SPECIAL OF
CHEMOMETRIC

Collected from
CHEMOMETRICS AND INTELLIGENT LABORATORY SYSTEMS, VOLUMES 1–5
An International Journal Sponsored by the Chemometrics Society
D.L. Massart, R.G. Brereton, R.E. Dessy, P.K. Hopke, C.H. Spiegelman and W. Wegscheider (Editors)

☐ As an Elsevier author I order one copy of this book with a 30% discount on the current price.

☐ As a member of the Chemometrics Society I order one copy of this book with a 30% discount on the current price.

☐ I order five or more copies of this book with a 50% discount on the current price. Please send me ...... copies.

The above discounts are not cumulative and only one discount is applicable per order.
The list price is 130 Dutch guilders or US$ 66.75 as at October 1990. These prices will remain firm until 31 December 1991.

All orders subject to the above discounts must be prepaid, made on (a copy of) this special order form and sent directly to the publisher.

☐ I enclose payment in the form of
  ☐ Bank Draft ☐ Personal cheque ☐ Eurocheque ☐ International Money Order
☐ Charge my
  ☐ Master Card ☐ Eurocard ☐ Access ☐ Visa ☐ American Express

Card Number.. Valid until ..../........ Signature:

Name:..

Address:..

....

....Zip/Postal Code ....

No postage is added to prepaid book orders. Customers in the Netherlands please add 6% BTW. Customers in New York State please add applicable sales tax.

Send to:
ELSEVIER SCIENCE PUBLISHERS
P.O. BOX 330
1000 AH Amsterdam
The Netherlands
FAX 31-20-5862 845

Customers in the USA and Canada
please send to:
Elsevier Science Publishing Co., Inc.
Attn. J. Weislogel
P.O. Box 882
Madison Square Station
New York, NY 10159, U.S.A.
FAX 212-633 3880

CHEMOMETRICS AND INTELLIGENT LABORATORY SYSTEMS

An International Journal Sponsored by the Chemometrics Society

which includes "Laboratory Information Management"

Editor-in-Chief:	**D.L. Massart** *(Brussels, Belgium)*
Editors:	**P.K. Hopke** *(Potsdam, NY, USA)*
	C.H. Spiegelman *(College Station, TX, USA)*
	W. Wegscheider *(Graz, Austria)*
Associate Editors:	**R.G. Brereton** *(Bristol, UK)*
	R.E. Dessy *(Blacksburg, VA, USA)*
	D.R. Scott *(Research Triangle Park, NC, USA)*

This international journal publishes articles about new developments on laboratory techniques in chemistry and related disciplines which are characterized by the application of statistical and computer methods. Special attention is given to emerging new technologies and techniques for the building of intelligent laboratory systems, i.e. artificial intelligence and robotics. The journal aims to be interdisciplinary; more particularly it intends to bridge the gap between chemists and scientists from related fields, statisticians, and designers of laboratory systems. In order to promote understanding between scientists from different fields the journal features a special section containing tutorial articles.

The journal deals with the following topics: Chemometrics; Computerized acquisition, processing and evaluation of data: Robotics; Developments in statistical theory and mathematics with application to chemistry; Intelligent laboratory systems; Laboratory information management; Application (case studies) of statistical and computational methods; New software; Imaging techniques and graphical software applied in chemistry. The research papers and tutorials are complemented by the **Monitor Section** which contains news, a calendar of forthcoming meetings, reports on meetings, software reviews, book reviews, news on societies and announcements of courses and meetings. This section also contains the "Chemometrics Newsletter", official bulletin of the Chemometrics Society.

Abstracted/Indexed in:
Analytical Abstracts, ASCA, BioSciences Information Service, Cambridge Scientific Abstracts, Chemical Abstracts, Chromatography Abstracts, Current Contents, Current Index to Statistics, Excerpta Medica, INSPEC, SCISEARCH

Subscription Information:
1991: Vols 10-13 (12 issues) US$ 692.00 / Dfl. 1232.00 including postage
ISSN 0169-7439

A free sample copy of the journal is available on request.

Elsevier Science Publishers
P.O. Box 211, 1000 AE Amsterdam, The Netherlands
P.O. Box 882, Madison Square Station, New York, NY 10159, USA

Chemometrics Tutorials

Collected from *Chemometrics and Intelligent Laboratory Systems – An Internal Journal,* Volumes 1–5

Chemometrics Tutorials

Collected from *Chemometrics and Intelligent Laboratory Systems — An International Journal*, Volumes 1–5

edited by

D.L. Massart
Farmaceutisch Instituut, Vrije Universiteit Brussel, Laarbeeklaan 103, B-1090 Brussels, Belgium
R.G. Brereton
School of Chemistry, University of Bristol, Cantock's Close, Bristol BS8 1TS, U.K.
R.E. Dessy
Department of Chemistry, Virginia Polytechnic Institute, Blacksburg, VA 24061, U.S.A.
P.K. Hopke
Department of Chemistry, Clarkson University, Potsdam, NY 13699-5810, U.S.A.
C.H. Spiegelman
Statistics Department, Texas A&M University, College Station, TX 77843, U.S.A.
W. Wegscheider
Institut für Analytische Chemie, Mikro- und Radiochemie, Technische Universität Graz, Technikerstrasse 4, A-8010 Graz, Austria

ELSEVIER
Amsterdam — Oxford — New York — Tokyo   1990

ELSEVIER SCIENCE PUBLISHERS B.V.
Sara Burgerhartstraat 25
P.O. Box 211, 1000 AE Amsterdam, The Netherlands

Distributors for the United States and Canada:

ELSEVIER SCIENCE PUBLISHING COMPANY INC.
655 Avenue of the Americas
New York, NY 10010, U.S.A.

ISBN 0-444-88837-3

© Elsevier Science Publishers B.V., 1990

All rights reserved. No part of this publication may be reproduced, stored in a retrieval system or transmitted in any form or by any means, electronic, mechanical, photocopying, recording or otherwise, without the prior written permission of the publisher, Elsevier Science Publishers B.V./ Physical Sciences & Engineering Division, P.O. Box 330, 1000 AH Amsterdam, The Netherlands.

Special regulations for readers in the U.S.A. – This publication has been registered with the Copyright Clearance Center Inc. (CCC), Salem, Massachusetts. Information can be obtained from the CCC about conditions under which photocopies of parts of this publication may be made in the USA. All other copyright questions, including photocopying outside of the USA, should be referred to the publisher.

No responsibility is assumed by the Publisher for any injury and/or damage to persons or property as a matter of products liability, negligence or otherwise, or from any use or operation of any methods, products, instructions or ideas contained in the material herein. Because of rapid advances in the medical sciences, the Publisher recommends that independent verification of diagnoses and drug dosages should be made.

Although all advertising material is expected to conform to ethical (medical) standards, inclusion in this publication does not constitute a guarantee or endorsement of the quality or value of such product or of the claims made of it by its manufacturer.

This book is printed on acid-free paper

Transferred to digital printing 2005

CONTENTS

Foreword VII

Computers in the laboratory

Chapter 1. R.E. Dessy: Scientific word processing 1
Chapter 2. R.D. McDowall, J.C. Pearce and G.S. Murkitt: The LIMS infrastructure 12
Chapter 3. E. Flerackers: Scientific programming with GKS: advantages and disadvantages 21

Expert systems

Chapter 4. N.A.B. Gray: Dendral and Meta-Dendral – the myth and the reality 26
          B.G. Buchanan, E.A. Feigenbaum and J. Lederberg: On Gray's interpretation of the Dendral project and programs: myth or mythunderstanding? 48
          N.A.B. Gray: Response to comments by Buchanan, Feigenbaum and Lederberg 51
Chapter 5. T.V. Lee: Expert systems in synthesis planning: a user's view of the LHASA program 53
Chapter 6. G.J. Kleywegt, H.-J. Luinge and B.-J.P. Schuman: PROLOG for chemists. Part 1 67
Chapter 7. G.J. Kleywegt, H.-J. Luinge and B.-J.P. Schuman: PROLOG for chemists. Part 2 92

Experimental design and optimization

Chapter 8. E. Morgan, K.W. Burton and P.A. Church: Practical exploratory experimental designs 104
Chapter 9. K.W.C. Burton and G. Nickless: Optimisation via Simplex. Part I. Background, definitions and a simple application 124
Chapter 10. J.C. Berridge: Chemometrics and method development in high-performance liquid chromatography. Part 1: Introduction 139
Chapter 11. J.C. Berridge: Chemometrics and method development in high-performance liquid chromatography. Part 2: Sequential experimental designs 153

Signal processing, time series and continuous processes

Chapter 12. R.G. Brereton: Fourier transforms: use, theory and applications to spectroscopic and related data 166
Chapter 13. A.G. Marshall: Dispersion vs. absorption (DISPA): a magic circle for spectroscopic line shape analysis 181
Chapter 14. G. Kateman: Sampling theory 196

Multivariate and related methods

Chapter 15. S. Wold, K. Esbensen and P. Geladi: Principal component analysis 209
Chapter 16. M. Mellinger: Multivariate data analysis: its methods 225
Chapter 17. M. Mellinger: Correspondence analysis: the method and its application 233
Chapter 18. P.J. Lewi: Spectral map analysis: factorial analysis of contrasts, especially from log ratios . 250

Chapter 19. A. Thielemans, P.J. Lewi and D.L. Massart: Similarities and differences among multivariate display techniques illustrated by Belgian cancer mortality distribution data 262
Chapter 20. O.H.J. Christie: Some fundamental criteria for multivariate correlation methodologies 286
Chapter 21. W. Windig: Mixture analysis of spectral data by multivariate methods 293
Chapter 22. O.M. Kvalheim: Interpretation of direct latent-variable projection methods and their aims and use in the analysis of multicomponent spectroscopic and chromatographic data 306
Chapter 23. N.B. Vogt: Soft modelling and chemosystematics 321
Chapter 24. H.J.B. Birks: Multivariate analysis in geology and geochemistry: an introduction 340
Chapter 25. R.A. Reyment: Multivariate analysis in geoscience: fads, fallacies and the future 354
Chapter 26. M. Mellinger: Interpretation of lithogeochemistry using correspondence analysis 367
Chapter 27. H.J.B. Birks: Multivariate analysis of stratigraphic data in geology: a review 383

Fuzzy methods

Chapter 28. M. Otto: Fuzzy theory explained 401

Author Index 421

Subject Index 423

FOREWORD

In the 1990 fundamental review issue of the journal *Analytical Chemistry*, the editor outlines four ways in which scientists keep abreast of new developments. The first, and conventional, approach is by reading the primary research literature: although this is possible in well established disciplines, it is much harder in rapidly evolving, multidisciplinary areas such as chemometrics and laboratory computing. The second method involves computerized literature searches. There are several excellent databases available and these supply the practising researcher with keyword searches, but there still remains much development in this field. Especial difficulties arise in multidisciplinary fields: in chemometrics we need to refer to the chemical, statistical, computational and even medical and geochemical literature for sources of information and, as yet, comprehensive, interdisciplinary, databases require much further investment of time and effort. A third method is the review: such articles are regularly published by many established journals and provide a concise summary of the literature in well defined fields. Interestingly, the editor of *Analytical Chemistry* suggested a fourth medium for keeping abreast of scientific developments, namely the tutorial paper. These articles are especially important in multidisciplinary areas. Many similar concepts occur throughout several subdisciplines of science and are referred to in different contexts by different groups of investigators. A good tutorial will explain, exemplify and compare these methods and ideas, drawing on key applications in different areas. To workers in the field of chemometrics and the related area of laboratory computing these tutorials are of particular importance: concepts and methods developed by computer scientists, mathematicians and statisticians are first applied and refined by chemists and then used by investigators in fields as diverse as pharmacy, geochemistry, environmental chemistry and so on, so there is a strongly defined need for interdisciplinary communication.

In the journal *Chemometrics and Intelligent Laboratory Systems* we have a specific policy of publishing tutorial papers, solicited from leading experts in the varied disciplines relating to this subject. This book is of reprints of tutorials from the first 5 volumes of this journal, covering the period from late 1986 to mid 1989. The papers have been reorganized into major themes, covering most of the main areas of chemometrics. In addition to chemists, this book will be of interest to works in all areas of laboratory computing. The authors of the papers include analytical, organic and environmental chemists, statisticians, pharmacologists, geologists, geochemists, computer scientists and biologists. It is remarkable that people from such diverse backgrounds can speak the same language and contribute to the same journal. We feel that the tutorial papers in this journal have strongly contributed to this interdisciplinary communication, and helped workers develop a common ground and common terminology. Many of the papers are at different levels reflecting the diverse background of users of chemometric techniques: some users have a sophisticated knowledge of statistical and computational methods and are principally interested in how to apply these methods, whereas others have virtually no statistical background but need to understand what methods are available for improving instrumental performance.

There are several excellent texts available on chemometrics or subareas of this discipline, and at first sight it may be wondered why we should be adding yet another book to the marketplace. A key feature of publishing tutorial papers is that new concepts can be reported fairly rapidly and that diverse ideas, a feature of chemometrics, can be included under one cover. Texts written by one author, or a small group of authors, have the advantage of uniform presentation, common notation and level, but the disadvantage is that the text would inevitably cover only a selected number of topics in detail and takes time to be written and prepared. In this book we hope to catch a snapshot in time of the diversity of modern chemometrics whilst educating the users of these methods, and provide a valuable bridge between research papers in *Chemometrics and Intelligent Laboratory Systems* and related journals and between textbooks. This book is intended both as a personal reference text and as useful background for courses in chemometrics and laboratory computing.

In preparing this text we especially thank the many referees who contributed helpful and rapid comments on the tutorial papers, to the guest editors of the special issues *Multivariate Workshop for Geologists and Geochemists* (O.M. Kvalheim) (chapters 15, 16, 17, 20, 24, 25, 26, 27) and *Expert Systems in Chemistry and the Chemical Industry* (P.B. Ayscough) (chapter 4), and to the editorial board of *Chemometrics and Intelligent Laboratory Systems* for regular advice.

Richard Brereton, on behalf of the Editors

Richard Brereton is Associate Editor of
Chemometrics and Intelligent Laboratory Systems
and coordinates the Tutorials Section of the journal

Chapter 1

Scientific Word Processing

RAYMOND E. DESSY

Chemistry Department, Virginia Polytechnic Institute, Blacksburg, VA 24061 (U.S.A.)

(Received 9 March 1987; accepted 18 May 1987)

CONTENTS

1 Prologue	1
2 Introduction	2
3 A panorama of software tools	2
3.1 Word processors	2
3.2 Scientific word processors	2
3.3 Desk top publishing	2
4 An historical perspective	3
4.1 PC graphic adapters	3
4.2 Printers	3
4.3 Time and space	4
5 The three bears	5
6 The scientist versus graphic artist	5
7 A future perspective	6
7.1 The graphic artist	6
7.2 The market place	6
7.3 PC mismatches	7
8 The next generation	7
8.1 Graphic chips	7
8.2 Packaging and the market	10
8.3 Graphic standards	10
9 The SCSI connection	11

1 PROLOGUE

This Fall the Chemical Institute of America, with the support of the Federal Bureau of Information, will begin publication of the Electronic Edition of its Journal of Chemistry. Each subscriber will initially receive a CD-ROM disk for use on a personal computer. The CD-ROM will contain a scientific word processor to be used for manuscript preparation. Separate style sheets for Introduction, Experimental, and Results and Discussion sections will allow prospective authors to develop their material in the format required by the Journal. Style sheets for references, tables, forms and graphs will also be included. With the built-in scientific word processor, the user can enter the appropriate text and data, and be assured that the result is editorially acceptable. The CD-ROM will also contain templates for all of the normal chemical and biochemical core structures, including the CAS Ring Index. The scientist can use these to sketch out desired structures and reactions. IUPAC names will be generated automatically.

Articles may be submitted to the editors on diskette or by modem. The review process will be "on-line". Reviewers will receive their copy by modem and will append their comments on electronic "PostIt^{TM}" notes for return to the editor.

Accepted manuscripts will appear in a monthly interactive CD-ROM version. The interactive format will allow readers to scan the electronic documents using their own customized access program. This format will also let authors use photographs and images within their text. The CD-ROM will include a "current awareness" file containing all of the Titles and Abstracts for chemical and biochemical papers published in the preceding period. A full-text relational database will allow subscribers to search for pertinent references.

Real? NO. Technically possible today? YES.

2 INTRODUCTION

Today's scientist, instead of putting pen to paper, is more likely to use a mouse on a screen. For efficiency and clarity, the lab worker turns to scientific word processing software to prepare manuscripts and reports. In this laboratory world, the message is the medium, rather than the reverse. Adapting to this new class of tool, and selecting the right one for your environment, requires a historical and future perspective.

3 A PANORAMA OF SOFTWARE TOOLS

3.1 Word processors

The first word processors (WPs) gave us an electronic typewriter with entry, deletion, copy, and replace capabilities. Second generation WPs added header, footer, footnote, and perhaps outlining capabilities. Third generation WPs installed interactive spelling checkers, thesauri, and alternate fonts. Are they useful? Many scientists no longer use any secretary support for letters, memos, and manuscripts.

3.2 Scientific word processors

Scientists reveling in the freedom and speed of WPs often reach out and try to touch something — and it's not there. They reach out for ways to convey their symbolism to the screen and hardcopy. Their symbolic language, using molecular structures, mathematical formulae, and strange alphabets is rarely spoken by WPs intended for the business and office environment. The need for a scientific word processing (SWP) tool is obvious.

Imagine incorporating the following into your text without scissors and paste:

Or automatically assembling molecules into a reaction chain:

This manuscript was prepared on a SWP system whose total cost was less than US$2500: an IBM-PC/XT clone with 640 Kbytes of memory, 20 Mbytes of disk, a high resolution monochrome display, a mouse, a dot-matrix printer, and Chem-Text (MDL) software.

3.3 Desk top publishing

Another descendant of WP packages is called desk top publishing (DTP). More oriented to the problems facing graphic artists and layout editors,

DTP packages generate a product that often rivals typeset documents.

Both SWPs and DTP packages draw on advances in software, silicon, and laser technology that can deliver services at prices the lab can afford. Let's examine the elements behind this revolution.

4 AN HISTORICAL PERSPECTIVE

Apple's MacIntosh computer, based on the Motorola 68000 computer chip, coupled with the mouse concept developed at Xerox, gave us a new way to look at computer screens. No mystical incantations, no Hebraic-like words lacking vowels; just a screen and cursor that let us pull-down menus and highlight the desired operations. Or icons that conveyed even to the novice its message and capabilities: pencils, clocks, desktops, filing cabinets, and garbage cans. The key hardware elements in this success were a 32-bit computer driving a screen with a resolution of 512 × 342 picture elements, or pixels. And software that provided windows that you could "peek" through for simultaneous views of two operations.

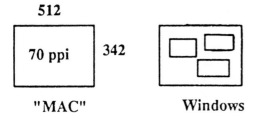

The MAC's screen pixel density of slightly less than 70 pixels/inch (ppi) allowed image displays that entranced users.

4.1 PC graphic adaptors

The closed architecture of the Mac, its intense use of software solutions that led to sluggish responses, and a scattered marketing strategy opened a window of opportunity for the IBM-PC which is based on the 16-bit Intel 8088. This mediocre architecture rapidly became the de facto standard. Its monochrome display adapter permitted only character based graphics. Although the expansion to an eight-bit ASCII code opened up 128 new characters, many of them useful in creating borders, the 80 × 25 character display matrix precluded true graphic displays. The color graphic adapter (CGA) interface also had poor graphic quality (320 × 200, or about 40 ppi). And certainly the PC's operating system, MS-DOS or PC-DOS, intimidated many users. But onto this inappropriate engine creative software writers began to bolt some enticing programs.

User and software demands rapidly led to a graphic board with better resolution: the Hercules 720 × 348 monochrome graphic adapter (HGA). And eventually IBM produced the IBM enhanced graphics adapter, a color 640 × 350 display. These correspond to a resolution of about 80 ppi.

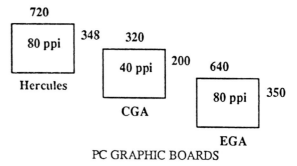

PC GRAPHIC BOARDS

4.2 Printers

Concurrently dot matrix printers were improving. Nine-pin printers became common. These formed alphanumeric characters by using a vertical set of closely spaced wires to create a matrix of dots. The human eye's lack of resolving power merged these into a recognizable object. Clever techniques employing multiple print-head passes over the original character matrix with a slight horizontal shift (called dithering) resulted in a near letter quality (NLQ) resembling impact font balls. Dot densities of 75–200 dots per inch (dpi)

became common. This allowed the generation of images on paper to exceed the resolution of the screen they were created on. Many dot matrix printers today have 20 or more wires. Promising astounding resolution, much available software and emulation routines use only 9 wires or can only print at half the advertised density. Inkjet printers give resolutions of about 250 dpi.

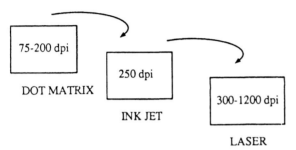

PRINTER RESOLUTION

The drive toward a print quality rivaling set type, however, was achieved with laser printers using Xerographic technology. The digital data corresponding to a character or image are used to modulate a laser beam striking a photosensitive drum, creating an electric charge. This electric charge is subsequently used to attract toner particles to the drum. This image is then rolled onto paper, and the toner particles fused into the cellulose by heat. Resolutions of 300 dpi are common, and are the reason laser printer output appears so crisp. 600 and 1200 dpi are possible. The latter produces characters that equal typeset quality.

4.3 Time and space

As we look at preparing printed pages, usually the last item of concern to the user is time and available memory. It should be the prime concern. Text generation at a fixed size of type, the simplest level, uses a character generating read-only-memory (ROM) that contains the individual matrices nedded to print each character set. No calculations are needed to print a character. Its ASCII equivalent code, actually generated when you stroke a key, is stored and used as a pointer into the ROM. The ROM memory cell so identi-

fied contains the correct dot pattern for the desired character.

For other fonts, or variable size letters, separate ROMs may be used. Or one can store templates and expand and reduce them accordingly. This requires some math capability. Generating images also requires considerable math as the lines and polygons used to construct the figures are created. In some systems the final image is stored as a bit (or pixel) map. For a 640×480 monochrome display this entails over 300,000 bits (40,000 bytes) per screen. In other systems the basic vectors are stored, and then manipulated everytime the display is required. With stored vectors the viewed image can easily be zoomed, compressed or panned. In either case graphics requires fast computer chips that can address large amounts of memory.

Screen update times often become a problem on PC based equipment. Most workers expect subsecond response in the iterative process of creating a document. Poorly written programs and slow processors lead to an unusable system. Graphic engines are becoming available that can solve these time problems. We will examine them shortly.

Printing is another time bottleneck. At 300 dpi an 8.5×11 image occupies about 1 Mbyte of RAM memory, and at 9600 Baud it would take 1 hour to transmit the image to a printer. Special high-speed computer–printer links and large print-buffer memories are obvious solutions. The plethora of standards among text preparation programs and printer control languages makes compatibility a serious concern. We'll examine this issue also.

5 THE THREE BEARS

Now we can look at important factors that differentiate WPs, SWPs, and DTP packages from the user's point of view. As you might expect full implementations of each differ in: what they do, what they do it on, and what you need to know.

Word processors are intended to handle the standard business and correspondence chores that require text entry, mail-merge, and limited database access. They can implement the basic emphasis features of **boldface**, underline, and *font change*. Minimum configuration PCs can handle these chores. Low resolution monochrome displays are usually sufficient. Any ASCII printer will suffice. And if it takes more than an hour to learn how to use 85% of the system's features, it has been designed poorly. But they are too small for most scientific applications.

Scientific word processors provide extensions that include a rich set of the fonts used in manuscript preparation, including greek, continental, fraktur, script, chemical and math symbols. Sketch pad facilities should include squares, rectangles, circles, ellipses, and a variety of etch-a-sketch lines. For chemical environments construction tools for 2-D and 3-D molecular structures and reactions are a requirement. Tables and forms utilities are necessary. The art work they can create needs to be placed in-line with the textual material. This requires an ability to compress or expand the drawn materials. The lines and vectors should not appear excessively "jaggy", and this implies screens and printers with medium resolution. Since most scientists use such a program sporadically, it must be easy to learn and remember. Learning to "drive" a new system should not take more than a morning.

The archetype of SWPs for chemical applications is certainly ChemText. Examples servicing other areas are T^3, Volks Writer Scientific, and The Egg.

Desk top publishing packages are currently targeted toward corporate publishing offices, smaller companies with intensive page layout requirements such as instant printing and copy shops, designers and free-lance artists. House organs, newsletters, advertising copy, and flyers are ideal targets. With the need to emulate set type quality, the systems require high resolution displays and printers. The computing burden is large. Therefore, since most current personal computer systems are based on Motorola 68000 or Intel 8086/80286 chips, speed/performance suffers. Laser printers are required to take advantage of all the detailed layout features. The software systems are powerful and complex. It takes several days to become acquainted with the basic features. In summary, at the present time:

DTP programs are very hardware specific;

DTP programs are very front-end software specific;

DTP program needs do not match current PC architectures.

The learning curve is steep going from SWPs to DTP. And current desk top publishing programs may be a little too big for the scientific desk.

In summary, current SWPs help reduce the iterative proofing, technical drafting liaison, and scissor-and-paste tasks that make communicating scientific results such a burden. Like the fable dealing with the three bears, they are currently just right.

6 THE SCIENTIST VERSUS GRAPHIC ARTIST

But, change is coming.

The graphic artist creates a page like a fine Japanese chef creates a dish to satisfy the eye and stomach.

This product differs markedly from today's scientific reports and manuscripts where content and uniformity are more important than aesthetics. Scientists have experience in what constitutes an acceptable report or manuscript. The scientist keeps adding material in a fixed style format until the task is done. Currently, scientific authors are not concerned with the squeezing, cutting, flowing, and moving of material till it all fits into a pleasing arrangement. The scientist need only satisfy the stomach.

But more user-friendly programs, vendor pressure, the desire to match what peer group members are creating, and rapidly changing publishing policies will force SWPs to evolve towards DTP characteristics. Consider the trend toward camera-ready copy, the social pressure to move from draft to NLQ or "letter quality", the fact that some journals are encouraging manuscript submission on diskettes, and the increasing costs of traditional printing. Today's luxuries will become tomorrow's necessities. It is therefore worth considering the world of the graphic artist and DTP.

7 A FUTURE PERSPECTIVE

DTP packages currently address a market that expects, consciously or subconsciously, a product that has been crafted by a graphic artist, a person familiar with what makes the printed page live, breath, and communicate. Unfortunately we take this skill for granted. Like fine music, most of us can appreciate the result of the graphic artist, yet we cannot produce it ourselves. No software package can substitute for lack of understanding or ability.

7.1 The graphic artist

What are some of the basic issues facing a graphic artist? What services does the package you might consider need to provide?

7.1.1 Page layout
Good page layout must be concerned with:
kerning — the spacing between letters;
leading — the spacing between lines;
call out — the specific style features for a section;
text wrap — the flow of text around figures;
threading — the flow from one column to another;
continuation page — the flow from one page to another.

Both kerning and leading improve the readability and appearance of text. Kerning takes cognizance of the adjacent letters in determining spacing. For example, spacing between the letter-pairs "AW" and "MN" should be different. Since kerning depends on the preceding and following letters it is a time intensive step. Kerning may be done automatically by look-up tables or algorithms, or may have to be done manually. Carrying full style-sheets or call-outs with each section is memory intensive, but necessary. Fitting and flowing text around art work or photographs, and making everything fit within the allowed space is human and computer intensive. Both art work and print size must be continuously adjusted to encourage reader attraction and then assure readability.

7.1.2 What do you see, when, and how?
Is the package "what you see is what you get" (WYSIWYG)? Can you see on the screen what will be printed, and, if not, when can you see it?

Is a "greeking" facility available to let you preview the results of art-work inclusion? Can you compress or expand a page to get a different perspective?

7.1.3 Import
Will the system accept formatted text from foreign WPs? The alternate is to import all materials in ASCII format. Can the system "frame" grab art work from foreign systems?

All vendors are struggling with import of text material from WPs and SWPs. Most will accept some subset of the following WPs; Word, Windows Write, WordStar, MultiMate, WordPerfect, Samna Word, XyWrite, DisplayWrite and DCA. Image imports may include Windows Paint, PC Paint, PC Paintbrush, Lotus, Symphony, Harvard Presentation Graphics, Dr. Halo, Freelance, and GEM Draw/Graph.

7.2 The market place

The early success of the MacIntosh, in part, was due to the coupling of a bit-mapped screen running a DTP package called PageMaker (Aldus).

In turn this was connected to the Apple LaserWriter by a standard page descriptor language (PDL) called PostScript (Adobe). We'll discuss bit-mapping in a moment. A PDL is a software interface between the page make-up software and the printer. It translates between the format needs generated by the DTP program and the opto-electro-mechanical vocabulary of the printer which implements those requests.

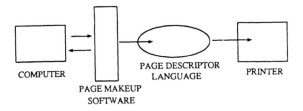

A new combination from Hewlett Packard includes the Intel 80286 based Vectra PC with PageMaker (Aldus), a PDL called DDL (Document Descriptor Language, Imagen), and the HP LaserJet printer.

Xerox, a pioneer in many aspects of this field, is offering a collection that includes the HP Vectra PC-AT clone, a DTP program of its own called Ventura, and a choice of PDLs including Interpress (Xerox), DDL and PostScript. The target printer is the LaserJet.

A plethora of other DTP packages are becoming available, including Spellbinder (Lexisoft), Page Perfect (IMSI), The Office (Laser Friendly), First Impression (Megahaus), SofType (Softest), Harvard Professional Publishing and Clickart (both Software Publishing).

7.3 PC mismatches

Mismatches between the heavy computer demands of these packages and the capabilities of current PC microprocessors are being addressed by add-in boards. Some, such as LaserVue (Sigma) and ConoVision (Conographic) provide enhanced screen resolution: 1600×1200. This matches the visual display to what is printed. Without such add-ins the call for special fonts often leads to the visual display of generic plain-vanilla fonts that can be supported by an 80 ppi display, and subsequent printing of the desired font at > 300 dpi by the laser printer. What you see is *not* what you get.

7.3.1 Time and space

Typical laser printing times for a single page range from 8 to 64 minutes, depending on the DTP package used. Therefore, add-in accelerator boards are available that improve the connection between the PC and the Canon engine found inside most laser printers. Typical examples are Turbo (Univation) and JLaser Board (Tall Tree).

Document file sizes increase dramatically as one goes from WPs to SWPs to DTP. WP document files take little more than one byte per character of text. A 24 page double spaced manuscript might take 64 Kbyte. Expect that to double for SWP documents. Watch DTP requirements carefully. Final storage for a two page document for a technically oriented PC magazine article, using several different DTP packages, required from 64 to 512 Kbytes! A factor of eight. Large Winchester add-ons are obviously essential.

8 THE NEXT GENERATION

We see that the demands of even current DTP packages exceed the capabilities of common PC hardware configurations. It is therefore relevant to examine what is likely to happen in this rapidly changing area so that users may decide what to do and when.

8.1 Graphic chips

A new generation of graphic display devices is emerging which recognizes the general need for an application specific integrated circuit (ASIC) that can draw lines, circles, and shade/fill or move and manipulate pixel arrays at very high speeds. The players include:

Vendor	Graphic engine
Advanced Micro Devices	AMD 95C60
Texas Instruments	TMS 34010
National Semiconductor	NS 8500
Intel	Intel 82786
Hitachi	HD 63484
Motorola	MC 68490

Each ASIC cited seeks to provide fast graphic support suitable for SWP and DTP applications. All promise to deliver screen resolutions of 1000 × 1000 or more at a reasonable price. They differ in approach, and the market they hope to address.

8.1.1 Who, me worry?

Why should I be concerned about these matters? Simple. Try performing the following reflection operations on your favorite PC. First reflect the top half of the screen to the bottom; then the left half to the right half. Which takes longer? The latter operation may take almost a hundred times longer. Why? The difference is caused by how images are stored and manipulated in the PC's memory.

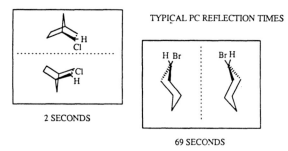

8.1.2 Architecture

Regardless of the implementation, these chips are intended to activate the basic operations of a graphic engine:

1. Take display instructions from a list created by the PC in its main memory and write data into a large display memory for eventual viewing on the screen.

2. Define a part of that display memory as the current bit-mapped display for viewing.

3. Transfer blocks of data into the current bit-map from other parts of display memory.

4. Operate on blocks of data using Boolean (logic) or arithmetic operations. These operations allow overlaying, bleed-through, and smoothing.

5. Continuously move the data to the display screen for viewing.

The display memory is a very high speed video RAM (VRAM) that is dual ported. One side gives the engine full parallel access for update. The other side feeds video data to the display via a fully buffered serial shift register. This allows concurrent operation-update and display. A typical graphic engine is shown in the accompanying diagram.

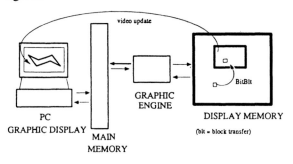

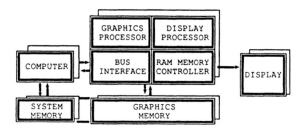

Let's examine a few of the alternate solutions:

1. How do you program the devices?

Software vs. hardware programming: Some graphic engines are oriented toward flexibility and programmability. They require extensive external software development support systems, but they can be tailored to fit a specific application. Other graphic engines employ a fixed set of self-contained macroinstructions; but their functioning is rigidly specified.

2. How is bit-mapped visual information stored?

Plane vs. packed memory architecture: Each pixel to be displayed may have one or more bits associated with it. Multiple bits are used to indicate grey-scale shades in monochrome or color in multichrome displays. Other attributes may be conveyed such as font and emphasis in text string displays.

In plane architectures, memory is constructed as a set of multiple planes each having a bit dimension equal to the pixel dimension of the screen. Each pixel may need 4 bits (or more) to describe its character — 4 bits allows 16 levels of grey or 16 different colors. The example shown requires four different bit planes. The X–Y coordinate position in each plane corresponds to the location of the associated pixel on the viewed display. This type of bit-mapped approach is conceptually simple. It does pose problems in accessing all the bits associated with one pixel (called the pixel depth). This is because all types of memory are divided into 8- or 16-bit chunks and addressed as a sequential collection. The bits associated with each pixel are located in different words in memory. Getting at all of the bits for a pixel requires accessing multiple words. Time is lost. As pixel depth increases, providing more gradation in grey-scale or color, the response time increases. However, if each plane has its own controller, and all controllers are activated simultaneously this disadvantage can be cancelled out. Cost is increased.

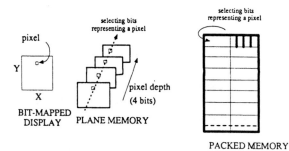

In packed architecture all of the bits associated with a pixel are stored in one word. Access is faster. However, this requires a larger word size and more speed, complexity, and cost in the engine. A trade off exists between parallel controllers or a more powerful engine.

3. How are small sections of the display moved around?

Bit block transfers (BitBlt): Manipulation of sections of the display involves arrays of bits that comprise a character, portion of a graph, or section of an image. These actions of moving and zooming a section, or rewriting it into a new location, usually require changes in the relative location of the array's boundries with respect to the fixed word boundaries of RAM storage. The solution is special hardware that defines an array of bits, pixels or characters and their manipulation as a discrete entity. This is the concept of Bit BLock Transfer (BitBlt). PiXeL BLock Transfer (PxlBlt) is the concept of BitBlt generalized to two dimensional operations on pixel arrays for use with Boolean and arithmetic operators. ChrBlt is the concept extended to text characters. Many graphic engines support all 16 two operand Booleans (AND, OR, XOR, etc.) and add/subtract math functions. These are essential in overlying one image on another.

Most BitBlt implementations require a hardware device that can shift bits rapidly as they move from one relative location within a RAM memory cell to another. This is called a barrel shifter.

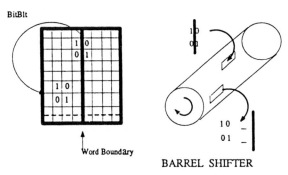

Text/graphics: Display of text and graphics often involves two distinct functions. Graphics is bit-mapped, while characters are generated from a special ROM or RAM look-up table to conserve time. In this approach text and graphics information are stored separately and the text overlayed on the graphic image at display time. True integration handles text exactly like graphic information.

4. How are windows implemented?

Windows: User interfacing is rapidly moving toward a windowed approach. This involves simultaneous display of two or more windows that allow the user separate views into the various concurrent tasks that may be executing within the computer. Movement, updating, or overlaying these windows requires rapid management of display memory parameters. This may be done in software, but is very time intensive. For example, try running Microsoft Windows on an 8088 based PC. Even an 80286 based PC-AT is sluggish! The software approach to window management does, however, allow flexibility because software drivers can accommodate a variety of configurations.

On the other hand, graphic engines can assume the responsibility of handling window management in silicon, providing phenomenal speed. But this forces a standard on vendors writing software.

8.2 Packaging and the market

The TI and AMD chip sets appear designed for the high-end market needing speed and resolution. The Hitachi and Intel products appear geared to a more moderate market. The accompanying table shows the multiplicity of approaches taken by typical vendors in their construction of graphic engines.

Characteristics of graphic engines

	CPU (bits)	Program	Memory	Barrel shifter	Graphic & text
Motorola	32	soft	packed	ucode instr.	int.
Texas Instruments	32	soft	packed	yes	int.
Advanced Micro Devices	16	hard	plane	yes	int.
Intel	16	hard	packed	no	sep.
National	16	soft/hard	plane	yes	int.

Each approach has its own unique strengths. Several of the plane architectures allow a vendor to increase pixel depth at no loss in speed. This opens up the world of color! Many hues are needed to represent three dimensional objects by shading. Eight bit-planes would allow 256 hues on screen at one time. Other architectures allow several chips to be "ganged" together, providing a "tiled" screen image with more pixels, higher resolution, and different screen aspect ratios. This is important since current screens not only have limited resolution, they are the wrong shape for document preparation. The printed page has greater height than width.

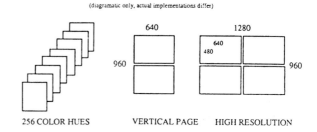

(diagramatic only, actual implementations differ)

256 COLOR HUES   VERTICAL PAGE   HIGH RESOLUTION

8.3 Graphic standards

These engines will be built into the next generation of PCs. They will begin to force consideration of the last element in the chain — a standard for graphic engine device drivers and graphics software. Recall the three page-descriptor languages mentioned above. Three window environment languages are also in common use, Windows (Microsoft), GEM (Digital Research) and X-Windows (for UNIX systems). A multiplicity of standards creates a healthy environment for technical progress, but it makes mixing hardware and software difficult.

In graphics a common protocol is needed to define the basic primitives common to all graphic programs: line draw, polygon generation, and shadowing/filling/shading operations. There are a plethora of "standards", and considerable emotion is evoked in any attempt to consolidate the situation. Graphics Kernal Standard (GKS) is a 2-D standard in a world rapidly becoming 3-D. A 3-D standard called Programmers Hierarchical Interactive Graphics System (Phigs) contends for attention. The more traditional and very widely used Plot-10 software uses the CORE standard. GKS and CORE and pre-bitmap standards and their performance on PC level equipment suffers.

If each graphic programming language and each graphic engine has its own unique standard the immense number of permutations and combinations will plague the user. A low level device interface protocol called the Computer Graphics Interface (CGI) has been suggested. It provides a standard bit-map oriented set of commands to drive graphic engines like those described above. The CGI provides a "virtual" terminal target that higher level graphic languages can aim at. Virtual is a term used to describe an object that doesn't exist, but which the user program perceives as a real entity. Virtual Device Interface protocols can solve the connection dilemma. One firm, Graphic Software Systems, has placed its generic version called Direct Graphics Interface Specification (DGIS) in the public domain to promote a de facto standard. DGIS is a ROM-able package for MS-DOS that translates the standard command set into the instructions required by whatever graphic engine is installed. DGIS has been endorsed by Ashton-Tate (dBase) and Lotus (1-2-3).

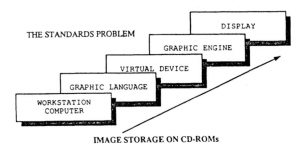

IMAGE STORAGE ON CD-ROMs

Compact disks store musical waveforms as a series of 16-bit binary numbers stored as two 8-bit bytes. This format can also be used to store ASCII text and graphic images. Each disk can hold 500 million bytes of information. This is the equivalent of 256,000 typewritten pages, thousands of templates, pictorials and icons, or 500 high resolution screen images. Standards are evolving for such storage promulgated by vendors such as Philips and Sony, or users such as the High Sierra Group. Simple text storage is covered by the CD-ROM protocols, while mixtures of text and graphics are covered by Interactive CD-ROM specifications. Chemical databases, spectral databases, or chemical structures in the form of wire-frame and space-filling models are obvious targets.

9 THE SCSI CONNECTION

Interconnecting all these components, PC, laser printer, and CD player, requires a high speed conduit. One emerging standard is the Small Computer Standard Interface bus, or SCSI. This is a 13-wire interfacing standard that consists of 8 wires for parallel transfer of a data byte and 5 "handshake" lines to coordinate transfer. It can easily operate at speeds up to 1.5 Mbytes/second. Many peripherals and computers are adopting the convention to make it easy for PC users to mix and match their purchases. Faster synchronous and 16-bit versions are being proposed.

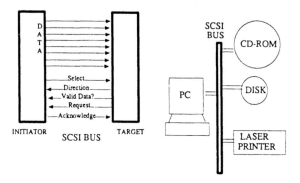

What an alphabet soup: SCSI, CD-ROM, CGA, EGA, CGI, GKS, GEM, Windows, X-Windows, DDL, Postscript, Interpress! One wonders if we can keep up. Many scientists have already encountered software/hardware incompatibilities as their needs and interests have grown. They can appreciate the tangled web that scientific word processing and desk top publishing may weave if standards are not established officially, or in the market place by the birth and death of product lines. The later will leave behind trails of lost time, energy and money. And since there is no correlation between elegance of design and success in the market place, we may end up with a standard that is less than best. The solution is to understand the area well enough to make intelligent investments.

Chapter 2

The LIMS Infrastructure

R.D. McDOWALL *, J.C. PEARCE and G.S. MURKITT

Department of Drug Analysis, Smith Kline and French Research Ltd., The Frythe, Welwyn, Herts. AL6 9AR (U.K.)

(Received 13 May 1987; accepted 4 February 1988)

CONTENTS

1 Introduction 12
2 The LIMS system manager 13
   2.1 Justification for a system manager 13
   2.2 Functions of a system manager 13
   2.3 Validation 14
3 The education and training of users 15
   3.1 Involving the users 15
   3.2 The role of management 15
   3.3 Introducing users to the system 16
4 Documentation 16
   4.1 User documentation 16
   4.2 System documentation 17
5 Problems and their notification 18
6 LIMS user groups 19
   6.1 Departmental user group 19
   6.2 External user groups 19
7 Conclusions 19
References 19

1 INTRODUCTION

Laboratory information management systems (LIMS) are databases tailored to an individual laboratory's needs. They are intended to integrate the analytical results produced by laboratory equipment with the sample information in order to reduce administration and speed the production of the final report [1]. There is great interest in LIMS as a means of meeting the challenges of greater efficiency in the laboratory, containing costs and reducing administration while complying with the requirements of good laboratory practice (GLP). These systems are complex to design and implement, so that very careful planning and forethought is required [2,3].

The areas discussed in this article are the need for, and functions of, a LIMS system manager, the education and the training of users, documentation of the system software, a procedure for the notification of problems and the setting up of a Departmental LIMS user group to discuss and guide the future development of the system. These aspects have evolved in our laboratories and now represent the infrastructure that we believe will allow the maximum productivity from a LIMS installation.

2 THE LIMS SYSTEM MANAGER

Computer systems, such as a LIMS, do not function on their own. The complexity and size of such systems requires that an increasing amount of time is spent overseeing their operation. Also, once operational, a great deal of reliance is placed upon a LIMS installation; it is vital that the system itself is highly reliable [4]. Therefore after the decision has been taken to acquire a LIMS, it is imperative that the first consideration is the appointment of a system manager. Ideally this person should be involved with the project from the start: from writing of the specification documents through to the acceptance testing, validation, implementation and operation of the system. Initially, this will be a job in name only as the installation may be some time in the future; but there will be much to do in developing ideas for the requirements of the system.

The system manager is instrumental in ensuring the success of the system so the following attributes are important and worthy of mention. For most installations, the position must be full-time, preferably with a deputy who can stand in for the manager in his or her absence or who can help during periods of heavy demand. It is essential that the manager has a good technical understanding of the computer and the operating system but he should also be a scientist, with a feel for the user's requirements because of a need to converse with people over a wide range of subjects. To be effective in this aspect of the job, good interpersonnel skills are necessary: on one hand being firm with the supplier, yet on the other showing understanding for the users.

2.1 Justification for a system manager

The proposed LIMS will cost a lot of money to purchase or develop. Additionally, much time and intellectual effort will have to be spent in specifying the system, which, in all probability, will not be accounted for in the initial costs. Thus, the system is a large investment of the organisations' resources; one that is worth protecting.

The recurring cost of a system manager will be relatively small in relation to the overall cost of the LIMS. However, the additional overhead will be repaid through the gains obtained from the speedy and efficient introduction of the system to the user community; when the benefits of the system are seen by both the users and management.

The organisation that is installing a LIMS is essentially dedicating a member of staff to work, at least part- or full-time, on the system. It will be the system manager who will be responsible for the introduction of the system to the users; therefore, for the maximum benefit and a relatively rapid implementation of the LIMS, a system manager is essential. These points must be made clear to senior management who will be responsible for authorising the position and the associated expenditure.

2.2 Functions of a system manager

The functions of a system manager are many and varied; the main duties are summarised in Table 1. Two of the main ones are the training of users and writing the documentation of the LIMS. The documentation includes not only the user guides for the software but also the standard operating procedures [5] used in regulated industries to ensure data integrity and good results. Once these are written, the system manager is responsible for their subsequent updating when new software enhancements are installed or new procedures are introduced.

Regular tending of the system is a major function. Old records should be deleted from the database once they are archived and the user accounts must be checked to ensure that files no longer used or required are removed. These tasks ensure that the computer can operate efficiently, since a disc that is full or fragmented will not operate efficiently and the response time of the computer will be slow.

Acceptance testing, validation and revalidation are areas where the system manager generates

TABLE 1

The main functions of a LIMS system manager

Operation of the system
    Acceptance testing
    Validation of the system
    Validation of software bug-fixes
    Periodic revalidation of the complete system
    Troubleshooting problems identified by users
    Liaison with the supplier of the system and the computer department

Documentation
    Writing system procedures
    Writing the user documentation
    Maintaining the user and system documentation to reflect current practices
    Maintenance of training records

Education and training
    Education of users and management
    Training and retraining of users
    Informing users of new upgrades, problems etc.
    Member of the LIMS user group

Computer housekeeping
    Archive and retrieval of data
    Removal of unwanted files from the system
    Deletion of unwanted data from the database

---

confidence in the LIMS for the users. The initial acceptance tests are designed to see if the system functions as it was specified. It is unlikely that a fully working system will be obtained at the first attempt [2] and therefore the system manager will be involved in respecifying parts of the software or devising solutions to obviate problem areas. Ideally the system manager should work in conjunction with an elite group of users to determine whether the LIMS as delivered is acceptable. If the system, in whole or in part, is not deemed acceptable negotiations must take place with the users, computer professionals, management and the supplier to work out a solution. Once the system is deemed acceptable then it must be validated.

2.3 Validation

In regulated industries there is a legal requirement to demonstrate the flow of information from the reported results back to the original sample. Computer systems are not exempt from this rule and validation is the process of producing the documented evidence that the LIMS, or any computer system, does what it purports to do [6,7]. Validation is similar to acceptance testing in that it must follow a predefined plan but differs in that the output must be kept and any discrepancies between the expected result and the actual result must be noted and explained. The end result of the validation is a certificate, signed by a management representative, that the system is authorised for use.

Can a computer system, including a LIMS, be completely validated? The answer is no. The reason for this is that there are many pathways through the software and it would take literally a lifetime to check them all. The pragmatic approach is to look at the software and decide the main pathways and concentrate the validation effort there.

Validation is achieved through a validation protocol, which is a written plan of the tests to be carried out on the system and the expected results of those tests. The tests should consist of three stages:

1. The first stage should be a series of relatively simple tests to 'exercise' the software. The rationale for this is that if the system does not work, then little resource has been spent finding problems. This case is of most use immediately after a system upgrade where the users need to know quickly if the system is operable.
2. The next stage is to test the system to its limits, for instance the largest batch size of samples possible should be taken through the procedures in the LIMS.
3. The final stage of testing is an attempt to 'break' the system; here entries should be different from or should exceed the expected values. (This stage could be combined with stage two above to reduce the time taken to validate the system.) Where entries are verified against data entries in the database, the logic of these should be examined.

In the authors' laboratories, the size and complexity of the validation protocol has grown as the system has been used more extensively and enhanced. This has had an effect on the time and effort needed to run the protocol: originally it took two days to execute when first devised, but now four days are required in its present format. Even so, it is unrealistic to expect that all pathways will have been tested and that the system will work without problems. Following a system upgrade or major software enhancement, problems that the validation protocol did not discover will surface during the subsequent use of the system; these should be dealt with under the procedure outlined in Section 5.

Once a LIMS has been validated, any enhancement or bug-fix should also be checked to ensure that it works as it was intended. This process is called revalidation. Whether the whole system or only the part of the software is revalidated is a decision left to the system manager or to the laboratory management.

In any case, the whole system should be revalidated at periodic intervals to ensure that it is still in control even if no enhancements or problems have occurred. This should detect any corruption of the software in test paths of the revalidation protocol. In a system that is undergoing change, the revalidation protocol must also change to reflect the current system.

3 THE EDUCATION AND TRAINING OF USERS

An essential component of a LIMS installation is the education and training of the users; this enables the implementation to proceed in a smoother manner than otherwise would be the case. The psychology of implementation of a computer system is an important factor to consider. The main reasons that computer systems fail are because the users are not motivated to use the system or because it does not live up to their expectations. The education and training process must overcome these problems [8].

3.1 Involving the users

Initial impressions of a LIMS are very important, since staff may well be daunted by the prospect of using new technology. The education process starts with the writing of the requirements specification and continues throughout the lifetime of the system. Time must be spent explaining the various benefits the system can offer, whilst being realistic about its limitations.

The enthusiasm of the users will be generated if they are involved in the specification process for the new LIMS; they should be consulted about what functions they would like to see and allowed to comment on the functional specification. It is not the intention that this consultation process should slavishly incorporate all suggestions into the LIMS, but is a process of generating ideas that may be incorporated. It is better to have asked for input in the first place and then have turned it down with a reason, than not to have requested input at all. This is an important psychological point that should not be overlooked.

Users should be made aware of the fact that the system will not be perfect when initially introduced. (The development of an efficient system is a process of iteration whereby the first attempt is modified in the light of use and the modifications are themselves modified, and so on.) It is therefore imperative that, before the users are exposed to a LIMS, it has been validated and the major bugs removed. Building credibility of the system in the users' eyes involves providing simple services that work well initially with more difficult tasks following later [9]. Where necessary, schemes to avoid problems already encountered should have been devised, pending enhancements to solve the situation permanently. If the system does not function efficiently the LIMS will not be accepted by the users.

3.2 The role of management

The role of management is crucial in the implementation of a LIMS. It must be seen by all concerned that there is a strong management commitment to the system. This is not to be confused with a system that is to be installed at the whim of management — there must be a rational argument that the LIMS will increase productivity of the laboratory to justify its installation [10].

Management support is crucial during the implementation phase: when tension, egos and problems surface, there must be clear direction and incisive action on the requests of the LIMS project team, the users and the system manager.

3.3 Introducing users to the system

Like all computer systems, the best way to learn how to use a LIMS is with 'hands-on' experience. Small, rather than large, groups of people should be trained, as this approach allows individual tuition if required. The best person to do LIMS training is the system manager, since he will have an in-depth knowledge of the software. Training should take place at an allotted time, and not be fitted in around the routine work. This is essential, since otherwise it places the analytical staff in a position of learning new technology whilst still being expected to produce work on time. Ideally the group learning to use the system should be relieved of normal work pressures while training.

Once fully trained the analysts can use the system; however, there may be instances when a problem occurs and work stops while it is investigated. It should be part of the training process that staff are aware that events like this may arise. These problems usually occur just after a system has been commissioned or when there has been a major software upgrade. However, in a system that has been operational for some time faults will still be reported when pathways that have not been used before are attempted. In general, the longer the system is used, the fewer the problems that will be reported.

4 DOCUMENTATION

Computer systems in the laboratory environment require very complex software in order to be capable of handling all the data formats and situations that can occur. Therefore it is mandatory that documentation is available to allow the system to be used to its full potential.

4.1 User documentation

As an adjunct to training, documentation describing the system for the users which is readable, self-contained and in a standard format must be available. The intention is that it can be referred to when appropriate, it can be understood and it is useful. Documentation improves efficiency by allowing users to understand the system with which they are working and to overcome any fears of the LIMS. It is also a stated requirement of the UK GLP monitoring programme [11]. It should be aimed at the first-time user and should be in tutorial format. Useful guides to the writing of computer documentation are available [12,13]. The user guides within our laboratory are modular, allowing easy update of specific areas without the need to re-issue the whole document. There is also an introduction to the system and the peripheral devices available for the novice user (see Table 2).

The reason user guides should be written is that the documentation supplied with the software will refer to a "standard" system which may be neither relevant nor tailored to the laboratory's specific needs; it may also be out of date. Therefore it is in the users' interest that the laboratory write and update its own documentation. As the LIMS expands or new software upgrades are added then retraining of staff will be necessary. Additionally, the LIMS documentation should also be updated at the same time.

TABLE 2

A modular approach to a LIMS user guide

Introduction to the system

Data set (dictionary) creation and maintenance

Sample receipt

Worksheet generation

Results entry
    Data transfer from chromatography integrators
    Balance data capture
    Manual results entry

Validation of results
    Quality control samples
    Calibration standards
    Sample results

Reporting results
    Graphical presentation of results
    Report generation
    Transfer of report to word processor

4.2 System documentation

Unlike the user documentation, which is intended for a wide readership, the system documentation has a relatively small circulation, namely the system manager, his deputy and any computer support staff. Any procedure carried out on the system should be written down in such a manner to allow an adequately trained person to carry out the procedure [5]. This has two purposes: rarely-used procedures can be carried out efficiently by reference to the documentation, and

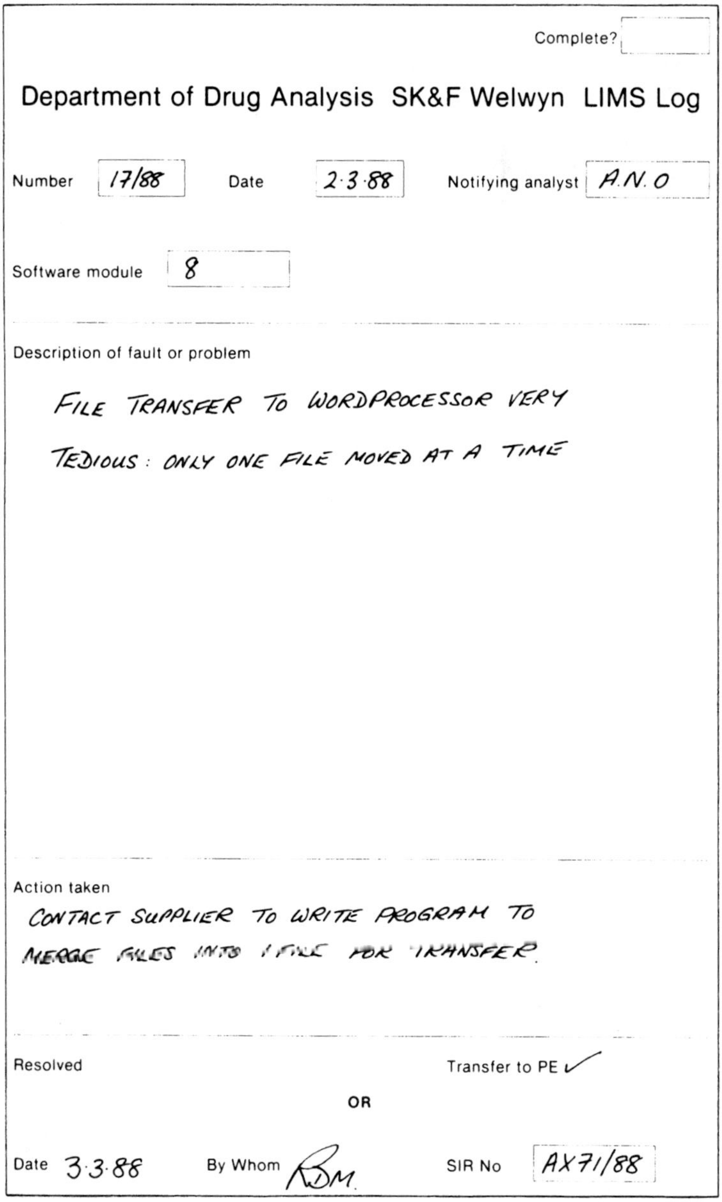

```
                                        Complete? [    ]

Department of Drug Analysis   SK&F Welwyn   LIMS Log

Number  [17/88]    Date  [2·3·88]    Notifying analyst  [A.N.O]

Software module  [8]

Description of fault or problem

    FILE TRANSFER TO WORDPROCESSOR VERY
    TEDIOUS: ONLY ONE FILE MOVED AT A TIME




Action taken
    CONTACT SUPPLIER TO WRITE PROGRAM TO
    MERGE FILES INTO 1 FILE FOR TRANSFER


Resolved                          Transfer to PE ✓
                         OR
Date  3·3·88      By Whom  RDM      SIR No  [AX71/88]
```

Fig. 1. Standard form for notification of problems.

any personnel changes, e.g. the system manager's leaving the company, can be overcome more easily than if there were no information available.

5 PROBLEMS AND THEIR NOTIFICATION

There should be a coordinated procedure for notification of system problems. In the authors' laboratories, a standard form, Fig. 1, is distributed to users to complete and return to the system manager, thus notifying him of any problems with the LIMS [3]. The form ensures that the problem is documented as an internal record and also demonstrated to regulatory agencies any fault and any subsequent action that has been taken to solve it. The form is in three parts; one is returned to

Fig. 2. Revalidation form.

the user when the problem is rectified, another is passed to the computer department for information on how the system is operating, and the last is retained by the system manager. Over a period of time the log of problems can reveal problem areas of the LIMS.

When any problem is passed to the supplier for solution, especially one involving software bugs, an additional form is used to check that the problem has been successfully remedied. The revalidation form, Fig. 2, outlines the tests to be undertaken to show that the solution is effective, along with any expected test results; these are compared to the actual results produced by the LIMS and the form is signed off when the problem is solved. In this way a complete appraisal documenting the history of the problem is available for inspection.

6 LIMS USER GROUPS

The formation of a LIMS User Group is very beneficial to all concerned: these user groups fall into two categories.

6.1 Departmental user group

The establishment of a departmental LIMS user group allows a formal channelling of information from the system manager to the users, laboratory management and vice versa. These meetings are formal, with minutes kept as a record of discussions [8]. The aim of these meetings is the improvement and expansion of the system, dissemination of information and canvassing opinion. Decisions affecting the development of the software can be made with the confidence that the users' needs and comments have been taken into account. All users can be informed of potential problems and new procedures are easily highlighted. Decisions affecting the system can be made with the confidence that the users' needs and comments have been taken into account. A typical user group agenda is shown in Table 3.

6.2 External user groups

Similar benefits are found with a user group affiliated to one company's product. This is the

TABLE 3

A typical LIMS user group agenda

1. Review of the last minutes
2. Action points
3. Review of problems
   (a) New problems since last meeting
   (b) Progress on existing problems
4. Enhancements
   (a) Those requested since the last meeting
   (b) Progress on existing enhancements
5. Documentation
6. Users' forum
7. Any other business

route to the long-term development and enhancement of the LIMS, since requests for new options are considerably strengthened when they are requested by many users. More informal in nature, they are a forum allowing fruitful discussions between users of similar systems and the supplier. It may be a useful idea, if a laboratory is contemplating the purchase of a particular vendors system, to attend a user group meeting before purchase to see what problems and advantages there are that the salesman has not mentioned.

7 CONCLUSIONS

In order to obtain the maximum benefit from a laboratory information management system (LIMS) an organisation within a laboratory to support and maintain the system is essential. The elements of this structure are a system manager, the efficient training of users together with good documentation and a procedure for the notification and documentation of problems. The establishment of a departmental LIMS user group is beneficial for monitoring progress and for the future development of the system.

REFERENCES

1 R.D. McDowall, Introduction to laboratory information management systems, in R.D. McDowall (Editor), *Labora-*

tory Information Management Systems — Concepts, Integration and Implementation, Sigma Press, Wilmslow, 1988, pp. 1–15.
2. M. Bertram and R.D. McDowall, The request for proposal and the functional specification, in R.D. McDowall (Editor), *Laboratory Information Management Systems — Concepts, Integration and Implementation*, Sigma Press, Wilmslow, 1988, pp. 154–174.
3. R.D. McDowall, J.C. Pearce and G.S. Murkitt, Laboratory information management systems — Part 2 Implementation, *Journal of Pharmaceutical and Biomedical Analysis*, in press.
4. G.E. Martin, LIMS with a strategic focus, *International Laboratory*, 16 (1986) 44–51.
5. E.H. Brown, Procedures and their documentation for a LIMS in a regulated environment, in R.D. McDowall (Editor), *Laboratory Information Management Systems — Concepts, Integration and Implementation*, Sigma Press, Wilmslow, 1988, pp. 346–358.
6. K.G. Chapman, A suggested validation lexicon, *Pharmaceutical Technology*, 7 (1983) 51–55.
7. S.S. Herrick, Validation of computer systems, *4th Annual Quality Assurance Roundtable, Houston, Texas, September 1983* (available from Beckman Instruments as a reprint).
8. G.S. Murkitt, Implementation, in R.D. McDowall (Editor), *Laboratory Information Management Systems — Concepts, Integration and Implementation*, Sigma Press, Wilmslow, 1988, pp. 67–73.
9. R.E. Dessy, Managing the electronic laboratory Part 1, *Analytical Chemistry*, 56 (1984) 725A–731A.
10. P. Johnson, Management considerations in the purchase of a laboratory information management system, in R.D. McDowall (Editor), *Laboratory Information Management Systems — Concepts, Integration and Implementation*, Sigma Press, Wilmslow, 1988, pp. 17–19.
11. *Good Laboratory Practice*, Department of Health and Social Security, London, 1986.
12. E.H. Weiss, *How to Write a Usable User Manual*, ISI Press, Philadelphia, 1985.
13. R.J. Brockman, *Writing Better Computer User Documentation, from Paper to On-line*, Wiley, New York, 1986.

Scientific Programming with GKS: Advantages and Disadvantages

E. FLERACKERS

Laboratorium voor Toegepaste Informatica, Limburgs Universitair Centrum, Universitaire Campus, B-3610 Diepenbeek (Belgium)

(Received 13 May 1987; accepted 29 September 1987)

CONTENTS

1 Introduction 21
2 Graphics standards 21
3 The Graphical Kernel System 23
4 Scientific programming with GKS 24
5 GKS: advantages and disadvantages 24
6 Sources of packages 25
References 25

1 INTRODUCTION

Computer graphics is the most versatile and most powerful means of communication between a computer and a human being. It is no wonder that scientists make an increasing appeal to computer graphics techniques. Standardization in the area of computer graphics has conferred portability upon graphical application programs as well as device independence. Besides these important advantages of standardization, the "programmer portability" is a real benefit, especially among scientists whose main interest is in their research and not in computer science.

In this paper a number of (de facto) graphics standards, will be discussed, viz.
— the Core Graphics System (Core)
— the Graphical Kernel System (GKS)
— Computer Graphics Metafile (CGM), formerly Virtual Device Metafile (VDM)
— Computer Graphics Interface (CGI), formerly Virtual Device Interface (VDI)
— Programmer's Hierarchical Interactive Graphics System (PHIGS)
— North American Presentation Level Protocol Syntax (NAPLPS)
— The Initial Graphics Exchange Specification (IGES)

Special interest will be paid to GKS. Since it is relatively simple and convenient GKS provides an interactive access to computer graphics for the scientist who likes to create his own graphics programs. The advantages and disadvantages of GKS will be discussed with examples of pros and cons drawn from our own experience.

2 GRAPHICS STANDARDS

Standardization in the area of computer graphics experienced a slow start, partly due to the rapid changes in hardware, but also due to the diversity of methodologies used. In the late 1960s and early 1970s, manufacturers of computer graphics hardware usually supplied software to their customers. These software packages from different vendors usually presented quite different interfaces to the application software level and, as a consequence, the application software was not portable. Among the first de facto standards the Tektronix PLOT-10 package is the best known. Real standardization originated from an IFIPS (International Federation for Information Processing) workshop on computer graphics held at Seillac, France in 1976 [1]. Here a distinction was drawn between the modelling side of computer graphics (where pictures are composed) and the viewing side (were composed pictures are displayed on a specified device in a particular orientation). Seillac resulted in the Core Graphics System (Core) [2] sponsored by the Association for Computing Machinery (ACM) Special Interest Group in Graphics (SIGGRAPH) Graphic Standards Planning Committee (GSPC) and in the Graphical Kernel System (GKS) [3] developed in the Federal Republic of Germany.

The Core system supports line-drawing computer graphics plus some raster graphics features. It is a three-dimensional system.

GKS has until now been limited to two-dimensional graphics (a GKS-3D standard is under development [4]). It is, however, an international standard, approved by the International Standards Organization (ISO), and is the most fully developed of the graphics standards. It specifies the interface to a graphics procedure library, which covers the graphical input/output needs of a wide range of applications. It also provides facilities for device-independent graphical input and output, for structuring and maintaining pictures, and for storing and retrieving pictures.

A consequence of the separation at Seillac of picture modelling and picture drawing is the single-level segmentation scheme common to Core and GKS. Segments provide a way to manipulate entire pictures or parts of pictures: they can be independently displayed, copied, deleted, or transformed. Once they have been created and closed their contents can no longer be changed. One cannot create a segment which in turn is itself composed of segments: only one level of segmentation is allowed. This means that the picture storage and manipulation facility (segmentation) is limited to a list of independent subpictures (segments). This is in conflict with the modern design principles of top-down design and stepwise refinement. The approach of allowing graphic entities to "call" other graphic entities is termed hierarchical. The Programmer's Hierarchical Interactive Graphics System (PHIGS) is a proposal for a three-dimensional system [5].

With GKS, a standardized application program interface to computer graphics has been defined. Apart from this there are two more interfaces: (1) the interface to the graphical devices; (2) the interface to the picture store.

The Virtual Device Interface (VDI) [6] is a standard being developed by ANSI (the American National Standards Institute). The objective of VDI is to provide a device-independent way to control graphics hardware; unlike GKS, it provides no picture-manipulation facilities. VDI is a lower-level standard, compatible with GKS in the sense that a GKS implementation can be built that uses VDI to communicate with hardware devices. The ISO nomenclature for VDI is Computer Graphics Interface (CGI).

The Computer Graphics Metafile (CGM), formerly known as Virtual Device Metafile (VDM) [7] has been approved as an American National Standard. It prescribes a standard way to record graphical information in a data file. A metafile is a means of permanent picture storage, and also a way to communicate graphical data between programs or between computers.

The Initial Graphics Exchange Specification (IGES) [8,9] is a data interface standard developed by the U.S. National Bureau of Standards to allow graphical engineering databases to be transferred between applications. IGES has become a de facto standard for CAD/CAM interfaces.

For years the use of graphics was limited to sophisticated mainframe systems due to the expense involved; complex graphic images require both extensive processing power and vast amounts of memory. Graphics quickly moved into the microcomputer environment when the size of processors and the price of memory decreased significantly. However, graphics for use with personal computers also needs standardization. Several graphics packages on microcomputers are based on GKS, VDI and VDM. Graphic Software Systems (GSS) has produced a VDI for the IBM Personal Computer. GSS-DRIVERS [10] provides a device-specific set of drivers based on the VDI standard for graphics peripherals. The programmer interface is based on GKS. GSS provides language support for C, Pascal and FORTRAN. The current graphics standards are well represented in a suite of software products available for the IBM-PCs called the Professional Graphics Series (IBM/PGS). These include GKS, a Graphics Development Toolkit which includes an implementation of Virtual Device Interface (VDI), a Graphical File System, for processing metafiles written according to the Virtual Device Metafile (VDM) standard, a Plotting System and a Graphics Terminal Emulator. Digital Research provides GSX, a GSS-like graphics package which runs under CP/M. For those who like to use a powerful, user friendly, interactive microcomputer graphic system there is the Graphics Environment Manager (GEM) from Digital Research.

The videotex industry has also been interested in microcomputer graphics. The standard used for videotex drivers is the North American Presentation Level Protocol Syntax (NAPLPS) [11]. A microcomputer with a NAPLPS implementation can be used for the creation, provision, and reception of videotex services [12]. It is perhaps not unreasonable that in the coming years videotex pages will be created mostly with microcomputers, accessed mostly through microcomputers and, to a great extent, stored in microcomputer-based databases. That being the case, NAPLPS may even become the dominant standard for computer graphics and applications.

3 THE GRAPHICAL KERNEL SYSTEM

The Graphical Kernel System (GKS) is a two-dimensional graphical system. GKS consists of a lot of user interface routines that give a programmer the ability to create graphics output and accept graphics input from a wide variety of graphical devices. These include monochrome and colour displays, printer, plotters and camera systems, as well as mice, data tablets, joysticks, and digitizers. A GKS implementation also typically provides programming access to GKS from several higher-level programming languages such as FORTRAN, Pascal, Ada and C.

GKS comprises six output primitives:
— POLYLINE draws a set of lines between a sequence of points.
— POLYMARKER marks a sequence of points with a specified symbol.
— FILL AREA fills an enclosed area with a solid colour or allows it to be filled by a specified pattern or hatch style.
— TEXT provides considerable flexibility in defining the quality of the text, its size and orientation, the origin, etc.
— CELL ARRAY is specifically aimed at the image processing community where the cell array defines the colour or grey level to be associated with individual elements of a rectangular array.
— GENERALIZED DRAWING PRIMITIVE defines a controlled method of adding more exotic primitives like circles, ellipses, etc.

Each of these primitives has a number of attributes and aspects such as colour, linestyle, character height, etc.

GKS input is defined in terms of a set of logical input devices, LOCATOR, PICK, TEXT, VALUATOR, CHOICE and STROKE which can be implemented on a workstation in a variety of ways. All input devices can work in the modes REQUEST, SAMPLE and EVENT.

GKS provides a one-level (segment) storage scheme for saving and manipulating parts of pictures. Thanks to the drivers GKS can be used on a wide range of devices of differing capabilities.

The capabilities of a GKS implementation are expressed using a level structure. The output level has three possibilities: 0: minimal output; 1: basic segmentation with full output; 2: full segmentation.

The input level also has three possibilities: a: no input; b: REQUEST input; c: Full input.

The simplest GKS implementation is, therefore,

level 0a, which only allows output to a single workstation at a time. The GSS implementation on microcomputers supports level 2b (full GKS, except for asynchronous input), most other PC implementation support only level 0b.

GKS also has a number of inquiry functions. Using these, an application may query GKS about the possibilities of a particular workstation. This is an important tool in writing device-independent software.

4 SCIENTIFIC PROGRAMMING WITH GKS

A lot of scientists need computer graphics only to plot results they get from their research. These line diagrams can be drawn using packages on microcomputers. The appearance of the plot, however, often does not fit with the individual's demands. GKS can provide a solution in these circumstances. The basis of GKS is easy to learn and is partly self-instructing. For the beginner, ref. 13 gives a good introduction, while the advanced user can find all information that he may need in ref. 14.

We have used GKS to write several software packages. One package can be used for drawing mathematical functions. The user can enter every function that can be written as a FORTRAN statement. Another package provides plotting possibilities for data which are given in x-y table. Both packages are fully menu-driven and consequently are very user friendly. The menu leaves the user to select the attributes and aspects of the drawing or the plot. A third package provides a graphical text editor. Texts are produced in different colour, size, orientation, expansion, etc. Several fonts are available. The position of the text is fixed using a graphical input device. We use the software mainly to make transparencies. Additional packages allow the creation of bar or pie diagrams. All these packages have been installed on a mainframe and produce output on such workstations as, for instance, the Tektronix 4107 terminal. Due to the standardization of GKS these packages can easily be transferred to other systems. Due to the device independence of GKS one can obtain a plot on paper or film in very little time. A self-written program using GKS allows the plotting of a spectrum recorded by a microcomputer in a realistic manner. To plot the intensity corresponding to a wavelength the observed colour at this wavelength is used. Some scientists in our institution have used GKS to draw molecules. The coordinates for the drawing were calculated using ORTEP. A self-written GKS-based program creates the actual plot.

We have had less success with fully interactive programs because the response times on the mainframe were too large. Such programs must be run on stand-alone graphics workstations or microcomputers based on the Intel 80286/80386 and Motorola 68020 processors.

We have also tried to do simulation and animation work on IBM-PC-like microcomputers using a GKS implementation. However, the overhead produced by GKS was a serious obstacle, so we have written our own driver which uses the low-level graphics possibilities of the microcomputer. A program written using our driver turned out to be an order of magnitude faster than GKS in executing raster instructions like moving windows. One can find interesting information about the use of GKS on microcomputers in ref. 15.

5 GKS: ADVANTAGES AND DISADVANTAGES

To conclude we summarize the advantages and disadvantages of GKS. The advantages of standardization are obvious; the portability of graphical application programs is very important. "Programmer portability", too, is a real benefit. It is sufficient to learn GKS to be able to write graphical programs on different systems. One can learn to use the most important procedures of GKS in, say, one day. The device independence of GKS makes it possible to obtain output on different workstations without changing the programs.

GKS also has some disadvantages, however. The implementation on small systems is usually a rather low-level one. As a consequence, for instance, a mouse may not be usable as graphics input device. Furthermore the interface function of GKS can produce a substantial overhead which makes GKS unusable for certain applications, especially on microcomputers.

A graphics system on its own can have interesting powerful primitive and operations which do not match with the structures of GKS. For instance, GKS supports virtually no pixel operations, although they may be very useful. Ruberbanding is also not well supported by GKS. For more demanding applications, especially interactive ones, the one-level segmentation of GKS can be a serious obstacle. Until now, GKS serves only 2D graphics. More and more applications, however, require 3D capabilities.

6 SOURCES OF PACKAGES

PLOT 10/GKS Graphics Subroutine Library (under license from Tektronix) is a GKS mainframe package. PLOT 10/GKS (Graphical Kernel System) Version 2.0 is a library of ANSI, FORTRAN 77 graphics subroutines. It is a level 2b implementation of the Graphical Kernel System. IBM supplies GKS for their microcomputers under ref. no. 6024203. Olivetti also supplies GKS drivers for the M24 (B&W and EGC version) and for the PE24.

The VDI (Virtual Device Interface) produced by Graphic Software Systems is available for the IBM Personal Computer. It can also be purchased from Olivetti.

GEM (Graphics Environment Manager) from Digital Research is an operating environment. It provides a variety of high-level functions whose purpose is to make it easier for the applications programmer to develop software that is both efficient and easy to use. GEM is currently available for the IBM-PC and compatible computers, running with the Intel 8086 microprocessor architecture. GEM is also available on the Atari ST, which uses the Motorola 68000 microprocessor.

Microsoft's Windows (MS Windows) and IBM's TopView are similar to GEM. Currently MS Window and TopView are only available for the IBM-PC and its clones.

Matrix Software Technology has recently put on the market three so-called Synergy products which offer interesting perspectives. These products are Toolkit, Layout and Desktop, all available on MS-DOS machines.

REFERENCES

1 R.A. Guedj and H.A. Tucker (Editors), *Methodology in Computer Graphics*, North-Holland, Amsterdam, 1979.
2 Status report of the Graphic Standards Planning Committee, *Computer Graphics*, 13 (Aug. 1979) 3.
3 *Graphical Kernel System (GKS), ISO Draft International Standard 7942*, International Organization for Standardization, Geneva.
4 *Graphical Kernel System for Three Dimensions (GKS-3D)*, ISO DP 8805, ANSI, 1430 Broadway, New York, NY 10018, U.S.A., 1985.
5 *Programmers Hierarchical Interactive Graphics System (PHIGS)*, Document X3H3/85-21, X3 Secretariat, CBEMA, 311 First St. NW, Suite 500, Washington, DC 20001, U.S.A., 1985.
6 *Computer Graphics Virtual Device Interface*, Document X3H3/85-47, X3 Secretariat, CBEMA, 311 First St. NW, Suite 500, Washington, DC 20001, U.S.A., 1985.
7 American National Standards Institute, Draft proposed American National Standard: *Information Processing, Computing Graphics, Virtual Device Metafile, Functional Description*, ISO document ISO TC 97/SC 5/WG 2 N 229 (1984).
8 American National Standards Institute, *Engineering Drawing and Related Documentation Practices — Digital Representation for Communication of Product Definition Data (IGES)*, American National Standard ANS Y14.26M (1981).
9 M.H. Liewald, Initial Graphics Exchange Specification: Successes and evolution, *Computers and Graphics*, 9, No. 1 (1985) 47–50.
10 *GSS-Drivers User Guide*, Ing. C. Olivetti & C., S.p.A., Direzione Documentazione 77, Via Jervis, 10015 Ivrea, Italy.
11 *Videotex/Teletext Presentation Level Protocol Syntax (North American PLPS)*, ANSI X3. 110-1983, CSA T500-1983, Dec. 1983.
12 K.Y. Chang, Microcomputer graphics and applications with NAPLPS Videotex, *IEEE Computer Graphics and Applications*, 1985.
13 F.R.A. Hopgood, D.A. Duce, J.R. Gallop and D.C. Sutcliffe, *Introduction to the Graphical Kernel System (GKS)*, Academic Press, New York, 1983.
14 G. Enderle, K. Kansy and G. Pfaff, *Computer Graphics Programming*, Springer-Verlag, Berlin, 1984.
15 R.F. Sproull, W.R. Sutherland and M.K. Ullner, *Device-Independent Graphics*, McGraw-Hill, New York, 1985.

Chapter 4

Dendral and Meta-Dendral — The Myth and the Reality

N.A.B. GRAY *

Department of Computing Science, University of Aberdeen, Aberdeen (U.K.)

(Received 26 October 1987; accepted 11 May 1988)

CONTENTS

1 The Dendral project 26
2 Origins 28
3 Heuristic Dendral 30
4 Diversification 33
5 Dendral 35
6 Meta-Dendral 39
7 Lessons from the Dendral project 44
References 46

1 THE DENDRAL PROJECT

The objective of the Dendral project was the development of computer programs that would aid the structural organic chemist in the process of structure elucidation [1,2]. The chemical structure elucidation process does not involve numerical computations; instead, the process is concerned with symbol-manipulation tasks such as the modelling of symmetries in graphs (i.e. in mathematical representations of chemical structures), the analysis of the implications of known structural constraints on the possible combinations of particular substructures, and the processing needed to infer structural constraints from patterns observed in spectral data. The tie between the chemical researchers on the Dendral project and workers in artificial intelligence (AI) originated from the need to work on symbolic computations, and from a requirement for a method for encoding the chemists' rules and procedures for spectral interpretation and prediction.

The structure elucidation process, as practiced by natural products chemists, involves several steps. Chemical and spectroscopic data are interpreted to yield structural constraints, that is specifications of substructures that must either be present in, or absent from the chemical structure under investigation. All possible candidate structures consistent with these constraints must be identified. These candidate structures are then evaluated and additional discriminating experiments planned so that the correct structure may be determined. Finally, the proposed "correct" structure must be confirmed either by synthesis or, possibly, by X-ray crystallography. Various

---

* Permanent address: Department of Computing Science, University of Wollongong, PO Box 1144, Wollongong NSW2500, Australia.

programs within the Dendral project were developed to assist in the first three stages — the interpretation of spectral and chemical data, the generation of candidate structures, and the evaluation of candidates and experiment planning. Other AI programs have been devised to assist in the processes of planning a synthesis as needed to confirm a proposed structure [3–5].

Many parts of the Dendral system are inherently purely algorithmic. For instance, an analysis of the stereochemistry of a chemical structure involves an iterative process for computing and investigating various symmetry groups of bonds and atoms. Other parts of the Dendral system are concerned with what is now regarded as the representation and use of knowledge.

The Dendral programs needed to represent structures and substructures — that is essentially factual knowledge concerning patterns of bonding relations that exist among atoms. The interpretation of spectral and chemical data requires either procedural knowledge or rules that can suggest possible substructures that are consistent with observed spectral patterns.

Constraints on the presence and absence of substructures (as derived in a spectral interpretation "Planning" step) must be used to guide the processes that generate candidate structures in the second step of the structure elucidation process. Thus, it is necessary to have procedures that can perform a form of constrained search through a space of possible structures.

Finally, Dendral's evaluation of candidate structures is based in part on the prediction of detailed spectral properties and such prediction processes again require knowledge of how, for example, particular combinations of structural features cause specific fragmentation processor to occur in a mass spectrometer.

Although the Dendral project was always focussed on specialized domain-specific problem solving, the project soon established and subsequently maintained a more general AI research component. The earliest ties between Dendral and AI arose from a need for recursive list-processing languages. A recognition of the structure elucidation problem as an instance of AI's Plan–Generate–Test paradigm led to the more substantial ties. The AI work focussed first on issues relating to the representation and use of domain knowledge in systems that could guide searches for appropriate candidate structures or could help in testing those candidates that were generated. Subsequent AI work was concerned with the capture and encoding of knowledge of the characteristic spectral properties of particular classes of compounds.

The AI workers associated with the Dendral project became concerned with issues of acquiring, representing, and using detailed domain-specific knowledge at a time when most AI work focussed on very general techniques of problem solving. The standard AI model of the time was resolution theorem proving — high performance AI programs were to be achieved using sophisticated inference systems based on theorem provers, and little attention was given to the need to encode the domain knowledge that was to be used by the theorem prover.

The Heuristic Dendral program constituted an empirical demonstration of an alternative approach to problem solving' - one that used detailed domain-specific knowledge to guide a relatively weak inference system. Heuristic Dendral illustrated how complex real-world knowledge could be represented in terms of discrete situation → action rules processed by a rule-interpreter. Subsequently, the Meta-Dendral work showed how search methods could be used as a basis for programs that attempt theory formation and learning tasks. This focus on knowledge and simple search techniques became a model for much subsequent work in applied AI, and led indirectly to the entire Knowledge-Based Systems industry of the 1980s.

The success of the Dendral project is now legendary, as can be illustrated by the quotations in Table 1. Dendral is presented as an AI success — a program that has "redefined the role of humans and computers in chemical research". Dendral's supposed success is cited as an example of AI applied to the real world and used as an argument justifying wider attempts to apply AI techniques.

However, this popular perception of Dendral is misleading. While the Dendral program had some success as a model for applied AI, most of the

TABLE 1

Some assessments of Dendral as an artificial intelligence system

---

The first expert system, produced at Stanford and called Dendral, was developed by a professor of computer science E. Feigenbaum and nobel laureate J. Lederberg. Dendral attempted to capture Lederberg's expertise in analyzing organic compounds from mass spectrometry. Issued under a Stanford licence and one of the most successful AI programs ever, Dendral can be found in many chemistry laboratories.
J. Shurkin, IEEE Expert, Winter 1986, p. 10.

Meta-Dendral, the first automated scientific discovery program, discovered previously unknown rules of mass spectrometry.
M. Walker, IEEE Expert, Spring 1987, p. 69.

Dendral surpasses all humans at its task and, as a consequence, has caused a redefinition of the roles of humans and machines in chemical research.
F. Hayes-Roth, D.A. Waterman and D.B. Lenat, Building Expert Systems, Addison-Wesley, 1983, p. 9.

Dendral is a success story. The results derived from its use are cited in over 50 scientific papers, which attests not only its usefulness but also its scientific credentials. It is in regular and routine use. The number of its users was expanding so rapidly that in 1983 a separate company was set up for its distribution and continued enhancement.
P.S. Steel, Expert Systems — A Practical Introduction, MacMillan, 1985.

An expert system can excel its human counterparts in terms of speed, completeness of knowledge, thoroughness, and its lack of human frailty. Existing expert systems such as DENDRAL and MYCIN have accomplished remarkable performance. The field of artificial intelligence is poised on the verge of a tremendous expansion of such systems into the marketplace. ART, the Automated Reasoning Tool, is literally a tool for building expert systems.
B.D. Clayton, Inference ART Programming Primer, Inference Corporation, 1985.

---

results of the project are in the form of algorithms and programs that have essentially no AI content. In fact, Dendral provides a further illustration of the way in which projects change as the problems become better understood. Just as the task of integration changed from an AI task based on heuristics (in programs like SAINT and SIN) to an algorithmic process (as in programs like MACSYMA) so Dendral changed from a demonstration system using heuristics, search and simple rules to a practical system using algorithmic methods.

2 ORIGINS

Dendral's origins date back to the early 1960s. From his work on the biochemical origins of life, Professor Lederberg was interested in the scope for chemical isomerism — that is in determining how many different compounds could exist with any given chemical composition. Since the 1870s, organic chemists have been working with a more or less correct model of molecular structures in terms of atoms, bonds and stereochemistry; but in all this time, the problem of chemical isomerism had never been tackled. In the 1930s, a small subset of the isomerism problem was addressed by Henze and Blair who used a "tree" representation of structures and determined the number of possible isomers for acyclic alkanes. However, the general isomerism problem was unsolved. In fact in the early 1960s, there was not even a way representing chemical structures uniquely, and chemical nomenclature was based largely on various essentially ad hoc schemes.

Lederberg introduced a new approach to the definition of (topological) structure based on trees and "vertex graphs" [1,2]. Lederberg's system was not simply a description of structure, it was inherently an algorithm for generating all structures of a given chemical composition. Used descriptively, Lederberg's system defined structures in terms of cyclic components (characterised as substituted vertex graphs) and acyclic parts (characterized as trees). A generator algorithm could create representations of all isomers of a given composition by partitioning a molecule's constituent atoms into cyclic and acyclic parts and subsequently building up, and joining together these parts (using a vertex graph generation algorithm for the cyclic parts, and a tree generation algorithm — the "Dendritic" algorithm — for the acyclic parts). In principle, these algorithms solved the problem of topological isomerism (stereoisomerism could be tackled once representations of topological structures had been created).

Lederberg's generation algorithms are recursive and consequently ill-suited to Fortran — the main scientific programming language of the period. Preliminary implementations of the algorithms were attempted in the naturally recursive Lisp language, and so a connection was established with AI researchers. The Dendral acyclic generator is basically a recursive combinatorial algorithm that finds all distinct partitionings of a given set of atoms (e.g. C and H atoms) into simple radicals such as $-CH_3$, $-CH_2-$, $>CH-$ and $>C<$, and further finds all distinct ways in which these radicals can be combined into larger units such as ethyl, propyl, or butyl radicals. Early versions of the Dendral (*Den* dritic *Al* gorithm) acyclic generator were running in Lisp in 1965. Results on the numbers of isomers of different empirical compositions were eventually published [6].

The "scope of chemical isomerism" is a rather esoteric topic. There are no practical applications that require information as to the total number of distinct ways of bonding together a particular set of atoms. However, Professor Lederberg had a second, more immediate application that also required computer assisted handling of chemical structural information. At this time NASA was planning the Viking missions that would land research probes on Mars. There was a possibility that Mars either still had, or at one time had had life forms; one of the tasks for the Viking probe was to perform chemical tests for molecules indicative of life. The kinds of molecules that might indicate life forms were expected to include organic acids and esters, ethers, alcohols, ketones and aldehydes, amines, amino acids, and possibly small peptides. Professor Lederberg had possibility for planning the experimental procedures for data acquisition and the analysis procedures for data interpretation that would resolve questions concerning life on Mars.

Of the analytic methods then available, only mass spectrometry (MS) constituted a viable candidate for inclusion in the Viking lander. A mass spectrometer (MS) could be built that would operate under the conditions that were expected to be experienced by the Viking lander. This mass spectrometer could be coupled to a simplified gas chromatography (GC) separation system and the combined GC–MS system used to obtain mass spectra from infinitesimally small amounts of individual compounds extracted from samples of Martian soil. The spectra acquired from the compounds were to be transmitted back to Earth for analysis.

Some of the compounds could be expected to be identifiable through the use of file-search techniques in which their mass spectra would be matched with reference data characterizing known compounds. However, although more than five million compounds have been reported, the largest files available (even in the 1980s) contain only around 80,000 mass spectra. It would obviously be unreasonable to presume that any Martian life forms would necessarily use the same compounds as identified in terrestrial life forms and characterized by data in available reference files. Consequently, many of the spectra might require other methods of interpretation and analysis.

It is unusual to have to attempt structure elucidation solely from mass spectral data. Normally, structure elucidation is only attempted when these is sufficient sample for many complementary spectral techniques to be applied. Further, a chemist generally has considerable information concerning the origin of a sample and data on the structure of other compounds as previously found in similar samples — such information can considerably help in the processes of spectral interpretation. Lederber's problem with his potential Martian samples was thus really novel, and he had to plan a new approach to structure elucidation.

By the 1960s, there had been a reasonable amount of theoretical work on organic mass spectrometry and the first texts were being published [7]. It was known that observed ions can be correlated with fragments of the original molecular structure that result from cleavage of individual acyclic bonds, or pairs of bonds in rings, or sets of three bonds in fused and bridged ring systems. The propensity for cleavage of different bonds was also partially understood, and it was often feasible to rationalize the observed mass spectral data of a known compound in terms of fragmentations of the bonds in its structure.

Other empirical analyses had established correlations between particular patterns of ions and intensities in a spectrum and the presence of a particular functionality in a structure. It was possible for an experienced spectroscopist to look at the spectrum of a simple monofunctional molecule, guess what functional group was present, and use this guess to help identify the molecular ion and, consequently, establish the molecular composition.

Lederberg's method for analyzing mass spectral data had to be based on such empirical methods for the identification of composition and functionality, and on procedures for predicting the spectral patterns that would be expected for given structures. If the composition and functionality of an unknown could be established, generation algorithms could be used to crease all of its possible isomeric forms.

Even if molecular composition and functionality are known, the numerous possibilities for different branching patterns in alkyl chains and for variations in the position of functional groups means that there can be hundreds, or even thousands, of possible candidate structures generated for quite simple empirical formulae. These various candidates can in principle be evaluated using procedures for spectral prediction. The bonds in a structure that are mostly likely to be cleaved can be identified, and consequently the mass of the most likely fragment ions can be calculated. Comparison of these predictions for ion patterns and the data actually recorded for an unknown should eliminate most candidate structures leaving just a few plausible structures that are consistent with all available data characterizing an unknown.

To make this approach possible, Lederberg required assistance from AI reseachers — assistance in the development of ways of representing and using knowledge concerning both the semi-empirical correlations between spectral patterns and substructures, and the more exacting prescriptions of how a given structure should cleave.

3 HEURISTIC DENDRAL

The mid-1960s are associated with one of those waves of enthusiasm for theorem proving approaches that have enveloped AI researchers. The favoured model for problem solving was to express any problem in terms of a formal, mathematical model. Knowledge of a domain was first to be axiomatized; then each problem was to be expressed as a theorem to be proven by inference from the given axioms (more accurately, a negated statement of the problem had to be expressed as a theorem that then had to be disproved). The main focus of research interest was on the inference methods that could be employed by such theorem provers. Stanford's AI laboratory was pursuing long-term research into general theorem proving methods, as well as carrying out work on robotics and vision, on problems associated with "understanding" natural language, on semantics of Lisp and so forth.

In addition to Stanford's main AI laboratory, there was a smaller research group, the Heuristic Programming Project (HPP) group, interested in the application of weak AI techniques, such as heuristic state-space search, to real-world problems. It was researchers from this HPP group, led by Feigenbaum and Buchanan, who cooperated with Lederberg on the problems of encoding and using chemical knowledge.

The Heuristic Dendral system was built up from the basic Dendral acyclic isomer generator. Dendral was extended by including a planning phase (the "Preliminary Inference Maker") that could perform a crude compound classification based on mass spectral data. Heuristic controls were incorporated in an attempt to filter the number of combinations of branched alkyl chains considered by the generator. An elaborate spectrum predictor was created to enable the final testing of candidates. Fig. 1 provides an illustrative summary of the kinds of mass spectral processing performed in the various parts of early versions of the Heuristic Dendral program.

In the first stages of its development, the mass spectral knowledge given to Dendral was bound closely into the code. The knowledge used in the Preliminary Inference Maker (PIM) was coded explicitly as a form of decision tree; an example tree is shown in Table 2 [8,9]. The PIM program worked through a hierarchy of compound classes (ketones, ethers, etc.) verifying that the molecular composition was compatible with a possible clas-

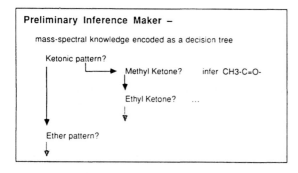

Fig. 1. Mass spectral processing techniques used in early versions of the Heuristic Dendral program.

atoms among those radicals that have still to be added to the partial structure. Goal-states correspond to complete structure representations that are consistent with structural and spectral constraints.

If at a given stage, one could perform a test that established that a particular operator (i.e. a particular partitioning of atoms among radicals) could not yield a satisfactory solution, then it would be possible to "prune" away an entire branch of the generation tree and so reduce the size of the search space. The heuristic test added to the Dendral generator checked for consistency between radical composition and mass spectral

sification and testing for specified ion patterns indicative of that class. A pattern of ions at specific masses (m/z values) was given for each class; if the pattern matched the observed spectral data, the Dendral generator would subsequently create structures incorporating the corresponding substructure.

Modifications of the Dendral generator were made in an attempt to exploit typical heuristic state-space search techniques from AI. The Dendral generator was now viewed as a "search" program. Dendral's "search" was really an exhaustive enumeration of all possible radicals that could be built with a set of available atoms and placed about a central atom or bond. If Dendral is treated as a state-space search, then states correspond to partially developed structures and operators correspond to the various (combinatorially generated) ways of partitioning the remaining

TABLE 2

Decision tree for classification of the spectra of ketones as used in the Heuristic Dendral program

Ketone –
Find two fragment ions, A and B, such that the sum of their masses is 28 atomic mass units (amu) greater than the molecular weight M.
Confirm the presence of ions at A − 28 amu and B − 28 amu
|Methyl-ketone3 –
| Require intense ions at m/z 43 and m/z 58
| Require ions at M − 43 and M − 15
| | Generate structures incorporating
| | $CH_3-(C=O)-CH_2-C(<)-CH<$
|Ethyl-ketone3 –
| Require intense ions at m/z 57 and m/z 72
| Require intense ions at M − 29 and M − 57
| | Generate structures incorporating
| | $CH_3-CH_2-(C=O)-CH_2-C(<)-CH<$
|NPropyl-ketone3 –
| Require intense ions at m/z 43, 71, and 86
| Require an ion m/z 58
| | Generate structures incorporating
| | $CH_3-CH_2-CH_2-(C=O)-CH_2-C(<)-CH<$
|IsoPropyl-ketone3 –
| Require intense ions at m/z 43, 71, and 86
| Require absence of any ion at m/z 58
| | Generate structures incorporating
| | $(CH_3)_2>CH-(C=O)-CH_2-C(<)-CH<$
|Generate structures with $>C=O$ but prohibit any
| occurrences of methyl-ketone, ethyl-ketone or
| propyl-ketone substructures
Ether
Verify absence of ions at M − 17 and M − 18
...
...

data. Each radical considered by Dendral was attached to the partial structure by a single bond; cleavage of that bond would lead to two fragments who masses could both be computed by the Dendral generator. A test required that at least one of these fragments was evidenced by an ion observed in the recorded mass spectrum. If the appropriate ions were not apparent, the "Heuristic" Dendral generator would discard that possible radical and all the structures that might have incorporated it. (These heuristic filters in the generator were abandoned in later versions of Heuristic Dendral — they involve assumptions about mass spectral processes that are not generally valid and so their use could lead to incorrect results.)

The Predictor program incorporated the most detailed mass spectral knowledge. The Predictor attempted to account for the relative intensities of the different fragment ions that could result from a given structure. The main part of the Predictor program was an iterative loop in which each bond of a structure was analyzed in turn to determine the expected intensities for the two fragment ions that would result from its cleavage.

This intensity analysis involved calculation of measures of the plausibility that a bond would break and of the relative likelihood of the charge being carried by each of the two fragments. The calculations involved numerous functions for assessing influence of neighbouring bonds, of the presence of hetero-atoms in fragments, of the relative sizes of fragments and so forth. Special case analyses were accorded to bonds in environments for which the general analysis proved inadequate. Thus, the knowledge of mass spectrometry in the Predictor program was initially coded in a conventional manner with tests, branches to special case code, and numerous functions. One the first reports on the problems of knowledge acquisition, and on possible methods of eliciting knowledge from an expert, concerns the difficulties of encoding a Lisp function that would constitute a correct description of β-cleavage fragmentation processes of ketones [10].

When Heuristic Dendral was extended to handle additional classes of molecules such as ethers [11] and amines [12], the interpretation and spectrum prediction functions were modified so that the system no longer relied solely on mass spectral data. With ethers it was found that mass spectral data did not provide sufficient discriminatory information to rank candidates reliably, so a proton nuclear magnetic resonance (NMR) spectrum predictor was added to the structure evaluation procedures. Proton NMR was subsequently used in the planning phase when Heuristic Dendral was extended to deal with amines. For acyclic monofunctional compounds, the possibilities for different types of branching in alkyl side chains constitute the major risk of a "combinatorial explosion" in the generator. Even a simple proton NMR interpreter could provide a count of the number of methyl groups in the molecule, and so could yield an extremely powerful constraint on the scope for branching in the alkyl chains.

These developments of Heuristic Dendral were accompanied by another more substantial change. Sections of the program, such as the Predictor, were rewritten into a form then described as "table driven programs" [10].

In early versions of the Predictor, mass spectral fragmentation behaviors described by chemists were encoded as individual Lisp functions. Thus, the programmer would invent a Lisp function, "VINYLIC", that contained special case coding to express the chemist's idea that "vinylic bonds, that is C–C single bonds adjacent to C=C double bonds, are generally not cleaved". The programmer's boolean VINYLIC function would involve hand crafted code to determine whether a C–C bond was adjacent to a C=C, and would be used in a special test that modified the plausibility of cleavage of that C–C bond.

The complexities of all such special case coding was severely limiting development of the system when it was perceived that, underlying all the special variations, there was a single form of processing. The chemists' descriptions of fragmentation processes were all simply definitions of substructures that had to be matched onto a given structure; if a matching was achieved, the plausibility of cleavage of a particular bond would be changed.

A "table-drive" version of the Predictor is illustrated in Fig. 2. The chemists' definitions of factors affecting bond cleavages are represented

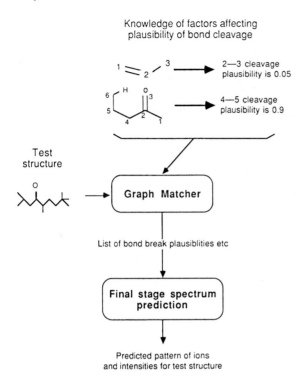

Fig. 2. A "table-driven" spectrum predictor.

using substructural templates; these templates define a bond environment and an associated change in break plausibility. (The term "table-driven" reflects the use of tabular record structures defining substructure connectivity and associated data.) A standard graph matching routine is used to find all possible matchings of each of the templates onto a structure whose spectrum is to be predicted. Each successful matching results in a change to a break plausibility. After the matchings are complete, the spectrum prediction process converts from break plausibilities to predicted masses and intensities.

The first gain from the "table-driven" revolution was that the knowledge of the experts, the chemists, was now expressed in a language they understood — the language of chemical substructures (instead of Lisp code). Errors in the encoded knowledge could be more readily identified. Frequently, these errors were errors of omission — the chemist's description of a substructure would be overly general resulting in its use in an inappropriate context. The chemist could be shown a trace illustrating how the substructure had been matched with a structure. This trace would immediately reveal any overly general attempt at matching and so allow the chemist to identify additional constraints that were omitted from the original substructure description. The chemist could then use a structure editor to add a constraint, e.g. a requirement that a particular neighbour not be part of a π-system, to the existing substructure definition.

A beneficial side effect of the whole coding process was that the domain experts often found that their own understanding of compounds' fragmentation behaviour was enhanced by the need to formulate a description that could be used by the programs. Further, the descriptions formulated for the program were sometimes useful when explaining the same mass spectral processes to students.

A second gain from a tabular representation of domain knowledge is that this knowledge becomes data capable of being manipulated by other programs. Programs can in some cases check the consistency of different "chunks of knowledge" from these tables. Analysis of particular examples may allow for the automated expansion of the tables with additional data.

Finally, the tabular representations of Dendral's mass spectral knowledge were equivalent to situation → action rules, or production rules. The fact that a production rule representation of knowledge had again proved effective was encouraging. It helped to unify the AI research of the Dendral project with other work. Waterman had recently completed a study using production rules in a system that learnt heuristics [13]. Newell and Simon had conducted lengthy studies in cognitive psychology that showed that a production rule formalism accounted for many human problem solving skills [14].

4 DIVERSIFICATION

Heuristic Dendral worked [15]. It could infer substructural constraints from simple features in mass and proton NMR spectra. It could generate structures compatible with composition and substructural constraints. It could predict the spectra of generated candidates in reasonable detail, and could rank candidates according to some measure of compatibility between predicted and observed spectra.

However, the Heuristic Dendral program was no more than a demonstration, an amusing toy. Its structure generation algorithms were limited to acyclic structures; its spectral interpretation and prediction procedures were limited to monofunctional compounds. Now monofunctional acyclic compounds are important to chemists — after all, these compounds serve as the solvents in which synthetic reactions are performed or with which natural products are extracted — but they do not constitute a class where structure elucidation problems are common. The chemists who had been drawn into the project had been promised computer aids for practical structure elucidation problems, not merely special case toys for simple mass spectra. The practical chemists needed computer systems capable of handling the large, polyfunctional, polycyclic structures that are of biochemical interest.

Thus by about 1970/71, it was time to take stock of the work of the Dendral project and consider various possible development paths. These development paths were as illustrated in Fig. 3.

For the programs to be useful in practical applications by chemists, the structure generation algorithms had to be capable of dealing with polycyclic, polyfunctional molecules. This extension to the generators was obvious a complex, largely mathematical task that could not be expected to achieve major results for some years; this work was continued as the "Dendral" project.

Work could continue on the application of rule-based systems to problems from particular well-defined subdomains of organic chemistry; such work could be expected to yield useful results in the short to medium term, and might also help identify automatic methods for acquiring and encoding the necessary domain-specific knowledge. This knowledge encoding/acquisition aspect particularly interested the AI researchers who wished to explore the processes of scientific theory formation [16]. The work on the use and formulation of knowledge, rules of mass spectra etc., formed the "Meta-Dendral" project.

Finally, Heuristic Dendral had demonstrated the utility of the production-rule formalism for encoding knowledge; there was interest in finding other practical domains where a similar encoding would also prove effective.

Rules for mass spectral prediction are complex. They require the matching of a subgraph representing chemical substructure onto a graph that represents the molecular structure. For each match found, the action component of the rule must be executed. This action component specifies bond cleavages, or cleavage plausibilities, that have to be interpreted in the context of a given match in order to derive the composition of the resulting ion. Thus, one requires a computationally sophisticated graph matcher to verify the premise of a rule and a quite specialized interpreter for the action part of a rule. Although these processes were elaborate, all that was really happening was that various attribute–value pairs of an object were being checked and, if certain conditions were satisfied, some weight was accorded to a particular expected result.

Heuristic Dendral made only rather limited use of situation → action rules. There were no real "chains of inference"; there was no need to apply one group of rules to establish the premise of

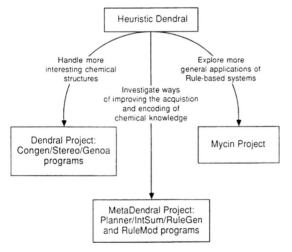

Fig. 3. Diverse developments from the initial Heuristic Dendral project.

some other rules. In the Predictor, there was a one step process — a substructure pattern (rule premise) was matched and spectral feature predicted (rule action). There was no need to perform any search for applicable rules; all rules in the program's tables had to be applied to a structure (many of course might fail to match and so not "fire"). Single-step inference and exhaustive rule-matching were appropriate for Heuristic Dendral's Predictor and for the corresponding interpreter — but obviously more general mechanisms of rule usage can be devised.

Researchers in the HPP group sought applications of production-rule systems in problems of classification or diagnosis (as in PIM's spectrum interpretation) and also in tasks that involved predictions of system behaviour (as in the Predictor). The newer systems focussed on rule usage. The rules themselves were simplified. A rule's premise became a statement of conjoined tests on the values of various fields (attributes) in a record structure that defined a particular object. The action parts of the rules were also simplified — actions simply updated the values in other fields of the object's record structure. Dendral's simple regimes of one-shot rule application and fixed classification trees were inappropriate to systems intended to demonstrate the power of knowledge-rich problem solving. HPP's more general systems utilized schemes in which chains of inference were built up dynamically through a search for those rules that could be applied in a given context and which would establish conditions for other rules to fire.

For the Dendral researchers, improvement in the expert's own understanding of a problem domain and the potential for use of the expert's rules in the context of instruction were both incidental. For the HPP group, these aspects of the use of knowledge were of much more central concern. The HPP group envisaged other knowledge based systems providing users with explanations as to how problems were being solved.

The complexity of the structure elucidation problems addressed by Dendral, and the rather specialized domain-knowledge that they involved, reduced the effectiveness of Dendral as a demonstration of the knowledge-based approach to problem solving. (It has been suggested that no other AI worker ever read a complete Dendral paper.) Practical demonstration of the knowledge-based approach to problem solving required a diagnostic task, whose complexity could be understood by those who were not specialists in the particular domain, that did involve fair amounts of specific knowledge and which occurred with sufficient frequency that it would be obviously advantageous to provide some automated assistance for human problem solvers.

Of the various projects at HPP in the early 1970s, Shortliffe's project on medical diagnosis constituted the best domain for further development of knowledge-based systems. Work on Shortliffe's project led eventually to the Mycin system for medical diagnosis [17]. Mycin solves a relatively simple problem of diagnosis and therapy recommendation. Such a problem could have been "solved" through the use of some fixed classification scheme working in a data-directed manner from a comprehensive data record for a patient that would be filled in prior to the attempted diagnosis. HPP's focus on the representation and use of knowledge resulted in a more flexible approach, and it is with Mycin that we have the first demonstration of goal-directed search for rules, consultative-style problem solving, explanations of rules to assist both expert and normal user, and so forth. However, the Mycin program lies outside the scope of this review.

5 DENDRAL

The Dendral researchers saw their task as involving (1) the implementation of algorithms for exhaustive isomer (structure) generation, (2) the extension of these topological structure generators to include stereochemistry, (3) the provision of a mechanism for applying structural constraints to the generator, and (4) the creation of systems that would help a chemist review large sets of candidate structures so as to identify additional discriminating experiments. The various structure generation systems developed by the Dendral group are noted in Table 3.

TABLE 3

Structure generation programs developed for the Dendral project

---

Dendral algorithm:
: Algorithm for generating acyclic, "tree-structured" graphs; as used in Heuristic Dendral.

StrGen program:
: Isomer generating system employing atom-partitioning, vertex graph catalogues, labelling algorithms, and the Dendral algorithm.

Congen program:
: Candidate structure generation system based on concept of superatoms, use of the StrGen system for intermediate structure generation, and additional labelling procedures for embedding of superatoms.

Congen-II:
: Congen — but using a more practical, if mathematically less sophisticated algorithm for the generation of intermediate structures.

Stereo:
: Module for the constrained generation of stereoisomers for candidate structures.

Genoa:
: Structure generation algorithm based on a generalization of standard graph matching techniques.

The Dendral group emphasized the importance of exhaustive, provably correct algorithms for isomer generation. The first proposed applications of the isomer generator were to be to the theoretical problem of the scope of chemical isomerism — the programs were to finally solve questions regarding the number of possible structural forms that could be built from a given set of atoms. However, exhaustiveness is also a prerequisite for a system that would satisfy the longer term objective of providing chemists with an aid for structure elucidation. For a structure to be "identified" on the basis of solely chemical and spectral evidence, it is necessary to show that no other structure is compatible with the evidence (even then, it is usually necessary for the proposed structure to be proved by some unambiguous process, such as synthesis). Therefore, all possible structures must be considered — and an exhaustive generator is required.

The first task of the Dendral researchers was to implement Lederberg's structure generation algorithms. Lederberg's approach involved combinatorial algorithms that found all possible distinct partitionings of atoms among cyclic and acyclic parts, and all possible divisions of the "cyclic" atoms among one or more separate ring systems. The forms of the acyclic parts could be explored using the Dendral generator. The cyclic parts were to be derived through an analysis of allowable vertex graphs.

Simple algorithms exist that can identify those vertex graphs that have to be used when generating cyclic structures for any given set of atoms of known degree. The forms of these reference vertex graphs can be obtained from a library. One of the first tasks of the Dendral researchers was to expand the pre-existing libraries of vertex graphs with those additional examples that would be needed by the structure generators [18]. (The various Dendral papers, like that of ref. 18, that report on the development of, or results obtained from specialized algorithms were all published under the imprimatur "*Applications of Artificial Intelligence for Chemical Inference*" because this provided a useful identifier for related research. This header was used irrespective of the fact that most of these papers did not address any AI related issues.)

The reference vertex graphs obtained from a library have to be built up into representations of complete cyclic parts of a molecule. This construction process involves an extensive sequence of "labelling" steps in which additional nodes are added to the edges of the vertex graph, and then atom names are attached to the nodes. The main difficulties in the structure generation process relate to problems of dealing with symmetries in the initial vertex graphs and in the partially labelled structures. The development of an exhaustive, irredundant generator required the development of new mathematical techniques for resolving the various labelling problems [19,20]. Developments eventually led to the StrGen program for structure generation [21,22]. StrGen was a purely combinatorial program; all its atom partitioning and graph labelling steps reduce to the selection of n objects from some set of m objects. StrGen involve no heuristics, no search, no artificial intelligence.

The StrGen system was used to solve specialized problems relating to the scope of isomerism [23,24]; however, application to practical structure elucidation problems required further developments. StrGen could treat known constituent parts of a molecule as atomic entities or "superatoms". Thus, if it was known that a compound incorporated a carbonyl group, then the empirical composition given to StrGen could be changed by eliminating one oxygen and one carbon atom and adding a bivalent "carbonyl atom". The "structures" generated would not represent actual chemical structures. A second processing phase was required to derive representations of chemical structures. In this second "Embed" phase, the superatom parts would be "embedded" — that is expanded out to reveal their internal structure. The general embedding process is complex and, once again, involves detailed analyses of symmetries [25]. Depending on the ways in which superatoms and other structural constraints were defined, it was possible for duplicate structures to result from the embedding step. Consequently, a final processing step required conversion of structures to canonical form [26] and filtering of any duplicates that this revealed. Additional substructural constraints could be applied through the use of "graph-matching" techniques; for example, if a particular substructure was known not to be present in a molecule, then a representation of this structure could be matched with all the final gen-

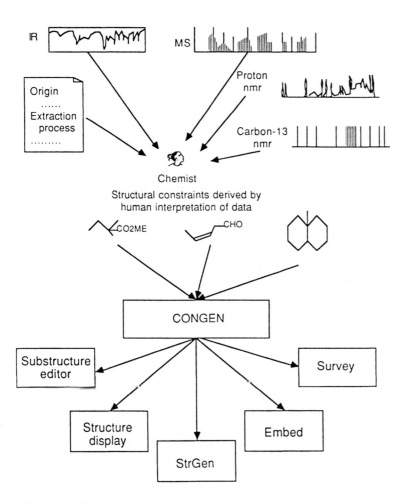

Fig. 4. In the Congen system, the human user performed the data interpretation tasks and the program handled the combinatorics of structure generation.

erated structures and those possessing the substructure could be discarded.

Dendral's first structure elucidation system was Congen — the CONstrained GENerator [27]. The Congen system, illustrated in Fig. 4, comprised an executive module that provided a convenient user-interface and which could invoke a number of other data-analysis and processing programs. The programs included a structure editor that allowed a chemist to define superatom parts, structure drawing routines, the StrGen program for intermediate structure generation, and the Embed program that derived representations of the final structures using a combination of a version of StrGen's labeller (to embed superatom parts) and a canonicalization routine.

Congen did not attempt any interpretation of spectra data. The structural constraints used by the program had be derived by the chemist user. The chemist had to interpret available spectral and chemical data and define any inferred structural constraints either as superatom parts or as restrictions on the bonding of particular superatoms. Thus, Congen did not have any AI "planning" component. Further, Congen did not attempt to apply any heuristic filters to the output of either the StrGen or embedder programs (for example, it did not discard "strained structures" with fused small rings). There were no heuristics, no domain rules, no knowledge in Congen — the program was based entirely on provably correct, graph-theoretic, combinatorial algorithms.

This lack of heuristic filters was not a consequence of any problems in implementing such procedures — it was an essential feature of a structure elucidation system. The structure elucidation process requires that all candidates be considered and that candidates only be discarded if there exists definitive evidence against them. Many heuristics could have been defined and used to constrain a generator (e.g. "cyclopropene rings are extremely rare, so don't create them") but such usage was inappropriate — the unknown whose structure was being determined might be one of the few naturally occurring cyclopropenes.

Once final candidate structures had been produced by Congen they could be evaluated by means of a number of structure surveying routines. Spectrum prediction routines, analogous to Heuristic Dendral's Predictor function, could be used as part of this surveying process. Mass spectra could be predicted either by means of detailed class specific rules [28] or through a more generally applicable semi-empirical theory [29]. ^{13}C NMR spectra could be predicted using a semi-empirical system that employed model substructures [30]. Candidate structures could be ranked according to a measure of the compatibility of observed and predicted spectra. However, it was not possible to exclude poorly ranked candidates — the semi-empirical theories used for spectrum prediction were too crude and could not be used in any "proof of structure" process. Structure ranking by spectrum prediction could at best help the chemist focus on the more plausible candidates.

Further development of the Congen program was mainly the work of Carhart. Carhart replaced the complex vertex graph analysis with an algorithm for generating intermediate structures that was related to the structure generation procedures originally developed by Kudo and Sasaki [31] and to the canonicalization procedures originally devised by Morgan [26]. The handling of symmetries in Carhart's algorithm was less sophisticated than the vertex graph approach but, overall, this new algorithm was more efficient in its processing of typical structural problems. Carhart's Congen-II became the service version of the Congen system as offered on the Sumex computer facility from about 1978. Nourse's Stereo module completed the structure generation process by generating all possible configuration stereoisomers of each candidate produced by the Congen system [32].

The Congen approach of first generating intermediate structures containing superatoms and then embedding these superatoms was eventually replaced by Carhart's Genoa algorithm [33]. Frequently, it is difficult for a chemist to derive distinct superatom parts from the available information on overlapping substructures inferred from spectral or chemical data. Carhart's Genoa algorithm provided a much more direct way of assembling structures from large, possibly overlapping, substructural components. Once again, the Genoa system contains no heuristics, no state-space search, no knowledge, no artifical intelligence; it is a purely algorithmic system based on a generalization of standard graph-matching procedures.

6 META-DENDRAL

In 1970/71, the Dendral structure-generator was known to constitute a long-term development project; in the meantime, practical, usable application programs were required. In order to justify further funding, the Dendral project had to be shown to be capable of delivering tools that could be used by chemists to help solve practical problems from the laboratory. The part of the Dendral project that led eventually to the Meta-Dendral work on theory formation started as a search for immediate applications of a rule-based system that could express correlations between substructural features and spectral processes.

As shown in Table 4, the Meta-Dendral project developed the programs Planner, IntSum, and the RuleGen/RuleMod system. IntSum and Planner were special purpose tools devised to solve immediate problems for working chemists — these programs were basically concerned with the identification of characteristic mass spectral fragmentation behaviours as exhibited in the spectra recorded for many members of some class of molecules, and the use of information on molecular fragmentations in restricted types of structure identification problem. The RuleGen/RuleMod system sought to take data, such as those produced by IntSum, and derive general rules describing mass spectral fragmentations; these rules would be similar in nature to those given by chemists to the original Heuristic Dendral program.

The Planner program [34] was intended to assist the chemist in the process of establishing the likely structure of a new isolated compound in a known family through the interpretation of its mass spectral data. The program's operation is suggested by the scheme shown in Fig. 5. Planar took as its inputs a definition of the common skeleton of the family of compounds, a set of rules characterizing the main fragmentation processes

TABLE 4

The Meta-Dendral programs

Planner:
A program for the "interpretation" of high resolution mass spectral data of a compound incorporating a known skeleton and substituent groups whose presence will not affect the fragmentations of the skeleton. Planner employs rules, developed by a chemist, that characterize the fragmentation processes of the skeleton. Planner can derive information concerning the placings of substituents on the known skeleton.

IntSum:
A program for analyzing a compound's high resolution mass spectrum in terms of possible fragmentations of its (given) structure. IntSum identifies all those fragmentations of a structure for which there is evidence in its spectrum, and can summarize data on related compounds in terms of equivalent fragmentations of a common skeleton. IntSum provides the information required by a chemist who is formulating rules for the prediction of mass spectra, or rules for use in an interpretation program such as Planner.

RuleGen/RuleMod (mass spectral version):
RuleGen automates the process whereby general rules (mass spectral "break processes") are inferred from descriptive data such as produced by IntSum. RuleMod refines the rules generated using the RuleGen procedure.

RuleGen (^{13}C NMR version):
A version of the RuleGen algorithm used to derive rules for predicting ^{13}C NMR shifts of atoms in specified substructural environments.

---

of that common skeleton, and composition and mass spectral data characterizing a particular unknown; the program's outputs were specifications of the probable nature and position of substituents groups.

Planner worked by calculating the compositions of the ions that would result from the specified skeletal break processes being applied to structures with particular assumed distributions of substituent atoms. If an ion of appropriate composition was found in the recorded spectrum, evidence would be accumulated towards establishing that particular assumed distribution of substituent atoms. Thus, in the example in Fig. 5, process B can be applied under the assumption that both oxygen are in the left most part of the molecule, or

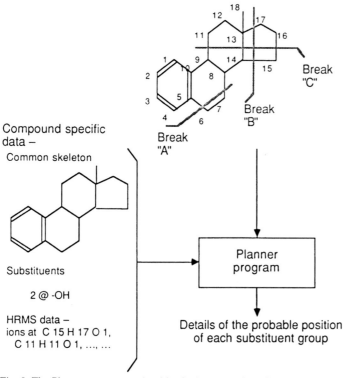

Fig. 5. The Planner program assisted in the interpretation of mass spectra of compounds from a known class.

that one oxygen is on either side of cleavage, or that both oxygens are attached to the three-carbon fragment that would be lost — these assumptions lead to predictions of ions at $C_{15}H_{18}O_2$, $C_{15}H_{18}O$, and $C_{15}H_{18}$. If $C_{15}H_{18}O$ was the only one of these ions observed, then this would constitute evidence that one –OH was attached to either C-15, or C-16, or C-17, and that the other –OH was bonded to one of C-1–C-14 or C-18. Through such analyses of each break process and a subsequent weighting of accumulated evidence, Planner could frequently identify and localize the substituent groups of a compound.

Programs such as Planner are only useful if there are sets of class-specific rules for the particular compounds of interest. Such rules are derived through analysis of similarities in the spectra of several closely related model structures. This analysis involves finding those break processes that seem generally applicable to molecules in a particular class and which account for most of the important ions in the observed spectra. The analysis process is basically a search — one has to explore some range of possible break processes and find those for which there is reasonable evidence among the available data.

There are some general constraints on the search. Thus, it is well known that bonds incident on tetravalent carbons are more easily broken than those incident on bivalent carbons, cleavages of bond β to π-systems tend to be favoured, and so forth. Using such constraints, the chemist postulates a small set of processes, tries these out on the available example compounds and checks whether they lead to observed fragments, possibly modifies a process by allowing for further loss of hydrogens etc., and finally stops when a particular set of break processes can be used to rationalize the majority of the intense ions in the spectra of each of the available example compounds.

Humans rarely perform effectively as such tasks involving a search through large numbers of data points and combinations of different interpretations. In particular, humans are inclined to stop when one set of break processes is found — even though there may well be alternatives of at least equal plausibility that have still to be considered. Of course such searches are relatively readily programmed, and the computer can be relied on to perform an exhaustive and unbiased search. A natural extension to the Planner system was a program that would perform analyses of sets of mass spectra and would identify all possible break processes that were evidenced in the data.

This program — called IntSum for Interpretation and Summary program [35] — took as inputs a definition of a common skeleton, a set of constraints on the nature and complexity of the fragmentation processes that were to be considered, and a set of structure definitions and details of recorded mass spectra for those structures. IntSum identified all possible skeletal fragmentations that were consistent with the given constraints and for which there was evidence in the observed spectra. The final outputs from the IntSum program took the form of tables identifying the various possible processes and their supporting evidence. From these output data, chemists could infer general rules of mass spectral behaviour for subsequent use either in the Planner program or in more conventional methods of spectral processing.

The next program in the Meta-Dendral sequence (the RuleGen/RuleMod system [36]) was devised to accomplish this rule generation or "theory formation" task. It is the RuleGen/RuleMod system that is commonly cited in claims that computers have discovered new theories for the physical sciences. The RuleGen/Rule Mod system was devised originally for the analysis of mass spectra data; Mitchell subsequently extended the RuleGen system to the analysis of ^{13}C NMR data [37]. Mitchell's system generated rules that relate precise ^{13}C shift ranges to specific environments for the resonating atom; Mitchell's ^{13}C RuleGen program provides the clearer illustration of the mechanism for rule generation.

Heuristic search techniques can again be applied. Here, the states in the search space correspond to rules of the form Substructural environment → Shift range. States differ primarily in their substructural descriptions. Operators, that define state–state transitions, actually change the substructure descriptions — for example by adding detail to the specification of a neighbouring atom's chemical type, or number of substituent hydrogens, and so forth. The shift range that is to be associated with a newly created substructure can be found by searching a reference file of structures and their assigned spectra for all instances of that substructure, and recording the minimum and maximum shift that have been observed for ^{13}C atoms in that specified substructural environment. Goal states can be characterized by various empirical criteria that define a useful prediction rule; for example, a rule may require a "sufficiently narrow"; associated shift range and an "adequate number" of supporting instances in the reference data file.

Fig. 6 illustrates RuleGen's search for ^{13}C prediction rules; these rules were developed from data on about 100 reference spectra of alkanes and alkyl amines. RuleGen began with a very general substructure and correspondingly weak prediction; the initial rule specifies that a carbon in an unrestricted substructural environment has to have a shift (or δ-value) in the range $-\infty$ to $+\infty$. New states were created by the application of operators that added detail to the substructure descriptions of existing states; thus, the direction of search was from general to specific. As illustrated in Fig. 6, RuleGen first added detail on the degree of the resonating atom (so distinguishing CH_3-, $-CH_2-$, $-CH<$, and $>C<$ resonances) and then explored elaborations such as CH_3-C, CH_3-N, CH_3-X_{sec} etc. (these representing all carbon bonded and nitrogen-bonded methyls and all methyls attached to any secondary center). The shift ranges assigned to these substructures were derived from the data characterizing the 100 alkanes and amines in the training set. Various heuristics determined whether (1) a state represented a rule of adequate quality for inclusion in the final rule set, or (2) a state represented an overly general rule that should be refined by

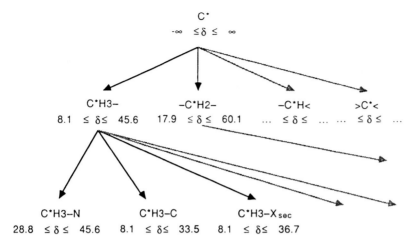

Fig. 6. RuleGen's search for ^{13}C NMR prediction rules.

the application of further operators, or (3) a state appeared to be on an unproductive path that could be abandoned.

MS-RuleGen's search for mass spectral rules was basically similar to the search for ^{13}C rules. However, the rules were inherently more complex and the substructure building operators were more diverse; both these factors act to enlarge the search space of possible rules. The search for rules was again from general to specific, starting with seed rule of the form X*X (interpreted as meaning "a bond between any pair of atoms is cleaved"). Operators could add atoms to this substructure, or add attribute values to existing atoms. First level elaborations of the seed rule included rules such as C*X (i.e. "cleave an atom between a carbon and any other atom"), XH_2-X ("cleave a bond between an atom with two hydrogens and another atom"), $X^*_{sec}X$ ("cleave a bond between a secondary atom and another atom"), and so forth. Evidence for a rule was found by matching these rule substructures against the substructural environments of those bonds cleaved in the fragmentation process that IntSum had identified.

Heuristics were again used to control the exploration of the space of rules defined by the substructure-elaboration operators. As in the case of ^{13}C data, the system frequently generated several apparently distinct rules that covered the same data points, and there were numerous partial overlaps among rules. A separate program, RuleMod, was needed to eliminate duplicates and merge overlapping rules.

The nature of mass spectral data limits the utility of the entire MS-RuleGen/RuleMod approach. The system can really only search for evidence for rules that apply to most members of a family of compounds. The result is a rather sparse set of rules; the rules rarely account for as much as 50% of the ion current. The rules are usually of low discriminatory power. They characterize fragmentations that are common to the class; but the rules don't always distinguish well among class members (and this discrimination task represents the only possible use of such rules).

In the original paper [36], it is noted that the Meta-Dendral program was "not discovering a new framework for mass spectrometry". Rather, the rules derived by the RuleGen/RuleMod system were compared with direct spectral ⇔ structural correlation rules such as were then being investigated by pattern recognition techniques. Contrasting their approach with the pattern recognition work, the authors of the Meta-Dendral papers emphasized their interest in providing the Chemist both with explanations as to the origin of rules and with methods of constraining a search for rules. In contrast to this rather conservative presentation of the Meta-Dendral results in the original scientific literature, reports in the secondary literature and in AI journals have been enthusiastic. Meta-Dendral is described as having "created rules for sub-families of molecules for which none existed before" and the general tenor of reports is to the effect that Meta-Dendral has achieved some significant advance in the theory of mass spectrometry.

Given these enthusiastic reports of the Meta-Dendral results, it is worth trying to assess the practical utility of Meta-Dendral's new rules or theories. One way of assessing the value of a new theory is to examine the frequency with which the work is subsequently cited. The evidence from the Science Citation Index confirms the supposition that the Meta-Dendral rules for keto-androstanes are of limited utility. The Meta-Dendral rules were cited in about twenty papers published in the next ten years; six of these citations were "continuity" references in subsequent Dendral project publications, seven citations were from general reviews along the lines of "Computer Techniques for Mass Spectroscopy", four citations were by authors of papers on alternative methods for interpreting mass spectra, and one citation was in a general review on mass spectral techniques for steroids. There was one reported attempt to use the keto-androstane rules — Gurst and Schrock noted that they failed to predict a fragmentation uniquely characteristic of a particular subclass of diketo-androstanes [38] (this results is exactly as one would expect, the Meta-Dendral rules characterize those processes that are common to a class of compounds).

Both the ^{13}C and MS versions of RuleGen are actually given a "theory" for the phenomena that they explore. In the case of ^{13}C RuleGen, this theory is a statement to the effect that shifts can be adequately explained in terms of topological models for substructures that can involve specifications of attributes (such as atom-type or number of hydrogens) on atoms up to two or three bonds from the resonating centre. This theory is built in; it is expressed through the forms allowed for substructures used in the rules and through the operators for modifying these substructures. The RuleGen programs can never transcend any limitations of the given theory; they can merely explore how well it can account for a given set of training data. The generated rules represent a particular instantion of the given theory as it applies to the specific training data.

The theoretical models given to Meta-Dendral RuleGen programs were in fact overly general, with significant constraints omitted. A spectral measurement (such as a ^{13}C shift) is a measure of how a particular subpart (a ^{13}C atom) is embedded in a whole structure. The properties of the subpart are going to be influenced most strongly by the character of its immediate neighbours, a second shell of neighbours will further influence the subpart's behaviour, other more remote neighbours may also influence the observed property to some measurable degree. There is an inherent hierarchy in these interactions; and, if the α-neighbors of that atom have not been fully specified, there is no point in seeking correlations between an atom's NMR shift and parameters such as the number of γ-hydrogens.

If the given theory embodies all the appropriate physical constraints, ^{13}C rule formation becomes essentially deterministic. There is no need for heuristic search, nor for empirical evaluation functions and so forth. The stereochemical environments of each carbon atom in each reference structure can simply be encoded, algorithmically, out to some maximum bond-radius. Each combination of substructure code and associated chemical shift either becomes a new, specific "rule" (with a point-prediction for shift rather than a range), or the code/shift combination provides new data to update the shift range associated with an existing rule with the same substructure code. Further, the coding procedure can be arranged so that the code (substructure definition) is canonical at each bond-radius (or shell level). If two atoms are in identical environments out as far as three-bond radii, then their 3-shell codes will be identical. Each code/shift combination can be used to update more general rules using 3-shell, 2-shell, and 1-shell descriptions (of course, these general rules are associated with increasingly broad prediction ranges). The final Dendral programs for ^{13}C spectrum prediction and interpretation [30,39] were based "rules" created through such a deministic approach. Once again, when the initial analysis is more complete, one moves from heuristics and search to algorithmic methods.

7 LESSONS FROM THE DENDRAL PROJECT

The achievements of AI and in particular of expert systems are mainly presented through secondary literature — reviews, books, product presentations and so forth. These presentations are normally designed to popularise some technique, or to encourage the wider use of particular approaches to problem solving, or to sell a product. They use data on previous AI systems to provide supporting evidence for their claims. The Dendral project has figured prominently in such reports — and, for the most part, is presented in a totally inappropriate way.

Even in those Dendral/Meta-Dendral programs that did utilize rules for spectral interpretation and prediction, the AI-related code for creating and using rules represented only part (often only a small part) of the total system code. Most of the code of these systems was involved in the user interface, in structure and substructure editors, in specialized file system, in graph-matcher and canonicalization routines, and so forth. It is likely that the rule creation/interpretation code will only represent a small part of the total code needed for problem solving in most complex domains. Systems as elaborate as the Meta-Dendral programs simply cannot be built from standard expert systems shells.

The Dendral programs of greatest practical value to chemists are the basic structure generators Congen and Genoa, and the Stereo module. Even these programs are of relatively limited use — they simply don't address a problem that arises with sufficient frequency as to really justify the use of computer aids. The chemical structural problems that do arise frequently (and that could utilise computer aids) are of the form "What is the nature of this pollutant?" — in these problems a rough classification of the compound (or mixture) as polynuclear aromatic, organo-phosphorus, chlorinated biphenyl etc. is required, the data available to the classification procedure concern the origin of the sample and the results of a few quick physico-chemical tests. These problems are insufficiently constrained for use of structure generators like Congen, and in any case there is no interest in the specific structural form of the pollutant.

Complete structure elucidation problems, such as can be handled by Congen, are relatively infrequent. Most of the time of the typical individual graduate student in a natural products laboratory is spent on the isolation and re-identification of previously known compounds; the three or four new compounds encountered each year represent the main results of that individual's work, and the elucidation of their structure is the only intellectual stimulating part of the entire exercise. The natural products researcher is unlikely to hand over the only "fun" part of the work to a computer program, particularly as in most cases biochemical constraints on skeletal form are such as to make the structural problem readily soluble with no need to exhaustively consider hundreds of potential candidates. Further, structure elucidation ultimately requires proof — proof either through X-ray crystallography or by means of one of the newer skeleton-tracing NMR techniques. Identifications based on Dendral-like procedures of candidate generation and ranking do not constitute proof of correct structure elucidation and so do not constitute complete solutions to a structure elucidation problem.

Although available for many years through the networked Sumex-Aim facility, Congen was never widely used (when an attempt was made to discover its usage, only about eight applications citing its use could be traced). The two main review papers that present the Congen/Genoa programs as aids for structure elucidation use contrived examples based on previously completed and published structure elucidation problems — there weren't any novel structural problems being solved by these programs that could have been used as the basis for illustration in such review papers [40,41]. The Genoa/Stereo program was made available through the company Molecular Design (San Leandro, CA, U.S.A.), but it represented only a minor and not very successful product line added to the repertoire of an already existing company. Claims of widespread use of the Dendral programs are simply without foundation. In any case, the programs Congen, Genoa, and Stereo have essentially no AI content and even if they had been widely used such usage would not constitute evidence for the successful application of AI techniques. (If Dendral is "one of the most successful AI programs ever" then god help the unsuccessful AI programs.)

Other misrepresentations of Dendral indicate how reports of the actual achievements of programs require much more careful phrasing. For example, original reports have compared the performance of graduate students and of the Heuristic Dendral program at the task of identifying a structure solely from composition and mass spectral data [10], and the correct results obtained from the structure generation algorithms of StrGen have been compared with the less successful attempts of post-doctoral students assigned similar structure-generation problems [21]. However these are not realistic tasks. A chemist normally interprets spectra in the context of considerable additional information concerning the origin and isolation of a compound. Structure generation problems, as actually encountered by chemists, involve small numbers of distinct multi-atom fragments and involve quite different conceptual problems from those associated with the generation of structures from large numbers of identical atoms. The lack of realism of these tasks is not stressed in the original reports — because these reports presume, substantial shared background knowledge. Those familiar with the domain understand that the tasks reported represent only a small part of the total process of structure elucidation, and that the examples chosen have been selected to give an easily comprehended measure of the scope of problems that can be tackled by the programs. However, those who do not share this presumed background knowledge tend to take these measures of performance quite out of context — and so the Dendral programs come to be reported as having greater expertise than post-doctoral scientists.

The Meta-Dendral systems had little potential for real application. In principle, the Planner system could be used to help identify structures that had some novel pattern of substituents on a standard skeleton. However, Planner was limited by its requirement that the fragmentation processes of a skeleton would not be substantially altered by the presence of the substituents — while this requirement was met in the case of the simply substituted estrogens, it is not a requirement that would generally be satisfied. In any case, such structural problems are much more readily solved through the use of other spectral techniques. The RuleGen/RuleMod analysis of mass spectral data could not generate rules that would be of significant value in structure elucidation. The spectral ⇔ structural correlations that these programs could identify were for compounds in well defined chemical classes and so were irrelevant for general structure elucidation problems. Further because the selection method picked those processes for which there was general evidence among the available reference data on example compounds, the rules were usually of low discriminatory power. These rules could not really be used as the basis of ranking candidates even when one did get a structure elucidation problem involving compounds in the class for which the rules had been developed.

Claims that "Dendral surpasses all humans at its task and, as a consequence, has caused a redefinition of the roles of humans and machines in chemical research" are made in ignorance. Such claims attempt to establish a myth — a myth that may in the short term help to sell some techniques. But, such myths only obscure the real achievements of the Dendral project.

The Dendral project has real achievements. Many are esoteric. Dendral led to advances in combinatorial mathematics, particularly graph labelling techniques [20,21,25]. Lederberg's Vertex Graph method for describing molecular structures, combined with methods for embedding complex subgraphs within other graphs led first to systems that could help enumerate isomers and then to practical structure generation systems. Although other methods are now used to establish unique identifiers for chemical structures [26], Lederberg's scheme for classifying structures is still of some current research interest. Brown, Carhart, and Nourse invented and refined many algorithms for the representation of canonical forms for structures, for subgraph matching, for analysing symmetry, and for handling stereochemistry. Few of these algorithms have been made as accessible to other researchers as one might have wished; however, these algorithms continue to be used and developed by specialists in companies such as Molecular-Design.

Possibly, the real achievement of Dendral was to help change the way people thought about computers and problem solving. The success of heuristic Dendral, and its stepchild Mycin, encouraged people to look beyond numerical computation and data processing tasks. Dendral and Mycin showed how complex real-world classification and diagnostic problems could be tackled provided that adequate domain-specific knowledge was given to a program that incorporated even a relatively simple inference procedure.

REFERENCES

1 R.K. Lindsay, B.G. Buchanan, E.A. Feigenbaum and J. Lederberg, *Applications of Artificial Intelligence for Organic Chemistry: The DENDRAL Project*, McGraw-Hill, New York, 1980.
2 N.A.B. Gray, *Computer Assisted Structure Elucidation*, Wiley-Interscience, New York, 1986.
3 E.J. Corey, A.K. Long and S.D. Rubenstein, Computer-assisted analysis in organic synthesis, *Science*, 228 (1985) 408–418.
4 A.P. Johnson, Computer aids to synthesis planning, *Chemistry in Britain*, (January 1985) 59–67.
5 W.T. Wipke, G.I. Ouchi and S. Krishnan, Simulation and evaluation of chemical synthesis — SECS: an application of artificial intelligence techniques, *Journal of Artifical Intelligence*, 11 (1978) 173–193.
6 J. Lederberg, G.L. Sutherland, B.G. Buchanan, E.A. Feigenbaum, A.V. Robertson, A.M. Duffield and C. Djerassi, Applications of artificial intelligence for chemical inference. I. The number of possible organic compounds: acyclic structures containing C, H, O, and N, *Journal of the American Chemical Society*, 91 (1969) 2973–2976.
7 H. Budzikiewicz, C. Djerassi and D.H. Williams, *Structure Elucidation of Natural Products by Mass Spectrometry*, Holden-Day, San Francisco, 1964.
8 B.G. Buchanan, G.L. Sutherland and E.A. Feigenbaum, Heuristic Dendral: a program for generating explanatory hypotheses in organic chemistry, in B. Meltzer and D. Michie (Editors), *Machine Intelligence, 4*, Edinburgh University Press, Edinburgh, 1969, pp. 209–254.
9 A.M. Duffield, A.V. Robertson, C. Djerassi, B.G. Buchanan, G.L. Sutherland, E.A. Feigenbaum and J. Lederberg, Applications of artificial intelligence for chemical inference. II. Interpretation of low resolution mass spectra of ketones, *Journal of the American Chemical Society*, 91 (1969) 2977–2981.
10 B.G. Buchanan, G.L. Sutherland and E.A. Feigenbaum, Rediscovery of some problems of AI in the context of organic chemistry, in B. Meltzer and D. Michie (Editors), *Machine Intelligence, 5*, Edinburgh University Press, Edinburgh, 1969, pp. 253–280.
11 G. Schroll, A.M. Duffield, C. Djerassi, B.G. Buchanan, G.L. Sutherland, E.A. Feigenbaum and J. Lederberg, Applications of artificial intelligence for chemical inference. III. Aliphatic ethers diagnosed by their low resolution mass spectra and NMR data, *Journal of the American Chemical Society*, 91 (1969) 7440–7445.
12 A. Buchs, A.M. Duffield, G. Schroll, C. Djerassi, A.B. Delfino, B.G. Buchanan, G.L. Sutherland, E.A. Feigenbaum and J. Lederberg, Applications of artificial intelligence for chemical inference. IV. Saturated amines diagnosed by their low resolution mass spectra and NMR data, *Journal of the American Chemical Society*, 92 (1970) 6831–6838.
13 D.A. Waterman, *Machine learning of heuristics*, Ph.D. thesis, Department of Computer Science, Stanford University, 1968.
14 A. Newell and H.A. Simon, *Human Problem Solving*, Prentice-Hall, Englewood Cliffs, 1972.
15 A. Buchs, A.B. Delfino, A.M. Duffield, C. Djerassi, B.G. Buchanan, E.A. Feigenbaum and J. Lederberg, Applications of artificial intelligence for chemical inference. VI. Approach to a general method of interpreting low resolution mass spectra with a computer, *Helvetica Chimica Acta*, 53 (1970) 1394–1417.
16 B.G. Buchanan, E.A. Feigenbaum and J. Lederberg, A heuristic programming study of theory formation in science, in *Proceedings of the 2nd International Joint Conference on Artificial Intelligence, September 1971*, W. Kaufmann Inc, Los Altos, CA, 1971, pp. 40–50.
17 B.G. Buchanan and E.H. Shortliffe (Editors), *Rule Based Expert Systems: The Mycin Experiments of the Stanford Heuristic Programming Project*, Addison-Wesley, Reading, MA, 1984.
18 R.E. Carhart, D.H. Smith, H. Brown and N.S. Sridharan, Applications of artificial intelligence for chemical inference. XVI. Computer generation of vertex graphs and ring systems, *Journal of Chemical Information and Computer Sciences*, 15 (1975) 124–130.
19 H. Brown, L. Masinter and L. Hjelmeland, Constructive graph labeling using double cosets, *Discrete Mathematics*, 7 (1974) 1–30.
20 H. Brown and L. Masinter, An algorithm for the construction of the graphs of organic molecules, *Discrete Mathematics*, 8 (1974) 227–244.
21 L. Masinter, N.S. Sridharan, R.E. Carhart and D.H. Smith, Applications of artificial intelligence for chemical inference. XII. Exhaustive generation of cyclic and acyclic isomers, *Journal of the American Chemical Society*, 96 (1974) 7702–7714.

22. L. Masinter, N.S. Sridharan, R.E. Carhart and D.H. Smith, Applications of artificial intelligence for chemical inference. XIII. Labeling of objects having symmetry, *Journal of the American Chemical Society*, 96 (1974) 7714-7723.
23. D.H. Smith, Applications of artificial intelligence for chemical inference. XV. Constructive graph labeling applied to chemical problems. Chlorinated hydrocarbons, *Analytical Chemistry*, 47 (1975) 1176-1179.
24. D.H. Smith, Applications of artificial intelligence for chemical inference. XVII. The scope of structural isomerism, *Journal of Chemical Information and Computer Sciences*, 15 (1975) 203-207.
25. H. Brown, Molecular structure elucidation. 3, *SIAM Journal of Applied Mathematics*, 32 (1977) 534-551.
26. H.L. Morgan, The generation of a unique machine descriptor for chemical structure — A technique developed at Chemical Abstracts Service, *Journal of Chemical Documentation*, 5 (1965) 107-113.
27. R.E. Carhart, D.H. Smith, H. Brown and C. Djerassi, Applications of artificial intelligence for chemical inference. XVII. An approach to computer-assisted elucidation of molecular structure, *Journal of the American Chemical Society*, 97 (1975) 5755-5762.
28. A. Lavanchy, T. Varkony, D.H. Smith, W.C. White, R.E. Carhart, B.G. Buchanan and C. Djerassi, Rule-based mass spectrum prediction and ranking. Applications to structure elucidation of novel marine sterols, *Organic Mass Spectrometry*, 15 (1980) 355-366.
29. N.A.B. Gray, R.E. Carhart, A. Lavanchy, D.H. Smith, T. Varkony, B.G. Buchanan, W.C. White and L. Creary, Computerized mass spectrum prediction and ranking, *Analytical Chemistry*, 52 (1980) 1095-1102.
30. N.A.B. Gray, C.W. Crandell, J.G. Nourse, D.H. Smith, M.L. Dageford and C. Djerassi, Applications of artificial intelligence for chemical inference. XXXIV. Computer-assisted structural interpretation of C-13 spectral data, *Journal of Organic Chemistry*, 46 (1981) 703-715.
31. Y. Kudo and S.I. Sasaki, Principle for exhaustive enumeration of unique structures consistent with structural information, *Journal of Chemical Information and Computer Sciences*, 16 (1976) 43-49.
32. J.G. Nourse, D.H. Smith, R.E. Carhart and C. Djerassi, Computer assisted elucidation of molecular structure with stereochemistry, *Journal of the American Chemical Society*, 102 (1980) 6289-6295.
33. R.E. Carhart, D.H. Smith, N.A.B. Gray, J.G. Nourse and C. Djerassi, GENOA: a computer program for structure elucidation utilizing overlapping and alternative substructures, *Journal of Organic Chemistry*, 46 (1981) 1708-1718.
34. D.H. Smith, B.G. Buchanan, R.S. Engelmore, A.M. Duffield, A. Yeo, E.A. Feigenbaum, J. Lederberg and C. Djerassi, Applications of artificial intelligence for chemical inference. VIII. An approach to the computer interpretation of the high resolution mass spectra of complex molecules. Structure elucidation of estrogenic steroids, *Journal of the American Chemical Society*, 94 (1972) 5962-5973.
35. D.H. Smith, B.G. Buchanan, W.C. White, E.A. Feigenbaum, C. Djerassi and J. Lederberg, Applications of artificial intelligence for chemical inference. X. IntSum: a data interpretation program as applied to the collected mass spectra of estrogenic steroids, *Tetrahedron*, 29 (1973) 3117-3134.
36. B.G. Buchanan, D.H. Smith, W.C. White, R. Gritter, E.A. Feigenbaum, J. Lederberg and C. Djerassi, Applications of artificial intelligence for chemical inference. XXII. Automatic rule formation in mass spectrometry by means of the MetaDendral program, *Journal of the American Chemical Society*, 96 (1976) 6168-6178.
37. T.M. Mitchell and G.M. Schwenzer, Applications of artificial intelligence for chemical inference. XXV. A computer program for automated empirical 13C nmr rule formation, *Organic Magnetic Resonance*, 11 (1978) 378-384.
38. J.E. Gurst and A.K. Schrock, Mass spectra of androstane-7,17-diones, *Journal of Organic Chemistry*, 45 (1980) 4062.
39. N.A.B. Gray, Applications of artificial intelligence for organic chemistry. Analysis of C-13 spectra, *Journal of Artificial Intelligence*, 22 (1984) 1-21.
40. C. Djerassi, D.H. Smith and T.H. Varkony, A novel role of computers in the natural products field, *Naturwissenschaften*, 66 (1979) 9-21.
41. C. Djerassi, D.H. Smith, C.W. Crandell, N.A.B. Gray, J.G. Nourse and M.R. Lindley, The Dendral Project: computational aids to natural products structure elucidation, *Pure and Applied Chemistry*, 54 (1982) 2425-2442.

 Invited Comments

On Gray's Interpretation of the Dendral Project and Programs: Myth or Mythunderstanding?

BRUCE G. BUCHANAN and EDWARD A. FEIGENBAUM *

Knowledge Systems Laboratory, Stanford University, Stanford, CA 94804 (U.S.A.)

JOSHUA LEDERBERG

Rockefeller University, New York, NY 10021 (U.S.A.)

INTRODUCTION

Gray's paper contains a description of the DENDRAL system and some of its subprograms, and opinions of the merits of this project. Previous overviews of the DENDRAL project include Gray's first references [1] and [2]. We feel that Gray has misunderstood several of the motives and achievements of this work and briefly comment on these below.

Gray's paper goes beyond the narrow and specific difficulties he finds with the DENDRAL programs. He broadens the scope to cast a shadow on the underlying symbolic-processing technology and methodology ("AI").

Since Gray has the perspective of a chemist, not a computer technologist, we feel it necessary to address this broader question as well.

EXAMPLES OF MISUNDERSTANDINGS

The first question is whether Gray has reported accurately on the DENDRAL programs. Because the DENDRAL work has been documented widely in the literature (see references in Gray's paper and in [1], it is neither useful nor necessary to do a thorough accounting in this brief note, but we present below two examples of misunderstandings:

1. The central element and conceptual foundation of the entire set of programs is the DENDRAL generator (later renamed CONGEN). Gray describes the generator in his Fig. 1 as a "heuristic filter" that checks for "some spectral evidence for each radical when building up structure". This is fundamentally wrong: the generator systematically enumerates possible structures within constraints.
2. Gray states that the Planner is part of Meta-DENDRAL (our learning program). In fact, it is the Plan part of the basic plan–generate–test method for producing solutions for structure elucidation problems, and has nothing to do with rule learning.

GOALS OF THE DENDRAL PROJECT AND GRAY'S MISINTERPRETATION OF THEM

1. Certainly intellectual assistance to chemists in doing the difficult and often error-prone task

of systematic structure elucidation was one of our goals. We did not regard structure elucidation as a task that was trivial and uninteresting for chemists, as Gray seems to believe it is.

2. Notwithstanding this, the DENDRAL work was motivated by the computer science question of how computer programs could be designed and built to assist with hypothesis formation in science. Chemistry and mass spectrometry were secondary to this question, although we also wanted to demonstrate the power of the design by building high performance programs. To illustrate this, when we published the major retrospective of the first decade and a half of this research in book form (Gray's first reference [1]), it was published in the McGraw-Hill Advanced Computer Science Series.

3. All work on expert systems, including our own in DENDRAL and many other projects, is motivated by the intent to provide assistance to human intellectual endeavor, not to provide replacements for skilled human problem solvers. It is unreasonable to assume, as Gray seems to, that Carl Djerassi and Joshua Lederberg believed structure elucidation chemists ignored all empirical data except mass spectra. The DENDRAL programs focused on mass spectra as exemplary of a source of information about structure, and in fact we explored uses of other sources as well.

4. University research projects almost never produce "industrial strength" products. They aim to pioneer proof-of-concept prototypes. The effort and expense to productize is an order-of-magnitude more than to do the basic research. In the usual way, the DENDRAL programs were licensed by Stanford University to an industrial firm for this productizing effort, but the firm (for whatever reason) never made the necessary efforts at product engineering, market education, or marketing. Hence market acceptance of DENDRAL was low. However, the firm used portions of DENDRAL in its other products, and indeed hired the three key young computational chemists of the DENDRAL Project (Smith, Nourse, and Carhart), one as Director of Research. The lack of market acceptance was not due, contrary to Gray's presumption, to lack of performance: DENDRAL's performance as a structure elucidation tool was demonstrated repeatedly to be high because of its systematic nature. The plan–generate–test paradigm used in DENDRAL was indeed adequate for solving complex hypothesis formation problems.

5. Gray does not appear to understand what kind of code is needed in practical expert systems. Gray remarks on how little "AI" code there is in DENDRAL (how does he count it? lines? concepts?), but apparently doesn't know that practical expert systems contain major chunks of code for "system", user interface handling, and routine symbol manipulation. The rule of thumb is: well over half.

CONCLUSION: THE CONTRIBUTIONS OF DENDRAL

The DENDRAL Project is regarded as the grandfather of knowledge based systems in AI. Today's commercially important sector of these is called expert systems. There are at least two thousand of these in use today in companies in the U.S.A., Europe, and Japan, and several thousand more prototyped or under development. The technology has become routinized, and the payoff in some cases has been in the tens of millions of dollars per year. This is not the place to tell that story; it is told elsewhere [3].

DENDRAL began at a time when the focus of AI research was on general problem solving methods (e.g. GPS, theorem-proving, etc.). DENDRAL research was the forcing function behind the so-called "shift to the knowledge-based paradigm" in AI, the paradigm that has since then dominated the field. DENDRAL experiments prior to 1968 led to the knowledge-is-power hypothesis, now called the knowledge principle by AI scientists, that the power of programs to solve complex problems derives not from the power of their reasoning methods but from what they know about their task domains.

The DENDRAL Project scientists were the first to formulate the rule-based representation of knowledge that today dominates the expert systems field. This representation method led directly to the famous MYCIN program at Stanford, whose clones fill the space of expert systems in use today.

The work on the DENDRAL generator, culminating in CONGEN, was also landmark work. It solved in a completely orderly, logical, and mathematically rigorous way the problem of the systematic generation of chemical structures, with and without constraints imposed by the user; and it offered canonical notational forms for the structures (not only important but necessary for chemistry).

The work on Meta-DENDRAL, the mass-spectral rule learning program that was our first attempt at theory formation from empirical data, revived the research area of machine learning research in the early 1970s. Its "test case" produced mass-spectral fragmentation rules interesting enough to be published by a major journal of the chemical literature. Much more important, its concepts led directly to the Version Space approach to the search problem in machine learning and through that path to the approach called Explanation-Based Learning.

Thus the DENDRAL legacy is rich in contributions to the stream of AI science and technology and to computational chemistry. Gray's paper tends to obscure that legacy and we wish to correct the misunderstandings arising from it.

REFERENCES

1. R.K. Lindsay, B.G. Buchanan, E.A. Feigenbaum and J. Lederberg, *Applications of Artificial Intelligence for Organic Chemistry: the DENDRAL Project*, McGraw-Hill, New York, 1980.
2. C. Djerassi, D.H. Smith, C.W. Crandell, N.A.B. Gray, J.G. Nourse and M.R. Lindley, The Dendral Project: computational aids to natural products structure elucidation, *Pure and Applied Chemistry*, 54 (1982) 2425.
3. E. Feigenbaum, P. McCorduck and H.P. Nii, *The Rise of the Expert Company*, Times Books, New York, 1988.

Response to comments by Buchanan, Feigenbaum and Lederberg

N.A.B. GRAY

*Department of Computing Science, University of Wollongong,
P.O. Box 1144, Wollongong, N.S.W. 2500 (Australia)*

THE DENDRAL STRUCTURE GENERATOR CONGEN AS AN "AI" PROGRAM

As a discipline in computing science, artificial intelligence (AI) is concerned with the explicit representation and use of "knowledge" for problem solving. The knowledge needed for any particular problem is obviously domain specific; AI researchers are concerned with general problems of knowledge representation and use. For some researchers the issue is simple – the only representation is logic, and all use must be through the inference procedures defined in the predicate calculus. Others, of a more empirical practical persuasion, utilise rules with confidence factors that are applied through production rule interpreters.

The purpose of a particular AI program may be to configure a computer system, generate a precis of a newspaper story, or diagnose an infection – the common element among such programs is the use of some inference system that can exploit extensive domain knowledge. How can one assess the AI content of a program? One can examine its use of domain knowledge.

The knowledge needed for chemical structure elucidation is varied – one requires an understanding of chemical stability, information on the extraction and work-up procedures through which the compound was isolated, and information on how to interpret spectral and chemical evidence. I have looked in Congen to find how it uses such domain knowledge (the versions I chose to ex-amine were the 1978 BCPL implementation based on Carhart's AMGEN algorithm and the 1980 version based on Carhart's constructive substructure search algorithm – code that should be familiar to Feigenbaum and Buchanan).

The program does indeed contain domain knowledge – there is a table of nine entries of the form "all graph nodes labelled 'C' have valence 4", "all graph nodes labelled 'O' have valence 2", and so forth. Apart from this use of domain knowledge, the code implements combinatorial algorithms. There are routines that encode algorithms that compare matrices to determine which is the lowest scoring on the basis of a given ordering criterion, there are routines that handle the problem of choosing all unique combinations of data elements chosen from some larger set. Some may consider such a program, based purely on combinatorial algorithms, to be an AI program and to possess knowledge ("it knows how to label a graph"). But on that criterion, a program implementing Hoare's Quicksort could be an AI program ("it knows how to order data elements") The notion of what is an AI program then becomes fatuous; the AI name can be accorded on the whim of the implementor.

USE OF KNOWLEDGE IN THE DENDRAL PROGRAMS

In addition to the core structure generator program, the various systems developed for the Dendral project possessed separate routines for spectral prediction and, to a much more limited extent, spectral interpretation. These spectral interpretation and prediction systems are empirical, and consequently have only limited value in a structure elucidation system. Spectrum prediction systems can at best help identify the most plausible candidate structures and so assist the chemist in the choice of further experiments that might discriminate among the remaining candidates.

These prediction routines use one-step inference – "if the molecule possesses this substructure, it will have a C-13 resonance at ... (or show ions that result from the cleavage of bonds A and B)". There are no long logical arguments to be pursued; one simply matches and asserts results. Performance depends solely on the size of the rule-library; implementation issues involve finding unique representations of the substructures, and providing fast look-up into the file of rules. Expertise becomes a glorified table-lookup. This form of expertise is closer to the model espoused by Dreyfus than to the model favoured by AI researchers.

Of the various Dendral systems, only one made a systematic attempt to thoroughly exploit chemical knowledge at all stages. As illustrated in my Fig. 1, Heuristic Dendral employed a decision tree classifier in its planning stage and a simple rule-based spectrum filter in its test phase. It also possessed a heuristic filter in its generator. The paper "Heuristic Dendral; a program for generating exploratory hypotheses in organic chemistry" (B. Buchanan, G. Sutherland, and E.A. Feigenbaum, *Machine Intelligence 4*, Edinburgh University Press, Edinburgh, 1969, pp. 209–254) describes the heuristic extensions to Lederberg's Dendral algorithm. These extensions implemented a "zeroth order mass spectral theory" that constrained the output of the generator. This paper states that "As each partition is generated it can be checked for plausibility before any attempt is made to generate the corresponding radicals. Each sub-composition is checked against the spectrum. If its weight is not present, the whole partition can be bypassed.". Lederberg's Dendral algorithm can be used as an exhaustive generator of acyclic structures and, as described in the comments by Buchanan, Feigenbaum and Lederberg, does systematically enumerate (acyclic) structures; but, the published description of the algorithm used in Heuristic Dendral is clear as to its use of heuristic mass spectral based filters.

THE PRESENTATION OF THE DENDRAL PROJECT

In the original scientific literature, the presentation of Dendral is correct. The Brown & Masinter papers describe the mathematical basis of graph labelling algorithms. Nourse's papers discuss the data structures and algorithms required for the computerized representation and manipulation of molecular stereochemistry. The Carhart & Smith papers on Congen and Genoa describe the algorithms, and provide worked examples that illustrate how such programs might be used to assist in the solution of a structural problem. The various papers on spectral prediction note any inherent biases of a prediction model, and review problems associated with the employment of empirically predicted spectra as aids for focussing on a likely candidate structure. This scientific work has great merit, irrespective of any implementation the algorithms and use (or neglect) of the programs. The work can be assessed by others in the appropriate peer-group of computational chemistry.

However, it is not the scientific papers that are read. Most people learn of Dendral through the AI literature. A standard text on building expert systems informs its readers that Dendral has caused a redefinition of the roles of humans and machines in chemical research; in the IEEE Expert journal, a member of the Stanford laboratory reports that Dendral ("based on a core AI concept, the use of heuristics") is one of the most successful AI programs ever and can be found in many organic chemistry laboratories. In their textbooks, computing science undergraduates learn of Dendral as a practical AI tool in use in chemistry. I am concerned by these presentations of the Dendral project. Professors Buchanan, Feigenbaum and Lederberg do not share my concern. It appears that they cannot perceive any grounds for concern.

Chapter 5

Expert Systems in Synthesis Planning: a User's View of the LHASA Program

THOMAS V. LEE

School of Chemistry, The University, Bristol BS8 1TS (U.K.)

(Received 13 May 1987; accepted 21 August 1987)

CONTENTS

1 Introduction 53
2 Retrosynthetic analysis and the LHASA program 54
3 The LHASA program 54
   3.1 Example 1 57
   3.2 Example 2 59
4 Evaluation of the LHASA program 62
5 Future developments of the LHASA program 65
6 Accessing the LHASA program 65
7 Conclusions 66
8 Acknowledgements 66
References 66

1 INTRODUCTION

For many years organic chemists have striven to apply computational chemistry to organic synthesis so enabling the computer-designed preparation of complex organic molecules to become commonplace. Although this has not yet been achieved, some notable advances have been made and it is now obvious that the computer does have a role to play in synthesis planning. This role will undoubtedly increase in future years but it is the purpose of this present article to show what can currently be achieved in synthesis planning by computer. The article is written from the point of view of a user of the most advanced of the synthetic programs which are commonly accessible. Since the impetus for advances in the development of any expert system is most likely to come from user demand such a view would appear to be the most valuable.

There are now a number of synthesis planning systems in use, the bulk of which utilise an empirical knowledge base. The first of these programs was initiated in 1968 by Corey and Wipke [1] being known as OCSS and is now under development in industry as SECS (Simulation and Evaluation of Chemical Synthesis). However despite there being about fifty other systems reported, including some such as EROS [2] and SYNGEN [3] which are not based upon an empirical knowledge base, there is no doubt that the most used and most advanced synthetic planning program is that developed from OCCS by Corey et al. at Harvard [4], viz., LHASA (Logic and Heuristics Applied to Synthetic Analysis). It is the use of this program that the current article will concentrate upon, prefaced by a simple description of the chemistry upon which the program is based.

2 RETROSYNTHETIC ANALYSIS AND THE LHASA PROGRAM

Retrosynthetic analysis is the modern method used by organic chemists to design the synthesis of complex molecules [5]. It involves working backwards from a target molecule via successive levels of precursors so arriving at recognisable potential starting materials. For example, in Fig. 1 the double arrow represents the theoretical retroanalytical step corresponding to the well known laboratory reaction of Wittig olefination. It is this simple and logical approach to synthesis that has been responsible for many of the major advances in synthetic organic chemistry during the last decade. Fig. 2 highlights the simplicity and power of the method by considering the synthesis of the compound **1**. A simple functional group interconversion (FGI) suggests that the acid **2** will be a good precursor of **1**. Such a decision depends upon some prior experience and upon the intuition of the chemist. The reason for this particular choice is to permit the selection of further design options later in the synthetic planning. This forward planning process is an integral part of retrosynthetic analysis and, as we will see, it can be mimicked by the expert system under consideration. The next stage in our synthetic design is to breakdown or disconnect **2** into simple precursors. One good option which all chemists (and any computer program aimed at synthesis) will recognise is to disconnect at the bond being 'cut' by the wavy line. Assuming a heterolytic cleavage (in which the pair of electrons represented by the bond both move to the same atom) we then arrive at four possible charged species, A–D, which are called synthons. Every bond in the molecule could be treated in this way but both chemists and the computer would in the majority of cases use their empirical knowledge base to disconnect this bond. This reasons for this are that most of the theoretical fragments A–D can be recognised as being equivalent to certain reagents or synthetic equivalents which are readily available and easy to use. The fact that fragment A does not have an immediately obvious simple equivalent does not preclude its use but suggests that the use of the reagents equivalent to C and D would be easier. Thus the synthesis would follow the sequence shown, involving a Michael addition, ester hydrolysis, decarboxylation and selective reduction with the exact conditions for each reaction being obtained from the chemical literature.

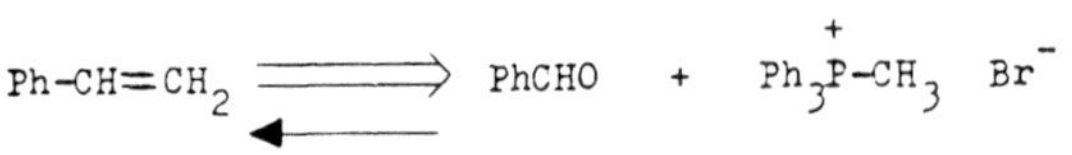

It is important to note that for simple molecules such as this, the approach tends to be a tactical one requiring a wide knowledge of chemical reagents and reactivities. For very complex molecules such a tactical approach often requires for an efficient synthesis, execution within an overall strategic framework in which one particular bond forming process is the key to the synthesis and all the tactical decisions are made to allow the use of this key process. As we will see the expert program we are considering allows the choice of a tactical approach or, more usefully, allows us to consider tactical bond forming processes within a strategic framework.

3 THE LHASA PROGRAM

As with most of the synthetic design systems LHASA is an interactive program using computer graphics [6]. After the user has entered the target molecule (the compound whose synthesis is required) the program enters a perception phase involving the system becoming aware of the presence of key structural features in the target such as rings or chiral centres. Following this the user now has a choice of either asking the program to generate all possible first level precursors (a process constrained only by the limits of the system's knowledge base) or the user can, by employing their own expert knowledge, impose certain constraints which limit the number of retrosyn-

$Ph-CH=CH_2 \rightleftharpoons PhCHO + Ph_3\overset{+}{P}-CH_3 \ Br^-$

Fig. 1. Retrosynthetic representation of the Wittig reaction.

thetic steps generated. The first of these options can generate an unmanageable number of alternative routes and is only of value for very simple molecules but in either case the user is now faced with a set of precursors each of which could now be processed in the same manner, but which normally are selected for processing by the user. This procedure is continued until precursors are suggested which the user recognises as being readily available. (One can envisage a future development

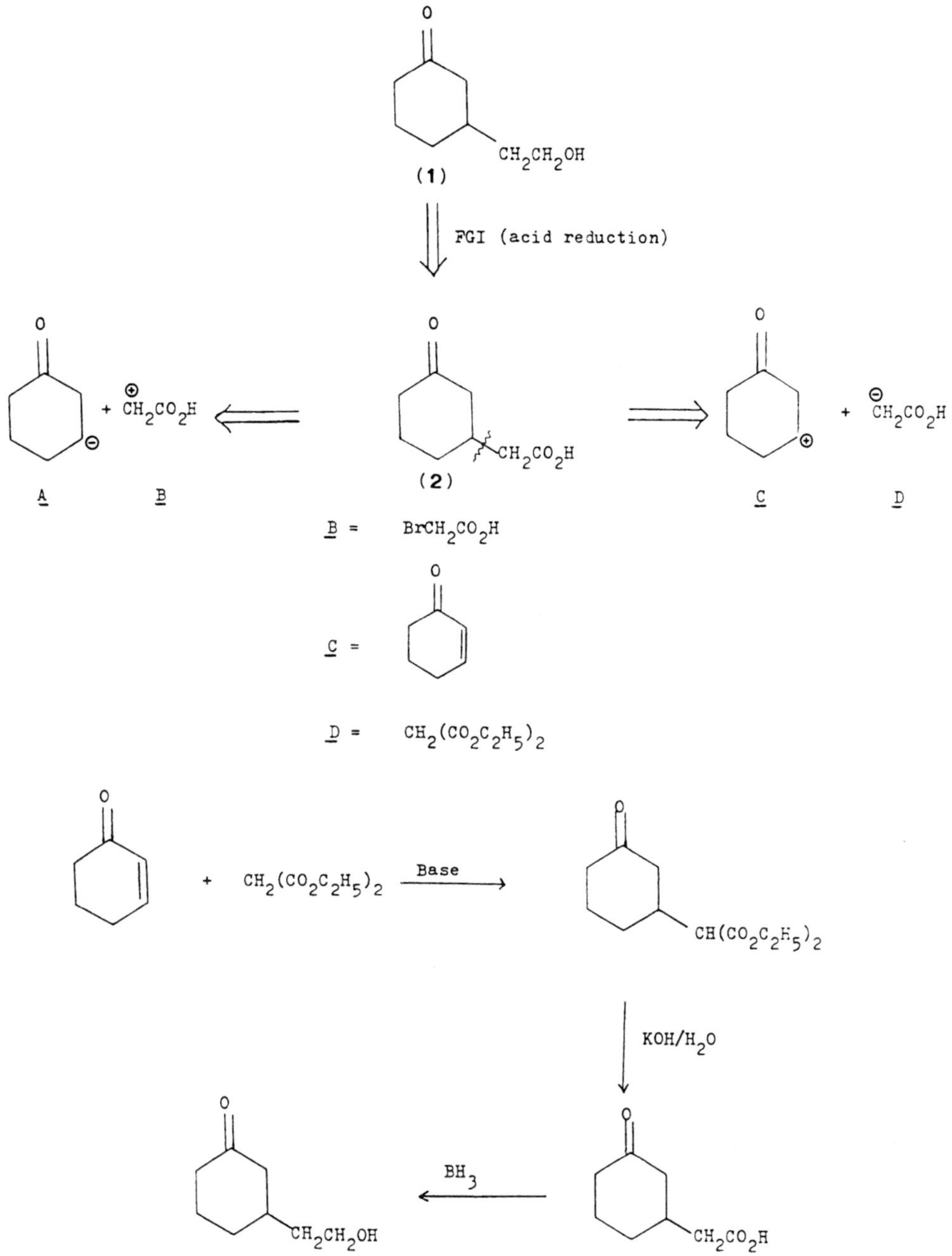

Fig. 2. Retrosynthetic analysis.

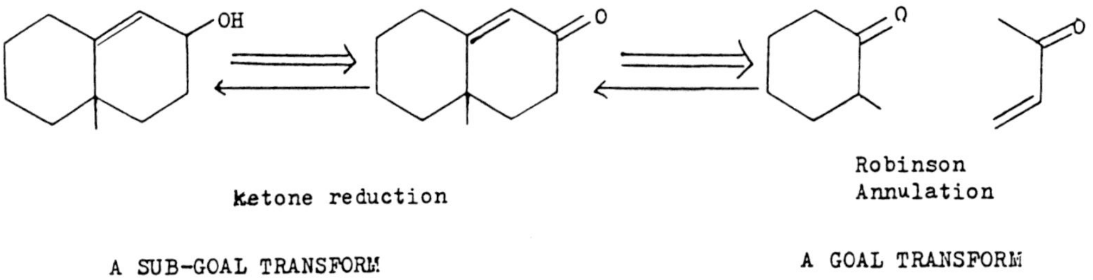

ketone reduction

Robinson Annulation

A SUB-GOAL TRANSFORM

A GOAL TRANSFORM

Fig. 3. Goal and sub-goal transformations.

whereby each precursor generated is automatically searched for in a data base of chemical suppliers' listings. However only a small proportion of "readily available starting materials" are commercially available.)

By use of the above constraints the user can

```
Transform 410
Name Claisen Rearrangement
...TL 3243(1969), JACS v.92 741,4461,4463(1970),v.95 553(1973)
Rating 30
...Kill Specifiers for Prescreen
Ringbond Trans2, Functional*group Transl
No*FG*within*alpha Trans3
Olefin*type Tetrasubstituted Z*Disubstituted
Spacing*1*5 Well*defined*T*first Well*defined*L*last
...
If there is a functional group on alpha to atom*3 &
        offpath then go to Block4
If there is a functional group on beta to atom*3  &
        offpath then go to Block3
If there is a functional group on gamma to atom*3 &
        offpath then go to Block4
Block3 Designate the group as group one
    If heterol*1 is nitrogen then go to Block6
    If the first group is carbonyl or: ester then go to Block2
    Go to Block4
Block6 If the first group is amide*3 then go to Block2
    Exchange the group for an amide*3 and *then go to Block5
Block4 Exchange the group for an ester on beta to atom*3 offpath
Block5 If unsuccessful then reject
Block2 Designate the group on beta to atom*3 offpath as group one
    If alpha to carbonl*1 is the same as alpha to atom*3
    Save as 1 the previous locant
    Kill if there is a hetero atom alpha alpha to saved*atom 1
Subtract 10 if atom*3 is a quaternary*center
....
        If there is a multiple bond in the first group &
            then break it
        If first group is carbonyl then join saved*atom 1 &
            and carbonl*1
        If first group is carboxyl then attach an ether  &
            to carbonl*1
    Attach an ether to carbonl*1
    Separate saved*atom 1 and atom*3
    Double bond*3
    Single bond*5
    Attach a hydroxyl to atom*5
```

Fig. 4. A LHASA transform (reprinted from ref. 7, with permission).

impose a specific strategy in the retrosynthetic analysis with several alternative strategies currently being available, e.g. one can limit the retrosynthetic step to Diels–Alder reactions or permit retrosteps which result in stereospecific alkene bond formation only. It is the wise choice of these strategies that is the essence of the use of LHASA. The imposition of a poor strategy will only generate poor synthetic pathways and it is at this point that the novice user realises that LHASA is a true expert system.

The generation of precursors in a retrosynthetic step is performed in the program by the use of "transforms". The LHASA database contains a set of transform specifications all of which possess a series of rules ensuring recognition of:

(a) the substructure which needs to be present in the target molecule if a transform is to be attempted;

(b) the range and restraints of the reaction being proposed, notably with respect to the effects of substituents in the vicinity of the reacting centres;

(c) the structural changes required to arrive at the precursor molecule.

Further to this there are two types of LHASA transform, viz., goal transforms and subgoal transforms; Fig. 3 illustrates examples of these processes. Goal transforms usually give precursors to the target of simpler structure whereas subgoal transforms generate functional group interconversions such as ketone reduction to a secondary alcohol. These latter transforms are used 'intelligently' being chosen by the system to allow a specific goal transform to be performed. Thus the computer is mimicking the logical thought processes employed by the chemist using retrosynthetic analysis (vide supra).

An example of a LHASA transform is shown in Fig. 4 [7] being written in a customised language called CHMTRN [8] which is easily understood by both chemist and computer.

It is of interest to discuss further some of the features of these transforms since they demonstrate clearly how they are a computer version of the logical decisions made, often intuitively, by the chemist. For instance they contain qualifiers (e.g. line 23 in Fig. 4) which are included to account for the fact that even though a target may contain the correct functionality a particular transform may not be appropriate. This may arise due to steric reasons for example or due to the proximity of interfering groups. These qualifiers thus express logically the many experimental and mechanistic features discovered only after many years of work on the more important organic transformations. Thus a synthesis design program such as LHASA which utilises a knowledge base, serves to provide a usable condensate of these many years' work rather than make it redundant (a fear which many organic chemists still express).

A further feature of the transforms are their 'ratings'. Each suggested reaction is given a numerical rating followed by an increment. The figure expresses how favourable the reaction should be with a positive increment indicating that the substrate is more favoured in the suggested reaction and a negative increment meaning the opposite. Finally transforms also contain literature references leading to precise experimental details which are of course the essence of preparative chemistry.

We will now work through two examples to demonstrate how LHASA is used in practice and how it correlates with the principles of retrosynthetic analysis.

3.1 Example 1

The ketone 3 (Fig. 5) has been synthesised and used as an intermediate in a natural product synthesis in the present author's laboratory [9]. The synthetic design of this molecule used retrosynthetic analysis but not LHASA so allowing a direct comparison of the program's suggestions with those used in practice.

The target molecule can be drawn onto the 'sketch pad' (screen) of LHASA using the graphical interface in about thirty seconds. On a Tektronix terminal the cross hairs cursor is placed at a desired location and the cursor activated by using any key to position an atom. The cursor is moved and activated at a second position so forming a bond and the process is repeated as necessary. Any atom can be entered by replacing the relevant symbols off the screen, with the default

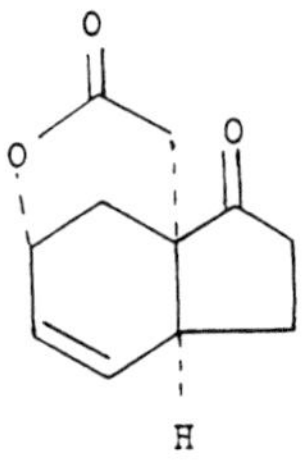

Fig. 5. Precursors to the ketone (3).

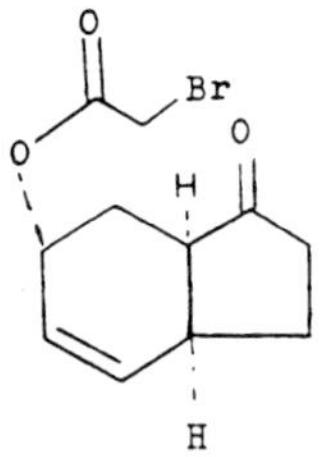

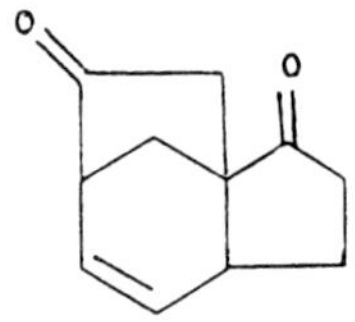

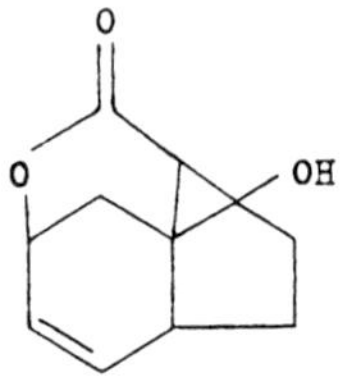

atom being carbon, and multiple bonds are formed by redrawing over single bonds. Various templates, e.g. a cylohexane ring, can be picked off the screen also to save the need for repetitive drawing. Stereochemistry can be denoted using WEDGE, SOLID and DASH 'buttons' present on the screen as part of the menu-driven LHASA system.

Hitting 'process' on the menu causes the program to enter its perception phase (any structural or stereochemical ambiguities are conveyed at this stage to the user for correction), the user is then faced with the process menu (Fig. 6) essentially consisting of short-range, bond-mode and long-range searches.

The simplest, and usually unwisest, course is to ask for an unconstrained analysis whereby all possible retrosteps will be considered. Doing this generates a whole range of bond breaking transforms most of which possess a very unfavourable rating. Thus we see suggested an oxy-Cope rearrangement or an intramolecular alkylation (with the precursor 4), or a Baeyer–Villiger reaction or a retro-aldol reaction (precursors 5 and 6). Each of these appears reasonable but none has generated a simplified precursor and so are of little value. However such an analysis does illustrate one important feature of LHASA. This is the retrosynthetic tree which is a graphical representation of the retroanalysis in which the target molecule is

SHORT-RANGE SEARCHES	BOND-MODE SEARCHES	LONG-RANGE SEARCHES
UNCONSTRAINED	CYCLIC STRATEGIC	STEREOSPECIFIC C=C
DISCONNECTIVE	POLYFUSED STRATEGIC	DIELS-ALDER
RECONNECTIVE	APPENDAGES	ROBINSON ANNULATION
UNMASKING	RING APNDG ONLY	CYCLOPROPANES
STEREOSELECTIVE	BRANCH APNDG ONLY	HALOLACTONIZATION
	MANUAL DESIGNATION	QUINONE DIELS-ALDER
		BIRCH REDUCTION

OPTIONS	PERCEPTION ONLY	CONTROL
ALIPHATIC ONLY	CYCLIC STRATEGIC	SKETCH
AROMATIC ONLY	POLYFUSED STRATEGIC	TREE
PRESERVED BONDS	6-RING CONFORMATIONS	DEBUG
PRESERVED STEREO	STEREOCHEMISTRY	TEST
STEREOCONSERVING	HUCKEL CALCULATION	EXIT
		HELP

Fig. 6. The process menu.

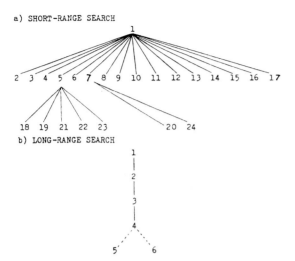

Fig. 7. Synthetic trees generated by LHASA.

connected to its precursor(s) via a solid line. Fig. 7a shows the tree (which was generated by the above unconstrained analysis with the precursors **5** and **7** being processed in a similar unconstrained manner). One useful point about this tree is that it demonstrates the tendency of short-range and bond-mode searches to generate broad trees. In contrast long-range searches usually give deeper trees in a procedure which, due to the automatic use of subgoals which allow a specific goal to be achieved, involves less user interaction (Fig. 7b).

If we now use a knowledge of chemistry in a preliminary analysis of the ketone **3** we can safely say that all organic chemists would recognise two features: (a) the lactone possessing an alkene one carbon removed from the C–O bond is a product of an halolactonization procedure and (b) a cyclohexene unit is most commonly derived from a Diels–Alder process. We now have an idea of what tactics we can employ in a long-range search for a retroanalysis of this simple molecule.

By stressing the use of halolactonization (from the process menu) we get the transforms shown in Fig. 8. Firstly an automatic subgoal transform introduces the iodide needed for the goal and then iodolactonization is suggested. This precursor (**7**) is then processed using the Diels–Alder long-range search which immediately gives the highly rated reaction shown. This involves a dienophile bearing electron-withdrawing groups which is a requirement for facile Diels–Alder reactions. It is worth noting that if one uses the Diels–Alder option on unactivated dienophiles then LHASA subgoals will automatically add electron-withdrawing groups prior to suggesting the specified reaction. Precursor **8** was then processed further using a bond-mode search via the manual designation button which allows the user to specify which bond is to be formed (the strategic bond). Specifying the alkene then gives a reasonable aldol reaction precursor. This strategy controlled analysis then gives the deep tree shown in Fig. 7b which constitutes a manageable synthetic sequence. Interestingly this strategy, with variations in detail, was essentially that employed in the synthesis of the ketone **3**. Although the above example is of minor value in a chemical sense it serves to illustrate the operation and some of the major features of LHASA. The remaining examples will emphasise some of the more specialised features of the system.

3.2 Example 2

The interesting natural product saudin (**9**) is currently a synthetic target within the present author's laboratories. Having already designed a synthetic route it was of interest to see how LHASA might handle this problem. The first step in our non-computerised analysis was to remove all of the relatively unstable functionality such as the two ketals and to open up the two lactones to give the cyclohexanone precursor **10** (Fig. 9). Most organic chemists would use this approach and LHASA which was written by chemists does have the facility for dealing with such functionality.

Thus, specifying an unmasking strategy from the short range search options results in LHASA suggesting the precursors shown in Fig. 10 derived from opening of the ketal function. The precursor **11** can then be analysed by choosing the short-range disconnection strategy, which eventually, as shown, resulted in a Michael addition reaction being suggested, so closely following the route under study in the laboratory. It is reasonable to say that any attempt to design a synthesis without this unmasking would most probably lead to a more complex synthetic route *. As a demonstra-

---
* Such statements are commonly disproved by the ingenuity of synthetic chemists.

Chapter 5

READY   DISCONTINUE   STOP
         REJECT RING   DEBUG
                        REFERENCE
                        COMMENTS
                        VIEW TREE
                        SUSPEND

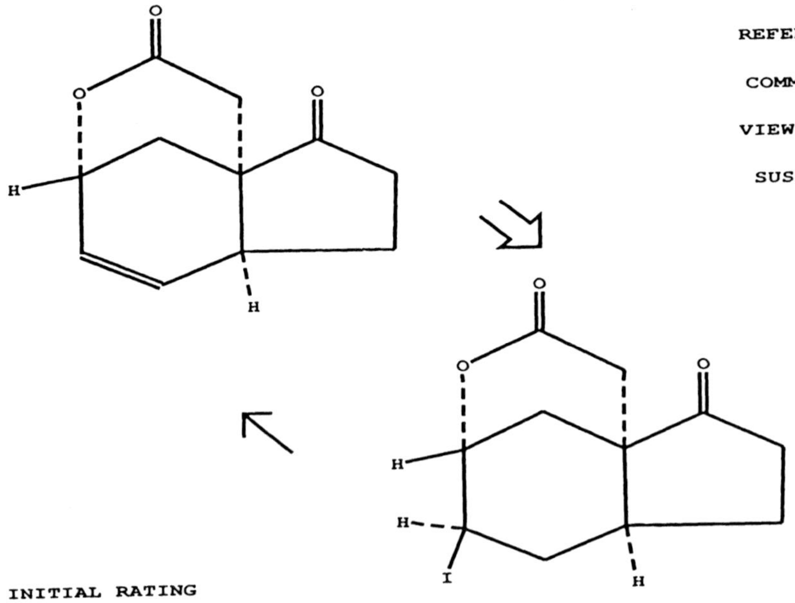

20 INITIAL RATING
 0 INCREMENT
 1 LINEAGE
 2

324 Dehydrohalogenation to Olefin
Prototype Conditions: R3N

         DISCONTINUE   STOP
         REJECT RING   DEBUG
                        REFERENCE
                        COMMENTS
                        VIEW TREE
                        SUSPEND

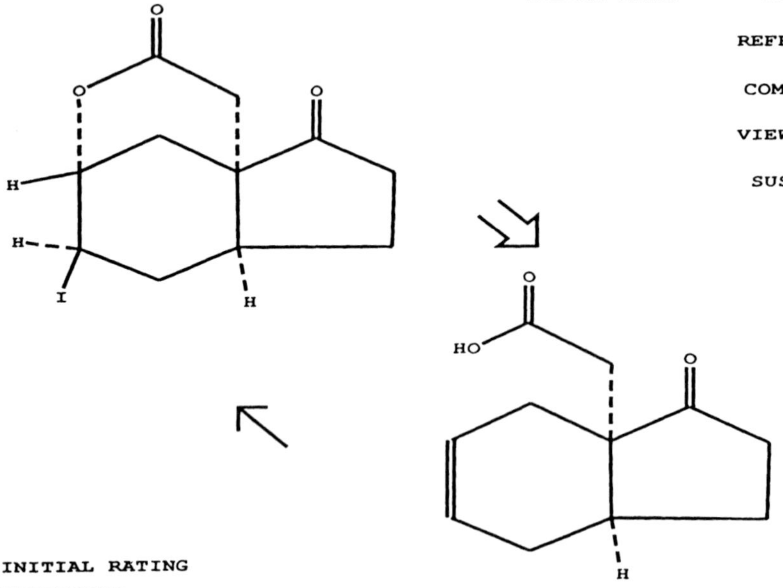

50 INITIAL RATING
40 INCREMENT
 1 LINEAGE
 2
 3

955 Halolactonization

Chapter 5

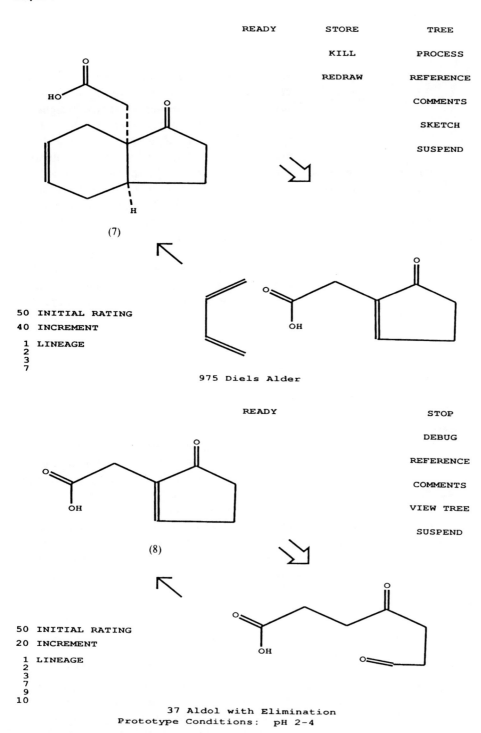

Fig. 8. Retroanalysis of the ketone (3).

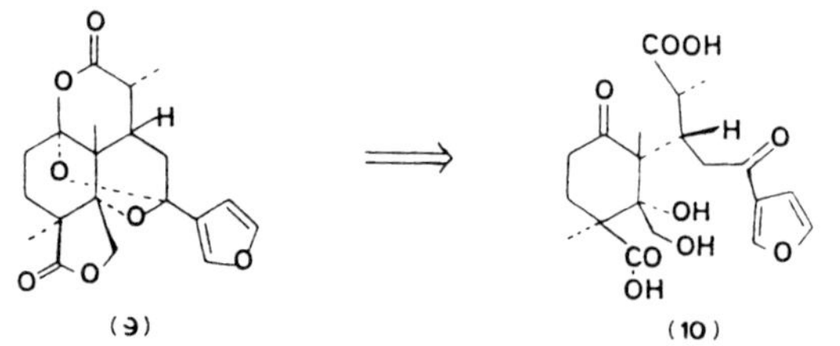

Fig. 9. Demasking of saudin (9).

tion if we ask LHASA to perform an analysis without prior unmasking by using an unconstrained strategy we find, as shown in Fig. 11, a variety of complex structures which are clearly not good precursors. Eventually the program could suggest the unmasking process as one of its options but such an untargeted approach clearly defeats the inherent strengths of the LHASA program.

The above examples have covered many of the major features within the LHASA program and will hopefully have provided some insight into its use and into the philosophy behind it. However we must now ask the critical question about any new technique or method: Is it useful? To address this question we must place the LHASA program within the context of modern organic synthesis and assess its limitations.

4 EVALUATION OF THE LHASA PROGRAM

One of the major reasons behind the dramatic advances in synthetic organic chemistry over the last two decades has been the rationalisation of the concept of retrosynthetic analysis. This has enabled chemists to explain fully their synthetic strategies within their own discipline and for the first time, has allowed non-specialists to understand and appreciate the synthetic planning involved. The LHASA program is an attempt to computerise this logical method in an interactive fashion in which the important decisions are still made by the chemist. The number of control decisions that can be made is limited by the size of the empirical knowledge base which at the moment is small relative to the body of knowledge that constitutes synthetic chemistry. Thus the strategic choices allowed by LHASA are, to the specialist, standard methodology offering very little new insight. However as LHASA grows and further strategies are added a point will be reached where strategic options that the chemist is very much less familiar with will be available for consideration so providing alternative insights.

Thus the first conclusion we can draw is that the actual structure of the program, with its emphasis upon an interactive regime, is potentially very powerful and to the specialist it is only the limited knowledge base that reduces the utility of the program. As we will discuss below the program is currently under development and one can foresee a point where the majority of today's synthetic problems could be solved by LHASA. However by that time such problems will be trivial and it is certain that synthetic chemistry will be solving problems of a different nature and different degree of complexity altogether. Therefore this author, at least, feels that LHASA, or similar programs, offer an opportunity for great developments in organic chemistry due to the removal of some of the more time consuming chores. Naturally the time scale for such predictions is open to debate.

However, what of the non-specialist? Organic synthesis underpins an enormous range of scientific disciplines since it involves, by definition, the preparation of materials and biologists, biochemists and geochemists, amongst others, regularly require access to usable quantities of organic substances. Does LHASA offer such non-synthetic specialists the opportunity to design and carry out organic preparations without reference to an experienced organic chemist? Presently, LHASA does not do this but not through any limitations of the program!

Organic chemistry courses to non-specialists rarely cover the topic of retrosynthetic analysis which would be a prerequisite to using the LHASA program. If they were to do so then LHASA in its present format could be used by non chemists, if their synthetic problems are not too complex. As the program develops it is likely to become of greater utility to the non-specialist long before innovative synthetic chemists themselves find it of use. To do this the geochemists, biochemists etc. wishing to make use of any synthetic design pro-

Chapter 5

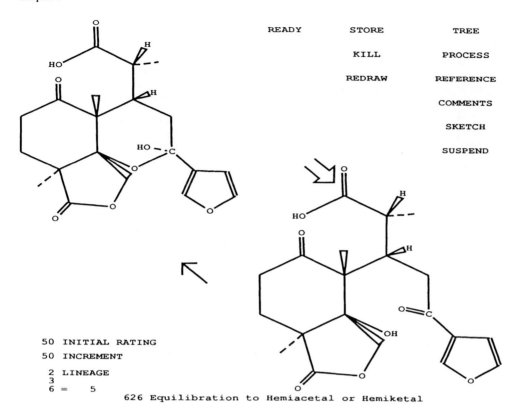

50 INITIAL RATING
50 INCREMENT
 2 LINEAGE
 3
 6 =   5

626 Equilibration to Hemiacetal or Hemiketal

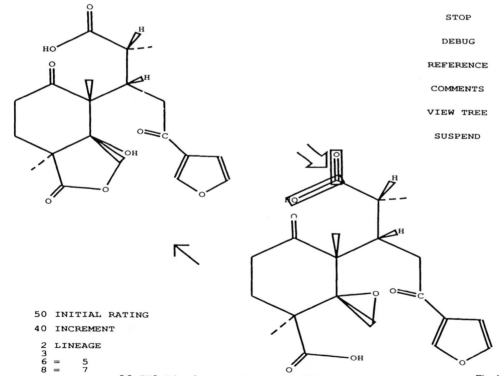

50 INITIAL RATING
40 INCREMENT
 2 LINEAGE
 3
 6 =   5
 8 =   7

96 SN2 Displacement on Epoxide or Episulfide
Prototype Conditions: NaOAc

Fig. 10. *(Continued on p. 64)*

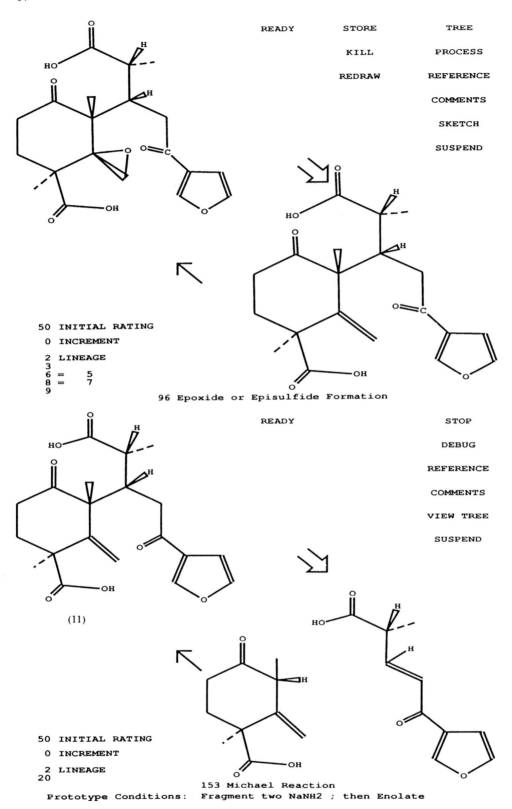

Fig. 10. Retroanalysis of saudin (**9**).

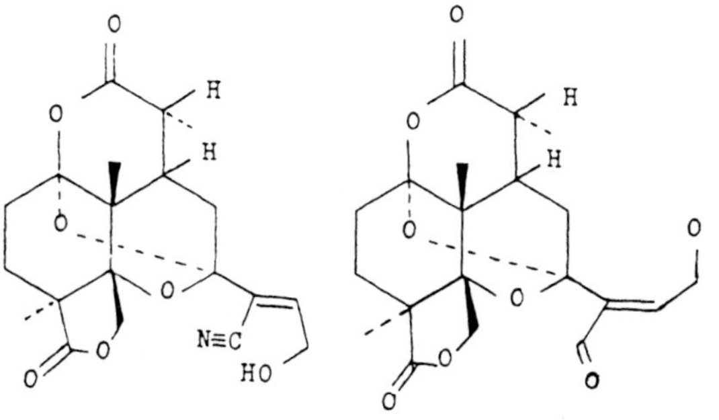

Fig. 11. Unconstrained precursors to saudin (9).

gram will need a good grounding in organic chemistry and notably, for LHASA, in retrosynthetic analysis, whilst demonstrating a willingness to adjust to the rigours of synthetic chemistry. Obviously they must also have the facilities and practical expertise to carry through the synthesis. Therefore I believe it is true to say that the existing LHASA program is potentially of use, serving to 'unlock' some of the mysteries of organic synthesis for the enterprising non-specialist.

Further to this, one often voiced criticism of organic chemistry, and synthesis in particular, is that as with all high technology sciences it is 'encased' in jargon which often appears impenetrable to the uninitiated. Any synthetic design program which hopes to be of use to non-specialists must try and avoid this. Currently LHASA is not as user friendly as it could be and does use a good deal of synthetic jargon. This reflects the state of development of the program rather than a lack of appreciation of these difficulties by the developers. These cosmetic but important considerations, will, I am sure, be dealt with eventually. For now though LHASA, which is a friendly system to the specialist, does require a good deal of patience on the part of non-synthetic chemists in their familiarising themselves with synthetic phraseology.

5 FUTURE DEVELOPMENTS OF THE LHASA PROGRAM

The LHASA program is currently under development in a number of areas [11]. One glaring weakness until recently was the program's difficulties in dealing with heteroaromatic and aromatic systems. This problem has been solved so well that even the preferred site of attack of an electrophile on a substituted benzene derivative can be accurately predicted and, if necessary, corrected for by use of subgoal transforms.

It would be fatuous to claim that retrosynthetic analysis is the only way to carry out synthetic design, especially since so many successful syntheses have used other approaches. A very common method is to relate the target molecule to a specific starting material and this strategy has now been incorporated into the LHASA program, by use of a mapping algorithm which searches for similarity matches. In a variation of this the target molecule can now be mapped against a 'pool' of readily available starting materials, with both a simple 'aromatic pool' and a 'pool' of chiral starting materials having so far been incorporated.

A further advance in LHASA, which is in the development stage, involves its interface with ORAC (Organic Reactions Accessed by Computer), a data base allowing organic reaction retrieval. By means of this the LHASA user will obtain rapid access to good literature precedents for reactions suggested by the LHASA program. Additionally it is hoped that eventually the information stored in the ORAC data base can be used as a basis for automatically writing transforms. Such a feature would expand the empirical knowledge base of LHASA enormously.

6 ACCESSING THE LHASA PROGRAM

LHASA can be run upon DEC VAX 11-750 or VAX 11/780 computers interfaced with graphics terminals that can emulate Tektronix 4010 or 4014 style graphics. The use of a data tablet or mouse to control a cursor is also essential as is a suitable hardcopy device. The majority of users access LHASA via a network which greatly reduces the initial cost of the system. In Europe the development of the LHASA system is done by an industrial consortium of major pharmaceutical and fine chemical companies. (Additionally it should be noted that SECS, the Princeton based system, is also being used by a group of Swiss and German companies.) Potential academic users of LHASA are often allowed free access to the system.

7 CONCLUSIONS

The LHASA program is the most advanced and well used synthetic design program and so serves as a good model for a discussion of the utility of such systems. This article is written by a user of LHASA and will hopefully result in a different perception and insight from that held by the program developers, so providing a more objective view.

It is clear that expert systems for designing organic synthesis are already accepted as a feature of modern chemistry, and will become of increasing importance. LHASA is a relatively simple system to use which, as it expands, will offer unique opportunities, initially more so to the non-specialist, and eventually to synthetic chemists, so alleviating some of the toil that is part of any research. Other synthesis design systems are also under continual development but none has yet reached the same degree of sophistication as LHASA and it is likely that this system will be the first to become widely accepted especially when linked with ORAC, the reaction retrieval system. The prospects for LHASA and/or similar systems are very exciting.

8 ACKNOWLEDGEMENTS

The author would like to thank Dr. A.P. Johnson and the University of Leeds' LHASA research group for providing access to the LHASA program and the University of Bristol for the purchase of computer hardware.

REFERENCES

1 E.J. Corey and W.T. Wipke, Computer-assisted design of complex organic syntheses, *Science*, 166 (1969) 178–192.
2 J. Gasteiger, Computer-assisted synthesis design, *La Chimica e l'Industria*, 64 (1982) 714–721.
3 J.B. Hendrickson, E. Braun-Keller and G.A. Toczko, *Tetrahedron*, 37, Supplement 1 (1981) 359.
4 E.J. Corey, A.P. Johnson and A.K. Long, Computer assisted synthetic analysis. Techniques for efficient long-range retrosynthetic searches applied to the Robinson annulation process, *Journal of Organic Chemistry*, 45 (1980) 2051, and references cited therein.
5 S. Warren, *Organic Synthesis: The Disconnection Approach*, Wiley, New York, Chichester, 1982.
6 The LHASA program was accessed via SERCNET from the University of Leeds' LHASA research group.
7 E.J. Corey and A.K. Long, Computer assisted synthetic analysis. Performance of long-range strategies for stereoselective olefin synthesis, *Journal of Organic Chemistry*, 43 (1978) 2208.
8 H.W. Orf, *Computer Assisted Synthetic Analysis*, Ph.D. Thesis, Harvard University, 1976.
9 T.V. Lee and J. Toczek, The use of selective lactonisation to achieve chemodifferentiation of two carboxylic acid functions. A novel entry into the elemanolide sesquiterpenes, *Tetrahedron Letters*, 26 (1985) 473.
10 J.S. Mossa, J.N. Cassady, M.D. Antoun, S.R. Byrn, A.T. McKenzie, J.F. Kozlowski and P. Main, Saudin, a hypoglycaemic diterpenoid from *Cluytia richardiana*, *Journal of Organic Chemistry*, 50 (1985) 916.
11 A.P. Johnson, Computer aids to synthesis planning, *Chemistry in Britain*, (1985) 59.

PROLOG for Chemists. Part 1

GERARD J. KLEYWEGT *

Department of NMR Spectroscopy, University of Utrecht, Padualaan 8, 3584 CH Utrecht (The Netherlands)

HENDRIK-JAN LUINGE and BART-JAN P. SCHUMAN

Analytical Chemistry Laboratory, University of Utrecht, Croesestraat 77A, 3522 AD Utrecht (The Netherlands)

(Received 29 March 1988; accepted 21 April 1988)

CONTENTS

1 Introduction	68
2 Scope	68
3 Implementation notes	68
4 Facts and queries	69
5 Conjunctive queries	71
6 Rules	72
7 Logic	73
8 Lists and recursion	75
8.1 Lists	75
8.2 Recursion	75
8.3 Examples	76
8.4 Tail recursion	78
9 Arithmetic	80
9.1 Addition	80
9.2 Multiplication	80
9.3 Comparisons	81
9.4 Example 1	81
9.5 Example 2	82
10 Input/output operations and supervisor programs	84
10.1 I/O operations	84
10.2 Supervisor programs	85
11 Control of evaluation	86
12 Database operations	88
References	90

1 INTRODUCTION

PROLOG stands for PROgramming in LOGic. The concept of using logic as a programming language emerged in the early 1970s, but has its roots in the 1950s [1]. The idea was developed by R.A. Kowalski, D. Kuehner and M. van Emden at Edinburgh and A. Colmerauer and P. Roussel at Marseilles. A major milestone was the efficient implementation of DEC-10 PROLOG developed at Edinburgh in the mid-1970s [2]. Three influential papers with respect to the development of logic programming are refs. 3, 4 and 5.

Nowadays, PROLOG is attracting a growing amount of interest from the entire scientific community [6]. The language is quite popular in its native continent, Europe, and the Japanese have selected PROLOG as the programming language for the core of their Fifth Generation Computer Systems Project [7,8]. Also, following a slow start, PROLOG now seems to be becoming more popular in the USA as well, challenging LISP [9,10] as the primary artificial intelligence (AI) programming language. Brief introductions to the syntax conventions of PROLOG have also begun to appear in the chemical literature [11–18]. Some of its applications in chemistry will be discussed towards the end of Part 2 of this tutorial.

PROLOG is a so-called declarative language, as opposed to Fortran, Pascal and others which are called imperative [19]. When programming in an imperative language, one has to specify procedures for solving problems. When using a declarative language, however, one provides a description of a problem, rather than a strategy for solving it.

2 SCOPE

This tutorial aims to provide newcomers to PROLOG with a first introduction into the main concepts of logic programming. Some programming experience with a conventional, third-generation language (e.g., Fortran, Pascal, Algol, C or Basic) will probably be required. For a fuller understanding, it is necessary that one actually does some programming in PROLOG, as is true for learning any other programming language. We have attempted to use "chemical examples" wherever possible and sensible: the reader will encounter examples from several chemical disciplines, from biochemistry to co-ordination chemistry. Thus we hope to show the reader the types of problems for which PROLOG might be better suited than a conventional language. Should the reader see an opportunity to put PROLOG to use, then we recommend that one of several available textbooks on Logic Programming be used for further study [2,20–23]. A more extensive list of book titles will appear in Part 2 of this tutorial.

3 IMPLEMENTATION NOTES

In this tutorial we will use the so-called standard syntax of Logic Programming Associates'

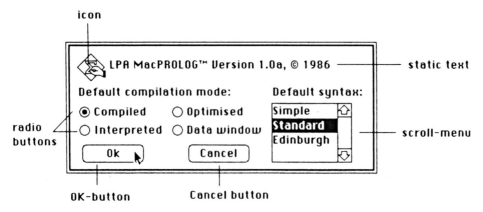

Fig. 1. A MacPROLOG dialog box showing the available syntaxes and compilation modes.

MacPROLOG (for Apple Macintosh personal computers) [24,25]. This particular PROLOG implementation provides an interpreter, compiler and an optimising compiler, and three different PROLOG syntaxes (standard, simple and Edinburgh syntax, see Fig. 1). There is as yet no internationally acknowledged standard PROLOG syntax. Although the international logic programming community tends to use the Edinburgh syntax most often, many chemists start out with the Simple micro-PROLOG syntax. We have opted to use the standard syntax. Conversion between the various PROLOG syntaxes, however, is usually quite straightforward (see the section about Logic). It is even possible (and not too complicated a task) to write small PROLOG programs which convert programs from one syntax into another. In MacPROLOG this feature is provided for the programmer.

MacPROLOG also provides the programmer with a wealth of built-in "functions". Wherever these are used in the examples, we will make mention of the fact that they might not be available in other syntaxes, or we will describe their functioning, so that users of other implementations may write "mimicking" functions themselves.

As for typography, parts of PROLOG code are printed in bold face; queries and output are printed in italics.

4 FACTS AND QUERIES

A PROLOG program may be viewed as a collection of knowledge. It consists of facts, rules, descriptions, relations etc. The simplest form of knowledge is a fact. In Standard syntax PROLOG a fact, for instance "copper is a metal", can be represented as:

((is_a_metal copper))

In other syntaxes, such a fact could even be written as: **copper is_a_metal**.

In this assertion, "copper" is the subject and "is_a_metal" is the so-called predicate. Both items are called terms of the assertion. Both must be represented as a single string of characters, which is why connecting underscores are used in "is_a_metal".

When we have entered ((is_a_metal copper)), what we in fact have done is to tell PROLOG that the fact that copper is a metal is known and true. This single fact already constitutes a "program" in PROLOG. This serves to show that the very notion of what a program is differs drastically from that in conventional languages. Note the absence of the program name, declarations of variables, a begin and an end. In addition, in PROLOG there is no strict separation between the program and its data.

Let us now add some other facts to our program, for example:

((is_a_metal nickel))

((is_a_metal iron))

((is_a_halogen fluorine))

((is_a_halogen chlorine))

Having a collection of facts is useful only if there is some mechanism to retrieve (or, more generally, manipulate) them. This shows up another difference between conventional languages and PROLOG. Having entered a Fortran or Pascal program, the next step would be to compile, link and run it (or to issue a run-command straight away, in the case of an interpreted language). PROLOG programs may be either compiled or interpreted as well. But, in order to obtain information or results, we issue queries.

For instance, we might wish to know whether copper is a metal. Hence, we issue the query: *(is_a_metal copper)*. What we thus do, is ask PROLOG to check whether the fact that copper is a metal is known to the program. Since there is indeed a fact in our program which states that copper is a metal, PROLOG will reply to our question with: *YES*. Analagously, the reply to our query *(is_a_metal chlorine)* will be: *NO*. Fig. 2 shows a screendump, with program and output windows, illustrating this program and a query/answer dialog in progress. Fig. 3 illustrates the results of a query session.

But there is more — suppose we want to obtain a list of all metals known to our program. We

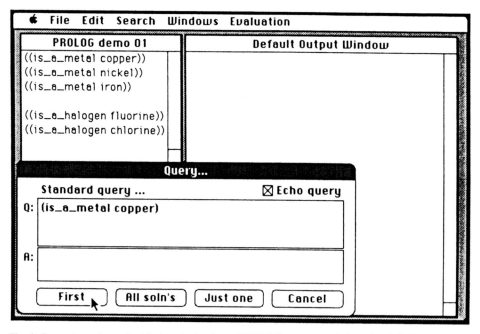

Fig. 2. Screendump from the Macintosh showing a PROLOG program window, the default output window and a query box.

would then issue the query: (*is_a_metal _metal*). In this query, "_metal" is a variable, as opposed to the constant "copper" in our previous query. Variables are strings of characters which commence with an underscore. (In some implementations, any unquoted string beginning with a capital letter

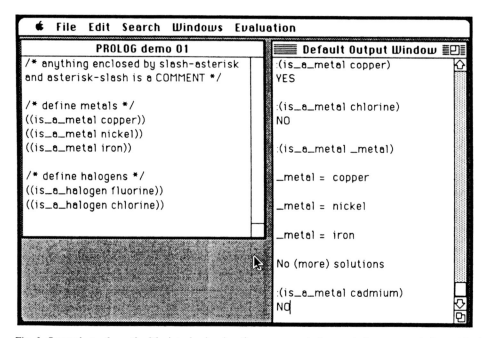

Fig. 3. Screendump from the Macintosh, showing the program window and the output window with the queries and their results.

is a variable too. In others, only strings that begin with one of the letters X, Y or Z may be used.)

When we issue this query, the variable _metal is unbound, i.e. it has no value assigned to it. What we ask PROLOG to do, is to find instances of the variable _metal, such that the query (with the variable _metal bound) represents a true fact. PROLOG has a built-in inference mechanism, which will subsequently attempt to match our query to facts known to the program. This is done in a top-down sequential manner, which implies that the facts which were entered first, will be retrieved first and so on. Hence, the first match occurs with the fact ((is_a_metal copper)), and PROLOG will display the answer: _metal = copper. The user will then be asked whether more solutions (if any) are desired. The manner in which this will happen depends on the particular PROLOG implementation. Suppose we do want more solutions, then PROLOG will come up with the next answer: _metal = nickel. Subsequently, it will find _metal = iron. Since there are no more is_a_metal-facts in our program, PROLOG will finally display the message: *No (more) solutions*.

There is a catch to all this, however. Suppose we were to issue the query: *(is_a_metal cadmium)*. Although this is perfectly true, there is no evidence in our program to support this thesis. Hence PROLOG will reply with *NO* to this particular query. This phenomenon is called negation by failure: everything the program does not "know" is not true (or, at least, not provable). A solution to this problem is the use of a so-called "catch all" clause [23], which is invoked only when no solution has been obtained and takes appropriate action (for instance, notifying the user that no evidence can be found for...).

5 CONJUNCTIVE QUERIES

Let us examine another example. Suppose we would like to create a program which enables us to reason about sequences of chemical reactions. We start by writing a program consisting of facts concerning chemical compound classes and chemical reactions. For instance, we know that primary alcohols can be converted into aldehydes and that these in turn can be converted into carboxylic acids. In PROLOG:

((Can_be_converted_into primary_alcohol aldehyde))

((Can_be_converted_into aldehyde carboxylic_acid))

Note that it is permitted to use any number of subjects in an assertion, even zero. Again, the program may be queried. For instance, the query ((*Can_be_converted_into_x aldehyde*)) will yield: _x = *primary_alcohol*. We may also introduce two variables in the query. Thus, the query ((*Can_be_converted_into_x_y*)) will yield two pairs of solutions, namely _x = *primary_alcohol*, _y = *aldehyde* and _x = *aldehyde*, _y = *carboxylic_acid*. (Of course, it would be more informative to use "sensible" variable names, such as _starting_material and _product, but this has not been done here for reasons of space.)

However, we might also be interested in answering such questions as: what will be the product(s) of any two-step reactions of primary alcohols. For formulating such questions, we may use conjunctive queries. A conjunctive query is merely a set of simple queries as those which we have used until now, with the provision that the entire query must now be satisfied. A conjunctive query will succeed if (and only if) all the individual queries are true (can be proved). Thus, the conjunctive query ((*Can_be_converted_into primary_alcohol aldehyde*) (*Can_be_converted_into aldehyde carboxylic_acid*)) will yield the answer *YES*. Variables can be used in conjunctive queries too—for example, the query ((*Can_be_converted_into primary_alcohol _x)(Can_be_converted_into _x _y)*) will yield the products of two-step reactions of primary alcohols. The variable _y will be bound to these products, and _x will be bound to the intermediate products. What we ask PROLOG is to find for us instances of the variables _x and _y, such that primary alcohols can be converted into _x, and these in turn can be converted into _y. The (only) solution to this query will obviously be: _x = *aldehyde*, _y = *carboxylic_acid*.

It is important to note, that the variable _x is used in both of the single queries constituting our conjunctive query. Therefore, only solutions in which this variable has the same value in both of the single queries are acceptable. Had there been another fact in the program such as

((Can_be_converted_into primary_alcohol
    chlorinated_alkane))

then _x = chlorinated_alkane would satisfy the first query, but not the second (since PROLOG cannot find evidence that a chlorinated alkane can be converted into a carboxylic acid).

PROLOG solves conjunctive queries by sequentially solving the individual queries. So, in our example, it will first attempt to solve (Can_be_converted_into primary_alcohol _x). A match will be found and _x will become bound to the value aldehyde. With this instantion of _x, the second query must be solved, namely (Can_be_converted_into aldehyde _y). Again a match will be found and _y will be bound to the value carboxylic_acid. Thus, the entire query has been solved and the values of the variables _x and _y will be printed.

If another solution must be found, PROLOG's built-in backtracking mechanism will be put to work. Variable _y will become unbound again, and another solution for the query (Can_be_converted_into aldehyde _y) will be sought. Since no such alternative solution can be found, PROLOG will further backtrack, and variable _x will become unbound again. Then, an alternative solution for (Can_be_converted_into primary_alcohol _x) will be sought. Binding _x to chlorinated_alkane will yield a new solution to this query, and the second query must then be solved again: (Can_be_converted_into chlorinated_alkane _y). Unfortunately, no instantiations of _y which satisfy this query can be found. Hence, PROLOG backtracks once again and attempts to find alternative solutions for the first query. Since these will not be found either, the attempt to find more solutions to the entire conjunctive query fails, and the message "No (more) solutions" will be displayed.

6 RULES

In the example of the previous section, typing conjunctive queries may be a rather time-consuming affair, especially if the same type of conjunctive query is required often. Also, up to now we have only used PROLOG as a sort of database querying language. We may, however, also add rules to our program. In the simplest form, these are predefined conjunctive queries. For instance, a general PROLOG formulation of a two-step reaction would be:

((Two_step_reaction _x _y)
 (Can_be_converted_into _x _z)
 (Can_be_converted_into _z _y))

We have defined a new predicate Two_step_reaction, with two variables _x and _y. The general format of a rule is: ((predicate + arguments)(...) (...)...). A call to a rule succeeds if all the subsequent conditions are true (or can be solved). In our example: (Two_step_reaction _x _y) is true (succeeds) if (Can_be_converted_into _x _z) is true AND (Can_be_converted_into _z _y) is true.

The rule states that there exists a two-step reaction for _x yielding product _y if _x can be converted into some intermediate product _z, which in turn can be converted into the product _y. Instead of issuing conjunctive queries, we may now directly call the rule, for instance by issuing the query: (*Two_step_reaction primary_alcohol _product*). The answer that PROLOG will provide us with is: *_product = carboxylic_acid*. Thus, the value of the intermediate variable _z in the rule is not printed. In fact, this variable has a so-called local scope: it only exists within the rule Two_step_reaction and its value is lost as soon as the rule has been applied. Hence, we may use the same variable names in separate rules, since there is no such thing as a global variable. Should we be interested in the intermediate products, then the format of the rule would have to be changed to

((Two_step_reaction _x _y _z)
 (Can_be_converted_into _x _z)
 (Can_be_converted_into _z _y))

Obviously, the format of the query must be altered accordingly. The result of the query (*Two_step_reaction primary_alcohol _y _z*) will now be _y = carboxylic_acid, _z = aldehyde.

Again, invoking rules in queries can be done with two objectives: to obtain confirmation (or negation) of facts, or to generate solutions. Hence, (employing the first format of the rule) the query (*Two_step_reaction primary_alcohol carboxylic_acid*) will result in PROLOG replying *YES*. Alternatively, the query (*Two_step_reaction _x _y*) will yield _x = primary_alcohol, _y = carboxylic_acid.

This also demonstrates another major difference between PROLOG and conventional programming languages. We may view a rule as a procedure or subroutine of a PROLOG program. However, whereas in a Fortran or Pascal program there is a clear distinction between input and output variables (i.e., variables which have a value when the procedure is entered, and those that are assigned a value within the procedure), this is not the case with PROLOG. In case of the query (Two_step_reaction _x _y), zero, one or both of the variables may have a value when issuing the query. Hence, one "procedure" does a job that would require three separate procedures in a conventional programming language:
- one that checks that there is a two-step reaction leading from one compound to another (both variables bound);
- one that generates products of two-step reactions of a compound (the first variable bound); and
- one that generates precursors of a compound, two retro-synthetic steps away (the second variable bound). Indeed, the result of the query (*Two_step_reaction _x carboxylic_acid*) will be _x = primary_alcohol.

When put to one of the latter two uses, rules can be used to derive knowledge that has not been explicitly coded (as facts). In fact, rules can be regarded as a generalisation of facts and help prevent one from storing knowledge as a large amount of separate facts, thus making knowledge more readily accessible.

Although the few, rather trivial examples presented so far may not yet convince long-time Fortran programmers, some favourable features of PROLOG are becoming evident. These include:
- enhanced readability of the programs. They can almost be written in plain English, or Swahili, if required. For example, renaming the Two_step_reaction predicate to Zwei_Stufen_Reaktion would make the corresponding rule transparent to German speakers;
- conciseness of programs. In our example, one rule served three different purposes;
- consistency of information representation and retrieval;
- the declarative nature of the language. The rule above only describes what a two-step reaction is.

We do not have to be concerned with the generation of candidate-values for the variables in the rule. It would be illustrative to write a program that does the same as our single Two_step_reaction-rule in a conventional programming language.

7 LOGIC

PROLOG is based on the clausal form of predicate logic. Clauses can be represented in general in the form [26]:

$$B_1, B_2, \ldots, B_n \leftarrow A_1, \ldots, A_m \qquad m, n \geq 0$$

where A_i is the ith condition of the clause, and B_j is the jth alternative conclusion. Its interpretation is: B_1 or ... or B_n is true if A_1 is true and ... and A_m is true. PROLOG uses a subset of these type of clauses, namely so-called Horn-clauses, which have at most one conclusion (i.e., $n = 0$ or 1). Three kinds of Horn-clauses are of interest:

assertions (facts): $\qquad B \leftarrow$
procedure declarations (rules): $B \leftarrow A_1, \ldots, A_m$
denials (queries): $\qquad \leftarrow A_1, \ldots, A_m$

When confronted with a denial (query), the inference engine will attempt to refute the denial by using other clauses.

The various terms used in predicate logic all have an analogue in PROLOG.

In predicate logic a term is:
- a constant symbol
- a variable symbol
- a compound term
  (a function with arguments)

Their respective analogues in PROLOG are:
- an atom, e.g. copper, 'CH3'
- a variable, e.g. _metal, _group
- a PROLOG structure consisting of a functor
  with arguments, e.g. (metal copper),
  (functional_group 'CH3')

Predicate logic has been used for the description of (a part of) the real world. In order to do this, it is necessary to state certain facts and define relations between them. In PROLOG there are, again, analogues of the concepts used in predicate logic. For example:

In predicate logic:
$\neg \alpha$ (meaning not α)
$\alpha \wedge \beta$ (meaning α and β)
$\alpha \vee \beta$ (meaning α or β)

Their respective analogues in PROLOG are:
(NOT termA)
(termA)(termB)
(OR(termA)(termB) or even ((clauseA))
((clauseB))

As can be seen from these examples, predicate logic and PROLOG are closely related, not only in principle, but also in practice. Even relatively complicated and implicating features of predicate logic are an essential part of PROLOG. As a last example we will show how quantifiers in predicate logic are related to PROLOG.

In predicate logic there is a so-called 'universal quantifier', normally written as: $\forall$. If one writes the following:

$$\forall x [\text{feathers}(x) \Rightarrow \text{bird}(x)]$$

then this means: everything ($\forall x$) which has feathers (feathers(x)) must be a bird (bird(x)). In PROLOG this is implicit in every clause and the PROLOG analogue,

((bird_x)
(feathers_x))

can be thought of as meaning exactly the same, because a clause is always true for all _x. Clauses in PROLOG are therefore said to be 'universally quantified'.

There are many more related subjects in this field and of course logic has a far broader scope than we can cover here. For a more extensive coverage the reader is referred to refs. 23 and 27.

From the symbolic notation introduced above, it is easy to show the differences between various PROLOG syntaxes. In the STANDARD syntax used in this tutorial, the following are examples of a fact, a rule and a query, respectively:

((metal_ion copper))

((salt _metal _other)
(metal_ion _metal)
(other_ion _other))

(salt copper _other)

In the SIMPLE (or micro-PROLOG) syntax, these are written as:

metal_ion(copper)

salt(_metal _other) if
   metal_ion(_metal) &
   other_ion(_other)

salt(copper _other)

In the Edinburgh syntax, this becomes:

metal_ion[copper].

salt[_metal,_other] :-
   metal_ion[_metal],
   other_ion[_other].

salt[copper,_other].

In the Simple and Edinburgh syntax rules, the two variables _metal and _other are enclosed in brackets. They thus constitute a list, which is the subject of the next section.

8 LISTS AND RECURSION

8.1 Lists

A very powerful feature of PROLOG is its use of lists (and, hence, as we shall see below, of recursion). Elements of a list are called atoms. Atoms can be anything, such as numbers, characters, strings or even other lists. Lists may be written as a sequence of atoms enclosed in brackets. An example of a list is:

(Woods_metal(Pb 0.25)(Cd 0.125)(Bi 0.50)

(Sn 0.125))

It contains information about the elemental composition of an alloy. A list can be regarded as consisting of a head and a tail. The head is the first atom of the list, the tail comprises the list of all other atoms following it. The above list may therefore also be written:

(Woods_metal|((Pb 0.25)(Cd 0.125)(Bi 0.50)

(Sn 0.125)))

Here, | is the so-called list constructor symbol. If we regard the tail of each list as a list in itself, having its own head and tail, we may eventually write:

(Woods_metal|((Pb|((0.25|()))

((Cd|((0.125|())))((Bi|((0.50|())))

((Sn|((0.125|())))

Obviously, this is a rather awkward notation for humans, although it is actually the format that PROLOG uses internally to store lists. Note that the empty list, (), is a special occurrence of a list containing no atoms at all. In fact, this is always the last element of a list.

Although in some respects lists resemble arrays as used in conventional programming languages, there are a number of important differences:
- lists need not be declared in advance, they may be expanded "infinitely";
- lists need not contain just one type of data (such as numbers), and may even contain variables;
- elements of arrays in conventional programming languages can be accessed by using an index, which makes these languages suitable for iterative procedures; list elements are most trivially accessed sequentially, by performing an operation with the head of the list and then with its tail, which makes PROLOG excellently suited for recursion.

8.2 Recursion

Some examples may serve to clarify the points made above. Suppose we want to write a procedure which checks whether an atom is part of a list. We will call the associated predicate **is_part_of_list**:

((is_part_of_list _el(_el|_Tail)))

((is_part_of_list _el(_head|_Tail))

(is_part_of_list _el _Tail))

The first clause states that an element _el is part of a list, if it equals the head of the list. The second clause states that an element _el is part of a list, if it is part of the tail of the list. This example demonstrates the use of recursion (in the second clause).

Now, if we issue the query: (is_part_of_list 2 (1 2 3)), the first clause will fail, since 2 is not the head of the list. Then the second clause is invoked, where _el will be matched to 2, _head to 1 and _Tail to the list (2 3). Now, is_part_of_list will be recursively called, but with the arguments 2 and (2 3), respectively. This time, the first clause produces a match, and PROLOG will display the answer YES.

Another possible use of this "routine" is for the enumeration of list elements, one by one. This can be effected by issuing the query: (is_part_of_list _x(1 2 3)). Matching the first clause produces the first hit: _x = 1. If we want more solutions, PROLOG tries the second clause, which will result in the call (is_part_of_list_x (2 3)). Again, the first clause produces a match: _x = 2. The second clause will yield the call (is_part_of_list_x (3)). Now there seems to be no tail any longer, and thus we might ask ourselves: will the first clause succeed? Indeed it does: remember that the last element of a list (internally) is always the empty list. Thus, matching produces: _el = 3 and _Tail = (), and the clause succeeds, yielding the final answer: _x = 3. The second clause will succeed, yielding the call (is_part_of_list_x ()). Now, no matching occurs with either clause, since the empty list itself can not be matched with (_head | _Tail). Table 1 gives a listing of the trace of this query's execution.

In most PROLOG implementations the is_part_of_list-predicate will be predefined. In MacPROLOG it is called "ON", although "member" is the name more commonly used.

This small example demonstrates an important feature of recursion: there is (most often) one special (boundary) case (in our example, the case where the desired element is the same as the head of the list), and one or more general cases (in our example, the case that the desired element is not equal to the head of the list).

8.3 Examples

Another example which demonstrates the use of lists and recursion is encountered when writing a procedure which will concatenate two lists:

((concatenate () _List _List))

((concatenate (_x | _Tail) _List (_x | _Rest))

(concatenate _Tail _List _Rest))

Let us see what happens, if we issue the query: (*concatenate (a b) (1 2) _N*); see also Fig. 4. The first clause produces no match (since the first argument is not the empty list), the second however does: _x = a, _Tail = (b), _List = (1 2) and _Rest is unbound. Note that _N is now (a | _Rest), where _Rest still has to become instantiated. This results in the recursive call: (concatenate (b) (1 2) _Rest. Again, only the second clause results in a successful match: _x = b, _Tail = (), _List = (1 2), and _Rest unbound again. The resulting recursive call: (concatenate () (1 2) _Rest) now only matches the first clause, which automatically binds _Rest to (1 2). Since this call succeeds, PROLOG returns to "unwind" the recursion. In the second call to the second clause, this will lead to _Rest becoming bound to (1 2) and in the first call it will become bound to (b 1 2). In our original query, _N will subsequently be bound to (a b 1 2), just as we were hoping it would be.

Again, it is instructive to consider the descriptive interpretation of the clauses: the first clause

TABLE 1

Trace of the execution of a query

:(is_part_of_list_x (1 2 3))

Starting new trace: TRACE 0
Call (0): (is_part_of_list_1928 (1 2 3)) ≪ E
Matches clause 1 head: (is_part_of_list_el(_el | _Tail))
Exit (0): (is_part_of_list 1 (1 2 3))

Leaving: TRACE 0

_x = 1

Continuing: TRACE 0
Redo (0)
Matches clause 2 head: (is_part_of_list_el(_head | _Tail))
Call (1 0): (is_part_of_list_1928 (2 3)) ≪ E
Matches clause 1 head: (is_part_of_list_el(_el | _Tail))
Exit (1 0): (is_part_of_list 2 (2 3))
Exit (0): (is_part_of_list 2 (1 2 3))

Leaving: TRACE 0

_x = 2

Continuing: TRACE 0
Redo (1 0)
Matches clause 2 head: (is_part_of_list_el(_head | _Tail))
Call (1 1 0): (is_part_of_list_1928 (3)) ≪ E
Matches clause 1 head: (is_part_of_list_el(_el | _Tail))
Exit (1 1 0): (is_part_of_list 3 (3))
Exit (1 0): (is_part_of_list 3 (2 3))
Exit (0): (is_part_of_list 3 (1 2 3))

Leaving: TRACE 0

_x = 3

Continuing: TRACE 0
Redo (1 1 0)
Matches clause 2 head: (is_part_of_list_el(_head | _Tail))
Call (1 1 1 0): (is_part_of_list_1928 ()) ≪ E
Fail (1 1 1 0)
Redo (1 1 0)
Fail (1 1 0)
Redo (1 0)
Fail (1 0)
Redo (0)
Fail (0)

No (more) solutions

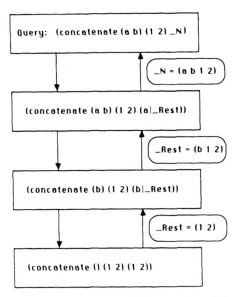

Fig. 4. Symbolic "trace" of the execution of the query (concatenate (a b) (1 2) _N).

((cleavage_site Eco_RI (G A A T T C)))
((cleavage_site Hae_III (G G C C)))
((cleavage_site Hind_III (A A G C T T)))

The following relation will determine whether a certain nucleotide sequence contains cleavage sites for restriction enzymes:

((can_cleave _res _nuc)
(cleavage_site _res _C)
(nucleotide_sequence _nuc _S)
(APPEND _X1 _X2 _S)
(APPEND _X3 _C _X1))

When this clause is invoked, it will first match the restriction enzyme's name with its corresponding cleavage site sequence _C; analogously, the nucleotide sequence _S that must be checked for cleavage sites is retrieved. The last two conditions state that a restriction enzyme can cleave a polynucleotide, if _S can be split into two separate lists _X1 and _X2, such that the former can be split into two separate lists, the latter of which equals the cleavage site sequence. Once again, there are several possible uses for this procedure:

– testing whether a restriction enzyme can cleave a polynucleotide: the result of the query (*can_cleave Eco_RI Thymidin_Kinase_gene*) will be *YES*.
– enumerating restriction enzymes that can cleave a given polynucleotide: the answer to the query (*can_cleave_res Thymidin_Kinase_gene*) will be: _res = Eco_RI, No (more) solutions.
– enumerating polynucleotides which can be cleaved by a given restriction enzyme: the query (*can_cleave Eco_RI _nuc*) will cause PROLOG to reply: _nuc = Thymidin_Kinase_gene, No (more) solutions.
– generating all sets of restriction enzymes and polynucleotides that can be cleaved by them insofar as these have been defined in the "knowledge base" (i.e., the relations "nucleotide_sequence" and "cleavage_site"): in this case the query should be issued with both arguments unbound: (*can_cleave_res_nuc*) will result in: _res = Eco_RI, _nuc = Thymidin_Kinase_gene; No (more) solutions.

states that concatenating the empty list to any other list yields that same list again. The second clause states that concatenation of a list (_x | _Tail) and a list _List, results in a list (_x | _Rest), where _Rest is the list resulting from concatenation of _Tail and _List.

Again, in most implementations, a predicate similar to "concatenate" will be predefined. In MacPROLOG it is called APPEND. As expected, the "concatenate" procedure can not only be used for appending two lists, but also for splitting a list into two sublists which, upon appending, will yield the original list.

An elegant example of list manipulation facilities is drawn from biochemistry. More particularly, it concerns the identification of cleavage sites for restriction enzymes in DNA sequences [28,29]. Suppose that we have information concerning nucleotide sequences and cleavage sites, for instance (for TK gene only nucleotides 320 through 340 have been used in order to conserve space):

((nucleotide_sequence Thymidin_Kinase_gene
(T C A T T G G C G A A T T C G A A C A C G)))

TABLE 2

Definition of some "differentiation"-clauses as well as the results of some queries

---

((differentiation x 1))

((differentiation _x 0)
 (INT _x))

((differentiation (_F + _G)(_dF + _dG))
 (differentiation _F _dF)
 (differentiation _G _dG))

((differentiation (_F − _G)(_dF − _dG))
 (differentiation _F _dF)
 (differentiation _G _dG))

((differentiation (_F * _G)((_F * _dG) + (_G * _dF)))
 (differentiation _F _dF)
 (differentiation _G _dG))

((differentiation (_F/_G)(((_G * _dF) − (_F * _dG))
  /(_G * _G)))
 (differentiation _F _dF)
 (differentiation _G _dG))

:(differentiation (x * (x * x)) _D)
_D = ((x * ((x * 1) + (x * 1))) + ((x * x) * 1))
No (more) solutions

:(differentiation (3/x) _D)
_D = (((x * 0) − (3 * 1))/(x * x))
No (more) solutions

:(differentiation (−3/(x * x)) _D)
_D = ((((x * x) * 0) − (−3 * ((x * 1) + (x * 1))))/((x * x) * (x * x)))
No (more) solutions

Another example of the power of list processing facilities is in symbolic differentiation. Table 2 lists a few "differentiation"-clauses as well as the results of a few queries. The first states that the derivation of x is 1; the second states that the derivation of a number is zero. Here (INT _x) checks that _x has been bound to an integer number. The other clauses apply rules for the differentiation of the sum, difference, product and quotient of two functions, respectively. It may require some effort to read the output, but the reader may verify then that the answers are correct. Alternatively, one could write a set of procedures that transforms the output list into a more readable format.

8.4 Tail recursion

As stated earlier, the order in which facts and rules are declared is very important, due both to considerations of execution speed and memory efficiency. If a query consists of more than one condition (as in the example above), then these conditions are placed on a so-called query stack by the compiler/interpreter. PROLOG's inference mechanism will then try to solve each of the conditions in turn (remember that PROLOG uses a depth-first search mechanism).

The way in which PROLOG treats variables gives the programmer a powerful programming tool, but also places a heavy burden on the programming environment. All sorts of temporary variables which are to be used for matching (and have to remembered for backtracking) will impose a large overhead on memory use.

In order to make efficient use of the available memory space (which is always limited, even with PROLOG systems that employ virtual memory), PROLOG must reclaim any storage space when it is no longer needed and remove the temporary variables from memory. Like many other programming environments, PROLOG uses a mechanism known as "garbage collection". To aid the compiler in producing efficient code (both in terms of memory and speed of execution), the programmer may use so-called "cuts" (to be discussed later; in effect, they control the evaluation, for instance by preventing a search for further solutions if only one is required) and a technique known as "tail recursion".

Recursion evokes a large overhead on memory usage which warrants some caution. Tail recursion, however, may be used to limit this overhead. Tail recursion entails putting the recursive call at the end of a rule. The classic example is the append-relation mentioned above:

((append () _resultlist _resultlist))

((append (_x | _list1) _list2 (_x | _list3))
 (append _list1 _list2 _list3))

Here, the recursive call in the second clause is the last (and, in this case, the only) call in the clause. If a relation is not tail recursive, it can usually be rewritten in a tail recursive form quite easily. For instance, the naive_reverse relation can be used to revert the order of the elements of a list. It could be defined as:

((naive_reverse ()()))
((naive_reverse (_head | _tail) _result)
 (naive_reverse _tail _temporary_list)
 (append _temporary_list (_head) _result))

Here, no tail recursion has been employed, since the recursive call to naive_reverse in the second clause is not the last call. Rewriting naive_reverse into a tail recursive form leads to:

((naive_reverse ()()))
((naive_reverse _list1 _list2)
 (reverting _list1 () _list2))

((reverting () _result _result))
((reverting (_head | _tail) _temp _list)
 (reverting _tail (_head | _temp) _list))

Now the order of the list is reverted using tail recursion only. It is instructive to try both versions for oneself. The difference will be especially clear if one uses a debugging facility known as the trace function (or puts so-called "spy points" on the predicates involved). See also ref. 23, Chapter 8, pp. 181–209.

Most PROLOG versions nowadays offer some very powerful tools for constructing and debugging programs. The main features of debugging in contemporary PROLOG environments is the ability to put so-called spy-points on any relation in the program and the use of the trace facility.

Spy-points give a programmer the ability to inspect the execution of his program very closely. As may have become clear from the examples given in this tutorial, most PROLOG programs consist of relations that call other relations. By placing so-called spy-points on any of these relations, it is possible to see what happens to those specific parts of the program. Even with a compiler such as MacPROLOG, placing spy-points enables the programmer to follow the execution of his program on a source level (so no more insertion and deletion of write statements) and in a very direct way. Because a spy-point can be placed on any relation, very direct debugging is possible. Relations which are (thought to be) correct can be left alone and the programmer does not need to worry about them. If, however, the programmer thinks that a certain relation is not working as it should, placing a spy-point on that relation will enable him to see when the relation is called by the program (an "entry" of a relation), when a relation is succeeding (an "exit" of a relation) or not succeeding (a "fail" of a relation) or even when the built-in backtracking mechanism of PROLOG is trying to use the relation again (a "redo" of a relation).

At all these stages of the debugging session, the programmer can see not only which relations are being called, but also what the value of every argument is, or is becoming, during the execution of a relation.

Since stepping through a program in this manner gives the programmer an enormous amount of information (some times a bit too much, see Table 1 for an example), debugging of PROLOG programs is most of the time much easier than debugging in a more conventional way. To lessen the amount of information during such a debugging session, PROLOG offers the possibility to watch only the entry points or exits points.

Up to now we have used lists in a way that resembles the use of arrays in conventional programming languages. There is, however, another important use for lists, namely for data representation, which more or less resembles the contemporary use of pointers and records in (a language like) Pascal.

Although many people tend to believe that data representations such a binary-trees can not be used in a language like PROLOG, the opposite is true. Binary-trees, and even the currently frequently-used B-trees, and the corresponding algorithms for searching, insertion and deletion in such trees, can be simply implemented and used in PROLOG. For extensive examples of the use of PROLOG and its relation with data representation, see for instance refs. 2 and 23.

9 ARITHMETIC

Of course, in practice, numerical manipulations will often be necessary. It should be noted that PROLOG is not particularly suited for fast or complex numerical computation. However, some basic operations are available, such as addition and multiplication, as built-in predicates (primitives). The number of available arithmetic primitives depends strongly on the particular implementation of PROLOG one is using. Also, some versions of PROLOG can only deal with integer numbers.

9.1 Addition

For additions we may use the SUM primitive, which has three arguments and three basic uses:
- checking that the sum of two given numbers is a given third number. For instance, (SUM 3 7 10) will succeed, since 3 + 7 = 10 is true.
- adding two numbers. For instance, (SUM 45 8 _x) will bind _x to 53.
- subtraction. For example, (SUM 34 _x 49) will force PROLOG to find a solution for the equation: 34 + _x = 49, and hence will result in the variable _x becoming bound to the value 15. Analogously, (SUM _x 12 6) will bind _x to −6.

Note that at least two of the three arguments of SUM must be bound when making the call. Thus, SUM can not be used to find all pairs of numbers whose sum is 56.

We shall demonstrate the use of SUM by means of a routine that determines the length of a list. (In most PROLOG implementations, this predicate will be predefined. In MacPROLOG it is called LENGTH.)

((length () 0))

((length (_head | _Tail) _n)

(length _Tail _m)

(SUM _m 1 _n))

The first clause states that the empty list has length zero. The second clause states that the length of a list is _n, if the length of the tail of the list is _m and _n = _m + 1. Hence, the query (*length (a b c d e f)_n*) will yield _n = 6. Note, that the conditions of the second clause can not be interchanged, since this would imply the use of two unbound variables in the call to SUM, which is not allowed. However, we can make the procedure tail-recursive:

((length () _n _n))

((length (_head | _Tail) _m _n)

(SUM _m 1 _m1)

(length _Tail _m1 _n))

This means, that we must use a start value of zero for the second argument in our query: (*length (a b c d e f) 0_n*). The second clause adds one to the second argument each time it is invoked. When all elements of the list have been processed, the first clause can be matched and this will force the third argument to become bound to the value of the second argument (which by then equals the number of elements of the list).

If one does not want to bother the user with the use of a second argument (the start value zero), one may use an intermediate procedure, which in turn calls the length procedure:

((determine_length _List _length)

(length _List 0 _length))

9.2 Multiplication

For multiplications we may use the TIMES primitive, which also has three arguments and three basic uses, analogous to those of SUM:
- checking that the product of two numbers is a given third number: (TIMES 3 −5 −15) will succeed.
- multiplying two numbers: (TIMES 5 2 _x) will bind _x to 10.
- dividing two numbers: (TIMES 10 _x 45) will bind _x to 4.5 and (TIMES _x 5 35) will bind _x to 7.

Again, at least two of the three arguments must have a value before calling TIMES.

To demonstrate the use of TIMES we shall write a tail-recursive procedure that computes the average of a list of numbers (where "length" is as previously defined):

((average _List _average)
 (compute_sum _List 0 _sum)
 (length _List 0 _length)
 (TIMES _length _average _sum))
((compute_sum () _sum _sum))
((compute_sum (_head | _Tail) _m _sum)
 (SUM _m _head _n)
 (compute_sum _Tail _n _sum))

Alternatively, we could write a procedure compute_sum_and_length which simultaneously determines the sum of the list elements as well as their number. In any event, the answer to the query (*average* (10 30 50 60 70 60 40 80 90 70) _average) is: _average = 56, No (more) solutions.

In cases where it is known in advance which of the arguments of TIMES or SUM is unknown, in MacPROLOG one may use any of the four predefined primitives $+$, $-$, $*$ or $\div$, which will result in faster execution.

For the evaluation of more complex structures, the is-primitive is quite handy. This allows the programmer to write such expressions as:

(is _x (* 3.14 (+ _y (* 10 _z))))

which evaluates the expression: _x = 3.14 * (_y + 10 * _z).

The availability of other arithmetic primitives is again critically dependent on the specific PROLOG implementation one is using. In MacPROLOG there are quite a number of them, such as SIN, COS, SQRT, ABS, INT, LN, PWR and TAN. If such primitives are not available, performing calculations may become quite cumbersome (and often staggeringly slow). Imagine having to write Taylor-expansion approximations in PROLOG for various trigonometric or exponential functions.

9.3 Comparisons

Another important aspect, not only of arithmetic operations, is the ability to perform comparisons. In MacPROLOG there are predefined predicates EQ (test for equality) and LESS (test for inequality). All other comparisons (less or equal, greater or equal, greater) can be programmed using these two primitives, for instance:

((greater _x _y)
 (LESS _y _x))
((greater_or_equal _x _y)
 (greater _x _y))
((greater_or_equal _x _y)
 (EQ _x _y))
((less_or_equal _x _y)
 (LESS _x _y))
((less_or_equal _x _y)
 (EQ _x _y))

(In MacPROLOG the primitives $<$, $\leq$, $\geq$, and $>$ are also predefined.) In the case of EQ, at least one of the arguments must have a value; in the case of LESS both arguments must be bound to a value. Hence, LESS can not be used to generate numbers that are greater than a certain number. The types of the arguments may be anything: numbers, characters, or strings. Hence the same primitives can be used for sorting numbers and for sorting a list of names into alphabetic order.

9.4 Example 1

We shall give another example of a deterministic procedure, in order to show the differences between programming such an application in PROLOG and in a conventional language. This example concerns the calculation of the number of rings plus double bond equivalents (RDBEs) on the basis of a stoichiometric chemical formula. For example, the number of RDBEs of benzene is 4. In general, for a formula $A_xB_nC_zD_y$ the number of RDBEs is given by:

RDBE = x + 1 + (z − y)/2

if y is the sum of the coefficients of the elements H, F, Br, Cl and I; n is the sum of the coefficients of O and S; z is the sum of the coefficients of N and P, and x is the sum of the coefficients of C and Si. Hence, in order to be able to compute the number of RDBEs, we need to establish the values of x, y and z.

First of all, we shall agree upon a notational formalism to represent chemical formulas. The most obvious choice is to use a list, which has as elements lists with two elements, namely an element symbol and the stoichiometric coefficient. Thus, the formula of benzene would be written as ((C 6) (H 6)).

For the computation of x, y and z, we will use a single predicate "get_xyz", with five clauses: one for each of the four types of elements, and one to terminate once all elements of the formula list have been treated.

One way to implement these clauses is:

((get_xyz ((_X _n)|_Tail) _x _y _z)
 (ON _X (O S))
 (get_xyz _Tail _x _y _z))

((get_xyz ((_X _x)|_Tail) _xs _y _z)
 (ON _X (C Si))
 (get_xyz _Tail _rx _y _z)
 (SUM _rx _x _xs))

((get_xyz ((_X _y)|_Tail) _x _ys _z)
 (ON _X (H F Br Cl I))
 (get_xyz _Tail _x _ry _z)
 (SUM _ry _y _ys))

((get_xyz ((_X _z)|_Tail) _x _y _zs)
 (ON _X (N P))
 (get_xyz _Tail _x _y _rz)
 (SUM _rz _z _zs))

((get_xyz () 0 0 0))

The main relation will be called "calculate_RDBE"; it will have two arguments, namely a formula list and the number of RDBEs. One possible implementation is:

((calculate_RDBE _formula _rdbe)
 (get_xyz _formula _x _y _z)
 (is _rdbe (+ 1(+ _x (÷ (− _z _y) 2))))))

Some examples of queries and their solutions are:

- (calculate_RDBE ((C 6) (H 6)) _rdbe) yields: _rdbe = 4, No (more) solutions;
- (calculate_RDBE ((C 6) (H 12) (O 6)) _rdbe) yields: _rdbe = 1, No (more) solutions;
- (calculate_RDBE ((C 16) (H 10)) _rdbe) yields: _rdbe = 12, No (more) solutions;
- (calculate_RDBE ((C 12) (H 18) (O 4) (N 2) (P 1) (S 3) (Si 1) (Cl 3)) _rdbe) yields: _rdbe = 5, No (more) solutions.

Note, that this procedure cannot be used to generate formulas which have a given number of RDBEs (i.e., issuing a query such as (calculate_RDBE_formula 4)). If this is done, the first clause of get_xyz will be invoked, which will result in a recursive call to get_xyz, with none of the arguments bound, etc. Hence, PROLOG gets into an "eternal" loop (which will continue until all the available memory has been used). Note, by the way, that the above program is not tail-recursive.

9.5 Example 2

A further example, using comparisons, takes us back to the use of PROLOG as an intelligent database query language. Suppose we have some facts concerning proteins:

((protein cytochrome_c (bovine heart)
 13370 11.4 1.71 1.19 0.71))

((protein myoglobin (horse heart)
 16900 11.3 2.04 1.11 0.74))

((protein chymotripsinogen (bovine pancreas)
 23240 9.5 2.54 1.19 0.73))

((protein β_lactoglobulin (goat milk)
 37100 7.48 2.85 1.26 0.751))

((protein serum_albumin (human)
 68500 6.1 4.6 1.29 0.734))

((protein hemoglobin (human)
 64500 6.9 4.46 1.16 0.75))

((protein catalase (horse liver)
 221600 4.3 11.2 1.25 0.73))

((protein urease (jack bean)
   482700 3.46 18.6 1.19 0.73))

((protein fibrinogen (human)
   339700 1.98 7.63 2.34 0.71))

((protein myosin (cod)
   524800 1.10 6.43 3.63 0.73))

((protein tobacco_mosaic_virus (tobacco plant)
   40590000 0.46 198 2.03 0.75))

These clauses, one for each protein, contain a predicate ("protein"), the name of the protein, its origin (as a list), the molecular mass (amu), the diffusion coefficient (10^{11} m^2 s^{-1}), sedimentation coefficient (10^{13} s), frictional ratio (dimensionless) and specific volume (ml g^{-1}). The physical data pertain to a temperature of 20°C and water solutions and were taken from ref. 30; background information concerning the entities used can be found, for instance, in ref. 31.

For convenience, we shall also use a rule to determine whether a certain number is contained within a certain interval, including the boundaries:

((in_between _number (_lower _upper))
 (greater_or_equal _number _lower)
 (less_or_equal _number _upper))

where the relations greater_or_equal and less_or_equal are as before. By issuing queries (coded as rules if required), we may obtain a wealth of information from our database.

For instance, to retrieve all information about proteins of human origin, we may issue the query: (protein_x (human) _m _x1 _x2 _x3 _x4), which will yield the following answers:

* _x = serum_albumin, _m = 68500, _x1 = 6.1, _x2 = 4.6, _x3 = 1.29, _x4 = 0.734;
* _x = hemoglobin, _m = 64500, _x1 = 6.9, _x2 = 4.46, _x3 = 1.16, _x4 = 0.75;
* _x = fibrinogen, _m = 339700, _x1 = 1.98, _x2 = 7.63, _x3 = 2.34, _x4 = 0.71;
* No (more) solutions.

If we want information concerning all proteins (in our database) with a frictional ratio between 1.1 and 1.2 and a molecular mass not greater than 50000, we issue the conjunctive query: (protein _x _s _m _x1 _x2 _x3 _x4) (in_between_x3 (1.1 1.2)) (less_or_equal _m 50000), which will yield:

* _x = cytochrome_c, _s = (bovine heart), _m = 13370, _x1 = 11.4, _x2 = 1.71, _x3 = 1.19, _x4 = 0.71;
* _x = myoglobin, _s = (horse heart), _m = 16900, _x1 = 11.3, _x2 = 2.04, _x3 = 1.11, _x4 = 0.74;
* _x = chymotripsinogen, _s = (bovine pancreas), _m = 23240, _x1 = 9.5, _x2 = 2.54, _x3 = 1.19, _x4 = 0.73;
* No (more) solutions.

Finally, if we wish to retrieve information about all proteins with a molecular mass greater than 100000, a sedimentation coefficient of less than 200 and a diffusion coefficient greater than 4, we issue the following conjunctive query: (protein_x _s _m _x1 _x2 _x3 _x4) (greater _m 100000) (LESS _x2 20) (greater _x1 4). This results in:

* _x = catalase, _s = (horse liver), _m = 221600, _x1 = 4.3, _x2 = 11.2, _x3 = 1.25, _x4 = 0.73;
* No (more) solutions.

By way of service to the reader, we also provide a relation that can be used to generate numbers, for uses much like that in for...next, for...do... and do...continue constructs:

((between _lower (_lower _upper _increment))
 (≤ _lower _upper))

((between _number (_lower _upper _increment))
 (+ _lower _increment _new)
 (≤ _new _upper)
 (between _number (_new _upper _increment)))

We leave it to the reader to figure out exactly how this relation works. To demonstrate its operation, consider the following queries and their solutions:

- (between _number (−4 2 1.3)) yields: _number = 4, _number = −2.7, _number = −1.4, _number = −0.1, _number = 1.2, No (more) solutions;
- (between _number (1 −33 −9)) yields: No (more) solutions;
- (between_number (1 33 9)), however, yields: _number = 1, _number = 10, _number = 19, _number = 28, No (more) solutions.

Obviously, the present relation can only be used to generate numbers in an increasing order. It should be a useful exercise to write a relation which generates numbers in a decreasing order. Of course, if the increment is known in advance (if, for instance, it is set equal to one), it need not be given as an argument. Instead, its value may be used directly in the summation in the second clause.

10 INPUT/OUTPUT OPERATIONS AND SUPERVISOR PROGRAMS

10.1 I/O operations

Obviously, it would be handy to have some more control over the information that is being printed. Hitherto, in answering our queries, PROLOG has only printed the instantiations of the variables in the query that belong to a certain solution. However, sometimes one may be interested in the value of only one or two variables, for instance in the case of the protein database. At other times, having an option available to input values may be quite convenient. Luckily, input/output operations are relatively straightforward in PROLOG, both where user-I/O and where file-I/O are concerned.

The predicate "P" may be used to display texts, numbers, values of variables etc. much as with the PASCAL "write" statement, whereas "PP" is comparable to the PASCAL "writeln" statement (i.e., with a carriage return added at the end). There is one difference, however: quoted constants are printed without quotes by "P" but with quotes when using "PP". As an example, the following relation will print all elements of a list consecutively and will print a carriage return when all elements have been printed:

((print_list (_head | _Tail))
 (P _head " ")
 (print_list _Tail))
((print_list ())
 (PP))

So, when the "condition" (print_list (Chemometrics and Intelligent Laboratory Systems)) is used in a clause, the result will be that the following line is printed on the screen:
Chemometrics and Intelligent Laboratory Systems

Alternatively, the following relation prints all elements on a new line and will number these lines:

((print_numbered_list (_head | _Tail) _old_counter)
 (PP List element number _old_counter is: _head)
 (+ _old_counter 1 _new_counter)
 (print_numbered_list _Tail _new_counter))
((print_numbered_list() _final_counter)
 (PP))

For example, if we issue the query: *(print_numbered_list (Chemometrics and Intelligent Laboratory Systems) 0)*, the result will be:

List element number 0 is: Chemometrics
List element number 1 is: and
List element number 2 is: Intelligent
List element number 3 is: Laboratory
List element number 4 is: Systems

YES

Hence, texts are printed just as such, whereas variables are used by their name. If a variable in a print statement has been instantiated (i.e., it has a value when the P or PP is evaluated), this value will be printed. If the variable has no value, the PROLOG will display its internal name (e.g., an underscore followed by a number).

For reading-in values for variables, PROLOG has the "R" primitive. This predicate has a single argument, which is a term that is typed on the keyboard, i.e. a number, constant, list or even a variable. In MacPROLOG there is a more advanced primitive, called PROMPT. This predicate has two arguments, both of which are lists. The first is a message to the user, which tells him or her what kind of input is expected; the other list will be bound to the user's response. When PROMPT is invoked, a dialogue box pops up on the screen in which the message is displayed and room is left for the user to type an answer. In other PROLOG implementations, the PROMPT primitive can be simulated (although the dialog box will be absent) by means of the following relation:

((prompt _message _answer)
(print_list _message)
(R _answer))

We use print_list here, rather than PP, since the latter would result in the entire list being displayed, including the outer parentheses. From now on, we shall use PROMPT for reasons of convenience.

Disk-I/O is also possible in PROLOG. Suffice it to say here, that one may read (and write) lists from (to) files, which in practice leads to two useful types of application:
- reading data from a file. For example, molecular spectra may be stored on disk for a large number of compounds and then be retrieved as required [32]. Writing data to a file may be useful, for instance, in the case of working with an expert system, if one wants to save all data (and, possibly, solutions) relating to a certain problem to disk for later use.
- reading parts of a program from disk. In case of applications which are too large to be held entirely in memory, a program may be stored in various files on disk which may then be loaded as required. This facility allows for building very flexible programs as well as extensive programs on computers with limited RAM.

10.2 Supervisor programs

Now we shall turn to a type of programming which gives PROLOG programs a somewhat closer resemblance to conventional programs. We may predefine, as it were, a set of queries which are typically performed in the same order many times: a "supervisor program". In fact, we are merely constructing a rule with a procedural nature. As an example, we will develop a small program which computes the molecular mass relevant to a certain user-input stoichiometric formula. First of all, we need some "knowledge" about elements and their masses:

((mass C 12))
((mass H 1))
((mass O 16))

Naturally, more of these clauses may be added. In case there are "sub-formulas" which often occur (for instance, amino acids, crystal solvent molecules, etc.), we may also add clauses about such entities, for example:

((mass water 18))
((mass methanol 30))

Subsequently, we write the supervisor program, which we call "compute_MW":

((compute_MW)
 (PROMPT (Enter your formula, please...)
 _Formula)
 (PP You entered the following formula:
 _Formula)
 (determine_MW _Formula 0 _MW)
 (PP The corresponding mass is _MW))

This program prompts the user to supply a formula, it prints it, calculates the molecular mass and prints its value. We will once again assume that formulas are represented as: ((C 6)(H 12)(O 6)) etc. All that is left to do is to write a relation which does the actual computation:

((determine_MW((_element _coeff)|_Rest)
 _old_mass _MW)
 (mass _element _mass)
 (is _new_mass (+ _old_mass (* _mass _coeff)))
 (determine_MW _Rest _new_mass _MW))
((determine_MW ()_MW_MW))

Note that the third argument of "determine_MW" does not become bound until all elements have been processed. At that stage, it is set equal to the accumulated (and actual) molecular weight. This strategy forces us to use an "intermediate" argument which must be assigned a start value (zero in this case) when "determine_MW" is first called. The major benefit, however, is that it enables us to use tail recursion which—as stated earlier—makes the program more efficient. The program is executed by issuing the query: *compute_MW*. (Since there are no arguments, the parentheses may be omitted.) An example session might look like this (refer to Figure 5 for a screen-dump which shows the PROMPT-box):

:compute_MW
You entered the following formula:
  ((C 6) (H 12) (O 6))
The corresponding mass is 180
YES

We will use this simple program in subsequent sections to demonstrate some other useful features of PROLOG, while making the program more efficient and versatile.

11 CONTROL OF EVALUATION

Let us take a closer look at the "determine_MW" relation of the preceding section. If we trace its execution, we will discover that many attempts to find other solutions are undertaken by PROLOG, although we know in advance that these will be futile. In particular, there may be many "mass" clauses, whereas at each stage we are only interested in one, namely the one that contains the element we are dealing with at any one moment. PROLOG does not "know" this, and will try all "mass" clauses in turn, even after the right one has been encountered.

Clearly, we need a mechanism to influence the execution of PROLOG programs, since such unnecessary (and futile) searching for alternative solutions will in general slow down execution of programs quite markedly.

Fortunately, there is such a mechanism, although there are some catches to its use. Its name is the "cut" an it is represented by a slash, "/". When inserted after a condition in a clause, PROLOG will not backtrack further than the "/". In practice, this means that once one solution has been obtained, PROLOG will not attempt to find any others (even if there are any). Hence, our determine_MW clauses need to be changed to:

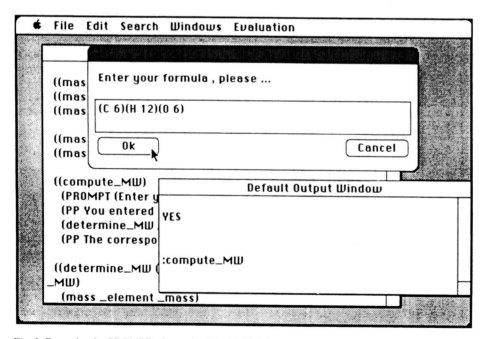

Fig. 5. Example of a PROMPT when using MacPROLOG.

((determine_MW ((_element _coeff) | _Rest)
  _old_mass _MW)
 (mass _element _mass) /
 (is _new_mass (+ _old_mass (* _mass _coeff)))
 (determine_MW _Rest _new_mass _MW))
((determine_MW () _MW _MW))

Typographically, the difference is scarcely noticeable, but in terms of program speed and efficiency the use of cuts may have quite dramatic results at times. Note that the use of a cut only makes sense in cases where condition may have more than one solution. Hence, inserting an additional cut after the "is" condition does not make the program more efficient or faster. However, adding a cut after the "is" condition instead of the one after the "mass" condition would also speed up the program.

In this example, another effect of cuts is demonstrated too. Not only does the cut prevent attempts to solve any of the preceding conditions; in addition no other clauses of the same predicate will be tried. In the original implementation of determine_MW, each call to it will result in the attempt by PROLOG to match both clauses, although the latter only applies in one special case (namely, when all elements of the list have been processed). Hence, although we known that attempts to match the second clause will fail in all cases but one, PROLOG still undertakes the operation. This is not done, however, in the second implementation, where the cut is used.

In our example, we have made the clause accounting for the empty formula-list the second one. In general, one should write the clauses in the most efficient order, i.e. the clauses of which one knows (or suspects) that they will be used most often should precede the less frequently used ones. This cannot always be effected though. In this example it works, since both clauses are mutually exclusive, i.e. a non-empty list as the first argument can only be matched in the first clause, whereas an empty list can only be matched in the second clause (since an empty list cannot be split into a head and a tail). Therefore, caution is warranted.

The use of cuts may also relieve the programmer of a lot of typing work, although such laziness carries several dangers. For instance, the relation:

((maximum _x _y _x)
 (LESS _y _x))
((maximum _x _y _x)
 (EQ _x _y))
((maximum _x _y _y)
 (LESS _x _y))

can be used to obtain the greater of two numbers. The first two arguments are these numbers and the third will become bound to the greater of the two.

Using cuts, we may make the program more efficient: if one of the three clauses can be satisfied, then we know in advance that none of the others will, since they are mutually exclusive. So a more efficient version could be:

((maximum _x _y _x)
 (LESS _y _x) /)
((maximum _x _x _x) /)
((maximum _x _y _y)
 (LESS _x _y))

In the second clause, we have used an additional trick, since this is the case where _x and _y have the same value. In the third clause, a cut is obsolete, since there are no more clauses succeeding it.

It is important to notice that the logical reading of the three clauses has not been altered: the cuts only influence the evaluation procedure. Cuts of this type are therefore called "green cuts". But what about the following implementation?

((maximum _x _y _x)
 (LESS _y _x) /)
((maximum _x _x _x) /)
((maximum _x _y _y))

The interpretation of this set of clauses is obvious to humans: if the first two clauses do not apply, then the conclusion must be that the second argu-

ment is greater than the first, and the test (LESS _x _y) is obsolete. But now consider the logical reading of the set of clauses:
- the maximum of two numbers x and y is x if y is less than x;
- the maximum of two equal numbers is that number;
- the maximum of two numbers x and y is y.

Clearly, in isolation the third clause is nonsensical from a logical point of view. It derives its validity in this paricular case purely from the fact that it appears in conjunction with two other clauses and in this particular order. Should another programmer write down these clauses in such an order that the third is not the final clause, then we are in trouble. This type of usage of cuts obviously changes the logical reading of the relation "maximum" and hence is considered to be bad programming practice. Cuts like these are therefore termed "red cuts".

A very useful feature of the Standard MacPROLOG syntax, that may entirely replace the use of cuts, is the exclamation mark. It may be used in conditions only and must precede the predicate. Its effect is that only one solution to the particular condition will be sought. Using this, yet another alternative to the determine_MW relation becomes:

((determine_MW ((_element _coeff) | _Rest)
 _old_mass _MW)
(! mass _element _mass)
(is _new_mass (+ _old_mass (* _mass _coeff)))
(! determine_MW _Rest _new_mass _MW))
((determine_MW () _MW _MW))

The "!" in the "mass" condition has the effect that only one matching "mass" clause will be retrieved. The "!" in the recursive call to determine_MW has the effect that only the first solution encountered will be elaborated upon.

Those readers interested in further practice are recommended to reconsider the demonstration program that calculates rings and double bond equivalents, and to make it more efficient, without altering the logical reading of the program. Use of the trace facility is recommended in order to obtain a fuller comprehension of the mechanism of using cuts. More about cuts, including listings of test programs to determine the exact behaviour of a particular PROLOG implementation, can be found in ref. 33.

12 DATABASE OPERATIONS

Let us return once again to the program that allows us to compute molecular weights, and add some "learning capability" to it. In order to be able to do this, we need PROLOG's capability to modify and extend PROLOG programs. PROLOG possesses a so-called dynamic database, which is a "pool" of memory in which we may store clauses of any kind. Clauses can be added to the database, used in programs, and deleted from it again. In Standard syntax MacPROLOG we may employ the primitives ADDCL, CLS, CL, DELCL, DELCLALL and KILL to this effect (in other PROLOG implementations, ADDCL may be called "assert" and DELCL/DELCLS may be called "retract"). Firstly, we will change the determine_MW relation into:

((determine_MW ((_element _coeff) | _Rest)
 _old_mass _MW)
(! get_mass _element _mass)
(is _new_mass (+ _old_mass (* _mass _coeff)))
(! determine_MW _Rest _new_mass _MW))
((determine_MW () _MW _MW))

The only difference between this and the previous version is that we introduce a new relation called "get_mass". This relation will consist of two clauses, one which attempts to retrieve a matching "mass" clause. If this fails, the program obviously does not "know" of the particular element or entity. Hence, a second clause will prompt the user to supply the mass of the entity, and will add this to the program:

((get_mass _element _mass)
 (mass _element _mass))
((get_mass _element _mass)
 (PROMPT (I do not know the mass of _element!
 Please enter it now...) _M)
 (EQ _M (_mass))
 (ADDCL ((mass _element _mass))))

Note that, due to the use of the exclamation mark in determine_MW a cut is unnecessary here. The first clause succeeds if there exists a "mass" clause pertaining to _element. The second prompts the user to supply the mass (since PROMPT always returns a list, we must extract the value of _mass in the succeeding condition) and adds a clause to the program.

As a matter of fact, both clauses, as they are now written, are not mutually exclusive. For them to be exclusive, we need to test (in the second clause) that indeed no "mass" clause pertaining to the element exists. We need to introduce negation here, "NOT" in PROLOG (see Part 2 of this tutorial), as well as CL, which may be used to retrieve a clause or to test whether any clauses exist. The second clause then changes to:

((get_mass _element _mass)
 (NOT CL ((mass _element _any_mass)))
 (PROMPT (I do not know the mass of _element!
 Please enter it now...) _M)
 (EQ _M (_mass))
 (ADDCL ((mass _element _mass))))

Since the variable _element is instantiated when get_mass is called, the first condition of the second clause will succeed if there is no "mass" clause pertaining to (the current value of) _element and fail if there already is one.

The primitives mentioned previously have the following functions:
- ADDCL adds a clause to the database;
- CL retrieves a clause;
- CLS retrieves a specified clause;
- DELCL deletes a clause;
- DELCLALL deletes all clauses of a certain relation that can be matched to the argument of this predicate;
- KILL kills an entire relation (i.e., all its clauses).

The ability to add, retrieve and delete clauses is a very powerful feature of PROLOG. In principle it enables one to write programs that write programs (etc.). Programming at this level of abstraction (i.e., manipulating programs) is called meta-level programming.

The CL primitive has another useful application, namely in checking whether a relation has been defined. If a relation is called when it does not exist, this will result in an error message. By using CL, we may test whether there is at least one clause (and hence the relation is defined) of the relation. This may be used to check whether a new program needs loading, or whether a knowledge base has been loaded, or whether a previous program terminated correctly or yielded solutions or....

In practice, this capability can be used to store the results of a certain operation for future use. The following example, taken from our own work, may help clarify this point, without getting into any technical details: PEGASUS [34] is a program that can be used to generate acyclic molecular structures. It works in three stages (see also Fig. 6):
- the first module (MolFormDed) prompts the user to supply a molecular weight and to select a set of elements, from which plausible molecular formulas are generated (see also ref. 35). Any formulas generated are stored in the dynamic database.
- the second module (Assemble) first checks to see if any formulas are present in the database. If this is not the case, the user is prompted to supply one. Otherwise, it displays the formulas and prompts the user to select one. Subsequently, the user must select a set of fragments (e.g., "CH3-", "-CH2-", etc.) and the program will generate linear combinations of these fragments which satisfy the molecular formula and store these in the database.
- the third module (Configure) retrieves all fragment sets from the database, prompts the user to select one, and then generates all possible different acyclic structures that can be built

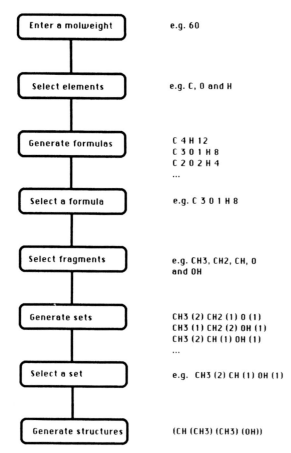

Fig. 6. Schematic representation of the use of PEGASUS, a program for the generation of acyclic molecular structures.

from the fragments in the set. These are stored in the database too, in order to be able to avoid the repeated generation of identical structures.

REFERENCES

1 R.A. Kowalski, The origins of logic programming, *BYTE*, (August, 1985) 192–193.
2 I. Bratko, *PROLOG Programming for Artificial Intelligence*, Addison-Wesley, Wokingham, 1986.
3 J.A. Robinson, A machine-oriented logic based on the resolution principle, *Journal of the ACM*, 12 (1965) 23–44.
4 D.W. Loveland, A simplified format for the model elimination theorem proving procedure, *Journal of the ACM*, 16 (1969) 349–363.
5 R.A. Kowalski and D. Kuehner, Linear resolution with the selection function, *Artificial Intelligence*, 2 (1971) 227–260.
6 E. Shapiro (Editor), *Proceedings of the Third International Conference on Logic Programming*, Springer-Verlag, Berlin, 1986.
7 E.A. Feigenbaum and P. McCorduck, *The Fifth Generation – Artificial Intelligence and Japan's Challenge to the World*, Pan Books, London, 1984.
8 K. Fuchi and K. Furukawa, The role of logic programming in the fifth generation computer project, in E. Shapiro (Editor), *Proceedings of the Third International Conference on Logic Programming*, Springer-Verlag, Berlin, 1986, pp. 1–24.
9 J. McCarthy, Recursive functions of symbolic expressions and their computation by Machine, Part I, *Communications of the ACM*, 3 (1960) 184–195.
10 S.E. Fahlman, Common LISP, *Annual Reviews in Computer Science*, 2 (1987) 1–19.
11 C.J. Rawlings, W.R. Taylor, J. Nyakairu, J. Fox and M.J.E. Sternberg, Reasoning about protein topology using the logic programming language PROLOG, *Journal of Molecular Graphics*, 3 (1985) 151–157.
12 A.J. Morffew and S.J.P. Todd, The use of PROLOG as a protein querying language, *Computers and Chemistry*, 10 (1986) 9–14.
13 G.J. Postma, B.G.M. Vandeginste, C.J.G. van Halen and G. Kateman, A comparison of LISP and PROLOG, *Trends in Analytical Chemistry*, 6 (1987) 2–5.
14 D.L. Massart and J. Smeyers-Verbeke, PROLOG: an artificial intelligence language, *Trends in Analytical Chemistry*, 4 (1985) 50–51.
15 D.L. Massart and M. De Smet, PROLOG: conjunctive queries and rules, *Trends in Analytical Chemistry*, 4 (1985) 111–112.
16 G. Musch, A. Thielemans and D.L. Massart, PROLOG: arithmetic relations, *Trends in Analytical Chemistry*, 5 (1986) 58.
17 R.E. Dessy, Expert systems, Part I, *Analytical Chemistry*, 56 (1984) 1200A–1212A.
18 D. Cabrol and T.P. Forrest, A different kind of language: PROLOG, Programming in Logic, *Journal of Chemical Education*, 63 (1986) 131–135.
19 S. Eisenbach and C. Sadler, Declarative languages: an overview, *BYTE* (August, 1985) 181–197.
20 J.R. Ennals, *Beginning Micro-PROLOG*, Ellis Horwood, Chichester, 1984.
21 C.J. Hogger, *Introduction to Logic Programming*, Academic Press, London, 1984.
22 K. Clark and F.G. McCabe, *Micro-PROLOG: Programming in Logic*, Prentice Hall, Englewood Cliffs, NJ, 1984.
23 W.F. Clocksin and C.S. Mellish, *Programming in PROLOG*, Springer, New York, 1984.
24 K.L. Clark and F.G. McCabe, *LPA MacPROLOG Reference Manual*, Logic Programming Associates, London, 1986.
25 P. French, *LPA MacPROLOG User Guide*, Logic Programming Associates, London, 1986.
26 R.A. Kowalski, Algorithm = Logic + Control, *Communications of the ACM*, 22 (1979) 424–436.
27 R.A. Kowalski, *Logic for Problem Solving*, North-Holland, New York, 1979.

28 R.C. Bohinski, *Modern Concepts in Biochemistry*, Allyn and Bacon, Boston, MA, 1979.
29 P. Friedland and L.H. Kedes, Discovering the secrets of DNA, *Communications of the ACM*, 28 (1985) 1164–1186.
30 A.L. Lehninger, *Biochemistry*, Worth Publishers, New York, 1975.
31 G.M. Barrow, *Physical Chemistry*, McGraw-Hill, Tokyo, 1979.
32 H.J. Luinge, G.J. Kleywegt, H.A. van 't Klooster and J.H. van der Maas, Artificial intelligence used for the interpretation of combined spectral data. Part III. Automated generation of interpretation rules for infrared spectral data, *Journal of Chemical Information and Computer Sciences*, 27 (1987) 95–99.
33 C. Moss, Cut&Paste – defining the impure primitives of PROLOG, in E. Shapiro (Editor), *Proceedings of the Third International Conference on Logic Programming*, Springer-Verlag, Berlin, 1986, pp. 686–694.
34 G.J. Kleywegt, H.J. Luinge and H.A. van 't Klooster, Artificial intelligence used for the interpretation of combined spectral data. Part II. PEGASUS: a PROLOG program for the generation of acyclic molecular structures, *Chemometrics and Intelligent Laboratory Systems*, 2 (1987) 291–302.
35 G.J. Kleywegt and H.A. van 't Klooster, Chemical applications of PROLOG. Interpretation of mass spectral peaks, *Trends in Analytical Chemistry*, 6 (1987) 55–57.

Chapter 7

PROLOG for Chemists. Part 2 *

GERARD J. KLEYWEGT *

Department of NMR Spectroscopy, University of Utrecht, Padualaan 8, 3584 CH Utrecht (The Netherlands)

HENDRIK-JAN LUINGE and BART-JAN P. SCHUMAN

Analytical Chemistry Laboratory, University of Utrecht, Croesestraat 77A, 3522 AD Utrecht (The Netherlands)

(Received 30 March 1988; accepted 21 April 1988)

CONTENTS

1 Logical primitives	92
2 List manipulation	93
3 Applications in chemistry	94
4 Example programs	95
4.1 Reasoning about ligand field splitting patterns	95
4.2 Transcribing DNA sequences into proteins	97
4.3 Calculating the Randic index for molecular graphs	99
5 Pros and cons	101
6 Concluding remarks	101
7 Acknowledgements	102
References	102

1 LOGICAL PRIMITIVES

In this section we will briefly introduce some additional (extra-) logical predicates. We have already used NOT for negation. It must be stressed here that NOT can only be used as a test but not for generating values for a variable. For instance, suppose we have in a program a sequence of conditions which prompts the user to input an element that is not equal to either C or H. This can be implemented as follows:

...
(PROMPT (Enter an element, but NOT C or H...) _V)
(EQ _V (_v))
(NOT ON _v (C H))
...

The following sequence of conditions, however, is erroneous:

...
(NOT ON _v (C H))
(PROMPT (Enter an element, but NOT C or H...) _V)
(EQ _V (_v))
...

---

* For Part 1 see: G.J. Kleywegt, H.J. Luinge and B.J.P. Schuman, PROLOG for Chemists. Part 1, *Chemometrics and Intelligent Laboratory Systems*, 4 (1988) 273–297.

In this case, the condition (NOT ON _v (C H)) will not be satisfied, since _v is unbound. One might imagine that PROLOG might start generating constants that are unequal to both C and H, but fortunately this does not happen. Since PROLOG does not have a clue as to what type of constants we are interested in, it could in theory continue generating constants eternally without ever generating the same symbol that the user inputs.

To introduce the OR predicate, we return to the "maximum" relation. In the original definition, we may contract the first two clauses into one clause as follows:

((maximum _x _y _x)
 (OR ((LESS _y _x)) ((EQ _x _y))))

The interpretation of this clause is: the greater of two numbers x and y is x if either y is less than x or y equals x.

An OR condition succeeds if either one of the lists of sub-conditions succeeds. The use of OR in general is not unavoidable and does not improve the readability of programs, but sometimes (especially in the case of clauses with a large number of conditions which would all have to be re-evaluated if we were to use two separate clauses) reasons of speed or efficiency will prevail.

We may even contract the entire "maximum" relation into one clause by making use of another primitive, namely IF:

((maximum _x _y _z)
 (IF (LESS _y _x)
     ((EQ _x _z))
     ((EQ _y _z))))

The interpretation of this clause is: the greater of two numbers x and y is z, where, if y is less than x, z equals x and otherwise z equals y. In general, if the condition following IF is true, then the first list of conditions will be evaluated, otherwise the second of list of conditions will be evaluated. Notice the following use of the maximum-relation: the query (*maximum Econometrics Chemometrics_z*) will succeed. The solution is: $\_z =$ *Econometrics, No (more) solutions.*

As is the case with OR, the use of IF does not contribute to writing programs which are the acme of elegant logic programming. Its usefulness stems solely from pragmatic considerations.

Two other very useful primitives are FORALL and ISALL (in other implementations they may be called bagof, setof, findall). Both primitives are related and can be used to perform one or a number of operations on several elements. Returning to the molecular weight computing program, suppose we want a listing of all elements known to the program, then one way of accomplishing this is:

...
(FORALL ((mass _el _m)) ((PP I know of element _el)))
...

The primitive ISALL can be used to generate all solutions to a (conjunction of) condition(s) and store them in a list. Solving the same problem as before, we could write:

...
(ISALL _list_of_elements _el
  (mass _el _m))
(PP I know the following elements: _list_of_elements)
...

The general format for FORALL is: (FORALL _list_of_conditions _list_of_actions), whereas that for ISALL is: (ISALL _list_of_solutions _solution _list_of_conditions). It will be clear that these primitives are rather high level concepts. Their use will often make the development of recursive routines obsolete.

2 LIST MANIPULATION

We have already dealt with a number of structure primitives, i.e. relations that manipulate lists. In particular, we have demonstrated the use of APPEND, ON and LENGTH. In this section we shall develop two other relations that are of general utility, namely "deletes", which deletes an element from a list, and "permutes", which generates permutations of a list. Another useful relation is "sort" but its implementation is beyond the scope of this tutorial. For those interested, it can be found in virtually any PROLOG textbook. In addition, MacPROLOG provides the SORT relation as a built-in primitive.

A relation that deletes just the first occurrence of a certain element of a list is:

((deletes_first _el _old_list _new_list)
 (APPEND _x1 (_el | _x2) _old_list) /
 (APPEND _x1 _x2 _new_list))

The use of the cut following the first APPEND ensures that no backtracking will occur. So the query: (deletes_first 3 (1 2 3 4 5 3 5 6 7) _new) will yield: _new = (1 2 4 5 3 5 6 7), No (more) solutions. Had we omitted the cut, then an extra solution would have been found, namely: _new = (1 2 3 4 5 5 6 7).

Notice that a call to this relation will fail if the element that must be deleted does not occur in the list at all. If we do not want this to happen, the following relation should be used:

((deletes_first _el _old_list _new_list)
 (APPEND _x1 (_el | _x2) _old_list) /
 (APPEND _x1 _x2 _new_list))
((deletes_first _el _list _list))

Now, the answer to the query: (deletes_first 3 (1 2 4 5 5 6 7) _new) will be: _new = (1 2 4 5 5 6 7), No (more) solutions.

A relation that deletes all occurrences of a certain element from a list is:

((deletes_all _el _old _new)
 (APPEND _x1 (_el | _x2) _old) /
 (APPEND _x1 _x2 _intermediate
 (deletes_all _el _intermediate _new))
((deletes_all _el _list _list))

The reader may easily verify that the query: (deletes_all 3 (3 1 2 3 4 4 3 3 5 5 3 6 3 7 3) _new) has one solution, namely: _new = (1 2 4 4 5 5 6 7).

A relation that generates permutations of a list can be implemented as follows:

((permutes (_head | _Tail) _list)
 (APPEND _V (_head | _U) _list)
 (APPEND _V _U _W)
 (permutes _Tail _W))
((permutes ( ) ( )))

The exact operation of this relation can best be examined by tracing its execution. Informally, the first clause may be interpreted as stating: a list consisting of a _head and a _Tail is a permutation of another _list, if the head element occurs somewhere on the other _list and if the remainder of the _list, when _head has been removed from it, is a permutation of the _Tail. In any event, the results of the query: (permutes _perm (1 A +)) will yield the following answers:

_perm = (1 A +), _perm = (1 + A), _perm = (A 1 +), _perm = (A + 1), _perm = (+ 1 A), _perm = (+ A 1), No (more) solutions.

The same relation may be used for checking purposes: (permutes (A + 1) (1 A +)) is true. Notice that the second argument must have a value. It the two arguments are interchanged, PROLOG will find one solution (the "non-permutation"), but will, on backtracking, be caught in an infinite loop.

3 APPLICATIONS IN CHEMISTRY

In recent years a number of applications of PROLOG in chemistry have been described in the literature. In this section we present a brief overview of some of these. Broadly speaking, these applications can be divided into three groups:
- 1 development of expert systems;
- 2 intelligent database query programs;
- 3 educational programs.

1. Several expert systems [1–8] written in PROLOG have been reported:

Fell and co-workers have developed a system that assists in luminescence analysis [9,10]. Gunasingham has developed two expert systems in PROLOG. One assists in the design of a cybernetic electroanalytical instrument [11], the other in the planning of HPLC separations of steroids [12]. In Massart's group a system was developed which assists in the development of procedures for pharmaceutical analysis [13,14], although at present an expert system shell rather than PROLOG is being used for its development. WIZARD [15,16] is an expert system that analyses the conformational space of molecules. It provides initial estimates of geometries and energies of likely conformers of a given molecule. These geometries may subsequently be refined by an existing molecular mechanics program, such as MM2. It thus serves as an intelligent front-end to conventional numerical software. In our own laboratories, we are engaged in the construction of EXSPEC, a microcomputer-based expert system for the interpretation of combined spectral information [17–19], and of ProNMR, a program that should take on the task of interpreting two-dimensional NMR spectra of proteins. Other applications relate to organic synthesis planning [20] and the selection of bacterial strains for bioconversions [21].

Usually, such expert systems are written directly in PROLOG. Due to the possibility of implementing meta-level programming, other methods can be used too, however. Rossi [22] briefly describes approaches to building on top of PROLOG (e.g., by implementing an explanation facility, or allowing for inexact reasoning) and to extending PROLOG itself. Such approaches are useful (and relatively easy to implement) when one needs control strategies or data-structures that differ from those standard to PROLOG (for instance, a breadth-first rather than a depth-first search mechanism).

2. A few applications of PROLOG have been described in which the language is used to develop an intelligent database query program. One such program [23,24] reasons about topological features of proteins. Another [25,26] employs PROLOG as a more general protein querying language.

3. Some approaches to using PROLOG for writing educational programs (CAI = Computer-Assisted Instruction, or ICAI = Intelligent CAI) have been published. Kawai et al. [27] have described a chemical reaction tutor. IRIS is a set of two programs (one in LISP, the other in PROLOG) that aid in teaching the interpretation of IR spectra [28]. Cabrol and co-workers are developing software that serves as an intelligent problem-solving partner in infrared spectroscopic analysis [29,30].

4 EXAMPLE PROGRAMS

4.1 Reasoning about ligand field splitting patterns

Let us now turn to inorganic for a first example program. The program deals with some aspects of co-ordination chemistry, namely the splitting patterns of energy terms of transition metal ions when these are subject to an octahedral (O_h) or tetrahedral (T_d) ligand field. Knowledge about such splitting patterns is of help when interpreting Visible-near IR spectra of transition metal complexes. For the theory, see ref. 31 and 32 for instance. For another application of symbolic computing (in LISP) to group theory, see ref. 33.

We shall commence by defining some knowledge about a number of first-row transition metal ions, in particular the number of their d-electrons:

((transition_metal "Ti3 + " 1))
((transition_metal "V3 + " 2))
((transition_metal "Cr3 + " 3))
((transition_metal "V2 + " 3))
((transition_metal "Mn3 + " 4))
((transition_metal "Cr2 + " 4))
((transition_metal "Fe3 + " 5))
((transition_metal "Mn2 + " 5))
((transition_metal "Co3 + " 6))
((transition_metal "Fe2 + " 6))
((transition_metal "Co2 + " 7))
((transition_metal "Ni2 + " 8))
((transition_metal "Cu2 + " 9))

The ground states for d^1 through d^5 ions are encoded as follows (where, for instance (2 D) is usually written as: ^2D):

((ground 1 (2 D)))
((ground 2 (3 F)))
((ground 3 (4 F)))
((ground 4 (5 D)))
((ground 5 (6 S)))

A general rule states that the ground state of a $d^{(5+N)}$ ion is the same as that for a $d^{(5-N)}$ ion (where $N = 1, 2, 3, 4$). Hence, instead of writing explicit facts for the ground states of d^6 to d^9

ions, we may write in more general terms:

((ground_state _d _T)
 (LESS _d 6)/
 (ground _d _T))
((ground_state _d _T)
 (SUM _dd _d 10)
 (ground _dd _T))

To determine the ground state of a given ion, we use the following rule:

((ground_term _metal _term)
 (transition_metal _metal _d)
 (ground_state _d _term))

The splitting patterns of free ion ground terms when brought into an octahedral ligand field are as follows (where (T 2 g), for example, would usually be written T_{2g}), where the terms are ordered from high to low energy:

((split_up (2 D) ((E g) (T 2 g))))
((split_up (3 F) ((A 2 g) (T 2 g) (T 1 g))))
((split_up (4 F) ((T 1 g) (T 2 g) (A 2 g))))
((split_up (5 D) ((T 2 g) (E g))))
((split_up (6 S) ((A 1 g))))

The splitting patterns for $d^{(5+N)}$ ions are the same as those for $d^{(5-N)}$ ions, but the terms are energetically inverted:

((octahedral_term _metal _Terms)
 (transition_metal _metal _d)
 (LESS _d 6)
 (ground_state _d _ground)
 (split_up _ground _Terms))
((octahedral_term _metal _Terms)
 (transition_metal _metal _d)
 (LESS 5 _d)
 (ground_state _d _ground)
 (split_up _ground _terms)
 (reverse _terms _Terms))

Here, the "reverse" relation is ued to invert a list (i.e., the result of reversing (1 2 3) would be the list (3 2 1)). To this effect we use a simple procedure (similar to the one introduced in Part 1 of this tutorial):

((reverse (_head | _Tail) _Rev)/
 (reverse _Tail _rev_tail)
 (APPEND _rev_tail (_head) _Rev))
((reverse ( ) ( )))

The splitting pattern for a tetrahedral ligand field is the same as that for an octahedral field, with energetical inversion of the terms. Also, since in a T_d ligand field symmetry labels (g for "gerade" or "even" and u for "ungerade" or "odd") do not apply, they have to be removed from the list of terms. To effect the latter, we use:

((transform_list (_h | _T) (_th | _tT))
 (!transform _h _th)/
 (transform_list _T _tT))
((transform_list ( ) ( )))

((transform (_t _m _s) (_t _m))
 (ON _s (g u)))
((transform (_t _s) (_t))
 (ON _s (g u)))

Finally, to determine the splitting pattern of energy terms in a tetrahedral ligand field, we use:

((tetrahedral_term _metal _Terms)
 (transition_metal _metal _d)
 (octahedral_term _metal _T1)
 (reverse _T1 _T2)
 (transform_list _T2 _Terms))

The relations have been programmed in such a way that various uses are possible:
- testing whether a given metal has a given ground term or splitting pattern;
- generating the ground term or splitting pattern of a given metal;
- generating all metals that have a given ground term or splitting pattern;
- generating all sets of metals and their ground terms of splitting patterns.

The results of queries of the last type have been collected in Table 1. For those interested, the program offers considerable scope for extension. For instance, regarding further splittings when going from O_h to D_3 (trigonal) or to D_{2h} (rhombic) via D_{4h} (tetragonal) symmetrical ligand fields. Alternatively, one could write a program that actually generates ground terms for a given number of d-electrons. PROLOG is excellently suited to such a task, which basically involves the generation of all possible permutations of d-electrons (or, if

Chapter 7

TABLE 1

Results of querying the program about ligand field theory, such that it produces lists of the ground terms and the splitting patterns in octahedral and tetrahedral field for all the first-row transition metal ions the program "knows" about. The lay-out of the output has been modified to conserve space

:(ground_term _ion _term)

_ion = "Ti3+"	_term = (2D)
_ion = "V3+"	_term = (3F)
_ion = "Cr3+"	_term = (4F)
_ion = "V2+"	_term = (4F)
_ion = "Mn3+"	_term = (5D)
_ion = "Cr2+"	_term = (5D)
_ion = "Fe3+"	_term = (6S)
_ion = "Mn2+"	_term = (6S)
_ion = "Co3+"	_term = (5D)
_ion = "Fe2+"	_term = (5D)
_ion = "Co2+"	_term = (4F)
_ion = "Ni2+"	_term = (3F)
_ion = "Cu2+"	_term = (2D)

No (more) solutions

:(octahedral_term _ion _term)

_ion = "Ti3+"	_term = ((E g) (T 2 g))
_ion = "V3+"	_term = ((A 2 g) (T 2 g) (T 1 g))
_ion = "Cr3+"	_term = ((T 1 g) (T 2 g) (A 2 g))
_ion = "V2+"	_term = ((T 1 g) (T 2 g) (A 2 g))
_ion = "Mn3+"	_term = ((T 2 g) (E g))
_ion = "Cr2+"	_term = ((T 2 g) (E g))
_ion = "Fe3+"	_term = ((A 1 g))
_ion = "Mn2+"	_term = ((A 1 g))
_ion = "Co3+"	_term = ((E g) (T 2 g))
_ion = "Fe2+"	_term = ((E g) (T 2 g))
_ion = "Co2+"	_term = ((A 2 g) (T 2 g) (T 1 g))
_ion = "Ni2+"	_term = ((T 1 g) (T 2 g) (A 2 g))
_ion = "Cu2+"	_term = ((T 2 g) (E g))

No (more) solutions

:(tetrahedral_term _ion _term)

_ion = "Ti3+"	_term = ((T 2) (E))
_ion = "V3+"	_term = ((T 1) (T 2) (A 2))
_ion = "Cr3+"	_term = ((A 2) (T 2) (T 1))
_ion = "V2+"	_term = ((A 2) (T 2) (T 1))
_ion = "Mn3+"	_term = ((E) (T 2))
_ion = "Cr2+"	_term = ((E) (T 2))
_ion = "Fe3+"	_term = ((A 1))
_ion = "Mn2+"	_term = ((A 1))
_ion = "Co3+"	_term = ((T 2) (E))
_ion = "Fe2+"	_term = ((T 2) (E))
_ion = "Co2+"	_term = ((T 1) (T 2) (A 2))
_ion = "Ni2+"	_term = ((A 2) (T 2) (T 1))
_ion = "Cu2+"	_term = ((E) (T2))

No (more) solutions

there are more than five, of "electron holes") with different sets of quantum numbers. Such a program might prove valuable as an aid to students in the teaching of this subject. Finally, it would be interesting to have a program that actually infers which terms will result if an ion is brought into a ligand field of known symmetry (by using so-called character tables).

4.2 Transcribing DNA sequences into proteins

Due mainly to the list-processing and pattern matching facilities inherent in PROLOG, it is an easy task to write a small program which "transcribes" a DNA sequence into a protein [34]. Let us first define our nucleotide sequence, using the predicate "sequence":

((sequence (G C C T T G T A G A A G C G C G T A T G G C T T C G T A C C C C T G C C A T C A A C A C G C G T C T G C G T T C G A C C A G G C T G C G C G T T C T C G C G G C C A T A G C A A C C G A C G T A C G G C G T G A A A C A C G)))

This particular example is an imaginary one, although interested readers may wish to know that is does contain a part of the Thymidine Kinase gene.

Subsequently, we need to tell PROLOG about the genetic code, hence about start codons, stop codons and actual coding codons:

((start_codon (A T G)))

((stop_codon (T A A)))
((stop_codon (T G A)))
((stop_codon (T A G)))

((codes_for (T T T) (Phe) 165))
((codes_for (T T C) (Phe) 165))
((codes_for (T T A) (Leu) 131))
((codes_for (T T G) (Leu) 131))

((codes_for (T C T) (Ser) 105))
((codes_for (T C C) (Ser) 105))
((codes_for (T C A) (Ser) 195))
((codes_for (T C G) (Ser) 105))

We shall not explicitly write down all 64 codes for reasons of space. They may, however, be found in any standard biochemistry textbook [35,36]. Incidentally, in each "codes_for"-clause we have added the molecular weight of the corresponding amino acid. Alternatively, we could have defined another relation which relates the amino acids to their molecular weights. However, because of the beautiful logic of the genetic code, most clauses can be typed by copying a previous one and altering a few characters — hence the gain in typing speed would have been negligible.

Now that we have defined our data structure, we may outline the algorithm. A possible way of performing the transcription is:
- retrieve the nucleotide sequence;
- find the start codon;
- find the stop codon;
- transcribe the fragment that is in between these two;
- do some bookkeeping in order to be able to compute the molecular weight of the resulting protein.

Notice that we will not incorporate so-called nonsense codons in our program. We leave this extension to the interested reader. A transcription of our algorithm into PROLOG might look like this:

```
((analyse)
/* retrieve nucleotide sequence */
  (sequence _T)
  (PP) (PP Polynucleotide: _T)
  (LENGTH _T _t1)
  (PP Length: _t1)
/* find start codon */
  (! get_start_codon _T _S)
  (LENGTH _S _t2)
  (SUM _t2 _sc _t1)
  (PP Coding starts at nucleotide _sc)
/* find stop codon */
  (! get_stop_codon _S _R)
  (LENGTH _R _t3)
  (SUM _sc _t3 _ec)
  (PP Coding ends at nucleotide _ec)
  (PP) (PP Effective coding part: _R)
/* "transcribe" sequence into a peptide */
  (! transcribe _R _A 0 _m)
  (PP) (PP Peptide: _A)
  (LENGTH _A _la)
```
  (PP) (PP Length: _la)
  (SUM _lac 1 _la)
  (* _lac 18 _dm)
  (SUM _mass _dm _m)
  (PP Mass: _mass amu)
  (PP) (PP))

We have used comments in order to show where the various algorithmic steps commence. Notice that, apart from the bookkeeping operations, the "program" does not differ very much from the description of the algorithm we gave in proper English. All we are left with now is the implementation of the lower-level routines. First, the following relation will locate the start codon and return that part of the sequence which starts there (including the start codon itself, which codes for methionine):

((get_start_codon (_x _y _z | _Rest) (_x _y _z | _Rest))
  (! start_codon (_x _y _z)))
((get_start_codon (_x | _Rest) _S)
  (! get_start_codon _Rest _S))
((get_start_codon ( ) ( )) /
  (PP (ERROR ! No start codon found !!))
  ABORT)

The first clause succeeds if three subsequent nucleotides form a start codon. The second shifts the sequence one nucleotide and recursively calls get_start_codon. The third clause is used only when no start codon has been encountered. ABORT is a MacPROLOG primitive which cancels the entire execution.

Second, the following relation will try to locate the first stop codon on the sequence that was returned by get_start_codon. It will return that part of the total sequence that actually codes for the protein (excluding the stop codon itself, since stop codons do not code for any amino acid). The use of the three clauses is quite analogous to that of the clauses of the get_start_codon relation:

((get_stop_codon (_x _y _z | _Rest)( ))
  (! stop_codon (_x _y _z)))
((get_stop_codon (_x _y _z | _Rest) (_x _y _z | _M))
  (! get_stop_codon _Rest _M))
((get_stop_codon _X _Y) /
  (PP (ERROR ! No stop codon found !!))
  ABORT)

Third, we need a relation that performs the actual transcription. It is a very simple one, involving only two clauses in which the sequence is transcribed and the sum of the molecular weights of the protein's amino acids is determined:

((transcribe (_x _y _z | _Rest) (_aa | _rA) _old _m)
 (! codes_for (_x _y _z) _aa _aam)
 (+ _old _aam _new)
 (! transcribe _Rest _rA _new _m))
((transcribe ( ) ( ) _m _m))

Executing the query *analyse*, finally, will yield the following sequence of output:

Polynucleotide: (G C C T T G T A G A A G C G C G T A T G G C T T C G T A C C C C T G C C A T C A A C A C G C G T C T G C G T T C G A C C A G G C T G C G C G T T C T C G C G G C C A T A G C A A C C G A C G T A C G G C G T G A A A C A C G)
Length: 110
Coding starts at nucleotide 17
Coding ends at nucleotide 101

Effective coding part: (A T G G C T T C G T A C C C C T G C C A T C A A C A C G C G T C T G C G T T C G A C C A G G C T G C G C G T T C T C G C G G C C A T A G C A A C C G A C G T A C G G C G)
Length: 84

Peptide: ((Met) (Ala) (Ser) (Tyr) (Pro) (Cys) (His) (Gln) (His) (Ala) (Ser) (Ala) (Phe) (Asp) (Gln) (Ala) (Ala) (Arg) (Ser) (Arg) (Gly) (His) (Ser) (Asn) (Arg) (Arg) (Thr) (Ala))

Length: 28
Mass: 3195 amu

The molecular weigth of the protein, by the way, is not merely the cumulative sum of the weights of the amino acids. Recall that for each amide-linkage formed, we must subtract 18. In total, there is one amide-linkage fewer than there are amino acids in the protein (provided it is not circular). This may help to explain the numerical operations involved in the "analyse" relation.

4.3 Calculating the Randic index for molecular graphs

There are several ways of assigning an index value to molecular graphs. Here we shall develop a small PROLOG program, which computes the so-called Randic index [37] for graphs of organic molecules. This index, and other indices as well, can be used in studies of structure–property relationships. The assumption is made that molecular structure can be characterised mathematically and precisely, and that the mathematically determined parameters of molecules can be correlated with the molecules' experimentally measured properties. A (molecular) graph can be regarded as a number of vertices (atoms) connected by edges (bonds). Each vertex has a "degree": the number of other vertices it is connected to. Similarly, each edge has a "value": the product of the reciprocals of the square roots of the degrees of the vertices it joins. The Randic index of a molecule is equal to the sum of the values of all the molecule's edges [37]. It appears that this topological index correlates well with the boiling points or the empirical Kovats index, derived from chromatographic retention data, for several alkane isomers.

Once again, the first thing to worry about is the manner of data representation. In this particular case we have to devise a representation formalism for chemical structures. Here, we shall use one which closely resembles that used in our work on the EXSPEC expert system [17–19]. An example should help clarify this formalism:

((STRUC propane ((s a b) (s b c)) ((CH3 a c) (CH2 b))))

Thus, we use a predicate called STRUC with three arguments. The first is the name off the molecule we are encoding. The second is a list that contains all bonds present in the molecule, whereas the third is a list that contains information about the atoms (with hydrogens atoms not considered separately: "superatoms"). The third argument defines the numbering scheme — in this example it implies that there are two methyl groups, "numbered" as a and c, and one methylene group, numbered as b. The second argument shows that there are single bonds between superatoms a and b and between b and c. Thus this clause indeed defines the structure of propane. Notice that any numbering system may be employed, provided it is unique. Also, our particular choice to designate single bonds by the letter "s" is in no way the only possible choice. It would have been equally valid if we had used "single", "1", "my_favourite_type_of_bond" or whatever. All of this is merely a matter of convention. Once decided upon, it should be used consistently throughout, though.

Now, an algorithm to obtain the Randic index from a structure could be:

- retrieve the structure and extract from it the necessary information (i.e., a list of all superatoms);
- determine the connectivity of all superatoms;
- determine the values that are to be summed and sum them.

The following relation can be considered the "main program":

((randic)
 (STRUC _name _bonds _atoms)
 (PP) (PP Structure _name)
 (PP _bonds) (PP _atoms) (PP)
 (KILL CONN)
 (! get_labels _atoms _list)
 (connectivities _bonds _list)
 (ISALL _values _val
  (ON _bond _bonds)
  (determine_value _bond _val))
 (PP _values)
 (! sum_of_list _values 0 _ind)
 (PP The structure of _name has Randic index
   _ind)
 (PP))

The following relation "extracts" all the superatoms from the third argument in the STRUC clause:

((get_labels ((_symbol | _atoms) | _left) _atomlist)
 (APPEND _atoms _rest _atomlist)
 (get_labels _left _rest))
((get_labels ( ) ( )))

For example, the query: *(! get_labels ((CH3 a c) (CH2 b)) _atomlist)* would yield: *_atomlist = (a c b), No (more) solutions.*

The following two relations determine the connectivity of each superatom. The first contains the "loop" over all superatoms, whereas the second does the job for one superatom at a time. Notice that the value of the connectivity is stored in the dynamic database, using the relation name CONN:

((connectivities _bonds _list)
 (PP Connectivities are as follows:)
 (FORALL ((ON _atom _list))
  ((! connectivity _atom _bonds 0 _n)
  (PP _atom has connectivity _n)
  (ADDCL ((CONN _atom _n)))))
 (PP))

((connectivity _atom ((_bondtype _atom _a2)
  | _bonds) _n _nfinal)
 (+ _n 1 _n1)
 (connectivity _atom _bonds _n1 _nfinal))
((connectivity _atom ((_bondtype _a1 _atom) |
 _bonds) _n _nfinal)
 (+ _n 1 _n1)
 (connectivity _atom _bonds _n1 _nfinal))
((connectivity _atom ((_bondtype _a1 _a2) | _bonds)
  _n _final)
 (connectivity _atom _bonds _n _nfinal))
((connectivity _atom ( ) _n _n))

Finally, determine_value will compute the value of a particular bond (this value must be multiplied, e.g., by 1,000,000 since MacPROLOG will only display three decimal places):

((determine_value (_type _a1 _a2) _val)
 (CONN _a1 _n1)
 (CONN _a2 _n2)
 (SQRT _n1 _root1)
 (SQRT _n2 _root2)
 (TIMES _root1 _inv1 1)
 (TIMES _root2 _inv2 1)
 (* _inv1 _inv2 _value)
 (* _value 1000000 _val))

Finally, the output from running the program on propane will look like this:

:randic

Structure propane
((s a b) (s b c))
((CH3 a c) (CH2 b))

Connectivities are as follows:
a has connectivity 1
c has connectivity 1
b has connectivity 2

The structure of propane has Randic index 1414213.562

YES

5 PROS AND CONS

In the previous sections and in Part 1 of this tutorial, we have already mentioned a number of strong and weak points of PROLOG. This section provides a brief recapitulation of these.

Elegant features and distinct advantages of PROLOG [38,39] are:
- The built-in backtracking mechanism, whichs forms an essential part of the inference methods used by PROLOG.
- Its specific use of variables.
- The ability for the programmer to concentrate more on the description of a certain problem, rather than on methods of solving it.
- Modularity is inherent to programming in PROLOG.
- The absence of such things as GOTO-statements and global variables.
- Its ability to handle symbolic information and data.
- Invertibility (a query can usually be employed either to discover a fact or to check whether a fact is true).
- Its inherent promise for parallel execution [40–43].

Disadvantages or shortcomings of PROLOG [38,44]:
- Negation by failure: if a fact is unknown and cannot be derived, then it is considered to be false (although this can sometimes be remedied by using "catch-all" clauses as mentioned earlier).
- Only Horn-clauses (first order predicate logic) can be used: so constructs such as $(A \vee B)$ if ... are not possible without some rewriting.
- (Obviously) its poor performance in number-crunching applications. Especially in 'real-life' scientific applications some degree of numerical computation will often be necessary. It is convenient then to have a number of extra-built-in arithmetic primitives or a link to procedural programming languages (such as Fortran or Pascal). Writing a Taylor expansion of the logarithm function in PROLOG is not an extremely pleasant passtime (for most people anyway).

Postma et al. have presented a brief discussion [45] in which they compare some aspects of PROLOG and LISP. Memory performance of LISP and PROLOG have been examined by Tick [46], who concludes that current hardware and compiler technology favours LISP, but that this may well change in the future. Much research is going into building special-purpose PROLOG machines or co-processors, an example of which is British Aerospace's Declarative Language Machine (DLM). The DLM is claimed to be the fastest AI-machine in the world [47], attaining a PROLOG execution rate of 620 kLIPS (where a LIPS is one Logical Inference Per Second).

6 CONCLUDING REMARKS

In this two-part tutorial we have presented the interested chemist with an overview of the most important features and characteristics of the programming language PROLOG. We hope that the use of examples from chemical contexts will have helped the reader to get a feel for the types of problems which are particularly suited for this language. Naturally, actually learning to program in PROLOG will require the reader to sit down at a machine and have a go at it. After a while, he or she might want to learn more about the language, its relation to logic or its use in specific applications. For such purposes, we gladly refer the reader to the existing literature. A list of books (which is by no means complete) concerning PROLOG and Logic Programming has been compiled in Table 2. Since we are currently considering expanding this tutorial into a book (specifically aimed at chem-

TABLE 2

Books about PROLOG and Logic Programming

T. Amble, *Logic Programming and Knowledge Engineering*, Addison-Wesley, Wokingham, 1987.

I. Bratko, *PROLOG Programming for Artificial Intelligence*, Addition-Wesley, Wokingham, 1986.

W.D. Burnham and A.R. Hall, *PROLOG Programming and Applications*, Macmillan, Basingstoke, 1987.

J.A. Campbell (Editor), *Implementations of PROLOG*, Ellis Horwood, Chichester, 1985.

K.L. Clark and S.A. Tarnlund (Editors), *Logic Programming*, Academic Press, London, 1982.

K. Clark and F.G. McCabe, *Micro-PROLOG: Programming in Logic*, Prentice-Hall, Englewood Cliffs, NJ, 1984.

W.F. Clocksin and C.S. Mellish, *Programming in PROLOG*, Springer-Verlag, New York, 1984.

J.R. Ennals, *Beginning Micro-PROLOG*, Ellis Horwood, Chichester, 1984.

S. Gararaglia, *PROLOG: Programming Techniques and Applications*, Harper & Row, New York, 1987.

F. Giannesini, *PROLOG*, Addison-Wesley, Wokingham, 1986.

C.J. Hogger, *Introduction to Logic Programming*, Academic Press, London, 1984.

F. Kluzniak and S. Szpakowicz, *PROLOG for Programmers*, Academic Press, London, 1985.

R.A. Kowalski, *Logic for Problem Solving*, North-Holland, New York, 1979.

J. Malpas, *PROLOG: a Relational Language and its Applications*, Prentice Hall, Englewood Cliffs, NJ, 1987.

C. Marcus, *PROLOG Programming*, Addison-Wesley, Reading, MA, 1986.

J.B. Rogers, *A PROLOG Primer*, Addison-Wesley, Reading, MA, 1986.

L. Sterling and E. Shapiro, *The Art of PROLOG*, MIT Press, Cambridge, MA, 1986.

M.J. Wise, *PROLOG Multiprocessors*, Prentice Hall, Englewood Cliffs, NJ, 1986.

ists), the authors would warmly welcome any comments or suggestions.

7 ACKNOWLEDGEMENTS

The authors would like to express their appreciation for the stimulating interest and useful suggestions extended to them by Professors R.E. Dessy and D.L. Massart as well as Dr H.A. van 't Klooster. In addition, they would like to thank Drs. P.J. Edauw MBA for giving them an update on molecular biology and various other persons who have acted as proofreaders and guinea pigs for this tutorial.

REFERENCES

1 R.E. Dessy, Expert systems, Part I, *Analytical Chemistry*, 56 (1984) 1200A–1212A.
2 R.E. Dessy (Editor), Expert systems, Part II, *Analytical Chemistry*, 56 (1984) 1312A–1332A.
3 J.W.A. Klaessens, G. Kateman and B.G.M. Vandeginste, Expert systems and analytical chemistry, *Trends in Analytical Chemistry*, 4 (1985) 114–117.
4 B.G. Buchanan, Expert systems: working systems and the research literature, *Expert Systems*, 3 (1986) 32–51.
5 D.H. Smith, Artificial intelligence: the technology of expert systems, *American Chemical Society Symposium Series*, 306 (1986) 1–16.
6 P.J. Denning, The science of computing. Expert systems, *American Science*, 74 (1986) 18–20.
7 G.J. Kleywegt, Artifical intelligence in chemistry, *Laboratory Microcomputer*, 6 (1987) 74–81.
8 J.W.A. Klaessens and G. Kateman, Problem solving by expert systems in analytical chemistry, *Fresenius Zeitschrift für Analytische Chemie*, 326 (1987) 203–213.
9 B.J. Clark, A.F. Fell, K.T. Milne and M.H. Williams, The use of an expert system based on total luminescence spectra for the identification of drugs separated by HPLC, *Journal of Pharmacy and Pharmacology*, 37 (1985) 129P.
10 K.T. Milne, M.H. Williams, B.J. Clark and A.F. Fell, Expert systems in luminescence analysis, *Analytical Proceedings*, 23 (1986) 157–160.
11 H. Gunasingham, Heuristic approaches to the design of a cybernetic electroanalytical instrument, *Journal of Chemical Information and Computer Sciences*, 26 (1986) 130–134.
12 H. Gunasingham, B. Srinivasan and A.L. Ananda, Design of a PROLOG-based expert system for planning separations of steroids by high-performance liquid chromatography, *Analytica Chimica Acta*, 182 (1986) 193–202.
13 M.R. Detaevernier, Y. Michotte, L. Buydens, M.P. Derde, M. Desmet, L. Kaufman, G. Musch, J. Smeyers-Verbeke, L. Dryon and D.L. Massart, Feasibility study concerning the use of expert systems for the development of procedures in pharmaceutical analysis, *Journal of Pharmaceutical and Biomedical Analysis*, 4 (1986) 297–307.
14 L. Buydens, M. Detaevernier, D. Tombeur and D.L. Massart, An expert system for the development of analytical procedures: UV spectrophotometric determination of pharmaceutically active substances in tablets, *Chemometrics and Intelligent Laboratory Systems*, 1 (1987) 99–108.
15 D.P. Dolata and R.E. Carter, WIZARD: applications of expert system techniques to conformational analysis. 1. The basic algorithms exemplified on simple hydrocarbons, *Journal of Chemical Information and Computer Sciences*, 27 (1987) 36–47.

16 D.P. Dolata, A.R. Leach and K. Prout, WIZARD: AI in conformational analysis, *Journal of Computer-Aided Molecular Design*, 1 (1987) 73-85.

17 H.J. Luinge and H.A. van 't Klooster, Artificial intelligence used for the interpretation of combined spectral data, *Trends in Analytical Chemistry*, 4 (1985) 242-243.

18 G.J. Kleywegt, H.J. Luinge and H.A. van 't Klooster, Artifical intelligence used for the interpretation of combined spectral data. Part II. PEGASUS: a PROLOG program for the generation of acyclic molecular structures, *Chemometrics and Intelligent Laboratory Systems*, 2 (1987) 291-302.

19 H.J. Luinge, G.J. Kleywegt, H.A. van 't Klooster and J.H. van der Maas, Artifical intelligence used for the interpretation of combined spectral data . Part III. Automated generation of interpretation rules for infrared spectra data, *Journal of Chemical Information and Computer Sciences*, 27 (1987) 95-99.

20 C.W. Moseley, W.D. LaRoe and C.T. Hemphill, Expert system rules for Diels-Alder reactions, *American Chemical Society Symposium Series*, 306 (1986) 231-243.

21 D. Lerner, P. Pingnand, C. Federighi, C. Maudelone and P. Meriaux, An expert system for the computer-assisted selection of bacterial strains for bioconversions, *Computers and Chemistry*, 11 (1987) 159-162.

22 G. Rossi, Uses of PROLOG in implementation of expert systems, *New Generation Computers*, 4 (1986) 321-329.

23 C.J. Rawlings, W.R. Taylor, J. Nyakairu, J. Fox and M.J.E. Sternberg, Reasoning about protein topology using the logic programming language PROLOG, *Journal of Molecular Graphics*, 3 (1985) 151-157.

24 C.J. Rawlings, W.R. Taylor, J. Nyakairu, J. Fox and M.J.E. Sternberg, Using PROLOG to represent and reason about protein structure, in E. Shapiro (Editor), *Proceedings of the Third International Conference on Logic Programming*, Springer-Verlag, Berlin, 1986, pp. 536-543.

25 J.M. Burridge, A.J. Morffew and S.J.P. Todd, Experiments in the use of PROLOG for protein querying, *Journal of Molecular Graphics*, 3 (1985) 109.

26 A.J. Morffew and S.J.P. Todd, The Use of PROLOG as a protein querying language, *Computers and Chemistry*, 10 (1986) 9-14.

27 K. Kawai, R. Mizoguchi, O. Kakusho and J. Toyoda, A framework for ICAI systems based on inductive inference and logic programming, in E. Shapiro (Editor), *Proceedings of the Third International Conference on Logic Programming*, Springer-Verlag, Berlin, 1986, pp. 188-202.

28 G.J. Postma, B.G.M. Vandeginste, C.J.G. van Halen and G. Kateman, Implementation of a teaching program for IR spectrometry in LISP and PROLOG, *Trends in Analytical Chemistry*, 6 (1987) 27-30.

29 R. Cornelius, D. Cabrol and C. Cachet, Applying the techniques of artifical intelligence to chemistry education, *American Chemical Society Symposium Series*, 306 (1986) 125-134.

30 D. Cabrol, J.P. Rabine and T.P. Forrest, An educational problem solving partner in PROLOG for learning infrared spectroscopic analysis, *Computers and Education*, 12 (1988) 241-246.

31 H.L. Schläfer and G. Gliemann, *Einführing in die Ligandenfeldtheorie*, Akademische Verlagsgesellschaft, Wiesbaden, 1980.

32 P.W. Atkins, *Molecular Quantum Mechanics*, Oxford University Press, Oxford, 1983.

33 G.D. Renkes, Symbolic computer programs applied to group theory, *American Chemical Society Symposium Series*, 306 (1986) 176-185.

34 P. Friedland and L.H. Kedes, Discovering the secrets of DNA, *Communications of the ACM*, 28 (1985) 1164-1186.

35 A.L. Lehninger, *Biochemistry*, Worth Publishers, New York, 1975.

36 R.C. Bohinski, *Modern Concepts in Biochemistry*, Allyn and Bacon, Boston, MA, 1979.

37 M. Randic, On characterisation of molecular branching, *Journal of the American Chemical Society*, 97 (1975) 6609-6615.

38 J.C. Emond and A. Paulissen, The art of deduction, *BYTE* (November 1986) 207-214.

39 G.J. Kleywegt and H.A. van 't Klooster, Chemical applications of PROLOG. Interpretation of mass spectral peaks, *Trends in Analytical Chemistry*, 6 (1987) 55-57.

40 S.J. Stolfo and D.P. Miranker, DADO: a tree-structured architecture for artifical intelligence computation, *Annual Reviews in Computer Science*, 1 (1986) 1-18.

41 T.J. Reynolds, A.J. Beaumont, A.S.K. Cheng, S.A. Delgado-Rannauro and L.A. Spacek, BRAVE — a parallel logic language for artifical intelligence, in R.P. van de Riet (Editor), *Frontiers in Computing. Preprints of the International Conference, Amsterdam, Dec. 9-11, 1987*, Elsevier, Amsterdam, 1987, pp. 221-235.

42 E.Y. Shapiro, *A subset of concurrent PROLOG and its interpreter*, ICOT Research Center Technical Report TR 003, Tokyo, 1983.

43 E. Shapiro (Editor), *Proceedings of the Third International Conference on Logic Programming*, Springer-Verlag, Berlin, 1986.

44 R.A. Kowalski, Algorithm = Logic + Control, *Communications of the ACM*, 22 (1979) 424-436.

45 G.J. Postma, B.G.M. Vandeginste, C.J.G. van Halen and G. Kateman, A comparison of LISP and PROLOG, *Trends in Analytical Chemistry*, 6 (1987) 2-5.

46 E. Tick, Memory performance of LISP and PROLOG programs, in E. Shapiro (Editor), *Proceedings of the Third International Conference on Logic Programming*, Springer-Verlag, Berlin, 1986, pp. 642-649.

47 A. Pudner, DLM — a powerful AI computer for embedded expert systems, in R.P. van de Riet (Editor), *Frontiers in Computing. Preprints of the International Conference, Amsterdam, Dec. 9-11, 1987*, Elsevier, Amsterdam, 1987, pp. 187-201.

Practical Exploratory Experimental Designs

EDWARD MORGAN *, KENNETH W. BURTON and PAUL A. CHURCH

Department of Science, The Polytechnic of Wales, Treforest, Mid-Glamorgan, CF37 1DL (U.K.)

(Received 29 July 1988; accepted 30 November 1988)

CONTENTS

Abstract 104
1 Introduction 105
2 Single-factor designs 105
    2.1 Student's *t*-test 106
    2.2 Analysis of variance 106
    2.3 ANOVA model 107
    2.4 Worked example of single-factor ANOVA 107
    2.5 Blocking 109
3 Factorial designs 110
    3.1 Interactive effects 110
    3.2 Two-level factorial designs 111
    3.3 Worked example of 2^3 factorial design 112
4 Fractional factorial designs 114
    4.1 Generating the fractional replicates 115
5 Matrix models 116
6 Response surface designs 118
    6.1 Worked example of central composite design 121
7 Conclusions 123
References 123

ABSTRACT

Morgan, E., Burton, K.W. and Church, P.A., 1989. Practical exploratory experimental designs. *Chemometrics and Intelligent Laboratory Systems*, 5: 283–302.

    Depending upon the particular aims of an experiment, different experimental designs may be adopted by the practical chemist. This article initially discusses the possible objectives that an experiment can take. Following this a discussion of mechanistic and empirical models leads into sections explaining some of the methods used in the analysis of experimental results. The technique of analysis of variance is explored by means of a worked example in which the effects of variables, blocks and interactions are discussed. When more than one variable is potentially important, factorial and fractional factorial designs are useful designs for quantifying their effects and possible interactions between them. Here again a worked example is presented. However, when it is desired to graph the response as a function of explanatory variables, such as when it is desired to find an optimum response, specific response surface designs can often prove more useful. These are presented by means of a matrix approach.

Chapter 8

1 INTRODUCTION

This tutorial is aimed at chemists who either have no clear ideas about experimental design or perhaps chemists who have some ideas but are interested in extending their knowledge in this area. The tutorial is not intended to be a comprehensive treatment of experimental design, but will hopefully show the reader its possibilities.

In general the aim of an experiment is to confirm or disprove something, discover some unknown principles or effects, or to test some suggested theory. There are two usual approaches to the solution of such a problem, each of which relies upon the formulation of a model. The first of these approaches involves building a mechanistic model by which the experimenter aims to increase his or her understanding of why certain things happen. In contrast to this, the second approach involves the formulation of an empirical model whereby a response may be predicted for a given set of experimental circumstances. Of course the mechanistic model may lead the experimenter to a deeper understanding of the experimental system but the effort required in both time and money may be too great. In addition the mechanism may not be understood sufficiently or be too complicated, thereby rendering the effort wasted. If all that is required is to decide whether a particular level of a variable has a significant effect upon the experimental response or what the experimental response is under a given set of experimental conditions, an empirical model will often suffice. However, it is important to remember that physical, chemical or biological meaning should not be ascribed to the individual terms in the empirical model unless there is some sound theoretical reason for doing so. Even if a model reasonably represents the response it does not automatically imply a causal effect of the stimulus or treatment upon the response.

The first step in designing an experiment is to state the objectives exactly. If this is done properly then the correct model, design and subsequent analysis of the data should confirm or disprove the experimenter's expectations. Important objectives in experimentation are the comparison of treatment effects, the estimation of parameters, the derivation of prediction equations for use as response surfaces and the optimization of operating conditions [1]. A response surface can be obtained where the response is graphed as a function of one or more explanatory variables. These may be used to predict future values as in a calibration curve. Also response-surface experiments may be used in the process of optimization to determine the values of the explanatory variables which give a minimum or maximum response. The optimization of laboratory procedures has traditionally relied on the one-variable-at-a-time approach. This can, however, produce incorrect results and is inefficient in that it may require many more experimental runs than correctly designed experiments. In this tutorial the empirical approach to experimentation is used and various types of exploratory experimental design adopted to suit several models.

2 SINGLE-FACTOR DESIGNS

In this section the effects of variation in single factors upon the response is considered. If the effect of the single factor is examined at only two levels then it is said to have two treatments, where a treatment is a stimulus that is applied to observe the effect on an experimental situation. In practice this refers to anything that may be controlled, such as the level of a variable (the factor is then a quantitative factor) or perhaps the absence or presence of a variable (qualitative factor) [2]. A change in the response due to a treatment is termed systematic variation. When assessing the likelihood that these treatments differ in their effect upon the response, replicates or repeated treatments are used so that treatment averages and estimates of how much these are likely to vary, known as variances, may be calculated. In contrast to systematic variation, this variation, termed residual variation, may arise from a number of sources which have not been taken into account or as a result of random errors in the measurement of the responses. In response surface experiments this residual variation can be further divided into two components known as lack of fit and pure error. This will be explained later.

2.1 Student's t-test

Where a comparison of the mean responses of two treatments, A and B, is desired and there are sufficient replicates, Student's t-test is often used. This test uses the variability of the data about a mean, as estimated by the standard deviation (the square root of the variance), to decide, at a certain probability level, whether the two treatments can be said to have different effects. The true or population standard deviation, σ, is often unknown and an estimated value, s, obtained from the treatment means is usually substituted.

At the outset it is usual to assume that the treatments have no effect upon the response. This is termed the null hypothesis and represented as $H_0:(\mu_A - \mu_B) = 0$. A two-tailed t-test may then be applied to the results since either treatment may induce the greater response. If the treatment is found to have a significant effect at a given probability level then one rejects this null hypothesis and accepts an alternative hypothesis in which the treatments have different effects, $H_1:(\mu_A - \mu_B) <> 0$. Often it is required to know whether treatment A produces a higher response than treatment B. In this case the alternative hypothesis is $H_1:(\mu_A - \mu_B) > 0$ and a one-tailed t-test is applied since one is only interested in whether treatment A gives a higher response than treatment B. However, the applicability of the t-test is limited to comparing the effects of two treatments by using their mean values.

2.2 Analysis of variance

Sometimes the experimenter wishes to compare more than two treatment means, to test whether they are the same or different. The null hypothesis to be tested is usually that the k treatment means are the same, and the alternative hypothesis is that they are not. Analysis of variance (ANOVA) is a useful technique for making decisions about these hypotheses. In analysis of variance it is actually the variation in the treatment responses which is used to decide whether or not treatment effects are significant or not. The techniques are usually justified on the supposition that the data can be treated as random samples from k normal populations having the same variance, σ^2, and differing, if at all, only in their means.

Analysis of variance aims to determine whether the discrepancies between the treatment averages can reasonably be said to be greater than the discrepancies within the treatments. Each treatment supplies an estimate of this within-treatment variation in the squared deviations of the responses from the treatment averages. These may be summed to form the within-treatment sum of squares or residual sum of squares (ResidSS)

$$\text{ResidSS} = \sum_{t=1}^{k} \sum_{i=1}^{n_t} (y_{ti} - \bar{y}_t)^2 \qquad (1)$$

where y_{ti} is the response obtained for the ith replicate in the tth treatment, $\bar{y}_t$ is the average of the replicates for the tth rreatment and n_t is the number of replicates for the tth treatment. Associated with each within-treatment estimate of the variance is a number of degrees of freedom (d.f.) which is in this case the number of replicates minus one. The reason for this is that the deviations from the treatment average must sum to zero, thus constraining the value of the last deviation. The number of degrees of freedom for the residual sum of squares is therefore the total number of experimental runs (N) minus the number of treatments (k). An average estimate of within-treatment variance, known as the within-treatment mean square or residual mean square (ResidMS), is obtained by dividing the residual sum of squares by $N - k$:

$$\text{ResidMS} = \frac{\sum_{t=1}^{k} \sum_{i=1}^{n_t} (y_{ti} - \bar{y}_t)^2}{N - k} \qquad (2)$$

On the assumption that each of the individual observations has been randomly taken from a population for the particular treatment then this within-treatment mean square supplies an estimate of the within-treatment error variance, σ^2, based on $N - k$ degrees of freedom.

The overall average $\bar{y}$ is defined as the sum of all the observations divided by the total number of observations (N). If there were no real differences between the treatment means then a second estimate of σ^2 could be obtained from the variation of the treatment means about the overall average. This estimate, known as the between-treatment mean square (TreatmentMS), is formed by dividing the between-treatment sum of squares (TreatmentSS)

$$\text{TreatmentSS} = \sum_{t=1}^{k} n_t (\bar{y}_t - \bar{y})^2 \qquad (3)$$

by the between-treatment degrees of freedom (number of treatments minus one).

$$\text{TreatmentMS} = \frac{\sum_{t=1}^{k} n_t (\bar{y}_t - \bar{y})^2}{k-1} \qquad (4)$$

This is most easily understood if one considers all the treatment groups to be of the same size ($n_t = n$). On the null hypothesis that there are no real differences between the treatment means, there are now two estimates of σ^2 available. If the between-treatment mean square is much greater than the within-treatment mean square, then there is little chance that the null hypothesis is correct and real differences between the treatments may have caused the between-treatment variation to be large.

A measure of the overall variation in the data could have been obtained by calculating the variance for the N results. This would been done by calculating the deviations of all the results from the overall average and squaring them, so providing a value known as the total sum of squares (TotalSS).

$$\text{TotalSS} = \sum_{t=1}^{k} \sum_{i=1}^{n_t} (y_{ti} - \bar{y})^2 \qquad (5)$$

This could then be divided by $N - 1$ to give the overall measure of the variation in the data. It is of value to the analysis of variance because it is easy to calculate and the within-treatment and between-treatment sum of squares add up to this total value;

$$\text{TotalSS} = \text{TreatmentSS} + \text{ResidSS} \qquad (6)$$

Thus it is possible to calculate the total and between-treatment sum of squares and obtain the within-treatment sum of squares by subtraction. The formula given in (5) may be simplified to

$$\text{TotalSS} = \sum_{t=1}^{k} \sum_{i=1}^{n_t} y_{ti}^2 - (N\bar{y})^2 \qquad (7)$$

in which the term $(N\bar{y})^2$ is sometimes called the correction for the mean, and the total of the squares of all the responses [the first term in (7)] is called the crude sum of squares. For this reason the quantity calculated in (5) and (7) is sometimes referred to as the total 'corrected' sum of squares.

2.3 ANOVA model

Analysis of variance is based upon a model which can give some idea of the changes to be expected in the response when treatment effects become important. If the data are random samples from a number of a normal population then a model

$$y_{ti} = \mu + \epsilon_{ti} \qquad (8)$$

can be written where each observed response, y_{ti}, the ith replicate for the tth treatment, is the sum of the true mean plus a residual error ϵ_{ti}. When a treatment such as a particular level of a factor is thought to be significant the model has to change to cope with this treatment effect and (8) becomes

$$y_{ti} = \mu + \gamma_t + \epsilon_{ti} \qquad (9)$$

where each observed response is now the sum of the true mean (μ), an effect due to each treatment (γ_t), and a residual error (ϵ_{ti}). For significance testing the residual errors are then assumed to be independently and identically distributed according to a normal population with mean zero and fixed variance σ^2.

2.4 Worked example of single-factor ANOVA

In an experiment to determine the effect of signal frequency on the hardness of a plate produced in an electroplating bath, four frequencies were investigated. These were 10, 55, 100 and 145 Hz. It was decided that it would be possible to run four replicates of each, giving a total of sixteen experimental runs. The order in which the individual experimental runs were carried out was randomized so as to avoid any possibility of systematic changes in the conditions with time influencing the responses.

Treatment	Hardness (Vickers numbers)	Mean
A 10 Hz	72, 74, 72, 74	73
B 55 Hz	74, 78, 76, 72	75
C 100 Hz	68, 66, 68, 70	68
D 145 Hz	63, 68, 65, 60	64

Obviously the mean treatment values are different but, because of random error, even if the true value is unchanged, the treatment means may vary from one treatment to another. If the null hypothesis adopted is that the treatments have no effect upon the response, the variance in the data may be estimated in two ways; one involving the variation between the treatments and the other within the treatments. The overall average has been subtracted from each response to simplify the calculations.

Treatment	Hardness (Vickers number)	Mean
A 10 Hz	2, 4, 2, 4	73
B 55 Hz	4, 8, 6, 2	75
C 100 Hz	−2, −4, −2, 0	68
D 145 Hz	−7, −2, −5, −10	64

$$\text{TotalSS} = 2^2 + 4^2 + 2^2 + 4^2 + 4^2 + \ldots$$
$$+ (-2)^2 + (-5)^2 + (-10)^2$$
$$= 362$$

$$\text{TreatmentSS} = 4 \times \left[(73-70)^2 + (75-70)^2 \right.$$
$$\left. + (68-70)^2 + (64-70)^2 \right]$$
$$= 4 \times [4 + 25 + 4 + 36] = 296$$

Using eq. (6) the ResidSS is obtained by subtraction: $362 - 296 = 66$.

The results of the above partitioning of variance is usually presented in an analysis of variance table (Table 1) from which inferences about the relative magnitudes of the variances may be drawn. In this case the ANOVA is referred to as single-factor ANOVA because the effect of one factor is being examined.

TABLE 1

One-factor ANOVA; plating example

Source of variation	d.f.	SS	MS	Variance ratio
Treatment	3	296	98.67	17.94
Residual	12	66	5.5	
Total	15	362		

Each sums of squares term in an ANOVA table has its own degrees of freedom (d.f.). The 3 between-treatment d.f. are the number of treatments minus one. The within-treatment d.f. are the total number of experimental runs minus the number of treatments ($16 - 4 = 12$). The mean square (MS) terms in the table were obtained by dividing the sums of squares terms by their degrees of freedom. These mean square terms are then two estimates of the variance of the data. If the null hypothesis is correct then it is to be expected that the estimates should be the same. A ratio of these variances is then calculated. In general a large value signifies that a treatment has had a significant effect upon the response. The residuals are assumed to be normally and independently distributed with zero mean anc onstant variance. The calculated variance ratio of 17.94 is much greater than the value of 3.49 derived from a one-tailed F distribution for 3 and 12 d.f. at a probability of 95%. The null hypothesis that the treatments are the same is therefore rejected and the alternative hypothesis that they are different is accepted. Therefore varying the applied frequency may be said to have a significant effect upon the hardness of the plate.

If on the other hand the calculated value had not exceeded the tabulated value at the given probability level, the alternative hypothesis could not have been accepted. However, this does not imply acceptance of the null hypothesis since there is a possibility that, if repeated, the experiment could actually show a significant difference.

If it is established that there is a significant difference between the treatments in the ANOVA at a given probability level, that is to say there is a high probability of a factor causing differences between the treatment means, it is often useful to distinguish between the mean values on an individual basis. There are various procedures available for this such as the least significant difference (LSD) [3], Scheffe's test [4], Tukey's paired comparison procedure [5], Newmann–Keuls test [6,7] and others. These tests require a measure of the precision of the estimated effects and is supplied by the residual mean square. For this example the least significant difference was calculated to compare the treatments on an individual basis. In the LSD method a quantity known as the standard error of the difference (sed) is used to distinguish between the means.

$$\text{sed} = \sqrt{\text{ResidMS}} \times \sqrt{2/n} \tag{10}$$

$$\text{sed} = 2.345 \times 0.707 = 1.658$$

The test utilizes a reference t distribution with degrees of freedom given by the residual error. If the differences between the means are greater than this value they can be said to significantly different at that probability level.

$$\text{LSD} = \text{sed} \times t_{\text{resid d.f.}}$$

$$\text{LSD} = 1.658 \times 2.18 \, (P = 0.05, \, 12 \text{ d.f.})$$

$$= 3.61 \tag{11}$$

The treatment means were y_D 64; Y_C 68; Y_A 73; Y_B 75. Thus the responses for the D treatments were significantly lower than those for C, A and B, and the responses for C were significantly lower than A and B at a probability level of 95%. However, there were apparently no significant differences between the responses for treatments A and B, and C and D at this level.

2.5 Blocking

One of the assumptions of ANOVA is that the uncontrolled variation is random. This random variation is used to test whether the procedures are different and replication is usually used as the source of such variation. Sometimes, however, it is not possible to carry out all the experimental runs under exactly the same conditions. For example, the procedure may be lengthy and several days are required to carry out all the experimental runs. The runs may therefore have been conducted with different temperatures, humidities or voltages. Alternatively there may have been insufficient raw material for all the runs. To cope with these problems a randomized block design could have been used.

Randomized blocks are used where it is necessary to subdivide the experimental conditions into blocks of relatively uniform conditions [8]. With a randomized block design the resulting ANOVA can separate the variation due to the treatments, blocks and experimental error. Suppose the effects of the four frequencies upon plate hardness were to be investigated but because of the amount of raw material which was available at any one time only 8 runs could be carried out on any one batch. One possible design could have been

Batch 1 A A C A D A C D
Batch 2 C B C D B B D B

If B had turned out to be significantly different from A it would not have been possible to say whether it was due to an actual difference between procedures or because of differing material. A better design, actually used, which has each treatment tested twice in each block is that given below:

Batch 1 A B D C C A B D
Batch 2 C B A A D B D C

Note that now any differences between the effects of A and B will not be entirely due to differences between the blocks. Again the positions of the treatments were allocated randomly within the batches so that changes in the response with time as a result of changing conditions could not have unduly influenced the averaged treatment responses. Information on the effects of blocking and the factors is most easily obtained if the design is balanced, meaning that the treatments are examined an equal number of times in each block. This ensures that all the effects are estimated independently of one another, a condition known as orthogonality.

The randomization used in the blocked design is termed restricted randomization and attempts to prevent imbalance through overall randomization. In blocked designs the effect of the factors are of interest and the effect of the block is of no interest and is to be eliminated. The ANOVA model has to change to cope with the introduction of this block effect:

$$y_{ti} = \mu + \gamma_t + \beta_i + \epsilon_{ti} \qquad (12)$$

where β_i is a block effect.

The results of the ANOVA for the plate hardness experiment with the two batches are presented as before with the overall average subtracted from the individual responses to simplify the calculations.

Block	Treatments					
	A	B	C	D	$y_{.i}$	$y_{.i}^2$
Batch 1	2	6	−2	−10		
	4	8	0	−7	1	1 $\Sigma y_{.i}^2 = 2$
Batch 2	2	2	−2	−5		
	4	4	−4	−2	−1	1
Column $y_{t.}$	12	20	−8	−24	0 = Grand total	
$y_{t.}^2$	144	400	64	576		
$\Sigma y_{t.}^2 =$	1184					

TotalSS = 362
TreatmentSS = 1184/4 = 296
BatchSS = 2/2 = 1

The residual sum of squares was calculated by difference and an analysis of variance table constructed (Table 2). The variance ratio for the batches (0.86) is less than the tabulated F value of 4.84 for 1 and 11 d.f. ($P = 0.05$). Therefore variation between the batches cannot be said to have had a significant effect upon the hardness of the plate at a probability level of 95%. The treatments were again found to have significantly different effects upon the plate hardness at a probability level of 95%.

TABLE 2

One-factor ANOVA with blocking; plating example

Source of variation	d.f.	SS	MS	Variance ratio
Treatment	3	296	98.67	16.70
Batch	1	1	1	0.17
Residual	11	65	5.91	
Total	15	362		

3 FACTORIAL DESIGNS

When the effects of two factors are to be examined at the same time, e.g. effects of temperature and catalyst upon yield of a reaction, each of the experimental runs or units will have one temperature and one catalyst only. The subsequent analysis of variance is then called a two-factor ANOVA because each response is classified according to two factors and they are of equal interest. The ANOVA model for such a situation is similar to that used for one-factor ANOVA with blocking except that the β_i term, now represents the effect of the second factor. Two-factor ANOVA also allows the separation and estimation of the three sources of variation: between the two factors and residual error.

3.1. Interactive effects

The two-factor ANOVA model represents any response as a linear sum of an overall constant, the two effects and an error term. However, in many situations synergistic or antagonistic effects between the two factors are likely. For example, in a reactor an interactive effect upon yield might occur where there are different effects of temperature with different catalysts. The ANOVA model again has to be modified to account for these interactions. An expected response now becomes the sum of the true mean, the effects due to the factors, the interactions between the factors and the residual error. The model for two factors is thus given by

$$y_{ti} = \mu + \gamma_t + \beta_i + \Omega_{ti} + \epsilon_{ti} \qquad (13)$$

where the effects are as before with the additional interactive effect Ω_{ij}. Interactions between blocks and factors may also be important and the sums of squares for the one-factor ANOVA example with blocking, as given in Table 2, may be further divided to test for the significance of this interaction. The residual sums of squares was calculated using the differences between the replicates and the interaction sums of squares obtained by subtraction from the total. A modified ANOVA table was then prepared (Table 3). Comparison with tabulated F values shows that the interaction

Chapter 8

TABLE 3

Two-factor ANOVA with interaction; plating example

Source of variation	d.f.	SS	MS	Variance ratio
Treatment	3	296	98.67	37.59
Batch	1	1	1	0.38
Interaction	3	44	14.67	5.59
Residual	8	21	2.625	
Total	15	362		

between the batch and frequency effects was significant at the 95% probability level, even though the effect of the different batches was insignificant at this probability level.

A useful way of examining interactive effects is to calculate cell means, which for the example are the mean values of the replicated batch-treatment combinations. Fig. 1 shows the extent of this interaction and indicates that it would not be wise for the experimenter to choose to ignore this interaction.

Factorial designs are designs in which all the possible combinations of factors and levels are investigated. Thus when there are three factors, the first at two levels, the second at three levels, and the third at four levels, the number of treatment combinations required is $2 \times 3 \times 4 = 24$. Such a design is termed a mixed factorial design because the factors have different numbers of levels. The design for the plating hardness experiment in blocks was effectively a factorial design because it investigated the effects of two factors each at two levels.

3.2 Two-level factorial designs

One class of design which is of great practical importance is the two-level factorial design in which all the factors are investigated at two levels. These are used to determine whether the effects of the factors themselves are important and to detect and quantify interactions. The examples given in the previous section considered two factors, frequency and batch, and an interaction between them. When three factors (factors denoted by boldface letters), **M**, **A** and **C** are investigated in a factorial experiment at two levels, the main effects (all effects denoted in ordinary type face) M, A and C, the two-factor interactions MA, MC and AC, and a three-factor interaction, MAC, can be assessed in eight experimental runs. A three-factor two-level factorial design is a 2^3 design, where the number of levels is raised to the power of the number of factors and requires $2^3 = 8$ runs for all combinations of factors. It is often useful to represent the lower level of a quantitative factor in any run by a $-$ and its higher level by a $+$. For qualitative factors $-$ and $+$ can represent the absence or presence of the factor respectively. A display of the experimental treatments is termed a design matrix **D** as shown for a 2^3 factorial design:

$$\mathbf{D} = \begin{bmatrix} x_{11} & x_{21} & x_{31} \\ x_{12} & x_{21} & x_{31} \\ x_{11} & x_{22} & x_{31} \\ x_{12} & x_{22} & x_{31} \\ x_{11} & x_{21} & x_{32} \\ x_{12} & x_{21} & x_{32} \\ x_{11} & x_{22} & x_{32} \\ x_{12} & x_{22} & x_{32} \end{bmatrix} = \begin{bmatrix} - & - & - \\ + & - & - \\ - & + & - \\ + & + & - \\ - & - & + \\ + & - & + \\ - & + & + \\ + & + & + \end{bmatrix}$$

In fact the $+ -$ notation is very useful because it can relate to a geometric view of the design (Fig. 2) and can also be used in the construction of fractional factorial designs [9]. Where there are more than three factors, additional cubes may be used to represent the lower and higher levels of these factors, with the number of cubes doubling for each additional factor.

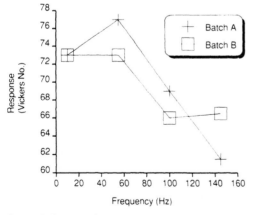

Fig. 1. Cell means of response (plate hardness, Vickers No.) for the two batches plotted against frequency of signal used.

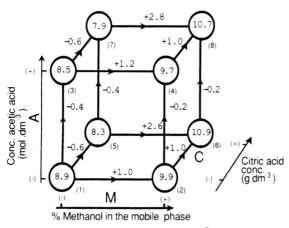

Fig. 2. Cube diagram for worked example, 2^3 factorial design.

3.3 Worked example of 2^3 factorial design

In the determination of phenols by reversed-phase liquid chromatography, several chemical factors are known to influence the separation of the phenol peaks. It was decided that three factors, percentage of methanol in the mobile phase (M), acetic acid concentration (A) and citric acid concentration (C) would be investigated in an isocratic HPLC system, the latter two factors having important influences upon peak shape. The response monitored was the chromatographic response function (CRF), a summation term of the individual resolutions between pairs of phenol peaks. The aim was to maximize the CRF since the higher the value the greater the resolution between the pairs of phenols. The levels of the factors used are presented below:

Methanol (M)
 $-$ = 50% $\qquad$ + = 70%
Acetic acid (A)
 $-$ = 0.002 mol dm^{-3} $\qquad$ + = 0.005 mol dm^{-3}
Citric acid (C)
 $-$ = 5 g dm^{-3} $\qquad$ + = 10 g dm^{-3}

Table 4 shows the treatment combinations for the eight runs necessary in an unreplicated 2^3 design and the responses (CRF values) obtained. Thus run 3 is represented by $- + -$ for M, A and C, respectively, indicating that in this particular run the percentage of methanol was 50%, acetic acid concentration was 0.005 mol dm^{-3} and the citric acid concentration was 5 g dm^{-3}.

Often the experimenter is not particularly interested in the statistical analysis but really wishes to estimate the effects of increasing or decreasing the levels of the factors upon the response. In factorial designs several estimates of the main effects are obtained, with the number depending on the exact design. In this example one estimate of the effect of changing the percentage of methanol upon the CRF, when both other factors are constant (A and C both $-$), is given by the difference in responses between experimental runs 1 and 2. Three other estimates of the methanol effect are obtained from the difference between runs 3 and 4 (A is +, C is $-$) and so on. These estimates are averaged to give the main effect of methanol, denoted by M. A simple method of calculating such a main effect is to subtract the average of responses in which a factor is at its low level from the average response with a factor at its high level. This may be viewed for methanol as the average of the responses on the front face of the cube subtracted from the average of the responses on the back face of the cube (Fig. 2).

$$\text{Methanol effect M} = \frac{r2 + r4 + r6 + r8}{4} - \frac{r1 + r3 + r5 + r7}{4}$$

$$= \frac{9.9 + 9.7 + 10.9 + 10.7}{4} - \frac{8.9 + 8.5 + 8.3 + 7.9}{4}$$

$$= 10.3 - 8.4 = 1.9 \qquad (14)$$

TABLE 4

2^3 design worked example, design matrix and CRF responses obtained

Run	M	A	C	CRF
1	$-$	$-$	$-$	8.9
2	+	$-$	$-$	9.9
3	$-$	+	$-$	8.5
4	+	+	$-$	9.7
5	$-$	$-$	+	8.3
6	+	$-$	+	10.9
7	$-$	+	+	7.9
8	+	+	+	10.7

The main effects of acetic and citric acid are similarly

$$A = \frac{8.5 + 9.7 + 7.9 + 10.7}{4}$$
$$- \frac{8.9 + 9.9 + 8.3 + 10.9}{4} = -0.3$$

$$C = \frac{10.7 + 7.9 + 10.9 + 8.3}{4}$$
$$- \frac{9.7 + 8.5 + 9.9 + 8.9}{4} = 0.2$$

A comparison of these main effects indicates that methanol had the greatest impact upon the CRF, with only minor influences by factors **A** and **C**. In this 2^3 design the three main effects were estimated with the same degree of precision as if there were four replicates of the two levels of each factor. This represents a great saving in the number of experiments required over the number that would have been required if the factors had been varied one at a time. The factor effects obtained above are not normalized for the change in the factors. To obtain a factor effect by 'regression analysis', this normalization must be carried out.

From Fig. 2 it is clear that the methanol effect was much greater with citric acid set to 10 g dm^{-3} than at 5 g dm^{-3}. This interaction between **C** and **M** is estimated in terms of the original responses as half the difference obtained when the methanol effect, estimated with **C−**, is subtracted from the methanol effect with **C+**.

Citric acid	Methanol effect
+ 10 g dm^{-3}	2.7
− 5 g dm^{-3}	1.1
	1.6

$$\text{MC interaction} = \frac{1.6}{2} = 0.8$$

It may equally be thought of as one half of the difference in the citric acid effect at the two levels of methanol. The remaining two-factor interactions are similarly calculated.

AC = 0.0
AM = 0.1

The MC interaction would seem to have a very important influence upon the CRF. Therefore the effects of methanol and citric acid should not be considered in isolation. Like the main effects, two-factor interactions are actually a difference between 2 averages, with half the eight in one average, half in the other. These may be viewed as contrasts between observations on the diagonal planes within the 2^3 cube [9].

The three-factor interaction, MAC, is estimated by half the difference obtained when an estimate of a two-factor interaction at the low level of a third factor is subtracted from the estimate of this two-factor interaction at the high level of the third factor. For example, considering the estimates of MA at the − and + levels of **C**:

	MA effect	Level of C
(10.7 + 8.3)/ −(10.9 + 7.9)/2 =	0.1	+
(8.9 + 9.7)/2 −(9.9 + 8.5)/2 =	0.1	−
difference =	0.0	

Therefore MAC = 0.0/2 = 0.0

The 2^3 design yields estimates of the three main effects, three two-factor interactions and a three-factor interaction, a total of seven effects in eight runs. This uses up all the available degrees of freedom and does not allow any estimation of the residual error for use in an analysis of variance unless the design is replicated. However, for most situations it is generally accepted that it is possible to combine the sums of squares of the higher-order interactions for use in the residual error mean square. The practical difficulty with this in a 2^3 design is that these higher-order interactions are often large. Therefore their sums of squares will tend to be larger than residual error sums of squares terms based purely on differences between replicate experimental runs. Also it is these higher-order interactions which are often of interest to the experimenter and combining their sums of squares into the mean square would obviously mean that their significance could not be tested. A useful graphical technique which gets around these problems is to plot the effects on normal probability paper [10–12]. This technique works because most of the effects will fall on a straight line, but

TABLE 5

Worked example, 2^3 design

Contrast coefficients for main effects and interactions.

Run	Mean	M	A	C	MA	MC	AC	MAC
1	+	−	−	−	+	+	+	−
2	+	+	−	−	−	−	+	+
3	+	−	+	−	−	+	−	+
4	+	+	+	−	+	−	−	−
5	+	−	−	+	+	−	−	+
6	+	+	−	+	−	+	−	−
7	+	−	+	+	−	−	+	−
8	+	+	+	+	+	+	+	+

any particularly large effects will not. This technique is especially applicable to screening experiments where the object is to discover which factors have important effects.

It would be extremely tedious to have to calculate the interactive effects from first principles whenever a factorial design is used and there is a short hand method available for the construction of the interaction contrasts from the original design matrix. Additional columns of + and − signs (contrast coefficients) are required and obtained by multiplying together, according to the usual algebraic rules, the signs of the rows for the main effects composing the interaction. Thus, for the MA interaction, the − sign for **M** in run 1 is multiplied by the − sign for **A** to give a + sign for the MA nteraction in run 1 and so on (Table 5).

The two-factor interaction MA is then simply calculated as the average of the runs with + signs minus the average of the runs with − signs or one-eight of the sum of the responses having applied the contrasts. The signs for the MAC three-factor interaction are likewise the product of the signs of the factors composing this effect. This method is quite general and can be extremely useful where three or more factors are under investigation. A tabular method has also been developed by Yates [13] to calculate the estimates and construct tables of analysis of variance for factorial designs [14]. This algorithm may be easily programmed into a spread sheet to provide automatic calculations after entering the responses and is thus very useful when sophisticated statistical design packages are unavailable.

Second-order effects in a 2^k factorial design are restricted to two-factor interactions, but it is also possible that there will be some curvature in the response with changing levels of the factors. This cannot be estimated using only two levels. 3^k factorial designs in which each factor is set at three levels are very useful for estimating this curvature, but require large numbers of experimental runs to estimate main effects and curvature even for relatively small numbers of factors. For example to estimate these effects for four factors would require at least $3^4 = 81$ experimental runs.

4 FRACTIONAL FACTORIAL DESIGNS

One of the disadvantages of full or complete factorial designs involving k factors is that the number of runs required by a full 2^k factorial design increases geometrically as k is increased. For example, two-level factorial designs with $k = 2, 3, 4, 5, 6,$ and 7 require $N = 4, 8, 16, 32, 64$ and 128 experimental runs respectively. However, when k is not small the desired information can often be obtained by performing only a fraction of the runs required for complete factorial designs.

In any complete factorial design for a large number of factors there are often high-order interactive effects which are negligible and experimental runs may be saved by not estimating these redundant effects. It is possible to carry out a half or a quarter of the number of experimental runs required in complete factorial designs and still have some estimates of the main effects and interactions.

Half-fraction factorial designs exploit the redundancy by deliberately confounding the highest-order interactive effects with the main effects. Any estimate of a main effect in this half-fraction is then a linear combination of the estimate of a main effect and a high-order interaction. The justification for this is that in terms of absolute magnitude, main effects tend to be larger than two-factor interactions, which in turn tend to be larger than three-factor interactions and so on. At some point higher-order interactions tend to become negligible and can be properly disre-

TABLE 6

Contrast coefficients for main effects and some interactions in a 2^{5-1} half-fractional replicate generated by **E = ABCD**

Run	A	B	C	D	E	ABCD	ABC	DE	BCDE
1	−	−	−	−	+	+	−	−	−
2	+	−	−	−	−	−	+	+	+
3	−	+	−	−	−	−	+	+	−
4	+	+	−	−	+	+	−	−	+
5	−	−	+	−	−	−	+	+	−
6	+	−	+	−	+	+	−	−	+
7	−	+	+	−	+	+	−	−	−
8	+	+	+	−	−	−	+	+	+
9	−	−	−	+	−	−	−	−	+
10	+	−	−	+	+	+	+	+	+
11	−	+	−	+	+	+	+	+	−
12	+	+	−	+	−	−	−	−	+
13	−	−	+	+	+	+	+	+	−
14	+	−	+	+	−	−	−	−	+
15	−	+	+	+	−	−	−	−	−
16	+	+	+	+	+	+	+	+	+

garded. The 'best' half-fractions of a 2^5 design, designated as 2^{5-1}, will have the 5 main effects confounded with the 5 four-factor interactions and will estimate these effects in $2^{5-1} = 16$ experimental runs. For example, in a 2^{5-1} design with factors **A**, **B**, **C**, **D** and **E**, the effect of **A** could be confounded with the **BCDE** four-factor effect. Should this be the case the two-factor interactions will be confounded with three-factor interactions.

4.1 Generating the fractional replicates

Suppose that a half-fraction replicate of a 2^5 factorial design is required, that is one wishes to estimate the 5 main effects and all possible interactions in 16 runs. A design matrix for the first four factors, **A**, **B**, **C** and **D**, is then written out in the standard order for a complete 2^4 factorial design. The column of contrast coefficients for the four-factor interaction (**ABCD**) is calculated and used to define the levels of fifth variable **E**. Thus **E = ABCD** so that where there is − for the **ABCD** interaction, variable **E** is set to its lower level (if a quantitative factor) and where there is + for **ABCD** then **E** is present at its higher level (Table 6). Notice also in Table 6 that the contrast coefficients for the effect of **ABC** have the same signs as **DE**. Therefore when these effects are estimated they will be the same i.e. confounded, and the effects are said to be aliases of one another.

Instead of laboriously calculating the individual column vectors of coefficients to decide the identities of the aliases, it is possible to do this more simply using what are termed defining contrasts or defining relations. In the above 2^{5-1} half-replicate, the design was obtained by leting **E = ABCD**. This is called the generator of the design. These are multiplied and the result is called the defining contrast or defining relation **I**. This indicates the effect with which the overall mean is confounded. Therefore

I = ABCDE (15)

The aliases are obtained by multiplying the defining contrasts with each of the effects. The rules are the usual algebraic rules with the additional condition that where a term appears an even number of times in the product it disappears. For example the alias of **A** is obtained by multiplying **A** by **ABCDE**. Therefore $A^2BCDE = BCDE$. For **AB**, the alias is $AB \times ABCDE = A^2B^2CDE = CDE$. This has already been confirmed earlier by multiplying the signs making up the effect.

In the above example the generator was **E = ABCD** and designated 16 runs out of the possible 32. It is also possible to obtain the other 16 runs by using the generator $E = -ABCD$ or $I = -ABCDE$. In such a case the estimates are no longer the sums of the confounding effects, but the differences, e.g. estimate of $A = A - BCDE$.

If another factor, F, is added to the 2^{5-1} design it becomes a quarter-replicate of a six-factor design. The number of experiments remains at 16, whereas a complete factorial with six factors would require $2^6 = 64$ runs. To obtain this quarter-replicate another defining contrast is obtained from an additional generator. Thus $E = ABCD$ and $F = ABC$. Therefore $I = ABCDE$ and $I = ABCF$ respectively. A third defining contrast is then obtained by multiplying the two together.

$$I = ABCDE \times ABCF = A^2B^2C^2DEF = DEF \tag{16}$$

The aliases for each effect can now be obtained in the usual way. For A this yields

$$\begin{array}{llll} A & = A^2BCDE & = A^2BCF & = ADEF \\ A & = BCDE & = BCF & = ADEF \end{array} \quad \text{or} \tag{17}$$

Thus in a quarter replicate of a 2^5 factorial, each effect has three aliases and the estimates of the main effects are free of each other and the two-factor interactions. This confounding alias structure can occasionally cause difficulty in interpretation but this is usually overcome by means of a few additional experiments.

The two-level fractional design is a powerful tool in the hands of the experimenter since it is possible, for example, to estimate as many as seven main effects in eight experiments (one-sixteenth fraction of a 2^7 design $= 2^{7-4}$). Such a design is known as a saturated design because the main effects are confounded by two-factor interactions, which must be assumed to be negligible or zero. When seven main effects are estimated the design does not allow statistical testing in the subsequent ANOVA because the seven degrees of freedom are used up. As a result of the high degree of confounding in highly fractionated designs it is very often necessary to use other fractions to resolve ambiguities. Appropriate generators are selected and the results combined with those of earlier fractions. Selection of appropriate fractions is also a useful means of running complete factorial designs in blocks, where insufficient time or experimental material is available to perform all the runs for the complete factorial design.

Fractional factorial designs are classified according to the order of assumed negligible effects. A fractional factorial is a particular resolution designated by a roman numeral placed after the design, and depends on the length of the shortest word in the defining relation [15]. For example, the 2^{5-2} design given above is resolution IV. Since fractional factorials may be generated in many ways, tables of 'best' designs have been worked out for different fractions of complete designs [9].

Plackett and Burman developed orthogonal saturated designs (resolution III) for cases when $k + 1$ is any multiple of four [16]. These are often useful since the usual fractions of complete factorials require 2, 4, 8, 16 etc. runs, whereas Plackett and Burman designs may be used, for example, to examine 11 factors in 12 runs or 19 factors in 20 runs.

Fractional factorial designs cannot, however, satisfy all experimental situations. Situations especially suited to these designs occur when interactions are negligible and when trying to establish which variables are important. It is then often possible to follow up a fractional design with a simplex optimization experiment once these important variables have been established [17].

5 MATRIX MODELS

Although the approach taken with the factorial designs is perhaps useful for the purposes of demonstrating main and interactive effects, it is time consuming and not really applicable to more complicated designs. In this section matrix models are introduced and applied to a 2^3 design with the intention of demonstrating their general applicability to the calculation of unknown parameters.

For each treatment combination in a design matrix a response is observed. A factor's effect is measured by response changes produced as a result of changing a factor's levels. When the exact relationship between factors and responses is unknown, a polynomial is often used to approximate the relationship. The order of this polynomial is limited by the type of design used. Two-level factorial designs can only estimate the main effects and interactions. Three-level factorial designs can also estimate the degree of curvature in the response because each factor is present at three levels. Other designs, known as response surface designs, have some of these properties but generally use fewer experimental runs to obtain these estimates. With each class of design it is possible to use matrices to estimate these effects from the design matrix. To demonstrate this the 2^3 design example will be used. For a 2^3 design in which only first-order effects are considered, any observation Y can be expressed as

$$Y = B_0 + B_1 x_1 + B_2 x_2 + B_3 x_3 + \text{error} \quad (18)$$

The expected response is a linear sum of a constant, B_0, and slopes B_1, B_2 and B_3, due to factors x_1, x_2 and x_3 respectively. The coefficients B_0, B_1, B_2 and B_3 are model parameters that are unknown and correspond to the main effects determined in the 2^3 design and may be represented as a column vector (B). An expected response in a first-order model expressed in vector notation is then

$$E(Y_i) = (1 \, x_{1i} \, x_{2i} \, x_{3i}) \begin{bmatrix} B_0 \\ B_1 \\ B_2 \\ B_3 \end{bmatrix} = \mathbf{x}_i \mathbf{B} \quad (19)$$

where i is a row in the design matrix. Extending the vector notation to all observations gives

$$\begin{bmatrix} E(Y_1) \\ E(Y_2) \\ \vdots \\ E(Y_N) \end{bmatrix} = \begin{bmatrix} \mathbf{x}_1 \mathbf{B} \\ \mathbf{x}_2 \mathbf{B} \\ \vdots \\ \mathbf{x}_N \mathbf{B} \end{bmatrix} = \begin{bmatrix} \mathbf{x}_1 \\ \mathbf{x}_2 \\ \vdots \\ \mathbf{x}_N \end{bmatrix} \mathbf{B} \quad (20)$$

or

$$E(\mathbf{Y}) = \mathbf{X}\mathbf{B} \quad (21)$$

The **X** matrix is called the model matrix and is derived from the design matrix by specifying the approximating model for $E(\mathbf{Y})$ e.g. first-order or second-order [15]. For a second-order model of a 2^3 factorial design this is the same as a complete table of contrast coefficients (Table 5). Having specified the model it is possible to move on to the process of estimating the coefficients in the model matrix by the method of least squares. These least squares estimators, $b_0, b_1, b_2, \ldots, b_N$ estimate B_0, $B_1, B_2, \ldots, B_N$ respectively. These are found by solving a series a simultaneous equations with b_0 to b_N as the unknowns. The solution to these simultaneous equations is given by the following matrix equation [18].

$$\mathbf{b} = (\mathbf{X}'\mathbf{X})^{-1}\mathbf{X}'\mathbf{Y} \quad (22)$$

where $\mathbf{b}$ is a vector containing all the least square estimates $b_0 \ldots b_N$, and is calculated from the model matrix $\mathbf{X}$ and the response vector $\mathbf{Y}$. Substituting values of $+1$ wherever a $+$ sign occurs in the model matrix and -1 for $-$ signs enables one to calculate an $\mathbf{X}'\mathbf{X}$ matrix, which for the 2^3 example is

$$\mathbf{X}'\mathbf{X} = \begin{bmatrix} 8 & 0 & 0 & 0 & 0 & 0 & 0 & 0 \\ 0 & 8 & 0 & 0 & 0 & 0 & 0 & 0 \\ 0 & 0 & 8 & 0 & 0 & 0 & 0 & 0 \\ 0 & 0 & 0 & 8 & 0 & 0 & 0 & 0 \\ 0 & 0 & 0 & 0 & 8 & 0 & 0 & 0 \\ 0 & 0 & 0 & 0 & 0 & 8 & 0 & 0 \\ 0 & 0 & 0 & 0 & 0 & 0 & 8 & 0 \\ 0 & 0 & 0 & 0 & 0 & 0 & 0 & 8 \end{bmatrix}$$

$$= 8 \begin{bmatrix} 1 & 0 & 0 & 0 & 0 & 0 & 0 & 0 \\ 0 & 1 & 0 & 0 & 0 & 0 & 0 & 0 \\ 0 & 0 & 1 & 0 & 0 & 0 & 0 & 0 \\ 0 & 0 & 0 & 1 & 0 & 0 & 0 & 0 \\ 0 & 0 & 0 & 0 & 1 & 0 & 0 & 0 \\ 0 & 0 & 0 & 0 & 0 & 1 & 0 & 0 \\ 0 & 0 & 0 & 0 & 0 & 0 & 1 & 0 \\ 0 & 0 & 0 & 0 & 0 & 0 & 0 & 1 \end{bmatrix}$$

$$= 8\mathbf{I} \quad (23)$$

where $\mathbf{I}$ is an identity matrix. The first row vector of $\mathbf{X}'$ is multiplied by the first column vector of $\mathbf{X}$. This multiplication is equivalent to squaring and then summing all the entries in the first column of $\mathbf{X}$. Because all the values are either -1 or $+1$, the value is 8. The product of the first row vector of $\mathbf{X}'$ with any other column vector of $\mathbf{X}$ is termed a covariance and is a measure of the dependency of the variables in the $\mathbf{X}$ matrix upon one another.

Thus it is possible to talk about covariances between the intercept, B_0, and any factor effects, or covariances between the factor effects themselves and with second-order coefficients.

An important property of an experimental design is orthogonality, in which there is no covariance among the factor effects. There might be covariance between B_0 and any or all of the factor effects, but as long as there is no covariance among the factor effects, the design is an orthogonal design. Therefore the 2^3 design is orthogonal because the off-diagonal elements of the $\mathbf{X'X}^{-1}$ are all zero. In addition all the coefficients or effects are free of confounding, a term applied when the coefficients cannot be estimated separately. This property simplifies the calculation of the inverse of the $\mathbf{X'X}$ matrix.

$$(\mathbf{X'X})^{-1} = \tfrac{1}{8}\mathbf{I} \quad (24)$$

Therefore the estimated parameter vector is

$$\mathbf{b} = \tfrac{1}{8}\mathbf{IX'Y} = \tfrac{1}{8}\mathbf{X'Y} \quad (25)$$

Thus

$$b_0 = \tfrac{1}{8}(+Y_1 + Y_2 + Y_3 + Y_4 + Y_5 + Y_6 + Y_7 + Y_8)$$
$$b_1 = \tfrac{1}{8}(-Y_1 + Y_2 - Y_3 + Y_4 - Y_5 + Y_6 - Y_7 + Y_8)$$
$$b_2 = \tfrac{1}{8}(-Y_1 - Y_2 + Y_3 + Y_4 - Y_5 - Y_6 + Y_7 + Y_8)$$

etc.

This matrix of parameter coefficients will obviously give the same estimates as the earlier method of calculating the effects. The variances of the estimates may be obtained from the diagonal elements of the $(\mathbf{X'X})^{-1}$ and the variance, σ^2, such that

$$\text{Var}(\mathbf{b}) = (\mathbf{X'X})^{-1}\sigma^2 = \frac{\sigma^2}{8}\mathbf{I} \quad (26)$$

The matrix $(\mathbf{X'X})^{-1}\sigma^2$, termed the variance–covariance matrix, contains estimates of variances on its main diagonal and covariances on its off-diagonal elements. Because the $(\mathbf{X'X})^{-1}$ matrix for the two-level factorial only has non-zero main diagonal elements, the covariances between the factor effects are zero.

6 RESPONSE SURFACE DESIGNS

Often the object of experimentation is to discover the combination of factors levels which maximizes the response. To accomplish this certain types of design are appropriate. The responses to be maximized may take many forms such as the yield of a reaction, less impurity or a faster rate of reaction. Whatever the response, the response surface designs are most applicable when the response and the factors vary in a continuous manner. For example, a response surface design may be applied to find the optimum combination of temperature of a reaction and amount of reactant added which maximize the percentage yield of a product. It is usual for only one response parameter to be optimized in any one experiment although it is possible to develop composite response functions for more than one response [19].

If only one factor is examined in a particular experimental design the response may be plotted against the level of the factor. If the factor is examined at two levels then it may only be modelled as a straight line. However, if the design allows three or more levels of the factor to be selected then the curvature in the response may also be estimated. For two factors, x_1 and x_2, examined in a 2^2 design, only the main effects and interaction between these factors may be estimated. This is equivalent to measuring the linear slopes in the directions of increasing levels of both factors and assumes that the response can be approximated by a plane in three dimensions. In a 2^2 design any response in any treatment combination is modelled as

$$Y = B_0 + B_1 x_1 + B_1 x_2 + B_{12} x_1 x_2 + \text{error} \quad (27)$$

where B_0 is the intercept, B_1 is the slope in direction x_1, B_2 is the slope in direction x_2 and B_{12} is the interaction term. Each B coefficient has its own least-squares estimator and these are denoted as b_0 to b_N. However, when curvature in the response is likely and an appropriate second-order design used, the model changes accordingly. The expected response when two factors are examined is then given by

$$Y = B_+ B_1 x_1 + B_2 x_2 + B_{11} x_1^2 + B_{22} x_2^2 + B_{12} x_1 x_2$$
$$+ \text{error} \quad (28)$$

where B_{11} and B_{22} are coefficients of curvature for x_1 and x_2 respectively. Once estimates of all these coefficients have been obtained it is possible to model the responses as a function of the explanatory variables and obtain a response surface [20].

An experimental design constructed to estimate coefficients for any approximating model should meet certain design criteria such as providing estimates for all the coefficients, require few experimental runs, provide a test for lack of fit, allow the experiment to be performed in blocks and allow specified variance criteria to be met for estimated coefficients and estimated responses. Perhaps the most obvious class of design to estimate second-order effects is the 3^k design. This satisfies several of the design criteria, but the number of experimental runs necessary is relatively large for small numbers of factors even when fractional replicates are used. To overcome this problem, Box and Wilson added star designs to 2^k factorials to form central composite designs. These designs do not require an excessive number of runs. They are two-level factorial designs or fractional factorial designs (N_c runs) which have been augmented with N_0 extra point(s) at the centre of the design and $2k = N_a$ (where $k =$ number of factors) extra points, one at either extreme of each factor and at the centre of all others [21]. Therefore composites of complete factorials require at least $2^k + 2k + 1$ runs. Another type of design which is useful for modelling response surfaces is the mixture design. These designs are for experiments where the response depends on the proportions of ingredients in a mixture rather than on their amount [22].

The minimum number of experimental runs required for a given number of factors by 3^k factorial designs and central composite designs is shown in Table 7. This table shows that using the composite designs can achieve a great saving in the number of experimental runs, especially when k is large. However, there are disadvantages in that fewer degrees of freedom are available for estimating the residual error and also some of the effects are estimated with unequal variances.

To construct a central composite design for a particular number of designs, values of N_c and 'a'

TABLE 7

Numbers of experimental runs required for 3^k factorial and central composite designs

Number of factors k	Treatment combinations	
	Three-level factorials 3^k	Composite $2^k + 2k + 1$
2	9	9
3	27	15
4	81	25
5	243	43
5	81 (1/3 fraction)	27 (1/2 fraction)
6	729 (1/3 fraction)	77
6 (1/3 fraction)	243	45 (1/2 fraction)

have to be determined. If a fractional factorial is used, it must be of resolution 'V' or greater to allow estimation of all second-order coefficients. The number of centre points should be greater than one to estimate pure error for a lack of fit test. How much greater than one is determined by the requirements for blocking and satisfying criteria for Var(**b**) and variance of the estimated responses. The value of 'a' for axial points is determined by variance criteria for estimated coefficients and responses. When working with central composite designs it is usual to scale the lower and higher values of the points in the factorial design to -1 and $+1$ respectively. The centre of all the points then takes a value of zero for each of the factors. A two-factor central composite design is shown in Fig. 3.

As mentioned earlier the variance–covariance matrix, $(\mathbf{X}'\mathbf{X})^{-1}\sigma^2$, determined by the design matrix **D**, has the variances of the parameter coeffi-

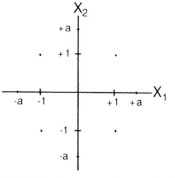

Fig. 3. Two-factor central composite design.

cients down the diagonal and the covariances between estimators as off-diagonal elements. For a central composite design it is the values for N_c, N_a, N_0 and the axial spacing (a) which determine this matrix and careful selection of these values can lead to certain desirable properties for the design. The variance–covariance matrix for a two-factor second-order model will be used to illustrate the dependency on parameters for a central composite design. For this design, $N_c = 4$, $N_a = 4$ with axial spacing of $+/- a = 1.267$ and $N_0 = 5$ centre points. The appropriate model matrix is

$$\mathbf{X} = \begin{bmatrix} x_0 & x_1 & x_2 & x_1^2 & x_2^2 & x_1 x_2 \\ 1 & +1 & +1 & +1 & +1 & +1 \\ 1 & +1 & -1 & +1 & +1 & -1 \\ 1 & -1 & +1 & +1 & +1 & -1 \\ 1 & -1 & -1 & +1 & +1 & +1 \\ 1 & +a & 0 & a^2 & 0 & 0 \\ 1 & -a & 0 & a^2 & 0 & 0 \\ 1 & 0 & +a & 0 & a^2 & 0 \\ 1 & 0 & -a & 0 & a^2 & 0 \\ 1 & 0 & 0 & 0 & 0 & 0 \\ 1 & 0 & 0 & 0 & 0 & 0 \\ 1 & 0 & 0 & 0 & 0 & 0 \\ 1 & 0 & 0 & 0 & 0 & 0 \\ 1 & 0 & 0 & 0 & 0 & 0 \end{bmatrix} \quad (29)$$

The matrix $\mathbf{X'X}$ is formed in terms of $N = N_c + N_a + N_0$ and the axial spacing as

$$\mathbf{X'X} = \begin{bmatrix} N & 0 & 0 & N_c + 2a^2 & N_c + 2a^2 & 0 \\ 0 & N_c + 2a^2 & 0 & 0 & 0 & 0 \\ 0 & 0 & N_c + 2a^2 & 0 & 0 & 0 \\ N_c + 2a^2 & 0 & 0 & N_c + 2a^4 & N_c & 0 \\ N_c + 2a^2 & 0 & 0 & N_c & N_c + 2a^4 & 0 \\ 0 & 0 & 0 & 0 & 0 & N_c \end{bmatrix}$$
(30)

The model matrix for this example is thus

$$\mathbf{X'X} = \begin{bmatrix} 13 & 0 & 0 & 7.21 & 0 & 0 \\ 0 & 7.21 & 0 & 0 & 0 & 0 \\ 0 & 0 & 7.21 & 0 & 0 & 0 \\ 7.21 & 0 & 0 & 5.154 & 4 & 0 \\ 7.21 & 0 & 0 & 4 & 5.154 & 0 \\ 0 & 0 & 0 & 0 & 0 & 4 \end{bmatrix} \quad (31)$$

The inverse of this matrix is found by solving equations from the relation

$$(\mathbf{X'X})(\mathbf{X'X})^{-1} = \mathbf{I} \quad (32)$$

by computational linear algebraic methods. For the above matrix this yields

$$\mathbf{X'X}^{-1} = \begin{bmatrix} 0.196 & 0 & 0 & -0.108 & -0.108 & 0 \\ 0 & 0.139 & 0 & 0 & 0 & 0 \\ 0 & 0 & 0.139 & 0 & 0 & 0 \\ -0.108 & 0 & 0 & 0.198 & 0.042 & 0 \\ -0.108 & 0 & 0 & 0.042 & 0.198 & 0 \\ 0 & 0 & 0 & 0 & 0 & 0.25 \end{bmatrix}$$
(33)

From the matrix $(\mathbf{X'X})^{-1}$ it can be seen that the covariances between the estimated intercept and first-order coefficients and the estimated second-order mixed coefficients will be zero [1]. However, covariances between the estimated intercept and estimated pure second-order coefficients, $\text{Cov}(b_o, b_{jj})$, will not be zero. Covariances between estimates of pure second-order coefficients, $\text{Cov}(b_{11}, b_{22})$, are zero if the design is set up so that

$$NN_c = (N_c + 2a^2)^2 \quad (34)$$

This condition for uncorrelated, estimated, pure second-order coefficients is true for any number of factors. A central composite design satisfying this condition is an orthogonal design. The axial spacing needed for orthogonality is

$$a^2 = \frac{\sqrt{(N_c + N_a + N_0)N_c} - N_c}{2} \quad (35)$$

For example, an orthogonal design for two factors with $N_c = 4$, $N_a = 4$ and $N_0 = 1$ has an axial spacing of $a = 1$. This design is equivalent to a 3^2 factorial design. If the number of centre points is increased from 1 to 5 then the axial spacing is increased to $a = 1.267$ and shows that the composite design for two factors given above is orthogonal.

Orthogonality eliminates covariances between estimated pure second-order coefficients. Rather than using a criterion for individual estimated coefficients, it is possible to use criteria based on the joint effect of all coefficients. One such criterion is based on variances of estimated responses for points that are an equal distance from the design centre. Designs that have points which are equidistant from the centre are termed rotatable designs. This criterion means that the variance of the predicted responses depends only on the distance from the design centre and not on the direction. A necessary condition for a design to be rotatable is that axial spacing for central composite designs be the fourth root of the number of cube points [23] or

$$a^4 = N_c \quad \text{or} \quad a^2 = \sqrt{N_c} \qquad (36)$$

Rotatability does not depend on the number of centre points. However, centre points can be added to orthogonal designs with a corresponding increase in the axial spacing to satisfy rotatability and orthogonality. For example an orthogonal design in two factors with one centre point has an axial spacing of 1 but is not rotatable. A rotatable orthogonal design for two factors must have 8 centre points with an axial spacing of 1.414.

6.1 Worked example of central composite design

In an investigation of the effects of burner height (H) and lamp current (L) upon the signal-to-noise (S/N) ratio response of an atomic absorption spectrophotometer, it was decided to use an orthogonal two-factor central composite design with five centre points. The minimum and maximum values of the factors were

	Units	−1	+1
Burner height H	(mm)	4	10
Lamp current L	(mA)	4	12

The composite design had $N_c = 4$ cube points, $N_a = 4$ axial points and $N_0 = 5$ centre points at an axial spacing of $+/- 1.267$. The design matrix and the responses obtained for the order in which they were carried out are given in Table 8. The **b** coefficients were then determined by means of eq. (22). These coefficients gave the second-order equation describing the response as a function of the coded factor levels

$$Y = 87.39 + 8.18H + 9.05L - 3.97H^2 - 18.92L^2 - 5HL$$

Predicted values of the S/N ratios were then calculated and subtracted from the observed S/N ratios to give the residuals. Reassuringly, the re-

TABLE 8

Two-factor central composite design worked example

Design matrix, responses (S/N ratio) and the residuals obtained from the regression equation.

Run	Factors		S/N ratio					
	H	L	Observed	Predicted	Residual	Min	0	Max
1	0	0	80	87.4	−7.4	x		
2	0	0	83	87.4	−3.4	x		
3	1	−1	71	68.6	2.4			x
4	0	1.267	75	68.5	6.5			x
5	0	0	97	87.4	9.6			x
6	0	0	75	87.4	−12.4	x		
7	1	1	74	76.7	−2.1	x		
8	−1.267	0	76	70.7	6.3			x
9	0	−1.267	44	45.5	−1.5	x		
10	−1	−1	41	42.3	−1.3	x		
11	0	0	100	87.4	13.6			x
12	−1	1	64	70.4	−6.4	x		
13	1.267	0	91	91.4	−0.4		x	

```
                                            xx    x
                                       x xxxxxx x x x
                                       |———|———|
                                       Min  0  Max
```

TABLE 9

Central composite design worked example

Analysis of variance for two-factor central composite design.

Source of variation	d.f.	SS	MS	Variance ratio
Total	12	3708.77		
Regression	5	3101.15	620.23	7.15
Residual	7	607.62	86.80	
Lack of fit	3	128.85	42.95	0.36
Pure error	4	478.77	119.69	

siduals seem to have occurred randomly and when plotted onto an axis are approximately normally distributed considering that there are only 13 runs (Table 8).

An analysis of variance table was then constructed (Table 9). The sums of squares due to the regression as a percentage of the total sums of squares was 83.6%, showing that a large proportion of the variance was explained by the regression equation. Unless a regression equation accounts for a large and significant part of the variation it will be a valueless representation of the data to which it has been fitted.

The variance ratio of the regression mean square to the residual mean square gave a value of 7.15 indicating a high probability that the regression equation is non-zero. However, as indicated earlier, the residual variation in response surface experiments can be broken down into two components, a lack of fit and pure error. The sum of squares due to pure error is calculated from the five centre points. The sum of squares due to lack of fit is then calculated by subtracting the pure error sum of squares from the residual sum of squares. The variance ratio of these mean squares was 0.36, a low value which indicates that the second-order model is an adequate approximation to the data. The residual mean square of 86.80 can thus be used as a reasonable estimate of the error variance. The square root of this value gives the standard deviation about the regression (9.32). This value indicates that there is a 68% probability that any single additional measurement of the S/N ratio within the region of interest would fall within plus or minus 9.32 units of the predicted response surface.

The estimated variances of the individual coefficients were then obtained by multiplying the appropriate value on the principal diagonal of the $X'X^{-1}$ matrix by the residual mean square.

$\text{Var}(b_0) = (0.196)(86.80) = 17.01$

$\text{Var}(b_1) = (0.139)(86.80) = 12.07$

$\text{Var}(b_2) = (0.139)(86.80) = 12.07$

$\text{Var}(b_{11}) = (0.198)(86.80) = 17.19$

$\text{Var}(b_{22}) = (0.198)(86.80) = 17.19$

$\text{Var}(b_{12}) = (0.25)(86.80) = 21.70$

The square roots of these variances provided the standard errors of the estimates. The regression coefficients were then individually tested to see if they were significantly different from zero by forming ratios of the **b** coefficients to their standard errors to give t-values. The probability that the coefficient is not zero is one minus twice the probability that a sample from a t-distribution should be at least as large as the observed t-value.

Coefficient	b_0	b_1	b_2
value	87.4	8.18	19.05
standard error	4.13	3.06	3.06
t-value	21.17	2.68	2.96
%Probability $< > 0$	> 99.5	> 95	> 95
Coefficient	b_{11}	b_{22}	b_{12}
value	−3.97	−18.92	−5.00
standard error	4.10	4.10	0.04
t-value	−0.97	−4.61	−125.0
%Probability $< > 0$	< 90	> 99.5	> 99.5

From the probability values, most of the coefficients would seem to be non-zero, confirming that

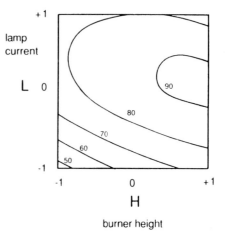

Fig. 4. Contour map of S/N ratio response as a function of the coded values for the factors, burner height and lamp current.

an adequate fit has been made. When an overall test of fit for the regression equation confirms that it is adequate, it is not very important whether an individual coefficient is significantly different from the zero, but may be useful when the regression equation fits the data poorly.

The stationary point of 0.96 for **H** and 0.11 for **L** was found by taking partial derivatives of the prediction equation with respect to each factor. These derivatives were equated to zero and solved for the factor levels. The response at the stationary point was then found to be 91.8 and by entering fixed contour levels for each of the factors, the contour map of the response over the region of interest was obtained (Fig. 4). This indicates that the stationary point in fact represents a maximum for the two factors and may be used as the optimum combination of burner height and lamp current.

7 CONCLUSIONS

Within the limitations of the actual objectives of an experiment, the experimenter may choose from a vast array of designs. When only one factor is of importance, the factor may be varied whilst keeping all others constant and a single-factor ANOVA applied to analyse the results. However, for many experimental situations, it is likely that several factors will be of interest. If quantification of the effects of these factors and possible interactions is required then full factorial or fractional factorial designs may prove useful. Full factorial designs provide large amounts of information on all the factors and interactions but, when the number of factors is large, tend to require too many runs to be practical. Thus when little is known about the chemical system, such as in the initial stages of an investigation, many factors could influence the response, fractional factorial designs may then prove more efficient at discovering the important factors. Once these have been established response–surface designs may be applied to graph the response as a function of the explanatory variables. However, when optimization of the response is required and the investigation lends itself to a sequential approach, simplex optimization may be more useful.

REFERENCES

1 C.K. Bayne and I.B. Rubin, *Practical Experimental Designs and Optimization Methods for Chemists*, VCH, Weinheim, 1986.
2 M.G. Kendall and W.R. Buckland, *A Dictionary of Statistical Terms*, Longman, New York, 1982.
3 D.B. Duncan, Multiple range and multiple F tests, *Biometrics*, 11 (1955) 1–21.
4 H. Scheffe, A method for judging all contrasts in the analysis of variance, *Biometrika*, 40 (1953) 87–91.
5 J.W. Tukey, Comparing individual means in the analysis of variance, *Biometrics*, 5 (1949) 99–108.
6 D. Newmann, The distribution of the range in samples from a normal population expressed in terms of an independent estimate of the standard deviation, *Biometrika*, 31 (1939) 20–23.
7 M. Keuls, The use of the Studentized range in connection with analysis of variance, *Euphytica*, 1 (1952) 112–119.
8 J.C. Miller and J.N. Miller, *Statistics for Analytical Chemists*, Ellis Horwood, Chichester, 1984.
9 G.E.P. Box, W.G. Hunter and J.S. Hunter, *Statistics for Experimenters; An Introduction to Design, Data Analysis and Model Building*, Wiley, New York, 1978.
10 C. Daniel, Use of half-normal plots in interpreting factorial two-level experiments, *Technometrics*, 1 (1959) 311–341.
11 D.A. Zahn, Modifications of and revised critical values for the half-normal plot, *Technometrics*, 17 (1975) 189–200.
12 D.A. Zahn, An empirical study of the half-normal plot, *Technometrics*, 17 (1975) 201–211.
13 F. Yates, *The Design and Analysis of Factorial Experiments*, Imperial Bureau of Soil Science, Harpenden, 1937.
14 O.L. Davies (Editor), *The Design and Analysis of Industrial Experiments*, Longman, London, New York, 1956.
15 P.W.M. John, *Statistical Design and Analysis of Experiments*, Wiley, New York, 1971.
16 R.L. Plackett and J.P. Burman, The design of optimum multifactorial experiments, *Biometrika*, 33 (1946) 305–325.
17 F. Vlacil and H.D. Khan, Determination of low concentrations of dibenzyl sulfoxide in aqueous solutions, *Collection of Czechoslovak Chemical Communications*, 44 (1979) 1908–1917.
18 N.R. Draper and H. Smith, *Applied Regression*, Wiley, New York, 1981.
19 E.C. Harrington Jr., The desirability function, *Industrial Quality Control*, 21 (1965) 494–498.
20 R.H. Myers, *Response Surface Methodology*, Allyn and Bacon, Boston, MA, 1971.
21 G.E.P. Box and K.B. Wilson, On the experimental attainment of optimum conditions, *Journal of the Royal Statistical Society, Series B*, 13 (1951) 1–45.
22 J.A. Cornell, *Experiments with Mixtures: Designs, Models, and the Analysis of Mixture Data*, Wiley, New York, 1981.
23 G.E.P. Box and J.S. Hunter, Multifactor experimental designs for exploring response surfaces, *Annals of Mathematical Statistics*, 28 (1957) 195–241.

Optimisation via Simplex

Part I. Background, Definitions and a Simple Application

K.W.C. BURTON

Department of Science, The Polytechnic of Wales, Pontyprydd, Mid Glam. (U.K.)

G. NICKLESS *

Department of Inorganic Chemistry, School of Chemistry, The University, Bristol BS8 1TS (U.K.)

(Received 1 December 1986; accepted 9 January 1987)

CONTENTS

1 Concept of optimisation	124
1.1 Introduction	124
1.2 Rigorous approach	125
1.3 Concept of Simplex	126
2 Method of Simplex operation	126
2.1 The quantity to be optimised	126
2.2 Selecting the factors to be optimised	127
2.3 Choosing the step size	128
2.4 Identification of the constraints	128
2.5 Location of the initial Simplex	128
3 Modified Simplex methods (variable-size Simplex)	135
3.1 Rules of modified Simplex method	135
3.2 Failure of contraction	136
4 Application of a simple Simplex	136
4.1 Long Simplex procedure	137
5 Conclusions	138
References	138

1 CONCEPT OF OPTIMISATION

1.1 Introduction

The requirement for methods of optimisation arises from the necessity to describe often in mathematical terms or rules the complexity of a system which occurs in practice. Even quite simple systems must sometimes be represented by theories which change with time or by parameters that vary in a random (possibly unknown) manner. For many reasons the theoretical concepts may be imperfect yet such concepts must be used to predict the optimum operating conditions of a system, so that some performance criterion, e.g. product strength or colour, is reached. At best, such a theory can only predict that the system is near to the desired optimum. Such a theory can often suggest that the system is near the desired optimum. So although "optimisation" is often taken literally to mean making something "as perfect, effective or functional as possible" [1] in chemical practice it usually means making something "acceptable". Thus optimisation can be regarded as "the collective process of finding the set of conditions required to achieve the best result from a given situation" [2].

Almost every chemical research and development project requires the optimisation of a system response (called the dependent variable) as a function of several experimental factors (called the independent variables). The optimisation of a chemical system is the process of adjusting these hopefully independent variables so that the response of the system achieves the best possible level (obviously within the limitations of the allowable modifications of the system). The primary aims can be:

1. the maximization of the yield of a reaction as a function of reaction time and reaction temperature;
2. the improvement of the stability of a product in solution;
3. maximizing the analytical sensitivity of a wet chemical method as a function of pH, reagent concentration, wavelength, etc.;
4. seeking the combination of levels for eluent variables that will produce a given level of separation in high-performance liquid chromatography (HPLC).

Such goals are usually attained by determining how each of the pertinent variables (sometimes called factors) affects the final result with the idea of acquiring data by systematic experimental design.

1.2 Rigorous approach

When the variables do not interact with each other, each variable may then be varied and optimised independently of the other variables. Regretfully however, variables do interact with each other and the one factor at a time approach will not always result in the best set of conditions. Such a univariate search procedure necessitates a very large number of experiments, particularly when more than two factors have to be considered. Therefore, univariate optimisation strategy wastes labour because it is inefficient since a larger number of experiments than is necessary is often carried out.

The traditional motivation underlying the theory of optimal design is that experiments should be designed to achieve the most precise statistical inference possible. The influence of optimal design has extended to almost all areas of experimental design. To apply optimal design theory in practice requires a criterion for comparing the performance over the set of possible experimental designs. The classical criteria are derived within the context of a linear model theory in which it is assumed that the experimental data can be represented by the equation

$$Y_i = f(x_i)'\beta + \epsilon_i \qquad (1)$$

Y_i is the measure of response from the ith experimental run.
x_i is a vector of predictor variables for the ith run.
f is the vector of p functions that model how the response depends on x_i.
β is a vector of unknown parameters.
ϵ_i is the experimental error for the ith run.

A natural way to measure the quality of statistical inference with respect to a single parameter is in terms of the variance of the parameter estimate. If the errors are uncorrelated and have constant variance σ^2, the variance-covariance matrix of the least square estimator, $\hat{\beta}$, is

$$\text{var}\{\hat{\beta}\} = \sigma^2 (X'X)^{-1} \qquad (2)$$

where X is the $n \times p$ matrix whose ith row is $f(x_i)$. It is assumed that X has full column rank.

Another useful way to measure the quality of inference is in terms of the variance of the estimated response at x, which, from eq. 1 is given by

$$d(X) = \sigma^2 f(x_i')'(X'X)^{-1} f(x_i) \qquad (3)$$

Both eqs. 2 and 3 depend on the experimental

design only through the $p \times p$ matrix $(\mathbf{X'X})^{-1}$ and suggest that a good experimental design will be one that makes this matrix small in some sense. Since there is no unique size ordering of the $p \times p$ matrices, various real-valued functionals have been suggested as measures of "smallness". Certain of the most popular of these optimal criteria are:

1. *D-optimality*. A design is said to be D-optimal if it minimises $\det(\mathbf{X'X})^{-1}$, where det is determinant.

2. *A-optimality*. A design is said to be A-optimal if it minimises $\text{tr}(\mathbf{X'X})^{-1}$, where tr is trace.

3. *E-optimality*. A design is said to be E-optimal if it minimises the maximum eigenvalue of $(\mathbf{X'X})^{-1}$.

4. *G-optimality*. A design is said to be G-optimal if it minimises $\max d(\mathbf{X})$, where the maximum is taken over all possible vectors $\mathbf{X}$ of predictor variables.

In practice, this optimisation problem may be difficult or impossible to solve analytically in a rigorous sense.

The general problem of experimentally optimising a function of several variables has been discussed both by Hotelling [3] and Friedmann and Savage [4]. The latter workers described a sequential process involving one factor at a time which leads to the sequential procedures of Box and Wilson [5]. These procedures were chiefly targeted at industrial processes, where very small changes to factors were made using very many measurements. Thus a model that fitted reasonably well over the limited region of the factor space covered can be made and used to predict the direction in which to move to obtain an improved response (evolutionary operation, or EVOP) [6]. But if the model, which is usually an empirical one, contains more than a few factors, then the number of experiments required to fit the model becomes impractically large, so it cannot be used further for synthesis or analysis procedures.

1.3 Concept of Simplex

The second and alternative strategy is to try to use an efficient experimental design that may optimise a relatively large number of factors in a small number of experiments. Thus although the Fibonacci search for a maximum along a line is the best method where the response depends on only one factor [7], for two or more factors the Simplex method is probably the most efficient and easily employed procedure. The method is not so rigorous mathematically but is very efficient, it does not use the traditional test of significance and is, therefore, faster and simpler than previous methods. Although the original Simplex method was introduced by Dantzig [8] it was limited in use, and it was not until Spendley et al. [9] developed the approach now referred to as the Basic Simplex method, which offered a general sequence for optimisation. The Basic Simplex procedure suffered from two disadvantages, firstly it is fixed in that it cannot adjust its size or shape to the surface to which it responds and secondly it can fail to optimise due to an unfavourable relative rotation of the initial Simplex. Therefore, an improved method called the Modified Simplex was soon proposed by Nelder and Mead [10]. The procedure contained two new concepts, expansion and contraction of the Simplex, so making it possible for the Simplex to adjust to the local gradient of the response surface. The Basic Simplex was first applied in the field of analytical chemistry by Long some five years later [11].

2 METHOD OF SIMPLEX OPERATION

Prior to optimisation the following aspects must be considered.

2.1 The quantity to be optimised

The decision as to which criteria should make up an optimum will depend upon the use intended for the final system. It may be a straightforward case of optimising only a yield, stability, linearity or it may be a complex function relating several such variables. An experiment often gives several responses of interest and a consideration of the auxiliary responses can shift the resulting optimum from the point which is maximum for a single principal response. A practical procedure is to follow the most important response with the Simplex system and record the secondary responses for each experimental point. After the maximum for the principal criterion has been reached, judgement of the optimum can be based upon a weighting of the relative importance of the responses. If a more systematic relationship is desired between several co-existing responses, a desirability coefficient may be derived by a weighting procedure and the resulting function maximised. Even if this procedure is used, the individual responses should be recorded for probable later consideration.

2.2 Selecting the factors to be optimised

To simplify the optimisation it is usually preferable to choose only the most important factors. The importance is determined by the comparative change in response caused by a change in level of each one of the factors, which may be based on a prior knowledge of the system (or even preliminary experimentation).

Factorial experiments are a good method of judging the relative significance of the possible factors and they offer a quantitative measure of the contribution of each factor to the overall response. Such a design specifies the proper combination of variables in each sample of the experiment; usually two-level factorials are adequate and their application is a well established procedure [12]. It is possible to include every conceivable variable in a factorial experiment for evaluation. The required number of samples increases exponentially as 2^k, where k is the number of factors; but this can be reduced by fractional replication, if interaction between certain factors can be assumed absent [13].

Because for the Simplex only those factors of obvious practical significance are generally of interest, it is usually unnecessary to perform a statistical analysis of variance on the data from factorial investigations. Instead, after the experiments employing the treatment combinations dictated by design have been carried out, the magnitude of the effects can be calculated by the standard method and their relative importance judged by inspection.

When ascertaining which factors yield the highest response, it must be realised that the apparent changes in responses are not an absolute measure of effect but will depend on the scales and differences as in levels selected for the experiments (i.e. the change or step size). Therefore, it may sometimes be desirable to normalise the responses by relating them to the total range possible for a factor. If there is doubt about the significance of a factor, it can always be included within the Simplex to ensure that due account is taken of the effect it creates.

Initially, however, consideration of two factors only is recommended because a graphical method can be used to locate the new Simplex or experiment to be carried out. The graphical procedure assists in visualising the optimisation of a greater number of factors, where calculations are to be used. Obviously, the graphical approach can only be employed where two factors are important or if the next factor in rank offers a much lower effect on response. Regardless of the number of factors used within the Simplex, any remaining factors are kept constant. The situation is usually achieved by maintaining uniform experimental conditions except for those factors being systematically varied according to the Simplex sequence.

Factors purposely omitted from the Simplex are usually those of lesser importance; however, in the region of the optimum reached using the Simplex, the slopes of the response surface can differ considerably from those in the region of the factor space determined by the first factorial. In turn, this may mean that the magnitudes of the effects due to a given change will differ and so may the relative order of the importance of the factors. Thus to locate the exact optimum, after the conditional maxima have been reached, re-run a factorial experiment to choose all the currently important factors and include those within a final multi-dimensional Simplex. The presence of any such interactions between factors in two sets of experiments makes it desirable to include *all* potentially significant factors in the original Simplex experiment but generally such interactions are not known at the onset of experimentation.

When optimising an analytical method, the concentration of the analyte being determined should not be taken as a variable for the Simplex, since the maximum response will proceed in the direction of increasing concentration. Instead, a concentration is chosen at a sufficiently low value where the range of the instrumentation will not be exceeded by the increase in response resultant from optimisation.

For the situation where two factors are interdependent with respect to the levels attainable (as in an eluent made up of two solvents, both of which affect the response but whose fractions must of a necessity always total unity) their combination may be located as a single factor with possible levels ranging e.g. from a mixture composed of fractions zero A, one B at one end of the scale, to a one A, zero B at the other.

2.3 Choosing the step size

Scales must be assigned to the factors being optimised and the spacing between successive experimental levels decided. The choice of the step size is arbitrary but it is of advantage if the step for each factor causes a comparable change in response. The effect of a factor upon the response determines the slope of the response surface and it usually changes as the maximum is approached. If one factor gives an effect which is small compared to the effects of other factors, it may be that the base level chosen for it is near a conditional maximum, the system is relatively independent of its level, or the unit adopted for it is disproportionately small. The situation can be checked by changing to a larger step size.

An initial large step size is usually an advantage since the maximum is approached more rapidly and error will have a proportionally smaller influence. The step size chosen should be large enough that the experimental error is a sufficiently small proportion of the total change in response due to the operation of the step. However, too large a step is bad if it causes excessive overshoot of the maximum when applied to a process already nearly optimised. If a peak is very sharp in comparison to the step size, it may well be missed. The situation can be checked (if at all suspected) by running the mid-point of the Simplex and if it yields a higher response than all of the apexes, the centre region must be explored with smaller steps. A large step may also make it difficult to manoeuvre between constraints or to stay on a high yield portion of a steep slope, but a reduction in size can be made after these problems are encountered.

When the levels of a factor cannot be assigned a definite ordering (qualitative parameters such as colour) the remaining (quantitative) factors are optimised for each level of the qualitative factor. The results of the separate optimisations are then compared to ascertain the acceptable version of the qualitative factor.

2.4 Identification of the constraints

Constraints are boundaries on the response surface which should not be crossed into regions of disallowed levels of one or more factors. These may be dictated by pressure or temperature limitations of the apparatus, solubility, concentration, etc. Undesirable features or results such as instability, slow reaction rates, can also be treated as constraints. If possible, these constraints should be set prior to optimisation, although in reality many constraints become only apparent during the actual experimentation. The region within the constraints may be considered as "the experimental region" and the Simplex will be seeking the best response available within this allowable portion of the factorial space.

2.5 Location of the initial Simplex

To establish the first Simplex of the series, it is necessary to decide upon the initial value for each of the factors (i.e., the experimental conditions at which the first experiment is to be performed dictated by the factors that have been selected for the optimisation procedure). These often will be the factor values in accepted use prior to the current investigation, unless preliminary experiments have indicated some better region of the factor space for commencing the Simplex series. Although the Simplex should finally reach the area where the optimum lies regardless of the starting position chosen, fewer iterations will be necessary if the starting point is at a point known to be close to the optimum.

Next, with the desired step sizes, the experimental points defining the initial Simplex are located by choosing values for the factors such that the points lie at the vertices of a regular Simplex of the required dimensionality.

Thus for a two-factor Simplex, the figure is an equilateral triangle, and so requires three experimental points. Since two factors can be followed graphically, all that is required is to lay out the two factors as the x and y coordinates and so locate the vertices of the Simplex on the graph. The initial experimental orientation of the triangle does not significantly affect the efficiency but it is helpful to orientate the Simplex with one side parallel to the axis of the primary factor if more factors are to be added at later stages.

Table 1 may be used to construct Simplexes of up to ten factors. For the initial location of each vertex, Table 1 specifies fractions of the step size, which are to be taken as the distance from the experimental origin. A triangular Simplex for factors A and B will require vertices 1, 2 and 3. Similarly a three-factor Simplex is a tetrahedron, so factors A, B and C require vertices 1, 2, 3 and 4. The four-factor Simplex is an analogous figure in four-factor space. Therefore, inclusion of factor D requires up to vertex 5.

Vertex 1 has coordinates of zero and so is located at the point chosen as the starting levels for the factors. This first vertex is the experimental origin and generally corresponds to the factor levels employed before the current attempt to optimise. For the remaining vertices, the step size for each factor is multiplied by the appropriate fraction in Table 1. Each result must then be added to the value which that factor has at the origin, i.e. at vertex 1. Repeating this process for each vertex gives the set of experimental coordinates needed for the initial Simplex. Thus the table produces the correct shape and orientation of the Simplex but its location is determined by what is picked as the starting point. The coordinates for each vertex specify the magnitudes of the factors to be used as parameters in the initial Simplex sequences.

Tables 2 and 3 offer examples of the calculations necessary for the initial experiments to be carried out for a two-factor and then a three-factor situation.

After the initial Simplex has been set up, experiments can be performed at the conditions obtained for each individual vertex. The worst experimental value is eliminated and a new vertex is located by reflecting the Simplex into the factor space in the direction opposite to the undesirable result. A single new vertex is required to form a new Simplex when taken together with retained vertices. For a two-factor case, this may be carried out by using a triangular mask cut to the size of the initial Simplex and laying it on the graph with two vertices coinciding with the two retained points. The third vertex will then be positioned opposite to the eliminated point, at the location for the next experiment. Optional with two but essential with three or more factors, the coordinates for the new point are found by a simple calculation. First the coordinates of the k retained vertices (from a Simplex of $k+1$ vertices) are tabulated; then these are summed over each factor and each sum is multiplied by $2/k$. Finally, the coordinates of the discarded point are subtracted to obtain the coordinates for the new point.

If a given vertex is retained after $k+1$ successive Simplexes, it is possible that it was a spurious value caused by error, so that the experimental value at these conditions should be rechecked by carrying out the experiment once more and noting the response. If the response is not reproducible it should be replaced by a new observation at that point.

If the response at a new point is also the lowest value in a new Simplex, causing it to reflect back onto the previous position, then the second lowest value should be eliminated instead of the lowest. The action will allow progress up a ridge which is being straddled or if a peak has been reached; this will cause the Simplex to circle the maximum, so that the cessation of progress will be verified and the region around the maximum defined.

As an alternative, the coordinates of the new vertex, the reflection vertex R, can be employed as follows.

The points W, N and B in Fig. 1 represent those factors which produce the worst, next-to-worst and best responses. Here, x_1 could represent column temperature and x_2 the carrier flow-rate for the separation of several components by

TABLE 1

Values of step sizes for up to ten factors

Vertex 1 has coordinates of zero and so is located at the point chosen as the starting level for the factors. This first vertex is the experimental origin and generally corresponds to the factor levels employed before the current attempt to optimise. For the remaining vertices, the step size for each factor is multiplied by the appropriate fraction in Table 1. Each result is then added to the value which that factor has at the origin, i.e. at vertex 1. Doing this for each vertex gives the set of experimental coordinates needed for the initial Simplex. (The table gives the correct shape of the Simplex and the position with respect to the experimental origin, but its actual location in factor space is determined by what is picked as the starting point.) The coordinates for each vertex specify the magnitudes of the factors to be combined as that sample or experiment.

Vertex No.	Factor No.									
	A	B	C	D	E	F	G	H	I	J
1	0	0	0	0	0	0	0	0	0	0
2	1.000	0	0	0	0	0	0	0	0	0
3	0.560	0.866	0	0	0	0	0	0	0	0
4	0.500	0.289	0.817	0	0	0	0	0	0	0
5	0.500	0.289	0.204	0.791	0	0	0	0	0	0
6	0.500	0.289	0.204	0.158	0.775	0	0	0	0	0
7	0.500	0.289	0.204	0.158	0.129	0.764	0	0	0	0
8	0.500	0.289	0.204	0.158	0.129	0.109	0.756	0	0	0
9	0.500	0.289	0.204	0.158	0.129	0.109	0.094	0.750	0	0
10	0.500	0.289	0.204	0.158	0.129	0.109	0.094	0.083	0.754	0
11	0.500	0.289	0.204	0.158	0.129	0.109	0.094	0.083	0.075	0.742

gas–liquid chromatography. High temperature and high flow-rate (e.g. point W in Fig. 2) may produce a very poor separation of a particular mixture. Slightly lower temperature and much lower flow-rate (e.g. point N) may give improved separation. A still lower temperature and an intermediate flow-rate (point B) may thus produce the best separation of all three points.

Symbols W, N and B are the vector notations for the factor combination of Fig. 2.

$$B = (x_{1b}\ x_{2b})$$
$$N = (x_{1n}\ x_{2n})$$
$$W = (x_{1w}\ x_{2w})$$

The point $\bar{P}$ in Fig. 2 is called "the centroid of the hyperface" remaining after the worst vertex (W) has been eliminated from the Simplex. It has simply the average coordinates of the remaining points. For the two-factor case

$$\bar{P} = \tfrac{1}{2}(N + B) = \left[\frac{(x_{1n} + x_{1b})}{2}\ \frac{(x_{2n} + x_{2b})}{2}\right] \quad (4)$$

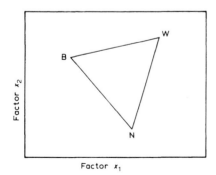

Fig. 1. Two-dimensional Simplex.

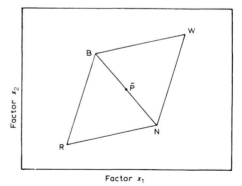

Fig. 2. Two-dimensional Simplex reflection.

Chapter 9

TABLE 2

Calculation of coordinates with respect to origin of initial Simplex

Consider a two-factor example, where temperature and pressure are the two factors dictating the optimisation procedure. Initial experiment conditions chosen: factor 1: temperature = 20°C; factor 2: pressure = 70 mm Hg. Step size chosen for factor 1: temperature (SSt) = 10°C; step size chosen for factor 2: pressure (SSp) = 10 mm Hg. As only two factors are being considered, the initial Simplex will be represented by an equilateral triangle (as the step sizes for both factors are the same). To obtain the experimental conditions for vertices 2 and 3 of the initial Simplex, Table 1 has been used. Thus:

Vertex number in initial Simplex	Temperature (°C)	Pressure (mm Hg)
Vertex 1	20	70
Vertex 2	20 + (SSt × 1) = 30	70 + (SSp × 0) = 70
Vertex 3	20 + (SSt × 0.5) = 25	70 + (SSp × 0.866) = 78.66

Hence the initial experimental conditions are now set.

TABLE 3

Calculation of coordinates with respect to origin of initial Simplex

Consider a three-factor example, where temperature, column size and carrier gas flow-rate are the three factors which dictate the optimisation procedure. Factor 1: temperature = 20°C; factor 2: carrier gas flow-rate = 40 ml/min; factor 3: column size = 200 cm × 0.125 in. I.D. Step size chosen for factor 1: temperature (SSt) = 10°C; step size chosen for factor 2: carrier gas flow-rate (SSf) = 5 ml/min; step size chosen for factor 3: column length (SSl) = 20 cm. As three factors are being considered, the initial Simplex will represent a tetrahedron. To obtain the experimental conditions for vertices 2, 3 and 4 for the initial Simplex, Table 1 has been used.

Vertex number in initial Simplex	Temperature (°C)	Flow-rate (ml/min)	Column length (cm)
Vertex 1	20	40	200
Vertex 2	20 + (SSt × 1) = 30	40 + (SSf × 0) = 40	200 + (SSc × 0) = 200
Vertex 3	20 + (SSt × 0.5) = 25	40 + (SSf × 0.866) = 44.33	200 + (SSc × 0) = 200
Vertex 4	20 + (SSt × 0.5) = 25	40 + (SSf × 0.289) = 41.445	200 + (SSc × 0.817) = 216.34

The vector $(\overline{P} - W)$ is added to $\overline{P}$ to give the coordinates of a new vertex, the reflection vertex, R. Hence $R = \overline{P} + (\overline{P} - W)$.

Similarly for the three-factor case (see Fig. 3) where x_1 and x_2 are unchanged but x_3 represents column length

$B = (x_{1b} \ x_{2b} \ x_{3b})$

$N_2 = (x_{1n2} \ x_{2n2} \ x_{3n2})$

$N_1 = (x_{1n1} \ x_{2n1} \ x_{3n1})$

$W = (x_{1w} \ x_{2w} \ x_{3w})$

$$\overline{P} = \tfrac{1}{3}(N_1 + N_2 + B)$$

$$= \left[\frac{(x_{1n1} + x_{1n2} + x_{1b})}{3} \ \frac{(x_{2n1} + x_{2n2} + x_{2b})}{3} \right.$$

$$\left. \frac{(x_{3n1} + x_{3n3} + x_{3b})}{2} \right] \quad (5)$$

and again $R = \overline{P} + (\overline{P} - W)$.

The movement of the Simplex, using a two-factor example, is illustrated in Fig. 4.

Point W is eliminated to give the reflected point R_1. If experimental conditions at R_1 give an experimental response which is better than the

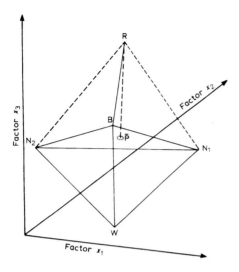

Fig. 3. Three-dimensional Simplex reflection.

experimental response at N then R_1 is accepted as a point in the new vertex BNR_1; but if worse than that obtained for N then R_1 is rejected. Subsequently, the experimental conditions giving the next best experimental response are eliminated in the *previous* Simplex (here the initial Simplex) giving the reflected point R_2. The response for the experimental conditions at R_2 is obtained and if it is greater than the value for the conditions at W then R_2 is the new point in the new Simplex BWR_2. However, if the experimental response for conditions R_2 is less than that obtained for condi-

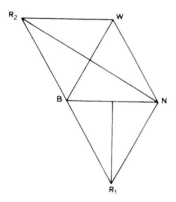

Fig. 4. Illustration of movement of Simplex; two-factor example (triangle). The initial Simplex is represented in BWN, where B = best experimental response, N = next best experimental response and W = worst experimental response.

tions W, R_2 is also rejected. There is no point in eliminating conditions at point B as this gives the best experimental response; hence contract the Simplex at B, that is taking B as the origin and setting up a new initial Simplex. However, if contraction is not possible due to the step size of each factor already being small, then the termination criterion has been reached.

Similarly, the movement of the three-factor Simplex, resulting in a tetrahedron shape, may be illustrated as shown in Fig. 5. Obviously, it is more difficult to visualise movement of the Simplex with three factors in comparison with the two-factor example in Fig. 4. But in fact the principle applies regardless of the number of factors involved. The following steps are followed:

(i) if R passes the criterion, then the new Simplex is obtained with vertices B, N_2, N_1, R.

(ii) if R fails then a new Simplex with vertices B, N_2, W, R_2 will be obtained provided that R_2 passes.

(iii) however, if R_2 also fails, then the new Simplex with vertices B, N_1, W, R_3 will be formed if R_3 passes the test.

But, if R_3 fails there is no virtue in reflecting the point B, hence the simplex should be contracted at B (that is taking B as the origin and setting up a new initial Simplex). If contraction is not possible due to the step size of each factor already being small, then the termination criterion has been attained.

To locate the maximum more exactly, the step size of the Simplex may be decreased to perhaps 0.25 or 0.10 of the previous step, but this depends on the term being optimised. (A large step should have been used initially, to decrease the effort required to reach the general region of the maximum; but then smaller steps should be used to increase the resolution in seeking the exact optimum.) To accomplish the goal, the coordinates of the Simplex giving the highest value are taken as the new experimental origin and the remaining vertices are located according to the design matrix by applying the smaller step. Decreasing the step size allows the maxima to be approached as closely as desired, although errors in adjusting small increments of the factors may prove to be the limiting problem.

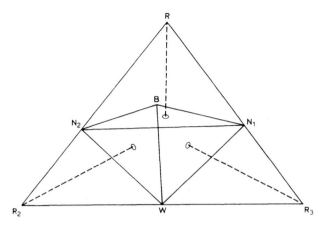

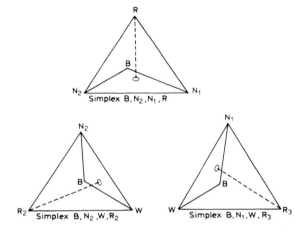

Fig. 5. Illustration of movement of Simplex; three-factor example (tetrahedron).

The objective of the sequential Simplex method is to force the Simplex to move to the region of optimum response. The decisions required to accomplish the process constitute the "Rules" of the Simplex method.

Rule 1. A move is made after each observation of response. Once the responses at all vertices have been evaluated, a decision can be made as to which vertex to reject. As has been seen, a new Simplex can be completed by carrying out only one additional measurement (R). Except in the initial Simplex, a move can be made after each measurement of response.

Rule 2. A move is made into that adjacent Simplex which is obtained by discarding the vertex of the current Simplex corresponding to the least desirable response (W) and replacing it with its image (R) across the centroid ($\bar{P}$) of the hyperface of the remaining points. This operation has been discussed above. Fig. 6 shows several moves of a Simplex. If the new vertex has the worst response in the new Simplex, Rule 2 would reflect the current Simplex back to the previous Simplex. The Simplex would then oscillate and become stranded. This situation is shown in Fig. 7. An exception to Rule 2 is necessary.

Rule 3. If the reflected vertex has the least desirable response in the new Simplex, do *not* reapply Rule 2, but instead reject the second lowest response in the Simplex and continue. Fig. 8 shows how Rule 3 prevents the Simplex from becoming stranded. Generally the conditions obtained as the new vertex *must* give a response which is not the worst in the new Simplex; if it is, it must be rejected and the next best conditions in the previous Simplex must instead be adopted. This procedure must be terminated when only the conditions giving the best response have not been eliminated (as there is no point in moving away from the conditions giving the best response). At this stage contract Simplex size if possible or else the termination criterion has been reached. If this occurs at the beginning of the optimisation procedure one must start the procedure again from a different starting point.

When dealing with experimentally measured responses, statistical fluctuations are to be expected. To lessen the uncertainty associated with the response in a Simplex, replicate measurements could be made at each vertex and the mean of these replicates could be used to assign a significance to differences in response at the vertices of a Simplex. This is, however, not always necessary since a single evaluation of response at each vertex is usually sufficient. This abandonment of traditional statistical procedures can be justified for two reasons. First, if the differences in the responses are large compared to the size of the indeterminate errors, the Simplex will move in the proper direction. Repetition of measurement would be wasteful. Secondly, if the differences in the responses are small enough to be affected by indeterminate errors, the Simplex might move in

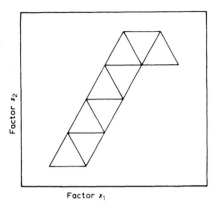

Fig. 6. Several Simplex moves (two-dimensional sequence).

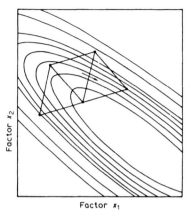

Fig. 8. Avoidance of oscillation by Simplex.

the wrong direction; however, a move in a wrong direction will probably (in a truly statistical sense) yield a lower response that would be quickly corrected by Rules 2 and 3. The Simplex, although momentarily thrown off course, would proceed again towards the optimum.

A special case that might cause a problem is that of a large positive error. As the Simplex moves, the less desirable responses are naturally discarded. The high responses, however, are retained. Thus it is possible that the Simplex will become fastened to a false high result and be mistaken for the true optimum. To help distinguish between an anomalously good response and a valid optimum, the following exception to Rule 1 is used.

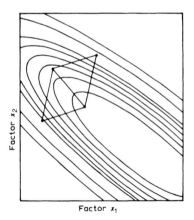

Fig. 7. Possibility of oscillation by Simplex.

Rule 4. If a vertex has been retained in $k + 1$ Simplexes and is not then discarded, before applying Rule 2 re-evaluate the response at the persistent vertex. If the vertex is truly near the optimum, it is probable that the repeated evaluation will be consistently high and the maximum will be retained. If the response at the vertex is high because of an error in measurement, it is improbable that the repeat measurement will also be high, and the vertex will eventually be eliminated. Occasionally, the Simplex might try to move beyond boundaries that have been placed on the factors.

Rule 5. If a new vertex lies outside the boundaries of the independent variables, do not make an experimental observation, instead, assign to it a very undesirable response (a negative response, as any valid response cannot be less than zero). Application of Rules 2 and 3 will then force the vertex back inside its boundaries, and it will continue to seek the optimum response.

The fixed-size sequential Simplex method suffers four limitations when it is used to locate a stationary optimum.

(i) In two dimensions, there is no difficulty in determining when the optimum has been located. The Simplexes will become superimposed because of the ability of triangles to close pack. Tetrahedra and higher dimensional Simplexes will not close pack. Thus, the first limitation is that it is not always clear when an optimum has been reached.

(ii) The original Simplex technique has no provision for acceleration. This is usually overcome by the technique of running several phases, the first is a large Simplex to roughly define the optimum response, the others to more precisely "home in" on the optimum.

(iii) One orientation of a Simplex will cause it to attain a false optimum. This problem can be overcome if another phase is run in which the new Simplex is rotated with respect to the previous Simplex.

(iv) The Simplex optimisation is susceptible to large random errors or noise inherent in the responses of the experiments; frankly the Simplex simply flounders about in the noise. Although any experiment with a large noise content is frustrating, the Simplex optimisation scheme is particularly susceptible because no functional relationship between the responses and the parameters is found. Thus the scheme lacks a model that would help reduce noise effects. In contrast, when a single parameter is varied to give responses, they are usually fitted to a model from which an optimum can be found by inspection or by calculation. Such a model will always have the effect of reducing noise in the optimum. There are Simplex algorithms that construct local models in order to reduce their noise susceptibility but they are not simple algorithms and have to be used as black-box computer programs [14,15].

There are applications in which none of these limitations are serious; these applications are truly "evolutionary operations" in which the Simplex method is used to attain and follow an optimum around a response surface that changes with time. Such applications are not rare in the physical sciences; keeping instruments in their best operating conditions as various of their components change characteristics with age is one example of such an application

Another problem in optimisation is that there is always the possibility that a local optimum has been located but not the global optimum. For the Simplex method (but also for other methods of optimisation) it is impossible to be certain that the global optimum has been found. Confidence can be increased if the same optimum is reached when the optimisation procedure is started from widely differing regions of possible domain.

3 MODIFIED SIMPLEX METHODS (VARIABLE-SIZE SIMPLEX)

As the name suggests the method is simply an improvement on the sequential Simplex method of Spendley et al. [9]. The modified Simplex method of Nelder and Mead [10] is a logical algorithm containing reflection, expansion, contraction and massive contraction rules. These rules can be understood by reference to Fig. 9 and the initial Simplex BNW (BNW representing different sets of experimental conditions) for a two-factor system.

3.1 Rules of modified Simplex method

Reflection is accomplished by extending the line segment $\overline{WP}$ beyond $\overline{P}$ to generate the new vertex R:

$$R = \overline{P} + (\overline{P} - W)$$

(Note equation for $\overline{P}$ has already been defined.)

Three possibilities exist for the measured response at R:

(1) The experimental response at R is more desirable than the experimental response at B. An attempted expansion is indicated, and the new vertex E is generated:

$$E = \overline{P} + 2(\overline{P} - W)$$

If the experimental response at E is better than the experimental response at B, it is retained as the new Simplex BNE. If the experimental response at E is not better than at B, the expansion is said to have failed, and BNR is taken as the new Simplex. The algorithm is restarted using the new Simplex.

(2) If the experimental response at R is between that of B and N, neither expansion nor contraction is recommended, and the process is restarted with the new Simplex BNR.

(3) If the experimental response at R is less desirable than the experimental response at N, a step in the wrong direction has been made, and the Simplex should be contracted.

Two types of contraction are possible.

(i) Firstly, if the experimental response at R is worse than that obtained at N but not worse than

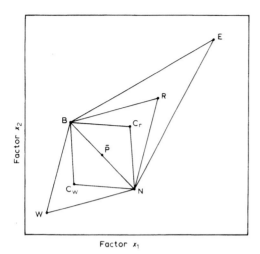

Fig. 9. Variable-size Simplex moves.

that at W, the new vertex should lie closer to R than to W:

$$\therefore C_r = \bar{P} + \tfrac{1}{2}(\bar{P} - W)$$

The process is restarted with the new Simplex BNC$_r$.

(ii) Secondly, if the experimental response at R is worse than the previous worst vertex W, then the new vertex should lie closer to W than to R:

$$\therefore C_w = \bar{P} - \tfrac{1}{2}(\bar{P} - W)$$

The process is restarted with the new Simplex BNC$_w$.

A failed contraction results if the experimental response at C$_r$ is worse that that at R, or if the experimental response at C$_w$ is worse than the response at W.

3.2 Failure of contraction

The problem of a failed contraction occurring causes a major dilemma, to which the ideas outlined below were suggested:

(a) Nelder and Mead [10] recommended a massive contraction in which the size of the Simplex is diminished ever further, which is shown graphically in Fig. 10. The idea, although effective, suffers from two distinct disadvantages. Firstly, it requires the evaluation of k new Simplex vertices before the algorithm can continue (where k equals the number of factors dictating the optimisation procedure). Secondly, the volume of the Simplex is contracted by (0.5^k) which might give rise to premature convergence in the presence of error.

(b) Ernst [16] recognised the second difficulty and recommended translation of the entire Simplex following a failed contraction (see Fig. 11). The Simplex does not contract but the process of translation does require the evaluation of $K + 1$ new vertices.

(c) Morgan and Deming [17] employing a suggestion of King [18] have examined the re-introduction of Rule 2 of the sequential Simplex case. If the contraction vertex is the worst vertex in the new Simplex, do not reject the vertex but rather reject the next-to-worst vertex N. The procedure is simple and does not require the evaluation of additional vertices. This latter method has proved successful in a number of circumstances.

Boundary violations are handled by assigning a very poor experimental response to the vertex whose location violated the boundary constraint (as for the sequential Simplex method). The new vertex is subsequently located by a C$_w$ contraction mid-way between the original point and the centroid and the optimisation is allowed to continue. Assignment of a negative value (say -1) for this response will cause the contraction giving C$_w$, as it is impossible to obtain a negative experimental response, hence the response will be the lowest in the new Simplex which is rejected.

Further discussion of more complex Simplex algorithms will be made in Part III of this sequence of tutorials.

4 APPLICATION OF A SIMPLE SIMPLEX

The following short section contains a description of the sequential fixed size Simplex to a familiar type of problem in analytical chemistry associated with environmental studies. In Part II of this sequence of tutorials more advanced and complex Simplex procedures will be described. It is hoped that these examples which have been taken from the literature and which have been calculated and presented in detail, will clearly illustrate the principles and practices which have been discussed in Sections 2 and 3.

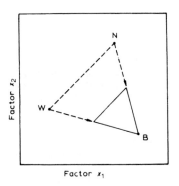

Fig. 10. Massive contraction in Simplex.

4.1 Long Simplex procedure

An example of a two-factor investigation is given in Fig. 12 for the method of determining sulphur dioxide (say in air) was presented by Long [11]. The method is based on the measurement of intensity of the magental colour formed by a Schiff reaction between pararosaniline, formaldehyde in mineral acid solution and sulphur dioxide. Factorial experiments had indicated that the two components in the method having the greatest effect on sensitivity were formaldehyde and hydrochloric acid for a constant concentration of sulphur dioxide. Hence these reactants were varied while all other factors were held constant. A solvent blank and a single concentration of sulphur dioxide absorbed in a mixture of sodium hydroxide and glycerine were compared under the experimental conditions represented by vertices 1, 2 and 3. The intial Simplex was started in a region somewhat removed from the usually suggested reagent conditions (point 0). In this particular example, the base of the Simplex is inclined to the x-axis but it is advisable to maintain the Simplex parallel to the x-axis to ease the later addition of extra factors and any decrease in step size with design matrix. Employing the experimental conditions corresponding to the first three points of the initial Simplex showed that point 2 gave the lowest response, so point 4 was located opposite point 2 across the hyperface of points 1 and 3.

After continuing the sequence, a decrease in response was noted when progressing from point 8 to point 9 (see Fig. 12). Fig. 13 is an enlarged part of this area which explores the area of the maximum. A smaller Simplex was then started from point 8 and included points 10 and 11. Since point 8 was the lowest value, the data for point 12 directly opposite were obtained. At point 13 where the response again decreased, the point giving the second lowest response (point 12) was instead eliminated to prevent the Simplex from oscillating. The process allowed the Simplex to advance and then to commence circling, which helped to define the shape of the response curve around the maximum. The maximum appears closest to point 10 and oculd have been more exactly located by a further decrease in step size. However, the formaldehyde had to be measured in increments of

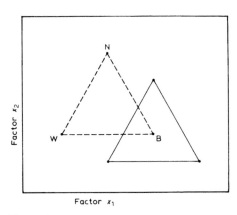

Fig. 11. Translation of Simplex.

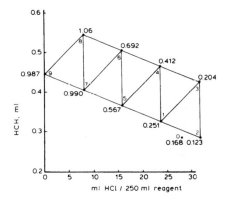

Fig. 12. Simplex optimisation of the reagent concentrations for the determination of sulphur dioxide. Apex number 1 ≡ 0.251 absorbance units.

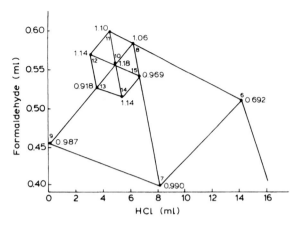

Fig. 13. Exploration of the maximum by Simplex.

0.01 ml and could not be more accurately dispensed in the practical application of this determination (a typical example of a constraint caused by experimental capabilities). At this stage more factors could also have been added and the investigation continued by the method of calculating the new vertex. The origin of such a set of experiments would have been located in the factor space at the level determined by the overall experimental conditions set by point 10. Notice, however, that the sensitivity has been increased by a factor of seven over the proposed original research method. At this point the experiment was terminated because an adequate response for the task set had been achieved.

5 CONCLUSIONS

The background, methods and rules appertaining to the use of the fixed-step and variable size step Simplex procedures have been discussed in detail. Additionally, a simple example of the Simplex procedure for the analytical determination of sulphur dioxide has been presented. In Parts II and III of these tutorials more advanced cases will be discussed and presented.

REFERENCES

1 S.N. Deming, Optimisation, *Journal Research National Bureau of Standards*, 90 (1985) 479–485.
2 G.S.G. Beveridge and R.S. Schechter, *Optimisation—Theory and Practice*, McGraw-Hill, New York, 1970, p. 367.
3 H. Hotelling, Experimental determination of the maximum of a function, *Annals of Mathematical Statistics*, 12 (1941) 20–46.
4 M. Friedman and L.J. Savage, in C. Eisenhart, M.W. Hastay and W.A. Wallis (Editors), *Techniques of Statistical Analysis*, McGraw-Hill, New York, 1947, Ch. 13.
5 G.E.P. Box and K.B. Wilson, On the experimental attainment of optimum conditions, *Journal of the Royal Statistics Society, Series B*, 13 (1951) 1–21.
6 G.E.P. Box and N.R. Draper, *Evolution Operation, A Statistical Method for Process Improvement*, Wiley, New York, 1969.
7 D.L. Massart, A. Dijkstra and L. Kaufman, *Evaluation and Optimization of Laboratory Models and Analytical Procedures*, Elsevier, Amsterdam, 1978.
8 G.B. Dantzig, *Linear Programming and Extensions*, Princeton University Press, Princeton, NJ, 1963.
9 W. Spendley, G.R. Hext and F.R. Himsworth, Sequential application of Simplex designs in optimisation and evolutionary operations, *Technometrics*, 4 (1962) 441–461.
10 J.A. Nelder and R. Mead, A Simplex method for function minimisation, *Computer Journal*, 7 (1965) 308–313.
11 D.E. Long, Simplex optimisation of the response from chemical systems, *Analytica Chimica Acta*, 46 (1969) 193–206.
12 O.L. Davies (Editor), *Design and Analysis of Industrial Experiments*, Oliver Boyd, London, 1984.
13 S. Addleman, Orthogonal main—effect plans for asymmetrical factorial experiments, *Technometrics*, 4 (1962) 21–58.
14 M.W. Routh, P.A. Swartz and M.B. Denton, Performance of the super modified Simplex, *Analytical Chemistry*, 49 (1977) 1422–1428.
15 P.F.A. van der Wiel, Improvement of the super-modified Simplex optimisation procedure, *Analytica Chimica Acta*, 122 (1980) 421–433.
16 R.R. Ernst, Measurement and control of magnetic field homogeneity, *Review of Scientific Instruments*, 39 (1968) 988–994.
17 S.L. Morgan and S.N. Deming, Optimisation strategies for the development of gas–liquid chromatographic methods, *Journal of Chromatography*, 112 (1975) 267–285.
18 P.G. King, *Ph.D. Dissertation*, Emory University, Atlanta, GA, 1974.

Chemometrics and Method Development in High-performance Liquid Chromatography

Part 1: Introduction

JOHN C. BERRIDGE

Analytical Chemistry Department, Pfizer Central Research, Sandwich, Kent, CT13 9NJ (U.K.)

(Received 27 April 1987; accepted 16 September 1987)

CONTENTS

1 Introduction and scope	139
2 The problems of method development in HPLC	140
2.1 The 'black box' of chromatography	140
2.2 The variables useful in separation development	141
2.3 Assessment of chromatographic separations	143
3 Approaches to separation optimisation	146
4 Definition and location of domain	147
5 Conclusions	151
References	151

1 INTRODUCTION AND SCOPE

Method development in high-performance liquid chromatography (HPLC) is a difficult and often bewildering task for the novice who will often approach the problem in a very much 'trial and error' manner. It is vital to recognise the limitations of such an attitude to method development and to appreciate the advantages offered by the application of chemometric principles. Comprehensive texts are available which deal with the instrumentation of chromatography [1], the optimisation of the chromatographic selectivity [2] and the implementation of automated methods for separation optimisation [3]. The objectives of this tutorial and the ones that follow are to explain why every chromatographer should be interested in applying chemometric concepts to method development, to discuss the ideas that are most likely to prove useful and to demonstrate how they might be applied in practice. It is not intended to review the current literature and so the references cited include only those which provide a more detailed discussion of the subject matter. The applications referred to are taken entirely from HPLC but the ideas and concepts are of wider interest to all interested in liquid chromatography or involved in experimental design.

2 THE PROBLEMS OF METHOD DEVELOPMENT IN HPLC

2.1 The 'black box' of chromatography

Chromatography, like other analytical methods, can be represented as the system shown in Fig. 1. At the centre is the chromatograph into which the sample is fed. Separation of the eluites is a function of the controllable variables (e.g. mobile phase flow-rate and composition, column chemistry, separation temperature etc.) and the uncontrolled variables (noise, drift, column performance). The eluites must be detected as they leave the column and their interaction with the detection system is presented as the chromatogram. Provided there is some means of objectively assessing the separation quality, optimisation of the separation can be achieved with little consideration of the fundamental aspects of these variables but the process will be more effective and efficient with a full consideration of their relevance.

The chromatographic separation between components is described by the resolution (R) between the peaks of the chromatogram:

$$R = \frac{(\alpha - 1)N^{1/2}k'}{2(\alpha + 1)(1 + k')} \qquad (1)$$

where k' is the average capacity factor between the adjacent peaks, α is the selectivity (k'_2/k'_1) and N is the number of theoretical plates provided by the column (a measure of the efficiency of the column). The capacity factor, k', is defined as

$$k' = (t - t_0)/t_0 \qquad (2)$$

where t is the peak retention time and t_0 is the column void time.

Fortunately the three terms of eq. 1 can be optimised independently although they are themselves not entirely independent. The plate count (N) is essentially fixed by the choice of column but it will be a function of the eluite, the temperature and consequent viscosity of the mobile phase and the performance of the chromatograph being used. If, at the end of the optimisation process, the plate count is either too high or too low, then the most practical way of providing an appropriate plate count is to shorten or lengthen the column. Simply increasing the column plate count is not a practical process for improving resolution because of the square root relationship it has with resolution (Fig. 2a) but a full discussion of its optimisation is outside the scope of this present tutorial.

The capacity factor is an expression of thermodynamic effects but, unfortunately, provides little scope for resolution optimisation as can be seen from Fig. 2b. The optimum range for the capacity factors of the most difficult peak pair to resolve is in the range 1-10 with 4 being an ideal compromise between resolution, speed and sensitivity:

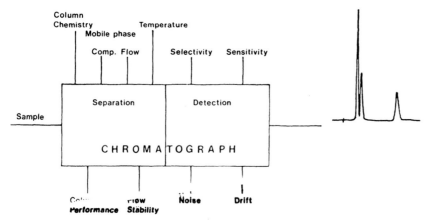

Fig. 1. A schematic representation for a chromatographic system with (above) the controlled (variable) factors and (below) the uncontrolled factors.

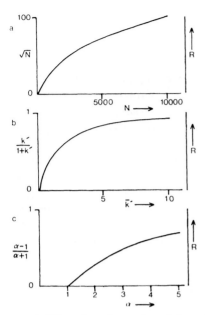

Fig. 2. Effect of (a) the number of theoretical plates (N), (b) the average capacity factor (k') and (c) the relative retention (selectivity) (α) upon resolution as defined by eq. 1.

found that the discrete variables are associated with decisions related to method development while the continuous variables are considered as part of separation optimisation. Fig. 3 [4] illustrates the major variables that are normally considered in separation development and optimisation. The discrete variables include the column parameters (length, diameter), the nature of the stationary phase (support properties, particle size and surface chemistry) and the components of the mobile phase (e.g. water and organic modifiers for reversed-phase separations). The continuous variables include the composition of the mobile phase (solvent strength, concentration of modifiers and/or ion interaction agents), the flow-rate and temperature. However, it will be recognised that this classification is, in itself, insufficient since there are some more fundamental decisions to be made about the mode of chromatography that will

extending retention beyond this will not provide significantly increased resolution. Contrary to the advice often found in texts and papers on separation optimisation, it does not mean that all peaks should be eluted in this range. If the retention of the most difficult peak pair has to be extended to $k' = 10$, then the chromatographic run time may have to be extended to $k' = 20$ or 30 if this peak pair happens to be due to the first eluted components.

The remaining term of eq. 1 is the selectivity term, α. While selectivity will be determined by the choice of the chromatographic mode and the column chemistry, it is the optimisation of the mobile phase composition that provides the greatest opportunity for maximising selectivity in liquid chromatography. From Fig. 2c it can be seen that effort applied to increasing selectivity gives an almost linear return in increased resolution so it is to α that most attention will be given.

2.2 The variables useful in separation development

Experimental variables are often classified as discrete or continuous. In practice it is most often

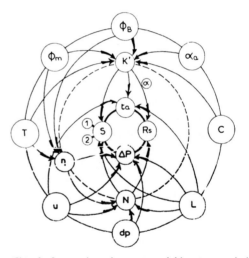

Fig. 3. Interactions between variables (outer circle), intermediate parameters (dotted circle) and separation quality criteria (inner circle). Variables: Φ_B = volume fraction of strongest solvent; Φ_m = volume fraction of a modifying solvent; T = temperature; u = flow (linear) velocity; d_p = particle diameter; L = column length; α_a = adsorbent activity; C = carbon content of alkyl-bonded stationary phase (or counter-ion concentration). Intermediate parameters: α = selectivity; k' = capacity factor; N = number of theoretical plates (column efficiency); n = solvent viscosity. Quality criteria: t_a = analysis time; R_S = resolution; P = pressure drop; S = peak height (sensitivity) for a low-concentration eluite available either (1) in a large amount or (2) in a limited amount. (Reproduced from ref. 4, with permission.)

be employed and the hardware (e.g. method of detection) that is most appropriate. Nevertheless, from this can emerge a more systematic approach to the development and optimisation of the separation: it follows the basic steps of good experimental design.

(i) Understand the problem. What information is available about the sample? Does this define the chromatographic mode (e.g. normal or reversed-phase, gel permeation etc.) that will have to be used? What hardware will be required (column, detection method)? What criteria will be used to judge a successful outcome of the method development and optimisation?

(ii) Devise a plan to solve the problem that is appropriate to the information available from step (i). What are the important variables to be considered and over what ranges?

(iii) Carry out the experimental optimisation to maximise the optimisation criterion.

(iv) Review the result. Can the separation be further optimised by considering the hardware (e.g. will a shorter column give a satisfactory result with a consequent saving in time and cost)? Has the problem been answered or do the data suggest a redefinition of the original problem is required: this then takes us back to step (i) above. Is it possible to extract further information by applying a different data processing technique?

It is practically impossible to develop a completely successful separation for a totally unknown sample. It is certainly impossible to fully optimise a separation without knowing exactly what is in the sample and what the desired outcome of the optimisation is to be. Some information on the nature of the sample must be gained before even the mode of chromatography is selected. For example, solubility in a polar solvent (water, methanol) suggests reversed-phase chromatography while solubility in hexane would suggest starting with normal phase. However, it is evident from the literature that more than 75% of separation problems are amenable to reversed-phase chromatography. This then can form the basis of the first decision with respect to a discrete variable — the column choice. In practical terms the obvious choice will be a reversed-phase column (C-8 or C-18) with 10,000 to 15,000 theoretical plates, i.e. a column of 15–25 cm packed with 3- or 5-μm particles. Chemometrics does not, however, always support the obvious choices. Having studied the relative discriminating power of stationary phases, Massart and co-workers [5,6] concluded that a cyanopropyl column was the preferred stationary phase. Coupled with a restricted set of mobile phases comprising n-hexane–methylene chloride–acetonitrile–propylamine (50 : 50 : 25 : 0.1) and acetonitrile–water–propylamine (90 : 10 : 0.01) the column can be used in reversed-phase and normal-phase modes respectively. Solvent strength is adjusted by varying the proportions of hexane or water. These combinations were applied to the separation of some 100 basic drugs but an extension to six solvents was needed to cope with acidic and neutral compounds. With the same column, normal- and reversed-phase eluents were made with hexane and water as solvent strength modifiers, with methanol, acetonitrile, tetrahydrofuran and dichloromethane as selectivity modifiers.

The choice of detection method is extremely difficult. Most modern liquid chromatographs are equipped with a single channel variable wavelength ultraviolet detector. These detectors are the mainstay although many involved in protein and peptide separations will choose a fixed wavelength detector since these offer the attractions of both lower cost and higher sensitivity. However, for method development, the extra advantages that accrue from the use of multi-channel ultraviolet detection cannot be overlooked and diode-array detection is most frequently cited. With multi-channel detection comes the possibility of tracking individual peaks during the separation optimisation and, as will be discussed in the next tutorial of this series, peak tracking is an essential part of several optimisation schemes. There are, however, exciting advances being made in artificial intelligence such that guidance is now available on the selection of detection conditions [7] and this is likely to be a rapidly expanding field as expert systems are applied to all areas of chromatographic method development [8].

The other hardware needed is simply that normally associated with HPLC. While fully automated schemes may require powerful computer controllers, autosamplers and pumps capable of delivering three or four solvents, much can be done with a simple isocratic system although, as a minimum, a two solvent system is recommended.

Having defined the chromatographic hardware, the mode of chromatography to be employed and the column to be used, the next step may appear to be to provide some solvents, inject the sample and experiment. This is the traditional 'trial and error' approach to separation optimisation which chemometrics so clearly demonstrates to be an inefficient and dangerous method. Before starting the first experiment it is necessary to decide by what criterion a satisfactory separation will be judged.

2.3 Assessment of chromatographic separations

An objective assessment of a chromatogram is essential in method development and optimisation since, without it, no judgement can be made as to a satisfactory outcome. As seen from eq. 1, resolution provides a description of the separation of peaks but, on its own, resolution may not be enough to describe the quality of the chromatogram in its entirety. There will normally be many peaks to be considered, each pair with its own resolution and some means needs to be found for describing the adequacy of their separations. It may also be important to consider just how many peaks appear in the chromatogram, their relative spacing and the time the total separation took. Furthermore, if some formal optimisation scheme is to be employed, it will be essential to express the quality of any separation as a single number which the scheme can use to establish optimum separation conditions.

As a major requirement in any separation development is to maximise selectivity, it may at first appear attractive to use selectivity itself (α) as a measure of chromatogram quality. Unfortunately, the converse usually turns out to be the case. From eq. 2 it was seen that selectivity is simply defined as the ratio of two capacity factors. If the capacity factor of the first peak is zero, then α immediately has a value of infinity — what better result could be desired? This is the pitfall of using α as a measure of separation since, for practical use, selectivity must be accompanied by retention. An additional drawback of using α is that it does not consider peak shapes. High α values accompanied by poor efficiency will not provide an adequate ultimate separation. Finally, the calculation of α requires knowledge of the column void volume: this is difficult enough in conventional reversed-phase chromatography. In complexation chromatography, such as when using columns employing cyclodextrins as the stationary phase, a method of accurately determining this parameter still has to be found. What is therefore required is measurement to reflect both separation and peak shape.

In Fig. 4, the main approaches to the determination of peak separation are shown. If the assessment is to be carried out manually, then the peak separation function of Kaiser [9] $P = f/g$ is easy to determine and robust. Unfortunately, P turns out to be very difficult to calculate automatically and an alternative needs to be found. An additional weakness of using P is that its value becomes progressively more difficult to determine, and its performance begins to fail, for values below 0.75 ($R < 1.0$) making it less satisfactory for the more difficult separations. Resolution would appear to be the ideal way of expressing the separation between peaks but, in practice, it is often very difficult to calculate. Use of eq. 1 requires that the column void volume and efficiency are known, both difficult to determine. From Fig. 4 it is necessary only to determine peak retention times and base widths but how wide is a chromatographic peak? This can only be found

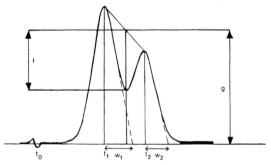

Fig. 4. Measurements of peak separation: P (peak separation) $= f/g$, S (separation factor) $= (t_2 - t_1)/(t_2 + t_1)$, R (resolution) $= 2(t_2 - t_1)/(w_1 + w_2)$, k'_1 (capacity factor) $= (t_1 - t_0)/t_0$, α (relative retention or selectivity) $= k'_2/k'_1$.

approximately after making assumptions about the shape of the peak. Some modern computing integrators do provide information of peak widths which can be used directly in determining resolution. For those which do not, it is possible to make some approximations which, while reducing the accuracy of the determination, are sufficient to provide a reproducible method of use in separation optimisation. The assumption made is that the chromatographic peaks are Gaussian. In this case the peak width is a function of both its area (A) and height (h):

$$w = 4A/(2\pi)^{0.5} h \qquad (3)$$

Eq. 3 assumes symmetrical, Gaussian peaks but these are rarely achieved in practice and so the base width will be underestimated. An approximate modification to this equation provides a readily obtained estimate of the peak width:

$$w = 2A/h \qquad (4)$$

Eq. 4 essentially treats the chromatographic peak as a triangle and will reflect variations in peak shape. It is not an accurate method of determining peak width but has successfully been used in resolution estimation in the fully automated optimisation of separations. It has the merits of being very easily obtained from most modern integrators which calculate both peak heights and areas.

The peak separation term, S, is a highly recommended method for assessing separation since it can be found without the need to know the column void volume. In a first approximation, $S =$ $R/\sqrt{\ }$ ... therefore a more easily derived measure of R for constant plate counts. However, especially in liquid chromatography, many optimisation problems will encounter far from constant plate counts. Nevertheless, under circumstances where plate counts are relatively constant, the use of S can be very convenient — either in its own right or as an alternative route to R.

Each of these terms considers only a single peak pair but the objective must be to devise a method for assessing the whole chromatogram. A simple way would be just to consider the separation between the worst separated pair, ignoring all others. If a set of chromatograms is to be compared, then this is a suitable approach but it does not provide a suitable criterion if it is desired to locate a single (global) optimum set of separation conditions since very different peak pairs may show the lowest separation in adjacent chromatograms and many optima will be indicated, rendering it impossible for a searching or modelling procedure to make effective progress.

The sum of all the resolutions will reflect gradual improvement in separations but, on its own, is of little value. Two peaks that are well resolved and easy to separate will dominate the sum and optimisation may result in their being over-separated while the most difficult pair to separate is ignored. One way of avoiding this is to sum the resolutions but to limit the maximum resolution that can be assigned to any peak pair. Thus once the resolution exceeds baseline ($R \geq 1.5$) any further separation is ignored so that attention can be focussed on lesser separations. This will certainly work better but will still suffer if there are a number of peaks which are poorly resolved. The maximum value assignable to R can be reduced (e.g. to 1.0) to reduce the number of peaks which continue to receive assessment as the separation improves.

Rather than add the individual resolution values calculated, a resolution product [10] can be used:

$$R_S = \Pi_i R_{S i+1, i} \qquad (5)$$

In this case the aim is to space the peaks evenly throughout the chromatogram since the lowest value of resolution has a dominant effect. A simple resolution product may still give a higher assessment to an inferior chromatogram but this can be overcome by using the relative resolution product:

$$r = \Pi_{i=1} R_{S+1,i} / \left[\left(\Sigma_{i=1} R_{S+1,i} \right) / (n-1) \right]^{n-1} \qquad (6)$$

where the denominator denotes the maximum possible value for ΠR_S in the given chromatogram [11]. The relative resolution product has the value of zero when a peak pair is coincident and a value of unity when all peaks are evenly spaced throughout the chromatogram. Unfortunately, even this function can produce inferior chromatograms since it takes no account of when the first peak appears in the separation. A final refinement is then to include an imaginary peak, indicated by $i = 0$, eluting at a desired point in the chromatogram. The equation then becomes:

$$r^* = \Pi_{i=0} R_{Si+1,i} / \left[(\Sigma_{i=0} R_{Si+1,i}) / (n-1) \right]^{n-1} \quad (7)$$

where $R_{Si,0} = (t_1 - t_0) N^{0.5} / 2(t_1 + t_0)$.

This final criterion then aims at spacing all peaks evenly early in the chromatogram. There is no consideration given to the ultimate duration of the separation. To consider the time a separation takes requires addition of a second term.

Multiple criterion decision making is a difficult subject and, in chromatography, is fraught with problems. It may seem a simple thing just to add a time-related term to an assessment of separation [a chromatographic response function (CRF)] [12], e.g.:

$$\text{CRF} = \Sigma_i \ln(P_i / P_0) + A(t_m - t_n) \quad (8)$$

where P_i is the actual and P_0 the desired separation between peak pairs, t_m is the desired analysis time, t_n the elution time of the last peak and A is a weighting factor. The function is constrained such that if $P_i > P_0$ then P_i is set equal to P_0. A second constraint is to set $t_m = t_n$ for all chromatograms for which analysis is completed earlier than the maximum time: this has the effect of neglecting time terms for all separations that meet the time criterion. A very similar criterion [chromatographic optimisation function (COF)], but which uses resolution, is [13]:

$$\text{COF} = \Sigma_i A_i \ln(R_i / R_{id}) + b(t_m - t_n) \quad (9)$$

where R_i is the actual resolution of the ith pair and R_{id} is the desired resolution. In this case the weighting A was included such that [...] between [...] could be made more [...] important.

Other terms can be added to assessment criteria such as noise (n_i) [14]:

$$\text{CRF} = 1/t_{95} \Pi_i f_i / (g_i + 2 n_i) \quad (10)$$

where t_{95} is 0.95 of the analysis time.

None of the terms presented so far takes into account the number of peaks found in the chromatogram — the objective may well be to produce the maximum number of peaks possible. A simple solution [15] would be to use:

$$\text{CRF} = \Sigma_i \ln(f/g) - 100(M - n) \quad (11)$$

where M is the maximum number of peaks expected (or a suitably high value if the precise number is unknown) and n is the number of peaks found. If resolution, time and peak numbers are all important then the following function has been successfully employed in unattended separation optimisation [16]:

$$\text{CRF} = \Sigma_i R_i + n^2 - a|t_m - t_n| - b(t_0 - t_n) \quad (12)$$

There are, unfortunately, some problems associated with these multiple term equations. Multiplying the final term of eq. 8 out, it can be seen that:

$$A(t_m - t_n) = \text{constant} - A t_n$$

so one term of the equation is essentially redundant. This is not the case for the first time term of eq. 12 (it is for the second!) and is not entirely the case for eq. 9 where time is ignored if it is less than the maximum specified. So some terms included in response functions are of little, if any, real value in objective assessment. A second problem is the one of mixing terms with very different units. All the multiple equations do this, for example dividing resolution by time or by subtracting the number of peaks from resolution. The net effect can be to cause multiple optima to be suggested depending upon the dominant term. This is a general point, not restricted to chromatography, that the shape and form of the response surface is a direct function of the objective criterion being used and different criteria will provide different respon[...]

The final selection of a criterion is a difficult decision. It must reflect the needs indicated in the original specification of the problem and the response surface that the chosen criterion generates must be valid both for the problem and for the optimisation procedure that will be used. For search procedures, the smoother the surface can be made, the better. Additionally, a criterion must be chosen that maintains a stable surface. Time normalised criteria or criteria normalised on the number of peaks detected, give rise to complex response surfaces, even when peak crossovers do not occur and are not well suited to sequential optimisation methods. Criteria based simply on the sum of separations must be avoided as they can be severely misleading but product criteria or criteria concerned with the minimum separation achieved for a given peak pair can be used. For methods that require response surface modelling (employing simultaneous experimental designs) the chromatographer is free to experiment with almost any response function amenable to the data available, bearing in mind the caveats discussed already.

3 APPROACHES TO SEPARATION OPTIMISATION

Separation optimisation is the next step in chromatographic method development after choosing the parameters and selecting an appropriate criterion. Parameter selection involves also choosing the ranges over which they can be varied: these limits define the parameter space or domain. Selecting two parameters defines a two-dimensional domain and a procedure suitable for at least two-dimensional optimisation should then be used. However, care needs to be taken that these parameters are not directly connected. For example, in mobile phase optimisation, a ternary mobile phase provides only two degrees of freedom (since the proportion of the third solvent is simply 100% − % solvent A − % solvent B) and is therefore a two-dimensional problem. While the response surface will be a function of the optimisation criterion selected, its shape will almost certainly not be simple. A simple, symmetrical surface (Fig. 5a) is amenable to univariate optimisation and can be tackled in the laborious way still advocated by many classical experimentalists. Thus all variables but one (a) are held constant and the single variable is optimised. Having located this optimum, the variable (a) is held constant and (b) is optimised: the global optimum will have been found. Unfortunately, HPLC optimisation is characterised by a high degree of interdependence between variables and the response surface will not be amenable to a single, univariate optimisation and several attempts will be needed (Fig.

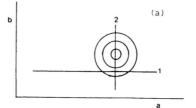

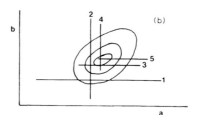

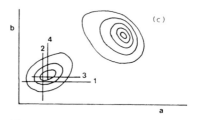

Fig. 5. Response surfaces and the application of univariate optimisation methods: (a) a simple surface with two independent parameters; (b) a simple surface with two dependent parameters; (c) a complex surface with two dependent parameters. Univariate optimisation locates the optimum in (a), requires several attempts for (b) and fails to encounter the area containing the global optimum at all in (c), locating a local optimum after several iterations.

5b). The real world of HPLC optimisation is more complex still and could well contain two or more optima which simple approaches will fail to reveal (Fig. 5c).

Most optimisation methods can be classified in one of two categories, sequential or simultaneous, although there is at least one method which combines elements of both. Sequential methods aim to approach optimum conditions in a stepwise manner, using the results of each experiment to guide the subsequent progression. Probably the best known method is the sequential simplex procedure which has been the subject of earlier tutorials and which will be discussed in more detail in the second tutorial of this series. Also in the following tutorial will be discussed a hybrid method in which the response surface is modelled and interpreted at each stage in the experimentation. This method is referred to as the method of 'iterative regression' or 'phase selection diagrams' [10,11,17]. It is generally more efficient than the simplex procedure for one or two variables but requires that eluites be tracked in subsequent separations so that the response surface can be correctly interpreted: peak tracking is not a requirement of the simplex procedure.

Simultaneous methods proceed according to a fixed experimental design, established prior to any experimentation. The objective is to conduct sufficient experiments in order to be able to model the response surface and predict the location of the optimum. Factorial and simplex lattice designs are well known examples of these and will be discussed in the third tutorial. Fixed experimental designs are also valuable to investigate the proposed parameters to establish their contribution to selectivity and hence to potentially reduce either the number of variables or the range over which they might vary. Reduction of parameter space is important if efficient optimisation is to be conducted and this will be considered next by illustrating the use of gradient elution to restrict the area that has to be searched for location of the optimum.

4 DEFINITION AND LOCATION OF DOMAIN

If separation optimisation is to be achieved efficiently and with a high chance of locating the global optimum, it is essential that the parameter space to be investigated is restricted to a manageable number of dimensions and to a realistically small size. The number of variables is a matter of judgement although factorial designs can be used to provide information on the variables contributing most to selectivity. Most frequently, gradient elution is used, particularly in reversed-phase chromatography, to limit the domain within which the optimum must lie.

Gradient elution is the term used to indicate that the composition of the mobile phase changes with time. It is usual to increase the eluting power of the mobile phase and this is usually achieved in reversed-phase chromatography by adding increasing amounts of methanol, acetonitrile or tetrahydrofuran to water and for normal phase by adding methylene chloride or chloroform to hexane. It is possible to use the results from an initial gradient separation to predict the appropriate isocratic conditions (and vice versa). A detailed discussion of the theory and properties of gradients is beyond the scope of this tutorial, we will concern ourselves simply with using gradients to predict isocratic requirements.

By definition, a linear solvent strength gradient obeys:

$$\log k_i = \log k_0 - b(t/t_0) \quad (13)$$

where k_i is the capacity factor the eluite would be expected to have with the isocratic mobile phase corresponding to that at the inlet at time t, k_0 is the capacity factor for the initial mobile phase composition, t_0 is the column hold-up time and b is the rate of change of the composition. The retention time (t_g) of a given eluite during the gradient can be found from:

$$t_g = (t_0/b) \log(2.3 k_0 b + 1) + t_0 + t_d \quad (14)$$

where t_d is the delay time of the system corresponding to the time taken for a change in composition to reach the head of the column. The instantaneous k' value of a solute as it leaves the column (k_f) is given by:

$$k_f = 1/(2.3b + 1/k_0) \quad (15)$$

Retention (k') in reversed-phase chromatography as a function of composition can be approximated by:

$$\log k' = \log k_w + S\Phi_b \quad (16)$$

where k_w is the capacity factor in pure water. If this linear relation is assumed, retention will change linearly with a linear change in elution power of the mobile phase, i.e.,

$$\Phi_b = \Phi_0 + \Phi' t \quad (17)$$

where Φ_b is the composition at time t, Φ_0 is the starting composition and Φ' is the rate of composition change.

The optimum value of Φ' depends upon the

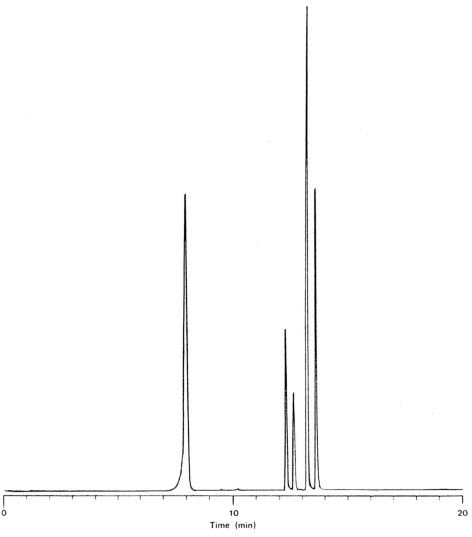

Fig. 6. Gradient separation of 6 components. Column, 11 cm Partisphere C-18, mobile phase A = water, B = methanol, linear gradient from A to B over 15 min. Flow-rate, 1 ml/min.

sample but it has been demonstrated that an optimum value for b is 0.2. It can also be shown that:

$$\Phi' = b/(St_0) \qquad (18)$$

For small molecules and a gradient of methanol-water, a value of 3 can be assumed for S although S does depend both upon retention and molecular weight.

The gradient (rate of change) can now be calculated as:

% methanol/min = $100 \times 0.2/(3t_0)$

giving a rate of change of 6.67%/min for a column with a hold-up time of 1 min.

Practical use of gradient elution as a 'scouting procedure' — a method of establishing the solvent strength(s) required to elute the compounds of interest — requires a linear gradient from 0 to 100% methanol at the rate described by eq. 18. Fig. 6 shows a gradient separation of six components under such conditions. A linear gradient from 0 to 100% methanol over 15 min was used, a slightly slower rate of change than is predicted from eq. 18 which suggests for the column used (hold-up volume of 0.8 ml, flow-rate = 1 ml/min) a 12-min gradient should be optimum. A rule of thumb is that a capacity factor, under isocratic conditions, of approximately 4 will be provided by the mobile phase composition being delivered at time $t_g - 2.5t_0 - t_d$. The third peak, having a retention of 12.6 min, therefore requires an isocratic composition of $(12.6 - 2.5 \times 0.8 - 0.7) \times 100/15 = 66\%$ methanol to provide a $k' = 4$. The separation using 66% methanol is shown as Fig. 7 and the third peak has a capacity factor of 3.5, an adequate correlation. Methanol is the preferred solvent to use for gradient to isocratic extrapolation since it has been found that the correlation achieved with acetonitrile gradients is less good. It is possible to calculate other requirements for different capacity factors using the equations given above and thereby obtain the limits of methanol–water that will provide elution of the detected components within a capacity factor range $1 < k' < 10$. An important point emerges from Fig. 7. The sample used is known to contain six components: all six are adequately resolved in the methanol–water isocratic separation. There is therefore no need to continue with any further selectivity optimisation if the chromatogram shown meets the requirements as originally specified (underlining the need to have a specification of the problem before experimentation commences). It may be that a faster separation is required or a separation with a different elution order would be preferable and in such cases further optimisation will be needed.

Unfortunately, if peptides and proteins are

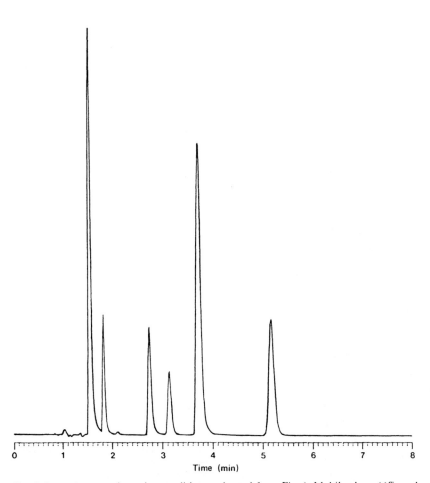

Fig. 7. Isocratic separation using conditions estimated from Fig. 6. Mobile phase 66% methanol in water.

being considered, then the above gradient model is subject to some errors. Eq. 14 needs modifying to account for the elution of the larger molecules at a time t_m before t_0. For small molecules, $t_s = t_0$.

$$t_g = (t_0/b) \log[2.3k_0 b(t_s/t_0) + 1] + t_s + t_d \tag{14a}$$

Similar arguments to those given above can now be used to predict isocratic retention times although much higher values of S are required. (For acetonitrile–water gradients, it has been demonstrated that $S = 0.48 M^{0.44}$ where M is the molecular weight [18]: values for methanol gradients will be similar.)

If two gradients are run with different durations (i.e. different values of b), since the ratio of b values is the ratio of the gradient times, the two equations analogous to eq. 14a can be solved by numerical methods to provide an estimate of b_1, b_2 and k_0. The isocratic retention can be predicted directly and with a greater degree of accuracy than before. Indeed, the running of two gradients is to be recommended for both small and large molecules and it has been demonstrated that this approach is applicable to reversed-phase and ion-exchange separations.

An alternative method of establishing this range is to use the graphical method developed by Schoenmakers et al. [19]. The exact mathematics of the procedure are described but, for manual estimation, the graphical method is easier to apply. Fig. 8 shows a plot of the volume fraction of methanol required to provide a given capacity factor from the net retention time under gradient elution. The assumption is made of a linear gradient of 0–100% methanol over 15 min and that the column hold-up time is 125 s. To apply the method with any column, the flow-rate must be adjusted to give this hold-up time. With the same solutes and column as before, a gradient separation at reduced flow (0.4 ml/min, $t_0 = 125$ s) is shown in Fig. 9. Note that the lower flow-rate has given an improved separation compared to Fig. 6 even though the rate of change was identical demonstrating the need for gradient optimisation and that the rate used was not optimum. In Fig. 10 two vertical lines have been drawn corresponding

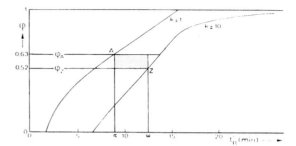

Fig. 8. Curves relating isocratic composition to the net retention under gradient elution for various values of the isocratic capacity factor. Assumes a linear gradient from 100% water to 100% methanol over 15 min with $t_0 = 125$ s. (Reproduced from ref. 19, with permission.)

to the net retention times $(t_g - t_0 - t_d)$ of the first and last eluted peaks and the range of methanol required to elute all components within that capacity range is indicated. The values show that all components can be eluted isocratically within a capacity factor range of 1–10 but only just. Had the values been reversed (i.e. a stronger eluent required to elute the final peak at $k' = 10$) then an

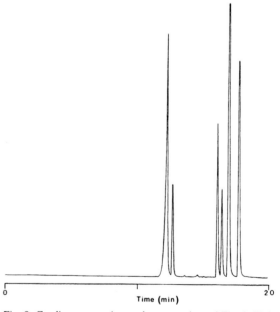

Fig. 9. Gradient separation under constraints of Fig. 8. Eluites as in Fig. 6; flow-rate, 0.4 ml/min. The first peak is eluted at 12.2 min, the last at 17.7 min corresponding to net (adjusted) retention times of 8.5 and 14 min respectively.

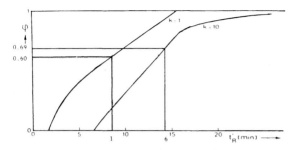

Fig. 10. Use of Fig. 8, relating isocratic composition required for $k' = 1$ and $k' = 10$ to the net retention achieved in Fig. 9.

isocratic separation in this range would not be possible. That the peaks are spread over approximately this capacity factor range is evident from Fig. 7.

Optimisation of the separation can proceed with 'fine tuning' of the modifier content of the mobile phase to achieve the desired retention and resolution. It will often be that further selectivity effects will need to be exploited, for example the use of ternary or quaternary mobile phases, the control of pH etc. In particular the use of ion-pairing will be frequently required and will need careful 'fine tuning' as a consequence of the high degree of interdependence between pH, the degree of ion-pairing and the concentration of the ion-interaction reagent.

Techniques for the systematic optimisation of multiple variables are the subjects of future tutorials.

5 CONCLUSIONS

Efficient method development and optimisation of HPLC separations demands an ordered and systematic approach. Any knowledge available about the sample (molecular weight, solubility, number of components) should be combined with the objectives of the separation to define the chromatographic mode to be used. An objective criterion is then needed for evaluation of separation qualities. Depending upon the sample, the goal and the criterion, optimisation can proceed by a parallel or serial route. However, for maximum efficiency, it is valuable to limit the parameter space to be investigated before optimisation commences — this is conveniently done with a gradient scan. The full capabilities of selectivity can then be exploited, if required, by use of one of the systematic schemes which will be discussed in the two following tutorials.

REFERENCES

1 N.A. Parris, *Instrumental Liquid Chromatography*, Journal of Chromatography Library Vol. 27, Elsevier, Amsterdam, 2nd ed., 1984.
2 P.J. Schoenmakers, *Optimisation of Chromatographic Selectivity. A Guide to Method Development*, Journal of Chromatography Library Vol. 35, Elsevier, Amsterdam, 1986.
3 J.C. Berridge, *Techniques for the Automated Optimisation of HPLC Separations*, Wiley, Chichester, 1985.
4 J.P. Bounine, G. Guiochon and H. Colin, A simple pragmatic optimization procedure for some parameters involved in high-performance liquid chromatographic separations: column design, temperature, solvent flow-rate and composition. *Journal of Chromatography*, 298 (1984) 1–20.
5 G. Hoogewijs and D.L. Massart, Development of a standardised analysis strategy for basic drugs using ion-pair extraction and high-performance liquid chromatography — I. Philosophy and selection of extraction techniques, *Journal of Pharmaceutical and Biomedical Analysis*, 1 (1983) 321–329 and II. Selection of preferred HPLC systems, 1 (1983) 331–337.
6 M. De Smet, G. Hoogewijs, M. Puttemans and D.L. Massart, Separation strategy of multicomponent mixtures by liquid chromatography with a single stationary phase and a limited number of mobile phase components. *Analytical Chemistry*, 56 (1984) 2662–2670.
7 G. Musch and D.L. Massart, Expert system for pharmaceutical analysis. II. Relative contribution of and rule validation for amperometric detection (oxidation mode), *Journal of Chromatography*, 370 (1986) 1–19.
8 Focus, Expert systems for liquid chromatography, *Analytical Chemistry*, 58 (1986) 1192A–1200A.
9 R.E. Kaiser, *Gas Chromatographie*, Geest und Portig, Leipzig, 1960.
10 P.J. Schoenmakers, A.C.J.H. Drouen, H.A.H. Billiet and L. de Galan, A simple procedure for the rapid optimisation of reversed-phase separations with ternary mobile phase mixtures. *Chromatographia*, 15 (1982) 688–696.
11 A.C.J.H. Drouen, P.J. Schoenmakers, H.A.H. Billiet and L. de Galan, An improved optimisation procedure for the selection of mixed mobile phases in reversed-phase liquid chromatography, *Chromatographia*, 16 (1982) 48–52.
12 M.W. Watson and P.W. Carr, Simplex algorithm for the optimisation of gradient elution high-performance liquid chromatography, *Analytical Chemistry*, 51 (1979) 1835–1842.
13 J.L. Glajch, J.J. Kirkland, K.M. Squire and J.M. Minor, Optimization of solvent strength and selectivity for re-

versed-phase liquid chromatography using an interactive mixture-design statistical technique, *Journal of Chromatography*, 199 (1980) 57–79.
14 W. Wegscheider, E.P. Lankmayr and M. Otto, Relationships between chromatographic response functions and performance characteristics, *Analytica Chimica Acta*, 150 (1983) 87–103.
15 J.H. Nickel and S.M. Deming, Use of the sequential simplex algorithm for improved separations in automated liquid chromatographic method development, *LC, Liquid Chromatography and HPLC Magazine*, 1 (1983) 414–417.
16 J.C. Berridge, Unattended optimisation of reversed-phase high-performance liquid chromatographic separations using the modified simplex algorithm, *Journal of Chromatography*, 244 (1982) 1–14.
17 A.C.J.H. Drouen, H.A.H. Billiet and L. de Galan, Practical procedure for the optimization of reversed-phase separations with quaternary mobile phase mixtures, *Journal of Chromatography*, 352 (1986) 127–139.
18 M.A. Stadalius, H.S. Gold and L.R. Snyder, Optimization model for the gradient elution separation of peptide mixtures by reversed-phase high-performance liquid chromatography. Verification of band width relationships for acetonitrile–water mobile phases, *Journal of Chromatography*, 327 (1985) 27–45.
19 P.J. Schoenmakers, H.A.H. Billiet and L. de Galan, Use of gradient elution for rapid selection of isocratic conditions in reversed-phase high-performance liquid chromatography, *Journal of Chromatography*, 205 (1981) 13–30.

Chemometrics and Method Development in High-performance Liquid Chromatography

Part 2: Sequential Experimental Designs

JOHN C. BERRIDGE

Analytical Chemistry Department, Pfizer Central Research, Sandwich, Kent, CT13 9NJ (U.K.)

(Received 6 June 1988; accepted 18 November 1988)

CONTENTS

1 Introduction 153
2 Response functions for sequential designs 154
3 The sequential simplex procedure 155
    3.1 Selection of algorithm 155
    3.2 Applications 156
    3.3 Summary — advantages and disadvantages 159
4 Iterative mixture designs 160
    4.1 Principles 160
    4.2 Applications 161
    4.3 Summary 163
5 Conclusions 164
References 164

1 INTRODUCTION

In Part 1 of this series of tutorials [1], the potential for the systematic application of chemometric principles to method development in high-performance liquid chromatography (HPLC) was introduced. In particular, the problems of selecting variables of value in the development process and assessing their impact upon the separation (by use of an objective function — an optimisation criterion) was discussed. It will shortly be seen that such optimisation criteria are essential for the successful use of sequential experimental designs and it is these designs that are the subject of the following tutorial. Two methods will be discussed in some detail, the use of the simplex algorithm and iterative mixture designs. As in the previous tutorial, selected examples from the literature are included for reference since this is a "how do I do it?" discussion rather than a critical review of available methods.

There are other sequential methods which readers may encounter but which will not be described in detail. These are the grid search method, as used in the Pesos development system (Perkin Elmer) [2] and the Optim method of 'adaptive intelligence' (Spectra-Physics) [3,4]. The Pesos method does not obviously use chemometric ideas, separation optimisation being achieved by searching, in a brute-force, stepwise manner, a pre-defined compositional space. No decisions are taken during the experimentation to guide the optimisation and data review is possible only at the end of the experimental sequence. Nevertheless, by virtue of the fact that the system explores the whole of the compositional space defined, optimum experimental conditions should be located. Separation quality is described by an optimisation function and the Pesos system is able to display the value of that function as a coded response map using computer graphics. The Optim approach to separation optimisation uses a decision making algorithm to guide the choice of experimentation. Unfortunately, since it is a proprietary system, about which full details have not been published, a discussion of its workings is not included in this paper.

2 RESPONSE FUNCTIONS FOR SEQUENTIAL DESIGNS

The selection of an appropriate function and method of assessment of chromatographic quality is paramount for the successful utilisation of sequential experimental designs. By their very nature, sequential designs rely on the results from previous experiments to provide guidance and direction for future experimentation. If those results are in any way ambiguous, it will be extremely difficult to reach a successful conclusion.

Implicit in this discussion is that response functions for sequential designs should be continuous and smooth. Climbing a mountain which is crisscrossed with yawning chasms is fraught with both difficulties and danger, especially if one is essentially ignorant of what lies ahead. The selected function should also be capable of directing the optimisation process towards the global optimum, not being distracted by local, less than ideal separations. It must yield a single number which closely reflects the chromatographer's assessment of separation quality, acknowledging that this assessment will almost certainly be a multi-criterion decision. Ideally, there should be some indication of occurrences of peak cross-over (changes in elution order during the optimisation process) but the combination of the function and the optimisation algorithm chosen should be able to exploit such cross-overs to search for the global optimum. Not surprisingly there is, unfortunately, no ideal response function which meets all demands for sequential optimisation schemes.

The assessment of separation quality was discussed in some detail in Part 1 of this tutorial series and a number of criteria were introduced. Extensive discussions of optimisation criteria are

TABLE 1

Chromatographic optimisation functions used with sequential simplex optimisation

Function*	Variables	Ref.
$F_{obj} = \Sigma[10(1.5 - R_i)]^2$	Ternary mobile phase	7
$F_{obj} = \Sigma 100 e^{1.5 - R_i} + (t_m - t_n)^3$	Ternary mobile phase	8
$CRF = \Sigma \ln(P_i/P_d) + a(t_m - t_n)$	Gradient parameters and flow-rate	9
$CRF = 1/t \, \Pi f_i/(g_i + 2N_i)$	Concentration of modifier and buffer, pH	10
$CRF = \Sigma \ln(f_i/g_i) - 100(M - n)$	Concentration of organic modifier, pH	11
$CRF = \Sigma R_i + n^a - b|t_m - t_n| - c(t_0 - t_1)$	Composition of ternary mobile phase, temperature, flow-rate, pH	12
$CRF = \Sigma r_i + n - (t_m - t_n)$ (for $t_m - t_n > 1$)	Composition of ternary mobile phase	13
$Y = p/M$	Gradient parameters (S/b)	14

* R_i (P_i) is actual resolution (peak separation) and R_d (P_d) is desired resolution (peak separation), t_n is retention time of last peak and t_m is desired retention time, t_0 the void time and t_1 the retention time of the first eluted peak, f and g refer to peak separation factors [1]. N is noise, M the number of peaks expected, n the number of peaks detected and p the number of peaks separated with a given resolution. CRF is Chromatographic Optimisation Function and Y is the extent of separation. The parameters a, b and c are selectable weightings.

also to be found in refs. 5 and 6. In general, functions used for truly sequential experimental designs (e.g. the simplex procedure) are based upon sums of terms, usually reflecting resolution and analysis time and a selection is presented in Table 1. For sequential designs involving iterative processes it is usual to use functions which require a greater knowledge of the retention and identity of individual solutes and two criteria have been most generally used. These are discussed in Section 4.1 where it will be seen that they are not complicated by the desire to introduce multi-criteria decision making into a single function.

Not taken into account in this discussion are criteria which can be successfully used when the separation is complicated by non-ideal peak shapes (e.g. a solvent peak or matrix peak) or when the objective is to separate a small number of components from a complex sample — except that this situation corresponds to a special case of assigning weighting factors to individual sample components within a criterion. Such criteria are the subject of current research [15].

3 THE SEQUENTIAL SIMPLEX PROCEDURE

The simplex method is a hill-climbing procedure whose direction of advance is dependent solely on the ranking of responses and not on any particular values on an absolute scale. The procedure is guided by calculations and decisions that are rigorously specified yet almost trivially simple. The great advantages of the simplex procedure in the optimisation of liquid chromatographic separations are that it assumes nothing about the mode of separation or the complexity of the sample. Nor does it require any pre-conceived model of the retention behaviour of solutes and so does not require that the solutes be identified or recognised in individual separations. The method has the further advantages of permitting the introduction of new variables during the optimisation process for the price of just one additional experiment per variable and one can also assess the progress of the optimisation during rather than at the end of the experimental sequence. The procedure is therefore efficient, multi-factor and has an empirical feedback enabling rapid attainment of the experimental optimum.

There are, however, negative aspects associated with simplex optimisation. The ranking of responses requires that the quality of the chromatogram from each experiment be assessed — hence the need to use a response function. As it is a 'blind' optimisation process it is generally unable to assess the ultimate quality of the optimum located. Thus where peak elution orders change in successive separations, the procedure would be expected to be unable to decide unambiguously which elution order should be pursued to provide the global optimum: one or more local optima may be located and the simplexes will locate the most favourable local optimum rather than continue searching for the global optimum. Also, in this new age of efficiency, it has to be recognised that the simplex procedure can require relatively large numbers of experiments to locate optimum separation conditions.

3.1 Selection of algorithm

Simplex optimisation has been discussed extensively in an earlier tutorial in this journal [16] and the theory will not be repeated here. The two most usual algorithms applied to liquid chromatography are the basic method of Spendley et al. [17] (Fig. 1) or the modified procedure of Nelder and Mead [18] (Fig. 2). The modified procedure is likely to be the algorithm of choice since it is able to more precisely locate the optimum and will usually be more efficient than the fixed step size simplex. Additionally, since the step sizes change, it is not necessary to make initial judgements on how large a step should be taken. Because the response surfaces generated by most of the commonly used response functions are often somewhat irregular, the newer algorithms such as the weighted centroid method [19] or the super modified simplex [20] may not offer significant advantages. They can be less efficient due to the noise often encountered in chromatography and they do require more complex calculations. There is a large number of commercially available software packages for simplex optimisation, for example COPS [21] or Simplex V [22] and these usually

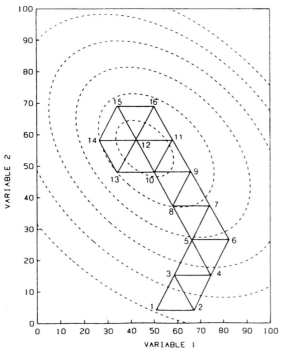

Fig. 1. Fixed step size simplex optimisation of two variables. Initial simplex is 123 and optimum region is at vertex 12. Reproduced from ref. 5 with permission.

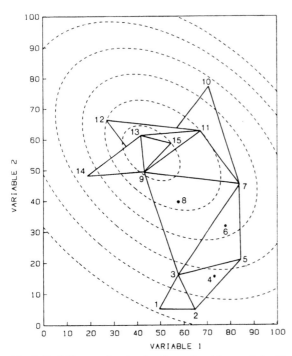

Fig. 2. Variable step size simplex. Reproduced from ref. 5 with permission.

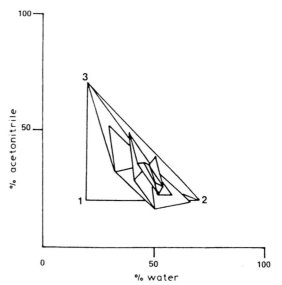

Fig. 3. Movements of simplexes in an unconstrained optimisation of a ternary mobile phase.

include most of the common algorithms. For those wishing not to use complex software, the procedure can be implemented purely graphically or easily programmed on one of the many spreadsheet packages [23].

3.2 Applications

The use of the modified simplex procedure in separation optimisation will first be discussed with reference to the six-solute mix described in Part 1 of this tutorial series. In this, the selection of column was made (reversed-phase, employing an octadecyl-bonded silica) and the experimental variables useful with such a column are water, as solvent strength adjuster and methanol, acetonitrile and tetrahydrofuran as selectivity modifiers. It is possible to exploit the simplex procedure with no further knowledge thant this. All that is required is to select a suitable response function (CRF, Table 1) to guide the procedure and to set up the initial simplex in an appropriate area. Fig. 3 shows the movements of a two-variable simplex which was allowed to investigate the whole of the compositional space available with three solvents present (water, methanol and acetonitrile). Starting from experiments 1, 2, 3, the simplexes have smoothly converged onto a region which indicates the optimum composition to be approximately 50% water, 13% methanol and 37% ac4-etonitrile; this being located after some 15 separations had been carried out. Note that the optimisation of a ternary solvent mix is only a two-variable experiment since the third solvent level is found by difference.

However, by using knowledge gained by first carrying out a gradient separation of the mixture under investigation [24], it is possible to constrain the simplex movements to a smaller area or volume of the factor space in order to provide faster and more reliable location of optimum conditions. From the gradient separation described in Part 1, a composition of 66% methanol provided an isocratic separation in which all peaks were eluted within the required time of five minutes. Using the "transfer rules" which will be discussed in detail in the third tutorial, it is possible to calculate that the optimum separation will require, for a combination of acetonitrile, metanol and/or tetrahydrofuran as modifiers, a water content in the range of 34 to 66%. Fig. 4 shows the movements of simplexes within a region bounded by 30 and 70% water and Fig. 5 shows the values of the CRF obtained during the optimisation process. The CRF used to guide the process is shown in eq. 1

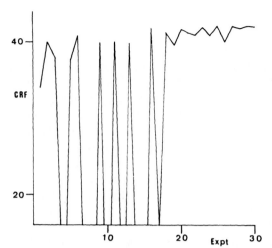

Fig. 5. Values of the CRF (Chromatography Response Function) produced during the optimisation shown in Fig. 4.

(see also Table 1). Note that resolution (R) is limited to a maximum value of unity.

$$\text{CRF} = \sum R_i + n^2 - b|t_m - t_n| - c(t_0 - t_1) \qquad (1)$$

Within 15 experiments the optimum region was very much more precisely located, being water–methanol–acetonitrile (50.5 : 14.5 : 34.1, v/v). The separation produced is shown in Fig. 6. Note, however, that experimentation continued through to experiment 30 since the stop criterion employed

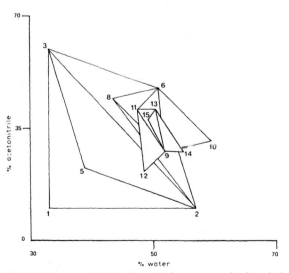

Fig. 4. Movements of simplexes during a constrained optimisation (30–70% water).

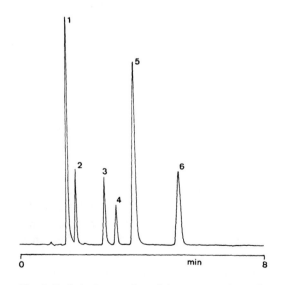

Fig. 6. Optimised separation of six-component test mix resulting from movements shown in Fig. 4.

in this particular implementation requires that the proportion of no individual component (e.g. water) vary by more than 3% in any three successive experiments. This is a severe criterion and, for a separation in which a 5% variation of methanol produces very little change in chromatogram quality, is not met in the current system. It would be greatly advantageous to know before optimisation started, or at least during the procedure, what the maximum value or the CRF should be and to stop the separation on achieving this value. For this example, the maximum CRF value attainable is 41 and (Fig. 5) it was achieved by experiment 17.

By using the extra information provided by multi-channel detection systems, and in the case of HPLC this will usually be a linear diode-array detector, it is indeed possible to estimate the number of peaks that are present in a mixture before a complete separation has been achieved. This information can then be used to provide a maximum possible value of the CRF such that, when this value is reached, the optimisation process can be halted. Rather than halting the procedure on the basis of the simplex size, which is a method frequently used, the stop criterion can now be based on response. However, it is not necessary to know the number of peaks in the present in the mixture prior to starting optimisation since the peak homogeneity can be checked for each chromatogram and the maximum number of possible peaks updated as more become discovered. Many chemometric techniques have been used to establish peak homogeneity and/or deconvolute peaks into their component elution profiles [25,26]. Fortunately, for simply estimating peak numbers, it is not necessary to call upon such complex software — instead use can be made of on-board capability usually provided by the detector's integration software to detect apparent shifts in peak retention time with detection wavelength. For example, where some separation between two components occurs, but without any observable resolution, the relative contributions of the two components will change throughout the elution profile. Where there is a sufficient spectral difference between the two components, both peak shape and retention time will be dependent upon detection wavelength. A sufficient difference in the shift of peak retention time with wavelength can provide an early indication of peak inhomogeneity [13]. To overcome some of the anomalies generated by using a multi-criterion CRF in simplex optimisation, the CRF (eqn. 1) can be simplified to that shown in eq. 2;

$$\text{CRF} = \sum R_i + n - (t_m - t_n) \qquad (t_m - t_n > 1) \qquad (2)$$

The time constraint for the first eluted peak is deleted and that for the last eluted peak is considered only if the difference between its retention time and the target retention time is greater than 1 minute. The term for the number of peaks (n) is not raised to a power to ensure that, when the time term is included, it has a significant influence on the CRF value. The net effect of this CRF is to force the maximum number of peaks to be searched for, with the last being positioned close to the target analysis time. Also this CRF is designed to give a maximum value when the separation meets or exceeds a pre-determined 'development' requirement. If the maximum value (CRF_{max}) is reached during the course of optimisation, then the simplex can be halted. CRF_{max} depends on the number of peaks in the sample and, for an unknown, is obviously unknown at the start of optimisation. However, by using a multi-channel UV detector, all UV active components can be detected and peak overlap frequently detected. As the simplex procedure progresses, updated estimates of the number of peaks present are calculated, used to determine CRF_{max} and, if the value of the CRF for the latest separation equals or exceeds CRF_{max}, the procedure is halted on the basis of response [13].

The algorithm, which includes the simplex procedure, is shown in Fig. 7. Note that the stop criterion used actually requires that CRF_{max} is attained at least three times. The optimisation procedure then goes on to check the peak elution order. As discussed above, a major drawback of the simplex procedure is its tendency to locate local optima. Several local optima can be generated by differing peak elution orders. If differing elution orders are detected, the chromatographer can be warned of the existance of local

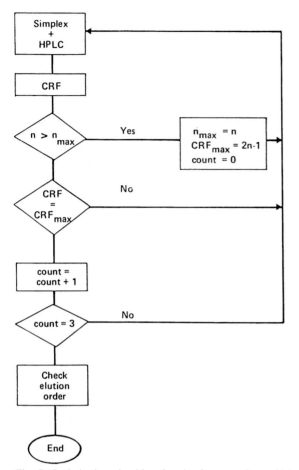

Fig. 7. Optimisation algorithm for simplex procedure which uses extra information provided by multi-channel detection to estimate the number of components present.

optima or new optimisation(s) can be started around each optimum. If all vertices indicate the same elution order, a greater degree of confidence can be attached to the result of the optimisation.

So far, the use of the simplex procedure for simple, isocratic optimisation of standard organic modifiers in reversed-phase chromatography has been considered. The method can be used without modification for the optimisation of other variables such as pH, ion-interaction (ion-pairing) reagent concentrations, flow-rate and/or separation temperature. Gradient separations can also be optimised. In this case it is necessary to find a descriptor of the gradient as the variable to optimise. This could be the starting composition and slope Φ' of the modifier content, or perhaps just a number corresponding to the shape of the gradient generated by the gradient programmer. A recent suggestion is to optimise the gradient steepness in terms of S/b (solvent strength/constant: $b = 0.2$ for 10-μm or 0.1 for 5-μm packings respectively). The actual gradient steepness is then calculated from Snyder's linear solvent strength (LSS) theory whose basic equation is

$$\Phi' = b/St_0$$

where Φ' is the volume fraction change per unit time. This approach can be used to optimise multi-segment gradient elution profiles [14].

3.3 Summary — advantages and disadvantages

It should be apparent from the discussion so far that the simplex procedure is a versatile, general optimisation method with wide applicability to HPLC separation optimisation. No assumptions need be made about the sample under investigation, either its composition or behaviour. Interdependent variables can be optimised, such as pH and ion-pair reagent composition, and it is easy to see how the optimisation is proceeding and halt it on the basis of separation quality (response) or when a pre-determined simplex size has been reached.

Furthermore, it is not necessary to identify or track peaks in successive chromatograms and thus, if computer assistance is being used, the software is both conceptually simple and easy to implement. This simplicity also means that the simplex method is one of the easier formal optimisation procedures to link to the on-line optimisation of HPLC separations now that microcomputer-controlled chromatographs are becoming more widely available. Following the demonstration of its feasibility in the early 1980's, at least two commercial manufacturers have made such an approach available on their instruments.

Such an apparently attractive procedure has to be accompanied by significant drawbacks. Most notable of these is the problem of locating local, rather than the global, optima. This will especially be the case when elution orders are changing during the optimisation although methods of at least altering to this have been introduced. Other difficulties associated with the simplex procedure are the often large number of experiments required (typically 15 to 30).

4 ITERATIVE MIXTURE DESIGNS

Iterative mixture designs [27, 28] seek to capitalise on the advantages offered by both simultaneous designs (such as the mixture design to be discussed in the third tutorial) and sequential procedures. In theory, the disadvantages of both might be overcome: in practice a universal procedure is still not realised.

4.1 Principles

Whilst being a procedure that is theoretically able to cope with many variables, the iterative mixture design is most easily understood by reference to just one. Also, as the number of variables increases, the computational demands become progressively higher. The procedure can be summarised as follows:

(a) The necessary solvent strength to achieve elution within a given time is determined. This is usually calculated from a gradient elution of the sample under study — a methanol–water gradient in the case of reversed-phase chromatography. The solvent strength for isocratic elution is then calculated by one of the two procedures described in the first tutorial.

(b) Isoeluotropic, binary solvent compositions are calculated — usually by using 'transfer rules' (to be discussed in Part 3). It may be that only two solvent mixes are to be investigated (producing a ternary mobile phase) in which case the optimisation is essentially a univariate problem. However, quaternary mobile phases can be optimised by considering three pseudocomponents but the visualisation and computation are more difficult.

(c) Chromatograms are obtained for each of the compositions under investigation (e.g. methanol–water and acetonitrile–water). The positions (retention times) of the eluites are established in each of the chromatograms.

(d) Linear plots of the logarithms of the capacity factors of each eluite versus composition are constructed for the entire range of ternary mixtures composed from the pairs of limiting binary pseudocomponents.

(e) The retention/composition data are then used to construct a graphical representation of separation quality versus composition. This can be a simple representation, such as minimum resolution, or can be more complex, for example using a chromatographic response function. To achieve the construction, it is assumed that the logarithm of the retention of each of the solutes will follow a linear dependence upon the mobile phase composition. There is no absolute reason why this should be the case and non-linearity will be dealt with in the iterative part of the procedure.

(f) From this response surface plot, an optimum composition is determined and the sample is chromatographed with this mobile phase.

These six steps represent the 'simultaneous' aspect of the iterative mixture design. The next steps are associated with the sequential, or iterative, part.

(g) The quality of the chromatogram from the predicted optimum composition is assessed against the original requirements of the separation. It is possible that the separation produced will satisfy these needs. More likely, however, is that the assumption of linear behaviour was erroneous and the separation is far from adequate. The new retention data for each solute are then used with the original data to refine the response surface plot. Linear behaviour between each data point is still assumed but it will be seen that curvature is gradually being introduced.

(h) The new plot is used to define a new, predicted, optimum composition and a separation is conducted with this mobile phase. The step (g) is repeated and the procedure continues until sufficient iterations have been conducted to produce a model that is sufficiently close to the true behaviour to enable an optimum separation to be achieved.

Referring once again to the six-solute mix, it can be calculated that isoeluotropic mobile phases to give a maximum retention time of ca. 6 minutes comprise 66% methanol, 54% acetonitrile and 44%

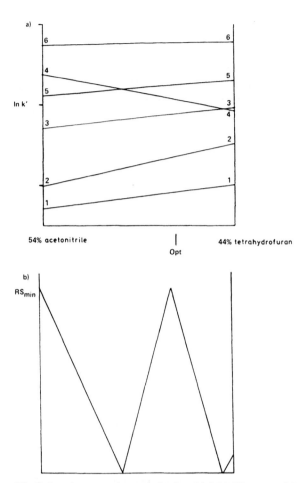

Fig. 8. Iterative regression optimisation. (a) Initial linear model describing retention as a function of mixing two binary solvents; (b) plot of minimum resolution between peaks as a measure of separation quality.

tetrahydrofuran. Fig. 8a shows the first step of the iterative procedure, using the acetonitrile and tetrahydrofuran pseudocomponents, with linear retention plots and Fig. 8b shows a representation of the minimum resolution between the limiting peak pairs. Fig. 9 is the chromatogram obtained from the composition predicted to produce an optimum separation. At least one problem is immediately apparent — there appear to be only five peaks! This problem highlights one of the difficulties that the iterative procedure shares with the simultaneous procedures, namely the requirement to track individual eluites in each separation. Fortunately, as can be seen from Fig. 9, the spectra obtained on the upslope and downslope of each peak reveal that the third eluted peak is a composite peak and there is sufficient dissimilarity between the spectra of the components in the mixture to allow positive identification of each peak.

The retention data from this separation are then fed back into the procedure to provide another prediction of an optimum composition. After these iterations, the original assumption of linear behaviour has been sufficiently modified towards the true, non-linear behaviour (Fig. 10) that the composition and accompanying separation are now sufficiently close to be able to halt the optimisation process: the optimum composition being 19% tetrahydrofuran, 31% acetonitrile and 51% water. The optimum separation is shown in Fig. 11: it can be seen that there are changes in elution order, brought about by the use of tetrahydrofuran, when compared with the simplex optimised separation.

This illustration represents a single variable optimisation. Other ternary mobile phases (e.g. methanol–acetonitrile) would be investigated in the same way. To extend the method to quaternary mobile phases requires the assumption that retention behaviour can be described as a series of planes. The three pseudocomponents that are to be considered simultaneously are positioned at the vertices of a triangle (rather as in the mixture design). The retention across the plane is then assumed to be linear and an optimum predicted. This optimum is used to divide the plane up into the three smaller, triangular planes. The iterative part of the procedure then takes over in a manner entirely analogous to that described above.

4.2 Applications

The majority of the applications of the iterative mixture design [27–31] result from the group of De Galan et al. who have pioneered this approach. To assess the quality of separations they have used a product criterion, a relative resolution product

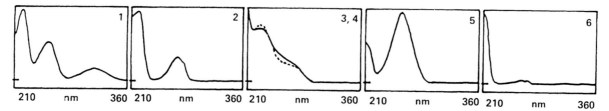

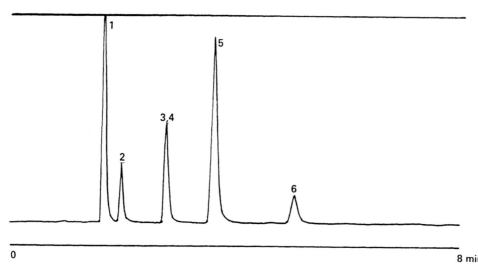

Fig. 9. Separation resulting from the composition predicted to be optimum from the first iteration shown in Fig. 8.

(r) which aims to achieve even spacing of all detected peaks throughout the chromatogram.

$$r = \prod_{i=1}^{n-1} Rs_{i,i+1} \bigg/ \left[\left(\sum_{i=1}^{n-1} Rs_{i+1,i} \right) / (n-1) \right]^{n-1} \quad (4)$$

where the denominator denotes the maximum possible value for ΠR in the particular chromatogram. However it was found that quite different separations produced the same values for this criterion, which was then modified to take into acount an imaginary peak ($i = 0$) which is considered to elute at the desired starting point of the chromatogram. The modified response function is now [28]:

$$r^* = \prod_{i=1}^{n-1} Rs_{i+1,i} \bigg/ \left[\left(\sum_{i=0}^{n-1} Rs_{i+1,i} \right) / (n-1) \right]^{n-1} \quad (5)$$

where $Rs_{1,0} = (t_1 - t_0) N^{0.5} / 2(t_1 + t_0)$. As well as evolving the relative response function, the method of predicting optimum composition was also improved. Rather than determine a separation at exactly the point predicted by the response surface plot, the concept of a 'shifted' optimum was introduced. The range of binary solvents is described by the parameter x which has the values $x = 0$ at one extreme and $x = 1$ at the other. The optimisation criterion is used to predict the optimum mobile phase composition but allowance is then made for the expected non-linear behaviour. It is argued that such an allowance reduces the possibility of experiments being conducted at positions too closely spaced and thus improves the efficiency of the searching. If it is assumed that two experiments have been conducted at compositions x_1 and x_2, the optimum must lie at a point x lying between these two points. However, instead of carrying out the next experiment exactly

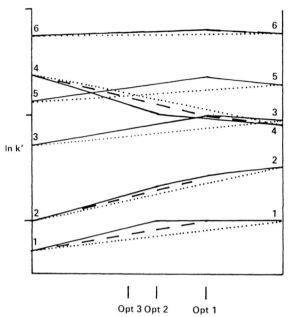

Fig. 10. Iterative regression optimisation showing three iterations: the third predicted optimum produced a satisfactory separation and the procedure was halted. ······ = First, ——— = second, ——— = third iteration.

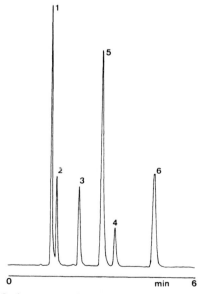

Fig. 11. Optimum separation resulting from the third predicted optimum: tetrahydrofuran–acetonitrile–water (19:31:51).

at this point, a new, 'shifted' composition x' is investigated as defined by:

$$x' = x + 0.4[0.5 - (x - x_1)/(x_2 - x_1)] \qquad (6)$$

This is surely entirely in accord with good chemometric principles where information about the system is used to improve the efficiency of the experimental design.

Some of the applications of iterative mixture design also address the problems associated with the tracking of elutes during the optimisation process [27,31]: diode-array UV detection appears to be the only practicable proposition at present. Simple absorbance ratioing is accompanied by too many problems to be sufficiently reliable an robust.

4.3 Summary

The iterative mixture design is an attractive approach to separation optimisation since it combines many of the advantages of conventional simultaneous and sequential designs. In doing so, however, it also has is own strengths and weaknesses. By assuming a simple model of retention behaviour (i.e. linear) progress towards an optimum solution can be made after only two experiments have been conducted (compare the minimum of seven experiments required to fit the special cubic model of the mixture design). Refining of the model is quickly accomplished without any involved mathematical or statistical requirements such that, with a univariate optimisation, a satisfactory solution can often be attained by the fourth or fifth experiment. The major drawback, as with all retention model-fitting methods, is the need to track peaks and correlate elution order. Until recently, this had to be done through the use of standard solutions but, with the advent of rapid scanning multi-channel detectors, in particular the linear diode-array UV detector, it is possible to use some of the newer chemometric techniques to follow peaks during the optimisation. A second drawback is the need to plot the response surface to calculate the new, predicted optimum. This is relatively straightforward with a single variable but the complexity rises exponentially as the number of variables increases. Without powerful micro/minicomputers, the simultaneous optimisation of more than two variables becomes a very time consuming and computing resource intensive method. Nevertheless, the iterative mixture design must still remain one of the more attractive methods for semi-automated optimisation.

5 CONCLUSIONS

In this second tutorial, examples of the sequential experimental designs have been discussed — they perhaps represent a design with which we find ourselves most comfortable since feedback upon the progress of the procedure is obtained after each experiment. While sequential designs may come under criticism, the simplex procedure in particular may be somewhat slow in comparison to other methods and may not locate the global optimum, it is relatively easy to automate [12]. One common feature of both simultaneous and sequential designs in HPLC separation optimisation is that both can use a gradient separation to establish the region of factor space within which further experiments should be located. (It turns out that such a wealth of information is available from a gradient separation that two gradient separations can provide sufficient data to enable off-line optimisation — at least of binary mobile phases — to be adequately achieved [32,33].) The movement towards a greater degree of off-line optimisation and calculation rather than experimentation is a general one. In the third tutorial, the benefits and disadvantages of the more calculation intensive simultaneous designs will be considered.

REFERENCES

1 J.C. Berridge, Chemometrics and method development in high-performance liquid chromatography. Part 1: Introduction, *Chemometrics and Intelligent Laboratory Systems*, 3 (1988) 175–188.
2 R.D. Conlon, The Perkin Elmer solvent optimisation system, *International Instrumentation-Research*, September (1985) 74–81.
3 M.P.T. Bradley and D. Gillen, Adaptive intelligence for automated separation optimisation: part I – Isocratic, *Spectra-Physics Chromatography Review*, 10 (2) (1983) 2–4.
4 M.P.T. Bradley and D. Gillen, Adaptive intelligence for automated optimisation: Part II – Gradient, *Spectra-Physics Chromatography Review*, 11 (1) (1984) 10–12.
5 J.C. Berridge, *Techniques for the Automated Optimisation of HPLC Separations*, Wiley, Chichester, 1985.
6 P.J. Schoenmakers, *Optimization of Chromatographic Selectivity. A Guide to Method Development*, Elsevier, Amsterdam, 1986.
7 A.S. Kester and R.E. Thompson, Computer-optimized normal-phase high-performance liquid chromatographic separation of *Corynebacterium poinsettiae* carotenoids, *Journal of Chromatography*, 310 (1984) 372–378.
8 D.L. Dunn and R.E. Thompson, Reversed-phase high-performance liquid chromatographic separation for pilocarpine and isopilocarpine using radial compression columns, *Journal of Chromatography*, 264 (1983) 264–271.
9 M.W. Watson and P.W. Carr, Simplex algorithm for the optimisation of gradient elution high-performance liquid chromatography, *Analytical Chemistry*, 32 (1979) 1835–1842.
10 W. Wegscheider, E.P. Lankmayr and K.W. Budna, A chromatographic response function for automated optimisation of separations, *Chromatographia*, 15 (1982) 498–504.
11 J.H. Nickel and S.N. Deming, Use of the sequential simplex algorithm for improved separations in automated liquid chromatographic methods development, *LC, Liquid Chromatography HPLC Magazine*, 1 (1983) 414–418.
12 J.C. Berridge, Unattended optimisation of reversed-phase high-performance liquid chromatographic separations using the modified simplex algorithm, *Journal of Chromatography*, 244 (1982) 1–14.
13 J.C. Berridge, A.F. Fell and A.W. Wright, Sequential simplex optimization and multichannel detection in HPLC: application to method development, *Chromatographia*, 24 (1987) 533–540.
14 F. Dondi, Y.D. Kahie, G. Lodi, P. Reschiglian, C. Bighi and G.P. Cartoni, Comparison of the sequential simplex method and linear solvent strength theory in HPLC gradient elution optimisation of multicomponent flavenoid mixtures, *Chromatographia*, 23 (1987) 844–849.
15 P.J. Schoenmakers, Criteria for comparing the quality of chromatograms with great variations in capacity factors, *Journal of Liquid Chromatography*, 10 (1987) 1865–1886.
16 K.W.C. Burton and G. Nickless, Optimisation via simplex. Part I. Background, definitions and a simple application, *Chemometrics and Intelligent Laboratory Systems*, 1 (1987) 135–149.
17 W. Spendley, G.R. Hext and F.R. Hinsworth, Sequential application of simplex designs in optimisation and evolutionary operation, *Technometrics*, 4 (1962) 441–461.
18 J.A. Nelder and R. Mead, A simplex method for function minimisation, *Computer Journal*, 7 (1965) 308–313.
19 M.W. Routh, P.A. Swartz and M.B. Denton, Performance of the super modified simplex, *Analytical Chemistry*, 49 (1977) 1422–1428.

20 P.B. Ryan, R.L. Barr and H.D. Todd, Simplex techniques for non-linear optimisation, *Analytical Chemistry*, 52 (1980) 1460–1467.
21 *COPS (Chemometrical Optimization by Simplex)*, Elsevier Scientific Software, Amsterdam.
22 *Simplex-V*, Statistical Programs, Rowlett, Houston, TX.
23 J.C. Berridge, Chemometrics in pharmaceutical analysis, *The Analyst (London)*, 112 (1987) 385–389.
24 J.C. Berridge and E.G. Morrissey, Automated optimisation of reversed-phase high-performance liquid chromatographic separations. An improved method using the sequential simplex procedure, *Journal of Chromatography*, 316 (1984) 69–79.
25 B.J. Clark and A.F. Fell, Multichannel spectroscopy in liquid chromatography, *Chemistry in Britain*, 23 (1987) 1069–1071.
26 G.G.R. Seaton and A.F. Fell, Multivariate analysis of non-homogeneous peaks in liquid chromatography, *Chromatographia*, 24 (1987) 208–216.
27 L. de Galan and H.A.H. Billiet, Mobile phase optimisation in RPLC by an iterative regression design, *Advances in Chromatography Vol. 25*, Marcel Dekker, New York, 1986, pp. 62–104.
28 A.C.J.H. Drouen, H.A.H. Billiet and L. de Galan, Practical procedure for the optimization of reversed-phase separations with quaternary mobile phase mixtures, *Journal of Chromatography*, 352 (1986) 127–139.
29 H.A. Cooper and R.J. Hurtubise, Mobile phase optimization for the reversed-phase liquid chromatography of complex hydroxyl aromatic mixtures, *Journal of Chromatography*, 324 (1985) 1–18.
30 H.A.H. Billiet, J. Vuik, J.K. Strasters and L. de Galan, Simultaneous optimization of reagent concentration and pH in reversed-phase ion-pairing chromaography, *Journal of Chromatography*, 384 (1987) 153–162.
31 A.C.J.H. Drouen, Computerised optimisation and solute recognition in liquid chromatographic separations, Ph.D. Thesis, Technische Hogeschool Delft, 1985.
32 L.R. Snyder, J.W. Dolan and M.A. Quarry, High-performance liquid chromatographic method-development using computer simulation, *Trends in Analytical Chemistry*, 6 (1987) 106–111.
33 J.W. Dolan and L.R. Snyder, Using computer simulation to develop a gradient elution method fopr reversed-phase HPLC, *LC–GC*, 5 (1987) 970–976.

Fourier Transforms: Use, Theory and Applications to Spectroscopic and Related Data

RICHARD G. BRERETON

School of Chemistry, University of Bristol, Cantock's Close, Bristol BS8 1TS (U.K.)

(Received 2 April 1986; accepted 26 August 1986)

CONTENTS

1. Introduction	167
2. Advantages of fourier methods	167
2.1. The signal to noise problem	167
2.2. Time averaging	167
2.3. The fourier spectrometer	168
2.4. Pulsed fourier methods	169
2.5. Interferometric fourier methods	170
3. The fourier transform	170
3.1. Time and frequency domains	170
3.2. The principles of fourier transformation	170
3.3. The discrete fourier transform	172
3.4. Time series	172
3.5. Sampling strategy	174
3.6. Digital resolution	174
3.7. Relationship between time and frequency domain parameters	176
3.8. Real and imaginary pairs	176
3.9. Phasing	176
4. Conclusion	179
References	179

Sections printed in small typeface contain details which may be omitted by some readers. Sections in small italic typeface contain mathematical details; those in small normal typeface contain explanations which assume familiarity with chemical spectroscopy.

Measurement units. Oscillations can be measured in units of radians per second (s^{-1} in notation used in this article) or cycles per second — Hertz (Hz in notation used in this article). These units are related by $1 \text{ Hz} = 2\pi \text{ s}^{-1}$.

Mathematical notation. The mathematical sections of the paper contain complex arithmetic, and i is used to denote the square root of -1.

1. INTRODUCTION

Comprehensive texts have been compiled by Griffiths [1] and Marshall [2] drawing together varied applications of the fourier transform (FT) in chemistry. The fast fourier transform (FFT) algorithm, originally developed by Cooley and Tukey [3] and applied to astronomic data, made possible rapid on-line computational analysis of fourier spectra. Two of the first major chemical applications were nuclear magnetic resonance (NMR) spectroscopy [4] and infrared (IR) [5] spectroscopy, but since then FT methods have been used in mass spectrometry (ion cyclotron resonance), dielectric and microwave spectroscopy, UV/visible spectroscopy, electrochemistry [1,2], naturally occurring chemical time series [6], and several other areas.

The objectives of this tutorial are to explain why fourier methods are useful to the chemist, how fourier transforms work, and to define terminologies associated with fourier processing. The applications referred to in this article are chiefly drawn from chemical spectroscopy, which in recent years has been revolutionised by the availability of fourier methods. However, an understanding of these techniques is of wider interest to the chemometrician.

2. ADVANTAGES OF FOURIER METHODS

2.1. The signal to noise problem

Astronomers need to extract information from weak sources in the sky. They cannot increase the intensity of such sources and are, therefore, dependent on optimising instrumental conditions. Infrared spectroscopy is one major tool, and early investigators were posed with severe signal to noise problems. Similarly early instrumental methods for structure elucidation in synthetic and natural product chemistry, such as NMR spectroscopy, were hampered by poor signal intensities.

We will illustrate the problem of noisy data by reference to NMR, which is an insensitive but very informative technique. The spectroscopist observes the difference between emission and absorption as transitions are induced between two energy levels. Because there is an excess of spins in the lower energy level, absorption will be greater than emission. However, typically, NMR energy levels are separated by 100 MHz (corresponding to about $4 \cdot 10^{-2}$ J mol^{-1}). Using the Boltzmann distribution we can show that there is an excess in the lower energy level of only a few parts per million at room temperature. Thus, originally, substantial quantities of compounds were required to obtain adequate signal strength for structure elucidation. Hence, unlike most other analytical techniques such as mass spectrometry and UV spectroscopy, NMR was of little use to chemists who had only small amounts available; 50–100 mg was typically required to obtain good ^{1}H NMR spectra in the 1960s. A further limitation came when taking the spectra of nuclei such as ^{13}C at natural abundance levels.

The enormous potential of many emerging instrumental techniques remained unrealised throughout most of the 1960s because of the limitations of weak signal strengths, and much interest centres on approaches for increasing signal to noise ratios.

2.2. Time averaging

One approach to solving the signal to noise problem is to add several spectra from the same source together. Providing spectroscopic and instrumental conditions are stable, the signal to noise ratio increases according to the number of scans added together.

Experimental noise consists of correlated noise (where each point in time is dependent on the previous point) and non-correlated noise. The latter type of noise is often modelled by a Gaussian distribution of mean (μ) and standard deviation (σ). Generally the experimental conditions are adjusted so that the mean is zero.

Random walk analysis [7] can be used to demonstrate that the variance of the noise (modelled as white noise) is proportional to the number of spectra added together, so that the standard deviation is given by

$$\sigma_M = \sqrt{M}\,\sigma \qquad (1)$$

where σ_M is the standard deviation of the noise after M spectra have been added together and σ is the standard deviation of the noise in a single spectrum. Signals arising from the source should, however, be additive, providing instrumental conditions are sufficiently stable, and their strength will increase linearly with the number of spectra added together.

The signal to noise ratio is statistically defined as the ratio of signal strength to root-mean-square (RMS) noise, for any given signal. Noise can be of various kinds, and we consider, here, only the case where noise is uncorrelated and independent of the signal: in other cases the analysis may differ. Random walk analysis shows that the *signal to noise ratio increases in proportion to the square root of the number of spectra added together*. This improvement is illustrated in Fig. 1. The signal to noise ratio doubles as 4 spectra are added together, triples as 9 spectra are added and so on.

Hence one solution to the problem of weak signals is to add several spectra together. However, to increase signal to noise ratio 10-fold would involve acquiring 100 spectra. If one spectrum takes 5 minutes to scan, the total acquisition time for 100 spectra would be 500 minutes or about 8 hours. Further improvements would involve several days, weeks or even months acquisition time, which would prove expensive, and often, because chemicals and scientific instruments are unstable over extended time periods, impossible. Hence the time averaged spectroscopy is limited in applicability.

2.3. The fourier spectrometer

Chemists are typically interested only in a small portion of a spectrum or chromatogram, i.e. the region where peaks occur. Frequently the majority of the data acquired by a scientific instrument are of no chemical interest (Fig. 2), and will consist of noise in the regions between peaks. The experimenter does not normally know in advance which regions of the spectrum consist of pure noise, so cannot predict how to make his sampling more efficient, using conventional methods.

The conventional approach to spectroscopy is via a scanning spectrometer. In such a case a slit is moved gradually across a spectrum, and at any one time only a small proportion of the spectroscopic information is recorded. A more efficient approach is to observe the entire spectrum all the time. In such a case, information about all the spectral frequencies will be contained in each observation, so the process of acquisition of useful information is speeded up. The approach is called a *multiplex* approach and the information about the spectroscopic frequencies must be encoded in

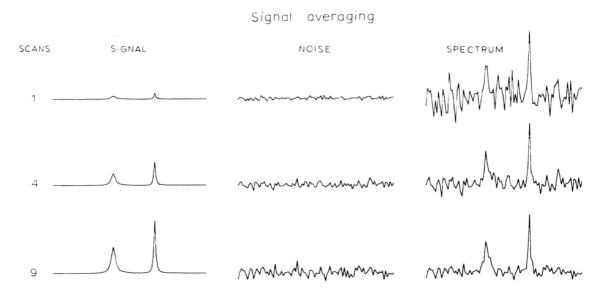

Fig. 1. The effect of signal averaging on simulated data. The absolute magnitude of the signal strength and the noise (modelled by a gaussian distribution) is illustrated (left and centre) as 1, 4 and 9 spectra (from top to bottom) are added together. The normalised spectra are illustrated on the right.

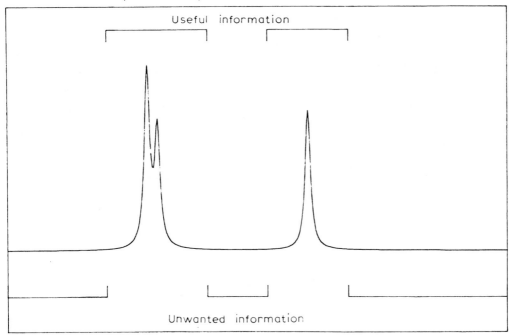

Fig. 2. Information content on an idealised spectrum. Areas between the peaks are of no interest to the spectroscopist.

a special way. To decode these data, a computer is needed, and fourier transform methods, as described below, are employed.

In order to acquire fourier data it is necessary to change the instrumental methods used to record information. There are two principal approaches, *pulsed methods* and *interferometers*, which are discussed below. In addition the advantages and need for fourier methods has recently been much more widely realised, so that data acquired by non-fourier methods and also naturally occurring cyclic chemical data (e.g. geochemical, environmental) can also be treated by methods described in this tutorial.

Clearly there will not be enough information in one datapoint to compute an entire spectrum consisting of several hundred or thousand significant points. Therefore it is necessary to acquire a large number of datapoints, which are obtained by sampling radiation (or magnetisation or other physically observable parameters) successively in time as described below. These data are stored in a computer and a fourier transform is used to convert this time series into a picture that is meaningful to the experimenter. Once a fourier transform spectrum is acquired, the signal to noise ratio can be increased still further by adding together several scans, just as in the case of conventional spectra.

2.4. Pulsed fourier methods

These are used in NMR, microwave and ion cyclotron resonance studies, inter alia. In such cases it is possible to excite the entire frequency (or mass) range using one pulse.

Because of the Heisenberg uncertainty principle

$$t \Delta \nu = 1 \tag{2}$$

where t is the excitation time, a pulse of t s long will excite a bandwidth of $\Delta \nu$ Hz in width. In NMR, spectra of interest are several hundreds or thousands of Hertz wide. It can easily be verified that a typical excitation pulse of a few microseconds is adequately short to excite a very wide range of spectroscopic frequencies, so a short pulse corresponds to a broad bandwidth, necessary for the fourier experiment.

Once the excitation is turned off the system will

decay back to equilibrium. This decay curve is recorded by the fourier spectrometer, and constitutes a *time series*, which contains information from the entire spectrum. The time series can be related to the conventional spectrum as described below.

2.5. Interferometric fourier methods

These are used in IR and UV spectroscopy, inter alia. In such cases two beams of radiation interfere with each other.

The most common method is via a Michelson interferometer. A beam of radiation is emitted from the sample. The beam is split into two via a splitter. These beams travel towards two mirrors and are, in turn, reflected back towards the beamsplitter where they interfere with each other. One mirror is fixed in distance from the beamsplitter and the other is varied in distance. The time series is obtained by varying the distance of the movable beam and recording the interference pattern as a function of time.

The interference pattern is recorded at different time delays between the two beams, and constitutes the *time series* used by the fourier spectrometer.

3. THE FOURIER TRANSFORM

3.1. Time and frequency domains

A conventional spectrum or chromatogram consists of several peaks. These peaks can be characterised by a shape (expressed as a distribution function), a position (normally the highest point in the peak), an integral, and one or more shape parameters (e.g. width).

A common spectroscopic lineshape is a lorentzian given by

$$I(\omega) = I_0 a / \left[a^2 + (\omega - \omega_0)^2 \right] \qquad (3)$$

where $I(\omega)$ is the intensity at ω s^{-1}, ω_0 is the centre of the peak, I_0 is an intensity parameter and $2a$ is the width at half height (as can be verified by the interested reader). This distribution is also called a Cauchy distribution.

The gaussian line shape distribution function is given by

$$I(\omega) = I_0 \sqrt{(\pi/4a)} \; e^{-(\omega - \omega_0)^2 / 4a} \qquad (4)$$

The width at half height is given by $4\sqrt{(a \ln 2)}$ as can be verified; ω_0 and I_0 are as defined for the lorentzian.

The constants used in the definitions of lorentzians and gaussians differ according to authors. We choose definitions in this tutorial to make all the algebra internally consistent. Thus the definitions of gaussian and lorentzian peak shapes can be algebraically derived from time series decaying at the rate of e^{-at^2} and e^{-at} respectively as discussed further below.

In both cases the line width depends on the parameter a alone; the position on ω_0, and the integrated intensity on the parameters I_0 and a.

In most experiments intensity and position parameters are of prime interest; occasionally width information is used. A spectrum or chromatogram could be described as the sum of lines characterised by different shapes, widths, positions and intensities. We will simplify the analysis below by restricting our discussion to spectra consisting solely of sums of lines of lorentzian (Cauchy) shape.

The raw data available from the fourier experiment are not, however, in such a convenient form, and fourier transforms are needed to convert the decay curve into an interpretable spectrum. The raw fourier data consist of readings of intensity versus time: these data are acquired in the *time domain*. The spectrum consists of readings of intensity versus frequency of radiation: this is displayed in the *frequency domain*. In some cases the raw data are not actually recorded against time (e.g. pH and concentration profiles) or else the desired end product is not necessarily a frequency profile (e.g. chromatographic data). The terminology time and frequency domains (frequency domain being used to describe the directly interpretable spectrum or chromatogram) should, nevertheless, always be used.

The transform extracts the parameters we are interested in from the raw data and converts a time domain series into a frequency domain spectrum (Fig. 3). Below we explain how frequency domain information is buried in the time domain data.

3.2. The principles of fourier transformation

There are several textbooks (referred to in the introduction) that deal with the mathematics of

Chapter 12

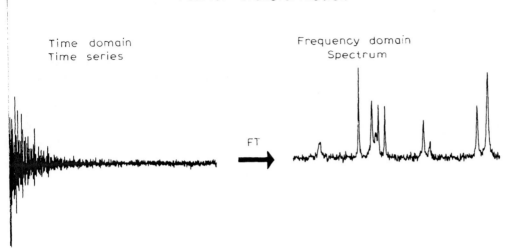

Fig. 3. The principle of fourier transformation. A time domain time series is recorded by the fourier spectrometer. It is fourier transformed into a frequency domain spectrum which is directly interpretable by the experimenter.

fourier transforms in detail. The aim of this section is to briefly summarise this area.

The calculation of intensity at frequency ω is performed by multiplying the entire time domain by a trigonometric function of frequency ω, and integrating the resultant waveform over time.

We mathematically define below various types of fourier transform. The conventions used in the literature vary according to author, so it is important, when studying papers in detail, to make sure which conventions are used. The main differences are the interchange of forward and inverse transform definitions, and the normalisation factor in front of the integral sign.

The cosine transform is given by

$$RL(\omega) = \int_{-\infty}^{\infty} f(t) \cos(\omega t) \, dt \qquad (5)$$

where $f(t)$ is the time domain response, and $RL(\omega)$ the resultant frequency domain intensity at ω s^{-1}.

Similarly, the sine transform is given by

$$IM(\omega) = \int_{-\infty}^{\infty} f(t) \sin(\omega t) \, dt \qquad (6)$$

These can be combined to give the complex transform

$$I(\omega) = \int_{-\infty}^{\infty} f(t)[\cos(\omega t) - i \sin(\omega t)] \, dt$$
$$= \int_{-\infty}^{\infty} f(t) e^{-i\omega t} \, dt \qquad (7)$$

We will call the above transforms forward transforms. They convert time domain data into frequency domain data. The corresponding inverse transform converts complex frequency domain data back to real time domain data.

$$f(t) = 1/(2\pi) \int_{-\infty}^{\infty} I(\omega) e^{i\omega t} \, d\omega \qquad (8)$$

The factor $1/(2\pi)$ is used to ensure that a forward transform followed by an inverse transform recovers the original data exactly.

In the case of typical spectroscopic data, a *real* time series is transformed into a *complex* frequency domain spectrum via a forward fourier transform. In certain types of spectroscopy, it is possible to acquire the time series in what is called the quadrature mode: this is particularly common in most modern NMR spectrometers. In such cases complex time domain data are acquired consisting of readings from orthogonal detectors which in turn are transformed into a complex frequency domain. For the sake of simplicity we will not consider this case here.

Several possible spectra can be displayed in the frequency domain.

1. The real spectrum, also called the absorption spectrum for ideally phased peaks (see below for a discussion about phasing), given by $RL(\omega)$.

2. The imaginary spectrum, also called the dispersion spectrum, for ideally phased peaks, given by $IM(\omega)$.

3. The power spectrum given by

$$P(\omega) = RL^2(\omega) + IM^2(\omega) \qquad (9)$$

4. The absolute value or magnitude spectrum given by

$$M(\omega) = \sqrt{P(\omega)} \qquad (10)$$

Some authors call the absolute value spectrum the power spectrum, so it is important to check conventions when the latter term is used in the literature.

3.3. The discrete fourier transform

We cannot sample an infinite number of datapoints, infinitely fast. Therefore, in practice, we use the discrete fourier transform (DFT).

The equations for DFTs can be derived from the equations above for continuous transforms. For example the discrete forward complex transform is given by

$$I(n\delta f) = \sum_{k=1}^{N} f(k\delta t) e^{-ik\delta t n\delta f} \delta t \qquad (11)$$

for a time domain consisting of N datapoints, acquired at regular intervals at δt seconds, transforming onto a frequency domain consisting of $N/2$ real, and $N/2$ imaginary datapoints at intervals of δf s^{-1}.

In practice, there are various differences between discrete and continuous transforms. First, the time series must be sampled at regular intervals, given by δt. Data spaced irregularly in time cannot be analysed by fourier methods. An interesting practical problem arises when there are uncertainties in the timing of the acquisition of each successive datapoint, which may occur when very rapid methods for acquisition are used. Second, in many cases data are acquired from time 0, not $-N$. In such circumstances, it is possible to envisage a curve at negative time values, which is identical to the curve at positive time values.

A further aspect that should be noted is the consequence of acquiring data over a finite time. The fourier integrals as derived in this tutorial and most other texts assume the limits of infinite time. If the time series has not decayed sufficiently over the period of acquisition then the approximation of infinite time will not hold. It can be verified algebraically, that this results in distortions both in lineshape and phase (see below for discussion of phasing). In practical spectroscopy, this can be a particularly serious consequence of discretely sampled fourier data acquired over a finite time, but could equally well apply to a continuously sampled time series if the limits of integration are finite.

In the discussion below we will use DFTs to illustrate experimental problems such as sampling strategy and phase errors, but we will use continuous FTs in the case of lineshape analysis, for convenience and ease of analysis.

3.4. Time series

Let us consider how a fourier transform acts on experimental data. Above we discussed how data are obtained for a fourier spectrometer, either by pulsed or interferometric methods.

The natural occurrence of a damped oscillator can be understood by reference to NMR spectroscopy [8,9]. A brief explanation is given here. Consider a spectrum consisting of a single peak. The initial intensity will correspond to the total integral of the spectral line, since the entire population of spins is excited by the broad pulse. The exponential decay arises from relaxation back to equilibrium: normally instrumental conditions can be adjusted so that the equilibrium is centred on zero, so that the average of the entire time series, ideally reached at infinite time, is zero. The oscillations are conceptually somewhat harder to understand. In NMR two models are interchangeably used to describe the process. In the quantum mechanical model, peaks arise from transitions between energy levels, separated by a given number of joules, which in turn can be related to a frequency, ν Hz. In the classical model spins can be thought of as magnets precessing around an applied field at ν Hz, these two frequencies being identical. However, the change in magnetisation with time is observed by a magnetic field precessing at a frequency close to the resonance of interest. If the peak is ν_1 Hz away from this observing pulse it will appear to oscillate in and out of phase at ν_1 Hz. Hence the frequency of the decay curve corresponds to the position of the resonance in the spectrum.

For the FTIR experiment, the frequency of sinewaves in the interferometer corresponds to the frequency of the radiation in the spectrum. The intensity parameter, as defined in this tutorial, should be intuitively obvious. The lineshape or decay method is somewhat more complex in such cases, and depends in part on instrumental factors, so the model of a damped oscillator will not adequately describe these data. In this tutorial, we will restrict our discussion to the NMR case, because the decay mechanism in FTIR does not lead to algebraically simple lineshapes. The interested reader is referred to more advanced texts [1,2,5]. However the overall mathematics and theory are identical to that of FTNMR.

A simple model for the time domain decay curve is the sum of damped oscillators (plus noise). In the case of a spectrum consisting of a single peak,

$$f(t) = I_0 e^{-at} \cos(\omega_0 t) + N(t) \qquad (12)$$

where a is a decay constant, ω_0 is the frequency of oscillation, I_0 is the initial intensity, and $N(t)$ is a noise function. These parameters are illustrated in Fig. 4.

The fourier integral has limits of infinite time. In order to approximate to this, the experimental time series should have decayed approximately to zero by the end of data acquisition. The experimenter normally adjusts conditions for pulsed data to ensure that acquisition is continued until the system reaches equilibrium which in turn is normally recorded as zero intensity. The situation with interferometric data is somewhat more complex and mathematical methods such as apodization (multiplying the data by a function which in itself ideally is zero at the end of the acquisition) often need to be employed to satisfy these conditions. This will be discussed in a subsequent tutorial.

For the damped oscillator, three parameters in the time domain can be linked to three parameters of interest in the frequency domain.

1. The initial intensity of the time domain signal corresponds to the overall integral intensity of the frequency domain peak.

2. The frequency of oscillation in the time domain corresponds to the position (frequency) in the frequency domain.

3. The decay rate constant corresponds to the line width in the frequency domain. This can be intuitively understood by reference to the uncer-

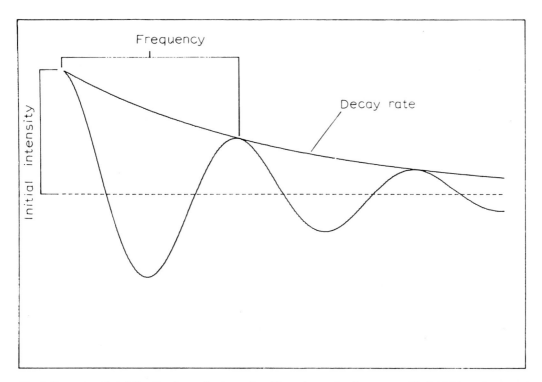

Fig. 4. Parameters that define the shape of a damped oscillator time series decay curve. The initial intensity is I_0, the decay envelope is given by e^{-at}, and the frequency is ω_0 s^{-1} or ν_0 Hz.

tainty principle. Mathematical proofs are available in a number of texts.

Table 1 illustrates the correspondence between parameters in the two domains. In practice there is likely to be considerably more than one peak present so fourier time series consist of a sum of time series. For simplicity we will restrict discussion in this tutorial, principally, to the case of a spectrum consisting of a single peak.

3.5. Sampling strategy

Above we discussed DFTs, and showed that the data sent to the computer must be sampled at regular time intervals. What is the effect of this sampling, and how are the sampled data related to the actual continuous process of decay? Consider a time domain series consisting of N datapoints sampled once every δt seconds. A peak oscillating at m Hz, in relation to the detector, will exhibit maxima of intensity at 0 s, $1/m$ s, $2/m$ s and so on. Consider the case of a peak oscillating at the rate of $1/\delta t$ Hz. The sine wave will exhibit maxima at 0 s, δt s, $2\delta t$ s ... $N\delta t$ s, i.e. every time the peak is observed, it is at a maximum. Hence this peak will appear to the recorder not to be oscillating, and will not be distinguishable from a peak at zero frequency. By extension of this argument, we can show that the computer will be able to analyse correctly the rate of oscillation of peaks that have frequencies between 0 and $1/(2\delta t)$ Hz. However, there is insufficient information to distinguish a peak oscillating at $1/(2\delta t) + x$ Hz, from $1/(2\delta t) - x$ Hz $[x \leq 1/(2\delta t)]$ (phase information sometimes helps, but this will be discussed below). This is illustrated in Fig. 5.

The frequency $1/(2\delta t)$ is called the *Nyquist* frequency. It is the maximum frequency of oscillation that can be distinguished from lower frequencies. The above analysis also demonstrates that the *rate of sampling in the time domain determines the distinguishable spectral width in the frequency domain*. The faster the sampling rate, the greater the range of frequencies that can be analyzed properly. A sampling rate of 1 datapoint per 0.5 ms is necessary if we wish to observe a frequency range of 1000 Hz. When peaks oscillate too fast for the experiment, they will be folded over (Fig. 6). A peak oscillating at 1200 Hz will appear at 800 Hz in the transform, if the sampling rate is 1 datapoint per 0.5 ms. Sampling strategy is important in the choice of acquisition parameters.

3.6. Digital resolution

The overall number of points in the time domain affects the digital resolution in the frequency domain. This can be understood as follows. If N

TABLE 1

Correspondence between time and frequency domain parameters, and common units used for measurement

Time domain	Frequency domain
Oscillation	*Position*
Frequency (Hz or s^{-1})	Frequency (Hz or s^{-1})
	Wavelength (nm, Å, cm etc.)
	Wavenumber (cm^{-1} etc.)
	Chemical shift (unitless)
	Energy (J, cal, eV etc.)
Decay	*Lineshape parameter*
Time constant to decay $1/e$ or $1/2$ initial intensity (s)	Width at $1/e$, $1/N$ height or standard deviation (units identical to frequency)
Rate constant (depends on mechanism: typically s^{-1})	Skewness and higher moments where appropriate
Intensity	*Intensity*
Intensity (physical units)	Integral (units same as time domain intensity units)
	Height (integral/position units)

Sampling rates and Nyquist frequency

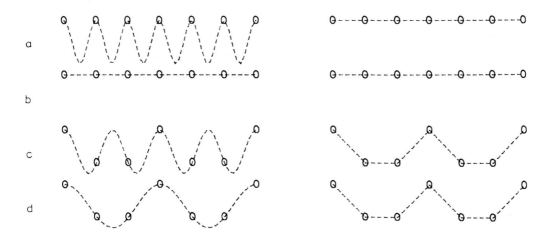

Fig. 5. Demonstration of sampling strategies and Nyquist frequencies. From top to bottom are illustrated, sine waves oscillating at (a) $1/\delta$ Hz, (b) 0 Hz, (c) $1/2\delta + x$ Hz, (d) $1/2\delta - x$ Hz, where $1/2\delta$ is the Nyquist frequency in Hz. On the left are the true sine waves, with sampling points indicated by circles. On the right is illustrated the information as recorded by the computer.

datapoints are collected, then there will be $N/2$ real and $N/2$ imaginary datapoints in the forward DFT. However, the range of frequencies is given by the Nyquist frequency ($1/2\delta t$), so the *digital resolution* in the frequency domain will be given by one datapoint per $\delta t/N$ Hz. Thus *the longer the acquisition time in the time domain, the greater the digital resolution in the frequency domain*. More advanced techniques such as zero-filling [10] will be discussed in later tutorials, and can be employed (with great care) to increase the digitisation rate in the frequency domain.

Demonstration of foldover

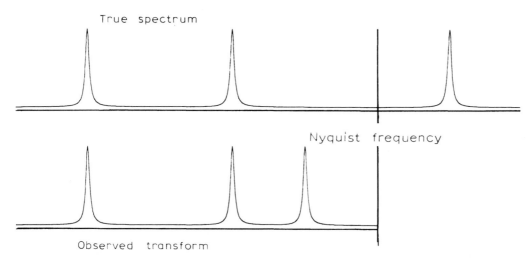

Fig. 6. At the top is a true spectrum, with the Nyquist frequency marked. Because one peak oscillates faster than the Nyquist frequency it appears folded over in the observed transform (bottom).

3.7. Relationship between time and frequency domain parameters

The above considerations lead us to an understanding of the relationship between parameters in the two domains. If N is the number of datapoints in the time domain, T is the total acquisition time, and S the spectral width in the frequency domain (related to the sampling frequency) then

$$N = 2 \cdot S \cdot T \tag{13}$$

3.8. Real and imaginary pairs

When we transform a real time series, we obtain both a real and an imaginary spectrum.

The cosine fourier transform is given by

$$RL(\omega) = \int_{-\infty}^{\infty} f(t) \cos(\omega t)\, dt$$

But a damped oscillator is described as follows

$$f(t) = I_0 \cos(\omega_0 t)\, e^{-at}$$

where ω_0 is the centre of the oscillator. We assume a symmetric even function about time zero. Hence,

$$\begin{aligned} RL(\omega) &= \int_{-\infty}^{\infty} I_0 \cos(\omega_0 t)\, e^{-a|t|} \cos(\omega t)\, dt \\ &= I_0 \int_{-\infty}^{\infty} [\cos(\omega_0 + \omega) + \cos(\omega_0 - \omega)]\, e^{-a|t|}\, dt \\ &= I_0 a / [a^2 + (\omega_0 - \omega)^2] \\ &\quad + I_0 a / [a^2 + (\omega_0 + \omega)^2] \end{aligned} \tag{14}$$

The interested reader should be able to derive eq. 4 from a time series of the form

$$f(t) = I_0 \cos(\omega_0 t)\, e^{-at^2}$$

thus verifying that the fourier transform of a gaussian is a gaussian.

The real fourier transform of an exponentially decaying sine wave consists of two peaks of lorentzian shape centred at frequencies ω_0 and $-\omega_0$ Hz. The negative frequency spectrum is a reflection of the positive frequency spectrum (this is not necessarily so for the noise distribution, but this aspect is not covered in this article). Providing the linewidth is significantly less than ω_0, we can throw away the negative frequency spectrum and keep exclusively the positive half.

In a similar way we can derive an expression for the peak shape in the imaginary spectrum and find that this is given for a lorentzian, by

$$IM(\omega) = (\omega - \omega_0) RL(\omega) / a \tag{15}$$

where ω_0 is the centre of the peak, $2a$ is the width at half height (frequency domain real spectrum) or a is the decay rate constant (time domain) as discussed elsewhere.

The appearance of the corresponding real and imaginary pairs is illustrated in Fig. 7. Note that the imaginary spectrum is not the derivative of the real spectrum: this can be verified algebraically. *For every algebraically defined decay mode in the time domain, we can derive corresponding real and imaginary fourier pairs in the frequency domain.* The FT of an exponentially decaying oscillator is a lorentzian. It can also be shown that the FT of a gaussian is, in turn, a gaussian.

3.9. Phasing

Above we modelled our decay curves as cosine waves. However, for pulsed methods, the computer must wait a short time after receiving a signal that the excitation is over, and before sampling can begin. This delay time has serious consequences.

If the trigonometric part of the cosine wave is given by $\cos(\omega_0 t)$, then if the delay time is t_0 s, the series becomes

$$f(t) = I_0 e^{-at} \cos(\omega_0 t + \omega_0 t_0) \tag{16}$$

where $\omega_0 t_0$ is a constant for a given peak. However,

$$\begin{aligned} \cos(\omega_0 t + \omega_0 t_0) &= \cos(\omega_0 t) \cos(\omega_0 t_0) \\ &\quad + \sin(\omega_0 t) \sin(\omega_0 t_0) \end{aligned}$$

$\omega_0 t_0$ *is constant for each peak, and can be replaced by a frequency dependent phase angle (ϕ), so that for a given peak*

$$\begin{aligned} f(t) &= I_0 e^{-at} \cos(\omega_0 t) \cos(\phi) \\ &\quad + I_0 e^{-at} \sin(\omega_0 t) \sin(\phi) \end{aligned} \tag{17}$$

Above we deduced that the lineshapes of pure cosine and sine waves transform into different peak shapes. The consequence of a time delay is that the real spectrum is a mixture of peak shapes, the amount of mixing being dependent both on the delay time between the end of the pulse and the beginning of the sampling, and the frequency of the peak. The amount of mixing depends on the phase angle. The above mechanism for variation

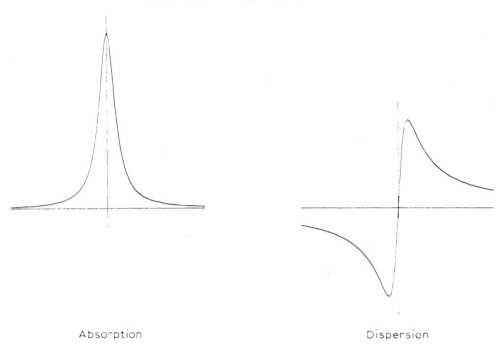

Fig. 7. The absorption (real for ideal phasing: left) and dispersion (imaginary for ideal phasing: right) lineshapes of a lorentzian are illustrated.

in phase angle accounts for a linear change across the spectrum, since the amount peaks are out of phase depends on how far the cosine wave has progressed at the beginning of sampling. The greater the frequency of oscillation, the further the sine wave has progressed (Fig. 8).

Normally, however, the experimenter is interested in pure peak shapes, and, therefore has to discover the phase angle from the data. We define the absorption spectrum by

$$ABS(\omega) = \cos(\phi)RL(\omega) + \sin(\phi)IM(\omega) \quad (18)$$

and the dispersion spectrum by

$$DISP(\omega) = \sin(\phi)RL(\omega) + \cos(\phi)IM(\omega) \quad (19)$$

where ϕ is the phase angle. If phasing is perfect (i.e. $\phi = 0$), then the absorption and real spectra are identical, and the dispersion and imaginary spectra are also identical. Peaks that have not been corrected for phase errors can be readily seen visually (Fig. 9). In addition to first order phase errors, as described above, some systems contain zero order phase imperfections, normally due to interferometer misalignment (this is common in interferometric spectrometers); higher order phase errors are also encountered. There are a large number of methods available for determining phase angles, and automatically calculating the resultant absorption spectrum from the real and imaginary components.

Sometimes determining phase angles and subsequent calculation of the phase corrected spectrum is difficult or not desirable, in which case the absolute value spectrum (see above) is used instead. It is important to note, however, that the absolute value spectrum is not quantitative. The integral of two overlapping peaks in the absorption spectrum is the sum of the integrals of each individual peak added together: this is not so in the absolute value spectrum as can be verified.

Consider two peaks A and B. The real and imaginary parts are given by

$$RL(\omega) = RL_A(\omega) + RL_B(\omega)$$
$$IM(\omega) = IM_A(\omega) + IM_B(\omega)$$

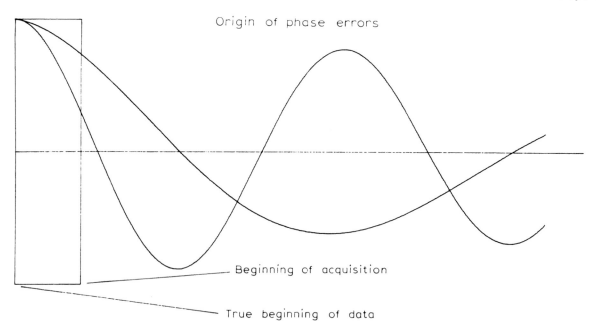

Fig. 8. Origin of first order phase errors in pulsed fourier spectroscopy. Two sine waves, oscillating at different frequencies (corresponding to two peaks in a spectrum), are illustrated. A delay between the signal to acquire data and the actual acquisition of data is reflected in the phase of the sine wave, which as is illustrated, is linearly proportional to frequency.

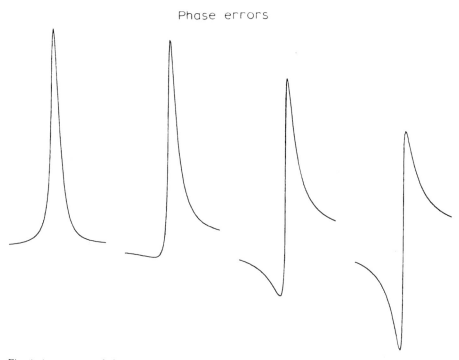

Fig. 9. Appearance of (from left to right) 0°, 30°, 60° and 90° out of phase lorentzian peaks.

Hence, the power spectrum is given by

$$P(\omega) = RL^2(\omega) + IM^2(\omega)$$
$$= RL_A^2(\omega) + RL_B^2(\omega)$$
$$+ IM_A^2(\omega) + IM_B^2(\omega)$$
$$+ 2[RL_A(\omega)RL_B(\omega) + IM_A(\omega)IM_B(\omega)] \quad (20)$$

The absolute value spectrum is the square root of the power spectrum. The sum of the absolute value spectrum of pure A and the absolute value spectrum of pure B is not the same as the absolute value spectrum of an equivalent mixture of A and B.

For ternary and higher mixtures, this non-additive property of the absolute value spectrum becomes even more severe, and it is, therefore, essential to phase correct complicated spectra if integrals are required.

4. CONCLUSION

Fourier transform spectroscopies were originally introduced into chemistry as a means of increasing signal to noise ratio, and their routine use was established when FFT algorithms became available. However, since then the special properties of fourier transforms have been exploited further. In particular, methods for filtering and resolution enhancement have been extensively studied [10]; two-dimensional transform methods have been developed [11]; image processing techniques such as maximum entropy [12] have been applied to fourier data; sophisticated lineshape analysis methods such as DISPA (dispersion vs. absorption plots) have been developed [2]; spectral analysis methods [13,14] involving fourier transforming the autocorrelogram of a time series have been used where strong correlated noise interferes with the signals of interest.

Many chemists and instrumental specialists now depend on the use of fast on-line fourier methods in the laboratory. These methods can be exploited further by understanding the underlying methodology. Future tutorials will be concerned with the problems of noise and how information theory can be used to estimate the amount of information available in chromatographic and spectroscopic data. Armed with this understanding we can use filters, spectral analysis and image processing methods to maximise the information we obtain from the raw data. Of great conceptual interest is how frequency domain methods (such as interpolation and curve fitting) compare to time domain methods (such as zero-filling and apodization); how reliably parameters such as integrals and peak positions can be estimated and how the instrumental operator can manually or automatically adjust conditions so that the parameters of interest can most accurately be measured. Finally many chemometricians deal with matrices of intensities which are normally obtained from instrumental data such as gas chromatography or NMR spectroscopy — it is vital to understand the problems and inaccuracies inherent in these data before using statistical methods for comparing samples. In many cases increasing the signal to noise ratio and the resolution of such data prior to analysis of matrices by multivariate methods can substantially increase the reliability of subsequent interpretation.

REFERENCES

1 P.R. Griffiths (Editor), *Transform Techniques in Chemistry*, Heyden, London, 1978.
2 A.G. Marshall (Editor), *Fourier, Hadamard, and Hilbert Transforms in Chemistry*, Plenum, New York, 1982.
3 J.W. Cooley and J.W. Tukey, An algorithm for the machine calculation of complex Fourier series, *Mathematics and Computers*, 19 (1965) 297–301.
4 R.R. Ernst, Sensitivity enhancement in magnetic resonance, *Advances in Magnetic Resonance*, 2 (1966) 1–137.
5 P.R. Griffiths, *Chemical Infrared Fourier Transform Spectroscopy*, Wiley, New York, 1975.
6 S.C. Brassell, R.G. Brereton, G. Eglinton, J. Grimalt, G. Liebezeit, U. Pflaumann and M. Sarnthein, Palaeoclimatic assessment through chemometric treatment of molecular stratigraphic data, in D. Leythaeuser and J. Rullkoetter (Editors), *Advances in Organic Geochemistry 12*, Pergamon, Oxford, in press.
7 C. Chatfield, *The Analysis of Time Series: An Introduction (Third Edition)*, Chapman and Hall, London, 1984, pp. 40–41.
8 D. Shaw, *Fourier Transform N.M.R. Spectroscopy*, Elsevier, Amsterdam, 1984.
9 T.C. Farrar and E.D. Becker, *Pulse and Fourier Transform Nuclear Magnetic Resonance Spectroscopy*, Academic Press, New York, 1971.
10 J.C. Lindon and A.G. Ferrige, Digitisation and data

processing in fourier transform nmr, in J.W. Emsley, J. Feeney and L.H. Sutcliffe (Editors), *Progress in NMR Spectroscopy*, Vol. 14, Pergamon, New York, 1980, pp. 27–66.
11 A. Bax, *Two dimensional nuclear magnetic resonance in liquids*, Delft University Press, Delft, 1982.
12 E.D. Laue, J. Skilling, J. Staunton, S. Sibisi and R.G. Brereton, Maximum entropy method in nuclear magnetic resonance spectroscopy, *Journal of Magnetic Resonance*, 62 (1985) 437–452.
13 G.M. Jenkins and D.G. Watts, *Spectral Analysis and its Applications*, Holden-Day, San Francisco, 1968.
14 G.E.P. Box and G.M. Jenkins, *Time Series Analysis, Forecasting, and Control*, Holden-Day, San Francisco, 1970.

Chapter 13

Dispersion vs. Absorption (DISPA): A Magic Circle for Spectroscopic Line Shape Analysis

ALAN G. MARSHALL

Departments of Chemistry and Biochemistry, The Ohio State University, Columbus, OH 43210 (U.S.A.)

(Received 18 September 1987; accepted 20 November 1987)

CONTENTS

1 Dispersion and absorption spectra 181
   1.1 Common origin of dispersion and absorption spectra 181
   1.2 Dispersion from absorption: the Hilbert transform 182
2 Dispersion vs. absorption (DISPA) 183
   2.1 Dielectric relaxation and the Cole–Cole plot 183
   2.2 DISPA for spectroscopy 183
   2.3 Extension of DISPA to Gaussian line shape 184
3 Effect of various line broadening mechanisms on a DISPA plot 184
   3.1 Peaks of different position or line width 184
   3.2 Effect of noise on a DISPA plot 185
   3.3 Other effects: baseline offset, time-domain truncation, zero-filling, modulation and power broadening, neighboring peaks, chemical exchange, magnetic field inhomogeneity 185
   3.4 Other DISPA-derived displays: linearized; radial-difference 188
4 Phasing: rotation of the DISPA plot 189
5 Applications 189
   5.1 Algorithms for on-line DISPA data reduction 189
   5.2 Nuclear magnetic resonance spectroscopy 190
   5.3 Electron paramagnetic resonance spectroscopy 192
   5.4 Ion cyclotron resonance mass spectroscopy 193
6 Conclusions and suggested future applications 193
7 Acknowledgments 193
References 194

1 DISPERSION AND ABSORPTION SPECTRA

1.1 Common origin of dispersion and absorption spectra

The most familiar spectral display is the absorption-mode spectrum. For a single Lorentzian line, the normalized absorption-mode spectrum, $A(\omega)$, is given by [1]

$$A(\omega) = \frac{\tau}{1 + (\omega - \omega_0)^2 \tau^2} \quad (1)$$

$A(\omega)$ is so named because it describes the

frequency-dependence of steady-state power absorption by a linear system (see below). Another spectral display is the dispersion-mode spectrum, originally named because it describes the frequency-dependence of refractive index, leading to the dispersion of white light into its component colors by a prism. The Lorentzian normalized dispersion-mode spectrum takes the form shown in eq. 2 [1].

$$D(\omega) = \frac{(\omega - \omega_0)\tau^2}{1 + (\omega - \omega_0)^2 \tau^2} \qquad (2)$$

Although originally identified from physically different experiments, dispersion and absorption in fact represent two complementary components of the same fundamental frequency-domain response in any spectroscopic experiment. For example, in Fourier transform spectroscopy [2], it can be shown that half of the frequency-domain spectral information is contained in $A(\omega)$ and the other half in $D(\omega)$ [3]. Their complementarity is most directly evident when $A(\omega)$ and $D(\omega)$ are assigned to be the real and imaginary parts of a mathematically complex vector, as shown in Fig. 1 [4]. $A(\omega)$ and $D(\omega)$ then appear as the projections of the complex spectrum, $C(\omega) = A(\omega) + iD(\omega)$, on its real and imaginary axes. (Of course, dispersion and absorption remain physically real quantities, but it is mathematically convenient to keep them separated as the real and imaginary parts of a complex number.)

It is worth noting that the dispersion spectrum is superficially similar to, but physically different from, the absorption-mode derivative spectrum. Absorption derivative spectra can appear either naturally, as from modulation in electron spin resonance (ESR) spectroscopy [5], or by computation from discrete absorption-mode spectra in order to better resolve overlapped broad peaks [6]. Computing the derivative of an absorption spectrum cannot increase its information content; the derivative simply makes it easier to visualize overlapped broad spectral peaks.

Dispersion and absorption spectra are sometimes detected in combination (e.g., in optical and X-ray spectroscopy) as the magnitude mode (also known as absolute-value mode), $M(\omega)$, or power

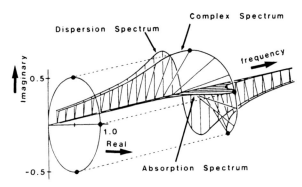

Fig. 1. Correctly phased Lorentzian spectrum, displayed in mathematically complex form. The real component vs. frequency is the absorption spectrum. The imaginary component vs. frequency is the dispersion spectrum. Viewed down the frequency axis, the real vs. imaginary component is the DISPA plot (see Fig. 2). (Reprinted with permission from ref. 4.)

spectrum, $P(\omega)$.

$$P(\omega) = (A(\omega))^2 + (D(\omega))^2 \qquad (3)$$
$$M(\omega) = (P(\omega))^{1/2} \qquad (4)$$

Historically, the absorption-mode spectrum has been the preferred display, because it is positive-valued (compared to the bimodal dispersion-mode) and is narrower than the magnitude-mode by a factor ranging from $\sqrt{3}$ (Lorentzian) to 2(sinc) [1]. However, one would like to recover the other half of the spectral information residing in the dispersion mode. For Fourier transform spectroscopy, one approach is to zero-fill the time-domain data before Fourier transformation, thereby doubling the number of absorption-mode data points and effectively transferring the dispersion information to the absorption spectrum [3]. Although useful for quantitating well-resolved peaks, zero-filling is not as helpful for analyzing highly overlapped spectra. In this tutorial article, I will try to show the advantages of an alternative combination of the dispersion and absorption spectra in a single display.

1.2 Dispersion from absorption: the Hilbert transform

As discussed in refs. 1 and 2, the same physical parameters (namely, resonant frequency, relaxation time, and number of spectroscopically active species) can be extracted from either the dispersion or absorption spectrum. Therefore, it is not surprising to find that there is a mathematical recipe, variously known as the Hilbert transform [7], Kramers–Kronig transform [8], or Bode relation [9], for interconverting between the two spectral modes.

$$D(\omega) = -\frac{1}{\pi} \int_{-\infty}^{+\infty} \frac{A(\omega)}{\omega - \omega'} d\omega' \tag{5}$$

We need to able to compute dispersion from absorption for two reasons. First, except for Fourier transform spectroscopy, only the absorption-mode spectrum may be directly (and/or easily) available experimentally, and the dispersion spectrum must therefore be computed from the absorption spectrum [10]. Second, some spectroscopic line shapes (including the relatively common Gaussian) are expressible in closed form only in the absorption-mode, and it is again necessary to use eq. 5 to evaluate the dispersion spectrum [11].

2 DISPERSION VS. ABSORPTION (DISPA)

2.1 Dielectric relaxation and the Cole–Cole plot

In 1941, Cole and Cole proposed a data reduction [12] which has since become the most highly cited paper in the field of dielectric relaxation [13]. As for spectroscopy, the dielectric experiment can be described by a complex frequency-domain line shape,

$$\epsilon(\omega) = \epsilon'(\omega) + i\epsilon''(\omega) \tag{6}$$

in which

$$\epsilon'(\omega) = \epsilon_\infty + \frac{\epsilon_0 - \epsilon_\infty}{1 + \omega^2 \tau^2}, \quad \omega > 0 \tag{7a}$$

$$\epsilon''(\omega) = \frac{(\epsilon_0 - \epsilon_\infty)\omega\tau}{1 + \omega^2 \tau^2}, \quad \omega > 0 \tag{7b}$$

represent the dielectric constant and dielectric loss [14]. Cole and Cole noticed the remarkable property,

$$(\epsilon'(\omega) - \epsilon_\infty)^2 + (\epsilon''(\omega))^2$$
$$= (\epsilon_0 - \epsilon_\infty)^2 = \text{independent of } \omega \tag{8}$$

A plot of $\epsilon'(\omega)$ versus $\epsilon''(\omega)$ for a system having a single dielectric relaxation time, τ, thus gives a semicircle centered on the abscissa. What made the Cole–Cole plot so useful was that for systems with two or more different dielectric relaxation times, the Cole–Cole curve was displaced inside the reference semicircle, thereby giving an immediate measure of the distribution in dielectric relaxation times [15].

2.2 DISPA for spectroscopy

In the mid-1970s, I noticed that the dispersion and absorption Lorentzian line shapes found in spectroscopy (eqs. 1 and 2) were mathematically quite similar to the relaxation line shapes of eqs. 7a and b. In particular, I found that the remarkable property of eq. 8 could be extended to the spectroscopic case [16].

$$[A(\omega) - (\tau/2)]^2 + [D(\omega)]^2$$
$$= [\tau/2]^2 = \text{independent of } \omega \tag{9}$$

Thus, a plot of dispersion vs. absorption (DISPA) for a single Lorentzian line is simply a circle, whose diameter is the absorption-mode peak height, centered at half the absorption-mode peak height on the absorption axis (Fig. 2). Compared to the dielectric case, the difference is that a spectroscopic peak is centered at $\omega_0 \gg (1/\tau)$, whereas the dielectric curves are centered at $\omega = 0$. Since dielectric relaxation experiments involve only positive frequencies, the Cole–Cole plot is limited to a semicircle ($\omega \geq 0$), whereas the spectroscopic experiment gives a full-circle DISPA plot ($\omega \geq \omega_0$ as well as $\omega \leq \omega_0$).

A subsequent literature search revealed that the circular nature of a DISPA plot for a Lorentzian line shape had been noted previously and independently [17,18], but no consequences of the DISPA plot were explored.

At this stage, it is worth noting that Fourier transform spectra are ideally suited to DISPA analysis for two reasons. First, discrete Fourier transformation of a time-domain in signal sampled at equal time increments automatically yields discrete frequency-domain absorption and dispersion spectra with the same vertical scaling. Thus,

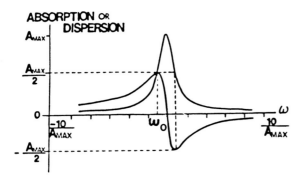

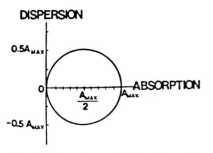

Fig. 2. Absorption and dispersion-mode spectra (top) and corresponding DISPA plot (bottom) for a correctly phased single Lorentzian line. The DISPA circle diameter is the absorption-mode maximum peak height.

dispersion and absorption spectra are automatically generated in a single experiment. Second, since Fourier transform spectra are sampled at equal frequency-domain increments, the corresponding DISPA curve will give widely spaced data points where the signal-to-noise ratio is highest (namely, near the center of the peak) and more closely spaced data points where the signal-to-noise ratio is poorest (i.e., far from the center of the spectral peak). Thus, the point spacing in a DISPA plot is well-suited for visual interpretation, since the data points of poorest signal-to-noise ratio are most tightly packed together in the display (see Section 3.2).

2.3 Extension of DISPA to Gaussian line shape

Before proceeding to the use of DISPA plots for line shape analysis, we will first consider the value of DISPA analysis for the other most common spectroscopic line shape, namely the Gaussian. Unfortunately, we soon discovered that although the absorption-mode Gaussian line shape is well-known, there is no analytic representation for the dispersion-mode Gaussian. Of the three most direct methods for generating a discrete dispersion-mode Gaussian spectrum — (a) Hilbert transform of an absorption-mode Gaussian; (b) Fourier transform, followed by even/odd conversion and inverse Fourier transform; and (c) infinite series formula — the infinite series method proved best [11].

The DISPA curve for a Gaussian spectrum is displaced outward from the DISPA circle for a Lorentzian of the same peak height. However, since the DISPA curve for a Gaussian of any line width is the same, one can restore the circular DISPA shape by simply shifting the DISPA curve back to the circle. The effect of various line-broadening mechanisms on such a 'normalized' DISPA plot for a spectrum with Gaussian component lines then proved to be the same as the effect of the same line-broadening mechanisms on the DISPA plot for a spectrum with Lorentzian component lines. In other words, if a spectrum is known to consists of Lorentzian components, then the DISPA plot is constructed directly from the dispersion and absorption spectra. However, if the spectrum is known to consists of Gaussian component lines, then the above normalizing procedure yields a DISPA plot which may be interpreted in the same way as for Lorentzian components [11].

3 EFFECT OF VARIOUS LINE BROADENING MECHANISMS ON A DISPA PLOT

3.1 Peaks of different position or line width

In spectroscopy, the frequency-domain response may represent a sum or average of two or more peaks of different position (i.e., different natural frequency) and/or different line width (arising from different relaxation time) [1]. Although these two types of inhomogeneous line-broadening are readily distinguished conceptually, they can prove quite difficult to distinguish experimentally.

For example, consider the two simple limiting cases shown in Fig. 3. Fig. 3a shows a spectrum

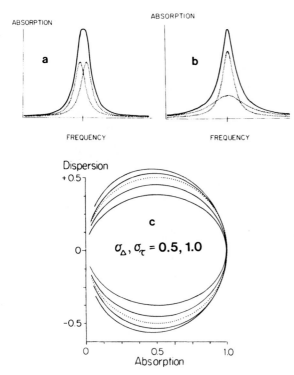

Fig. 3. Analysis of inhomogeneously broad spectra. (a) Absorption-mode sum of two Lorentzians of equal area and width but different position. (b) Absorption-mode sum of two Lorentzians of equal area and position, but different width. (c) DISPA plots for spectra consisting of a Gaussian distribution in position (with variance, σ_Δ) or log-Gauss distribution in relaxation time (with variance, σ_τ). Proceeding from outermost to innermost curves, $\sigma_\Delta = 1.0$, $\sigma_\Delta = 0.5$, $\sigma_\Delta = \sigma_\tau = 0$ (dotted circle), $\sigma_\tau = 0.5$, $\sigma_\tau = 1.0$. Note that the DISPA plot readily distinguishes a distribution in peak position (DISPA curve displaced outside the reference circle) from a distribution in peak width (DISPA curve displaced inside the reference circle).

In other words, what is needed is a diagnostic method for identifying the spectral line-broadening mechanism, so that the appropriate peak-fitting algorithm may be brought to bear on the problem.

The DISPA plot provides just such a diagnostic tool. As first observed empirically [16,19] and later derived under quite general conditions [20], a symmetrical superposition of two or more Lorentzians of equal height and width but different position will produce a DISPA curve displaced outside its reference circle (i.e., a circle whose diameter is the absorption-mode peak height), whereas a superposition of two or more Lorentzians of identical position but different width will generate a DISPA curve displaced inside its reference circle. Moreover, the displacement of a DISPA curve from its reference circle is directly related to the range of variation in component peak positions or widths [16,19,20], as seen in Fig. 3c for distributions in peak position or width. For superpositions of Gaussian peaks, the same conclusions apply, provided that the DISPA plot is first "normalized" as discussed in Section 2.3.

It is even possible to distinguish between an symmetrical doublet (DISPA curve displaced outside and to the right of its reference circle), triplet (DISPA displacement vertically outside the reference circle), and more than three lines (DISPA curve displaced outside and to the left of the reference circle) [16,19].

3.2 Effect of noise on a DISPA plot

The accuracy of any line shape analysis method is degraded by the presence of random noise in the spectrum. Fig. 4 shows graphically the effect of increasing levels of noise on a DISPA plot. It is clear that a signal-to-noise ratio of $\geq 20:1$ is necessary to decide whether a given DISPA curve is displaced (and, if so, by how far) from its reference circle [21].

3.3 Other effects

Any line-broadening mechanism which distorts the spectral line shape from that of a simple

composed of the sum of two peaks of identical width and amplitude, but different position, whereas Fig. 3b shows a spectrum consisting of the sum of two peaks of identical area and position but different width. Of course, if we know in advance that the spectrum consists of a sum of two peaks of different position (Fig. 3a), then we can devise an algorithm to determine the peak heights and positions by standard non-linear least squares fitting procedures. Unfortunately, such an algorithm will return best-fit values for those parameters even if the true line shape is actually the sum of two peaks of different width (Fig. 3b).

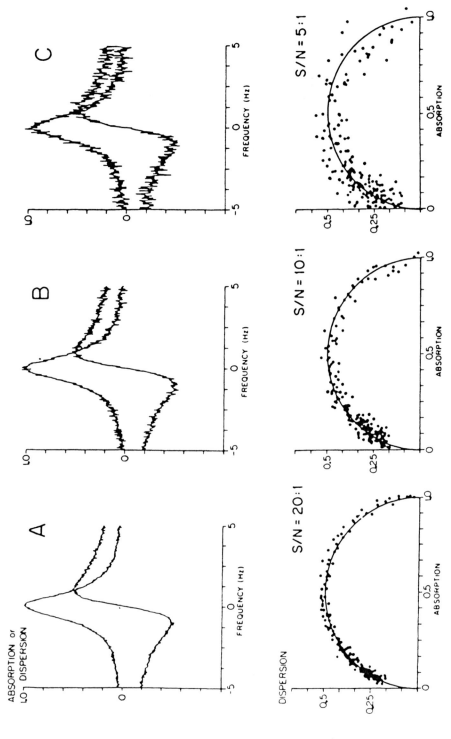

Fig. 4. Effect of random noise on the quality of the DISPA plot. For a single Lorentzian line, absorption and dispersion spectra (top row) and their corresponding DISPA plots (bottom row) have been computed for ratios of absorption-mode peak height to noise of (A) 20:1, (B) 10:1, and (C) 5:1. Clearly, a signal-to-noise ratio of ≥ 20:1 is desirable. (Reprinted with permission from ref. 21.)

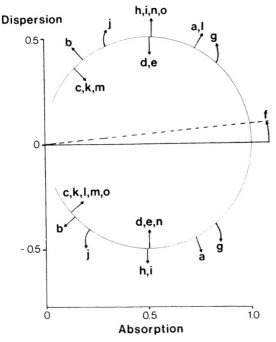

Fig. 5. Direction of displacement of a DISPA curve from its reference circle (diameter = maximum magnitude-mode peak height) for each of several line-broadening mechanisms. Note that the qualitative DISPA behavior suffices to distinguish between most of the possible mechanisms. (a) Unresolved doublet of Lorentzians of equal height and width; (b) Gaussian distribution in position of Lorentzians of equal width; (c) sum of two Lorentzians of equal area and natural (resonant) frequency but different line width; (d) log-Gauss distribution in relaxation time for Lorentzians of equal resonant frequency; (e) log-Gauss distribution in correlation time for Lorentzians of equal resonant frequency (NMR); (f) phase error (rotates DISPA plot); (g) spectrum obtained by Fourier transformation of a truncated time-domain in signal; (h) power broadening; (i) over-modulation; (j) chemical exchange between two sites of different frequency; (k) chemical exchange between two sites of different line width; (l) distortion produced by one peak of a doublet on the other; (m) effect of two outer peaks on the central peak of a triplet; (n) long recorder time constant (ESR derivative spectrum); (o) baseline drift (ESR derivative spectrum). Finally, for a simultaneous distribution in peak position and line width, the peak position distribution dominates the DISPA behavior. (Reprinted with permission from ref. 22.)

Lorentzian must also displace a DISPA curve from its reference circle. Fortunately, most line-broadening mechanisms distort the DISPA curve in characteristic ways, as summarized graphically in Fig. 5, and explained individually below.

Baseline offset. The effect of baseline offset in the dispersion or absorption spectrum (or both) is to shift the respective DISPA plot vertically or horizontally (or both), so that the DISPA curve is no longer tangent at the origin [4]. Baseline offset is readily recognized as the only spectral artifact which moves the DISPA curve away from the origin, and is easily corrected by subtracting the baseline average value from each data point in the original spectrum.

Time-domain truncation. One way to produce a Lorentzian frequency-domain spectrum is by Fourier transformation of an exponentially decreasing sinusoid of infinite duration [1]. It should therefore be clear that Fourier transformation of a time-domain signal which has been truncated after a finite acquisition period, T, must yield a spectrum which is no longer Lorentzian. The corresponding DISPA curve is displaced characteristically outward and to the right of its reference circle [21]. In such a case, truncation is usually evident from inspection of the time-domain transient signal. No line shape analysis of any kind should be attempted on a spectrum obtained from truncated time-domain in data.

Zero-filling. A common method for increasing the digital resolution of Fourier transform spectra is to 'fill' or 'pad' an N-point time-domain data set with $N(2^n - 1)$ zeroes, $n = 1, 2, 3, \ldots$. It is readily shown that the effect of the first zero-fill (i.e., the first N zeros) is to interpolate the true line shape without distortion, but that further zero-filling distorts the line shape and thus its DISPA plot [21].

Modulation and power broadening. When Fourier transform methods are not readily available (as in ESR spectroscopy), a steady-state spectrum is obtained by slow scanning across the spectrum, and only the absorption (or dispersion) is normally available. (In ESR, modulation is used to produce an absorption-mode derivative spectrum, which may be integrated to obtain the desired absorption-mode spectrum.) From the absorption spectrum, a dispersion spectrum may be generated by means of Hilbert transformation (Section 1.2), and a DISPA plot constructed. The only difference between a DISPA plot constructed from absorption alone rather than from independently measured absorption and dispersion is a reduction in signal-to-noise ratio by a factor of $2^{1/2}$ [10].

If the irradiation power is too high in such a scanning experiment, then the populations of the energy levels are perturbed by the spectrometer, and the resulting line shape is said to be saturated (if detected directly) or overmodulated (if detected by modulation of frequency or field strength). The effect of saturation [22] or overmodulation [10] is to displace the DISPA plot vertically outside of its reference circle. (Actually, since the absorption spectrum saturates before the dispersion spectrum, it would be more accurate to say that the right-hand edge of the DISPA curve is squeezed to the left.)

Neighboring peaks. Because the dispersion spectrum is so much broader than the absorption spectrum, a second peak as far as 20 line widths (measured at half-maximum peak height) away from the peak of interest can affect the DISPA plot for the latter peak [4,22]. For example, the effect of a single neighboring peak is to displace the DISPA plot diagonally, whereas the two outer peaks of a triplet displace the DISPA curve for the central peak rightward and inside its reference circle (see Fig. 5). For this reason, DISPA analysis (except for phase correction, as noted below) should be restricted to a spectral peak that is well-separated from other peaks in the spectrum.

Chemical exchange. Chemical exchange in NMR offers a particularly nice example of the value of DISPA analysis [16,19]. Fig. 6 shows the DISPA plots for a spin system undergoing chemical exchange between two sites of different chemical shift (i.e., different resonant frequency) in the intermediate- and fast-exchange limits. In the fast-exchange limit (33°C), the line shape is Lorentzian, as shown by the DISPA data points on the reference circle. At lower temperature (20°C), the exchange rate constant drops below the difference in resonant frequency between the two sites, and the resulting non-Lorentzian line shape is readily evident from the DISPA plot. Ordinarily, several experiments at different temperature would be needed to establish the exchange rate limit, but a single DISPA plot tells whether exchange is fast or not — a major advantage when (as for binding of small molecules to proteins) the temperature cannot be varied

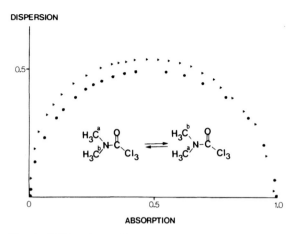

Fig. 6. DISPA characterization of exchange rate in NMR. Internal rotation about the C–N bond of N,N-dimethyltrichloroacetamide results in exchange between two sites of different proton NMR chemical shift. In the fast-exchange-limit (solid circles, 33°C), the line shape is Lorentzian and the DISPA data fall on their reference circle. At lower temperature (20°C, solid triangles) fast exchange no longer holds, as evident from the observed displacement of the DISPA curve from its reference circle. The exchange rate limit can thus be established from a single spectrum. (Reprinted with permission from ref. 19.)

without changing the properties of the system of interest.

Magnetic field inhomogeneity. In Fourier transform NMR spectroscopy, the line shape of naturally narrow (< 1 Hz at half-maximum peak height) resonances may be limited or even determined by the inhomogeneity of the applied static magnetic field. As might by now be expected, various magnetic field gradients will have characteristic effects on the DISPA curve. For example, odd-order z-gradients (e.g., z, z^3, z^5) produce symmetrical DISPA displacement (outside vertically, left, and right, respectively), whereas even-order gradients (e.g., z^2, z^4) produce asymmetrical DISPA displacements [19,23]. It has therefore been proposed that DISPA plots be used for systematic automated shimming of the magnetic field in NMR experiments [23].

3.4 Other DISPA-derived displays: linearized; radial-difference

The examples of this section demonstrate that the DISPA plot has obvious value in the identification of line-broadening mechanisms in spectroscopy, based upon the direction and extent of displacement of a DISPA curve from its reference circle. Nevertheless, it is natural to wonder the DISPA curve could somehow be transformed into a straight line, in order that displacements from the curve could be more readily evaluated.

Several methods for linearizing a DISPA plot have been considered [24,25]. First, since dispersion and absorption line shapes (eqs. 1 and 2) differ by a factor of $(\omega_0 - \omega)\tau$ in the numerator, a plot of $(\omega_0 - \omega) \cdot D(\omega)$ vs. $A(\omega)$ will produce a straight line of slope, τ, for a single Lorentzian line [24]. Unfortunately, such a plot is double-valued and fails to detect power broadening. Second, since the radius, $R(\omega) = [(A(\omega) - \tau/2)^2 + (D(\omega))^2]^{1/2}$, of a Lorentzian DISPA plot is constant with frequency (eq. 9), a plot of $(R(\omega))^2$ vs. ω gives a horizontal reference line from which to measure DISPA displacements [24]. An improved plot is $(R(\omega))^2$ vs. $\omega^{1/2}$, in which the frequency scale is expanded near resonance [24]. Finally, since a DISPA curve is normally evaluated according to its displacement from its reference circle, one could plot the 'radial difference' between a DISPA curve and its reference circle as a function of frequency [25]. All but the first of the above methods allow for judging DISPA data according to its deviation from a straight line, and none offers any demonstrable diagnostic advantage over the original circular display.

4 PHASING: ROTATION OF THE DISPA PLOT

A serious problem for analysis of Fourier transform spectral line shapes is the difficulty in distinguishing between asymmetric line shape and a small mis-phasing: i.e., the proportion of imaginary and real data combined to give the final absorption-mode spectrum (see Section 5.1 for an example). Fortunately, a unique and one of the most important features of a DISPA plot is that mis-phasing has the effect of rotating the DISPA curve by precisely the phase angle error [16]. This effect should seem intuitively reasonable, since a phase error of 90° effectively interchanges the dispersion and absorption axes, thereby rotating the DISPA plot by 90°; it therefore makes sense that the same effect should hold true for all other phase angles as well. Since no other line-broadening mechanism produces only a rotation of the DISPA curve about the plot origin, mis-phasing is readily identified and quantitated.

From a computational standpoint, three methods for finding the angle of rotation of a DISPA plot (and thereby the phase error at the natural ('center') frequency of that peak) have been proposed. The first method [26] employs iterative minimization of the difference between the two heights in a radial-difference DISPA plot, and has been applied to ESR spectra. The second method [27] uses a linear algebra construction to find the points at which a DISPA plot intersects its coordinate axes; a right-triangle construction then leads to the DISPA plot rotation angle. The first method is however iterative, and therefore computationally slow, and the second method does not inherently exclude data points of low signal-to-noise ratio. A third method [4] is based on the simple geometric property that the perpendicular bisectors of two chords of a circle must intersect at the center of the circle: the center of the rotated DISPA circle (and thereby the rotation (phase) angle) can be determined quickly and non-iteratively without ever constructing the DISPA circle itself, and by selecting only the data points of highest signal-to-noise ratio.

5 APPLICATIONS

5.1 Algorithms for on-line DISPA data reduction

Efficient algorithms for constructing a DISPA plot from stored discrete dispersion and absorption data sets are available in Fortran [25–28] and Pascal [28,29] for both mainframe and minicomputers (e.g., Bruker Aspect 3000, Nicolet 1280, Sun 3, MicroVAX II, Apollo). The advantage of the minicomputer algorithms is that they make possible on-line DISPA data reduction (e.g., for spectral phasing). As noted above, any DISPA algorithm should provide for baseline flattening of both adsorption and dispersion data sets before DISPA display. For speed in plotting, it may be desirable to plot only a decreasing fraction of the total number of data points in proceeding away from the center of the peak. Finally, in order that the initial DISPA display not be distorted by phase errors, the diameter of the DISPA reference circle should be determined from that magnitude-model (not the presumed absorption-mode) spectrum [4].

5.2 Nuclear magnetic resonance spectroscopy

In this and the following sections, discussion will be limited to DISPA analyses with provided answers not previously known (rather than simply confirming prior results). The first and most numerous applications of the DISPA method have involved Fourier transform NMR spectroscopy. Perhaps the most significant was the first experimental demonstration of the 'dynamic chemical shift' effect for quadrupolar nuclei of spin greater than 1. Predicted some ten years earlier, the second-order dynamic chemical shift produces two NMR peaks of different width and position, to give the somewhat asymmetrical peak shape shown in Fig. 7 [30]. However, in practice no one could distinguish such an effect from a slight mis-phasing of the spectrum. Since phase error and other line-broadening effects are readily distinguished by their effect (rotation or distortion, respectively) on a DISPA curve, we were able to demonstrate unequivocally the existence of the dynamic chemical shift in Fourier transform NMR spectra of the spin 3/2 ^{23}Na nucleus in aqueous soap solutions — see Fig. 7 [31]. Moreover, the detailed DISPA curve shape reveals the relative widths and separation of the two component peaks, from which the rotational correlation time could be established from a single experiment (rather than several experiments at different magnetic field strengths or temperatures, as previously required).

Several DISPA Fourier transform NMR applications have been based on the detection of two or more chemical shifts or line widths. For example, the proton NMR spectrum of benzene on dried silica gel revealed a distribution in chemical shift (rather than the previously assumed distribution in line width) [19]. On the other hand, the ^{31}P Fourier transform NMR signal from ribosomes

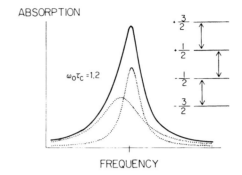

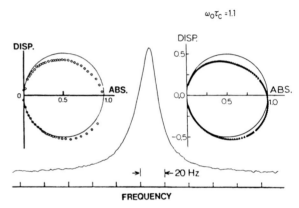

Fig. 7. DISPA demonstration of NMR dynamic chemical shift. Top: Theoretical NMR spectrum of the (narrow) $m_I = -1/2$ to $m_I = +1/2$ transition and the (broad and identical) $-3/2$ to $-1/2$ and $1/2$ to $3/2$ transitions for a slowly reorienting nucleus of spin 3/2. ω_0 is the Larmor frequency and τ_c is the rotational correlation time. Bottom: Experimental ^{23}Na Fourier transform NMR spectrum (middle) for an aqueous (15% D_2O) solution of 80 mM sodium laurate, 56 mM lauric acid, and 60 mM sodium chloride. The corresponding experimental DISPA plot (left) matches best to the theoretical DISPA plot for $\omega_0 \tau_c = 1.1$ (right), chosen from a DISPA library computed for various values of $\omega_0 \tau_c$. (Unpublished results kindly provided by T.-C.L. Wang.)

demonstrated a distribution in relaxation time (presumably arising from variations in flexibility along the phosphate backbone of the ribosomal RNA) [20]. Sykes et al. [32] used ^{19}F Fourier transform NMR DISPA to show that two fluorinated tyrosines in M13 coat protein had the same chemical shift — subsequent experiments based on solvent shifts confirmed that both tyrosines were in fact buried in similar environments in the cell membrane.

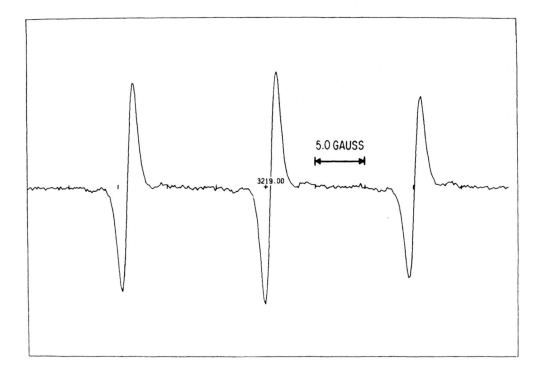
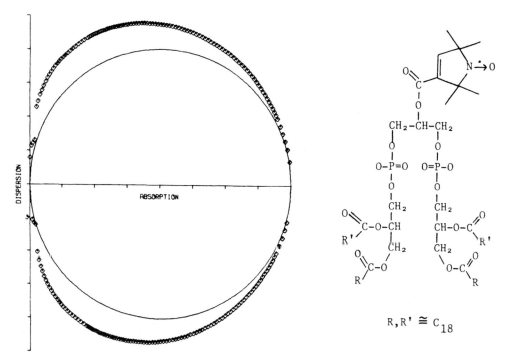

Fig. 8. ESR absorption-mode derivative spectrum (top) and corresponding DISPA plot (bottom) for the $m = 0$ transition of a nitroxide spin-labeled cardiolipin (structure at bottom right). The displacement of the DISPA curve outside and to the left of its reference circle is directly related to unresolved proton hyperfine splittings. (Reprinted with permission from ref. 2.)

^{13}C Fourier transform NMR spectra are particularly attractive candidates for DISPA analysis, because the peaks are well-separated (so that the dispersion-mode signals from neighboring peaks are less likely to overlap). For example, ^{13}C Fourier transform NMR DISPA plots have been applied to the analysis of segmental motion in polyethylenes and related polymers [33]. In other experiments, ^{13}C Fourier transform NMR DISPA analysis showed that the line shape is dominated by a relaxation time distribution which becomes broader as cross-linking increases (i.e., in proceeding from Dextran T40 and T2000 to Sephadex G100, G75, and G25 [28]).

It is well-known that tumors can often be distinguished from normal tissue by proton magnetic resonance (e.g., magnetic resonance imaging, based principally upon the H_2O signal). In that context, an intriguing recent DISPA NMR result is the observation that the water signal from normal mouse spleen is dominated by a distribution in line width (i.e., DISPA curve displaced inside its reference circle), whereas the water signal from AKR mouse spleen tumor cells is dominated by a distribution in chemical shift (i.e., DISPA curve displaced outside its reference circle) [34].

5.3 Electron paramagnetic resonance spectroscopy

DISPA plots generated from the ESR absorption-mode derivative line shape by means of numerical integration and Hilbert transform methods were initially applied to the rapid determination of of modulation amplitude and the detection of modulation broadening and other experimental artifacts in ESR spectra (e.g., amplifier time constant, baseline drift, and exchange broadening) [10]. The method was subsequently extended to the identification of line shape distortions due to unresolved hyperfine splittings, power broadening, and dispersion leakage [25].

Once the method had been generalized to incorporate Gaussian line shape, DISPA analysis was used to detect the presence of two different hyperfine ^{1}H coupling constants in the spin-label, 'Tempo' (2,2,6,6-tetramethyl-1-piperidinyloxy), compared to a single hyperfine ^{1}H coupling in another spin-label, 'Tempone' (2,2,6,6-tetra-

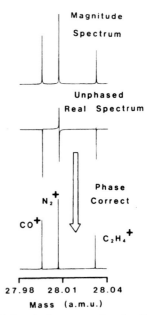

Fig. 9. DISPA-based phase correction for Fourier transform ion cyclotron resonance. From the angle of rotation of the DISPA plot (see text) for each of the three peaks of the uncorrected real spectrum (middle), phase-correction yields the absorption-mode spectrum (bottom), whose resolution (measured at half-maximum peak height) is better than that of the (unphased) magnitude-mode spectrum (top) by a factor ranging from $3^{1/2}$ to 2. (Reprinted with permission from ref. 36.)

methyl-4-oxo-1-piperidinyloxy) [11].

One of the most popular applications for ESR spin-labels has been the study of heat-induced phase-transitions in synthetic and natural phospholipid bilayer vesicles and membranes. A tacit assumption for such studies is that the hyperfine coupling constant is independent of temperature. Fig. 8 shows an ESR spectrum and the corresponding DISPA plot for a nitroxide spin-labeled cardiolipin molecule [35]. The unresolved hyperfine splittings are clearly evident from the pronounced DISPA displacement from its reference circle. Because the DISPA displacement is directly related to the magnitude of the hyperfine splitting, essentially independent of the absolute width of any one component line in this case, the temperature-dependence of the DISPA displacement from its circle should provide a way to test for the temperature-dependence of the hyperfine coupling constant.

5.4 Ion cyclotron resonance mass spectroscopy

The principal value of DISPA for Fourier transform ion cyclotron resonance (ICR) mass spectrometry is to facilitate phasing of the frequency-domain spectrum. Compared to Fourier transform NMR, in which the phase typically varies linearly across the spectrum by up to about half a cycle (i.e., 0 to π radians), Fourier transform ICR spectra can exhibit non-linear phase variation of > 1000 cycles across a single spectrum. In preliminary experiments shown in Fig. 9, we have used DISPA-based phase measurements [4] to phase simultaneously three peaks in a narrow-band Fourier transform ICR mass spectrum [36], and extensions to broad-band spectra are in progress.

6 CONCLUSIONS AND SUGGESTED FUTURE APPLICATIONS

For virtually every line-broadening mechanism, the extent of displacement of a DISPA curve from its reference circle is directly related to the line-broadening parameter of interest (e.g., the splitting in a doublet, the width of a distribution in peak position, the width of a distribution in peak width, etc.). Thus, it is in principle possible to determine such a parameter quantitatively from the DISPA displacement. However, even DISPA displacements can be similar for different line-broadening mechanisms. Therefore, the best use of DISPA analysis appears to be to establish qualitatively the line-broadening mechanism, and then use conventional non-linear least squares procedures to fit (e.g.) the absorption-mode line shape to that mechanism to determine the appropriate line-broadening parameter(s).

In all of the above discussion, we have considered only one line-broadening mechanism at a time. Actual spectral often exhibit several line-broadening mechanisms simultaneously, such as the proton or ^{13}C NMR spectrum of a polymer, for which a given peak may exhibit multiplet splitting, distribution in peak width (due to variation in flexibility and cross-linking at various locations in the polymer chain), and distribution in peak position (due to various magnetic environments at different locations along the chain). An ultimate goal of line shape analysis might therefore be to work backward from the DISPA plot to determine the relative contributions from two or more line-broadening mechanisms. As a preliminary example, we have considered a simultaneous distribution in peak position and width, and find that the two line-broadening mechanisms can to some extent cancel each other's effect [22]. Future work should aim at generalizing and quantitating such results.

It is important to remember that the information available from any spectral peak is inherently limited by its signal-to-noise ratio and the number of data points per line width [37]. Thus, another future effort should be to relate the validity of a DISPA plot to the signal-to noise ratio and digital precision of its component spectral data.

Finally, although the present discussion has been limited to magnetic resonance and ICR spectroscopy, the DISPA method should apply equally well to several other kinds of spectroscopy as well. For example, theoretical [38] and experimental [39] Cole–Cole plots have been generated in the optical frequency range from the wavelength dependence of the attenuated total reflectance from organic crystals, and used to study exciton surface polaritons. Dispersion and absorption for pure rotational (microwave) spectroscopy can be detected separately [40] or by Fourier transformation of the time-domain response [41–43], but there have as yet been no attempts to apply DISPA data reduction to the two spectra. Similarly, DISPA could readily be extended with the same advantages to any type of Fourier transform spectroscopy: e.g., infrared or visible interferometry, nuclear quadrupole resonance, etc. [2].

7 ACKNOWLEDGMENTS

This work was supported by grants from the U.S.A. National Science Foundation (8617244) and The Ohio State University.

REFERENCES

1. A.G. Marshall, Advantages of transform methods in chemistry, in A.G. Marshall (Editor), *Fourier, Hadamard, and Hilbert Transforms in Chemistry*, Plenum, New York, 1982, pp. 1–43.
2. A.G. Marshall, Transform techniques in chemistry, in T. Kuwana (Editor), *Physical Methods of Modern Chemical Analysis*, Vol. 3, Academic Press, New York, 1983, pp. 57–135.
3. E. Bartholdi and R.R. Ernst, Fourier spectroscopy and the causality principle, *Journal of Magnetic Resonance*, 11 (1973) 9–19.
4. E.C. Craig and A.G. Marshall, Automated phase correction of FT/NMR spectra by means of dispersion versus absorption (DISPA)-based phase measurement, *Journal of Magnetic Resonance*, 76 (1988) in press.
5. C.P. Poole, Jr., *Electron Spin Resonance*, Interscience, New York, 1967, Ch. 10.
6. T.R. Griffiths, K. King, H.V.St.A. Hubbard, M.-J. Schwing-Weill and J. Meullemeestre, Some aspects of the scope and limitations of derivative spectroscopy, *Analytica Chimica Acta*, 143 (1982) 163–176.
7. A.G. Marshall, Dispersion versus absorption (DISPA): Hilbert transforms in spectral line shape analysis, in A.G. Marshall (Editor), *Fourier, Hadamard, and Hilbert Transforms in Chemistry*, Plenum, New York, 1982, pp. 99–123.
8. J.A. Bardwell and M.J. Dignam, Extensions of the Kramers–Kronig transformation that cover a wide range of practical spectroscopic applications, *Journal of Chemical Physics*, 83 (1985) 5468–5478.
9. D.C. Champeney, *Fourier Transforms and their Physical Applications*, Academic Press, London, 1973, pp. 244–246.
10. F.G. Herring, A.G. Marshall, P.S. Phillips and D.C. Roe, Dispersion versus absorption (DISPA): modulation broadening and instrumental distortions in electron spin resonance line shapes, *Journal of Magnetic Resonance*, 37 (1980) 293–303.
11. T.-C.L. Wang and A.G. Marshall, Plots of dispersion versus absorption for detection of multiple positions or widths of Gaussian spectral signals, *Analytical Chemistry*, 55 (1983) 2348–2353.
12. K.S. Cole and R.H. Cole, Dispersion and absorption in dielectrics. I. Alternating current characteristics, *Journal of Chemical Physics*, 9 (1941) 341–351.
13. E. Garfield, Citation Classics, *Current Contents/Physical Sciences*, 3 (1980) 10.
14. A.G. Marshall, *Biophysical Chemistry: Principles, Techniques and Applications*, Wiley, New York, 1978, pp. 438–443.
15. R.H. Cole and P. Winsor, IV, Fourier transform dielectric spectroscopy, in A.G. Marshall (Editor), *Fourier, Hadamard, and Hilbert Transforms in Chemistry*, Plenum, New York, 1982, pp. 183–206.
16. A.G. Marshall and D.C. Roe, Dispersion versus absorption: a new spectral line shape analysis for radiofrequency and microwave spectrometry, *Analytical Chemistry*, 50 (1978) 756–763.
17. H. Kopfermann and E.E. Schneider, Nuclear moments, in H.S.W. Massey (Editor), *Pure and Applied Physics*, Academic Press, New York, 1958, p. 283.
18. E. Chovino, R. Sardos and R. Chastanet, Résonance paramagnétique électronique, *Comptes Rendus de l'Academie des Sciences, Paris, Serie B*, 273 (1971) 557–560.
19. D.C. Roe, S.H. Smallcombe and A.G. Marshall, Dispersion versus absorption: analysis of line-broadening mechanisms in nuclear magnetic resonance spectrometry, *Analytical Chemistry*, 50 (1978) 764–767.
20. A.G. Marshall, Spectroscopic dispersion versus absorption: a new method for distinguishing a distribution in peak position from a distribution in line width, *Journal of Physical Chemistry*, 83 (1979) 521–524.
21. A.G. Marshall and D.C. Roe, Dispersion versus absorption (DISPA): effects of digitization, noise, truncation of free induction decay, and zero-filling, *Journal of Magnetic Resonance*, 33 (1979) 551–557.
22. A.G. Marshall and R.E. Bruce, Dispersion versus absorption (DISPA) line shape analysis. Effect of saturation, adjacent peaks, and simultaneous distribution in peak width and position, *Journal of Magnetic Resonance*, 39 (1980) 47–54.
23. E.C. Craig and A.G. Marshall, Dispersion vs. absorption (DISPA) plots as an index of static magnetic field inhomogeneity. Use for adjustment of spinning shims in a superconducting magnet, *Journal of Magnetic Resonance*, 68 (1986) 283–295.
24. R.E. Bruce and A.G. Marshall, Linearized dispersion: absorption plots for spectral line shape analysis, *Journal of Physical Chemistry*, 84 (1980) 1372–1375.
25. P.S. Phillips and F.G. Herring, An evaluation of lineshape analysis in ESR using dispersion vs. absorption (DISPA), *Journal of Magnetic Resonance*, 57 (1984) 43–55.
26. F.G. Herring and P.S. Phillips, Automatic phase correction in magnetic resonance using DISPA, *Journal of Magnetic Resonance*, 59 (1984) 489–496.
27. C.H. Sotak, C.L. Dumoulin and M.D. Newsham, Automatic phase correction of Fourier transform NMR spectra based on the dispersion versus absorption (DISPA) lineshape analysis, *Journal of Magnetic Resonance*, 57 (1984) 453–461.
28. E.C. Craig, Application of dispersion versus absorption (DISPA) in Fourier transform nuclear magnetic resonance and Fourier transform ion cyclotron mass spectrometry, *Ph. D. Dissertation*, The Ohio State University, Columbus, OH, 1987.
29. T.-C.L. Wang, C.E. Cottrell and A.G. Marshall, Procedures for processing spectroscopic dispersion vs. absorption (DISPA) line shapes, *Computers and Chemistry*, 7 (1983) 183–197.
30. L.G. Werbelow and A.G. Marshall, The NMR of spin-3/2 nuclei: the effect of second-order dynamic frequency shifts, *Journal of Magnetic Resonance*, 43 (1981) 443–448.
31. A.G. Marshall, T.-C.L. Wang, C.E. Cottrell and L.G. Werbelow, First experimental demonstration of NMR dynamic frequency shifts: dispersion vs. absorption (DISPA) line shape analysis of sodium-23 in aqueous sodium

laurate/lauric acid solution, *Journal of the American Chemical Society*, 104 (1982) 7665–7666.
32 D.S. Hagen, J.H. Weiner and B.D. Sykes, Inaccessibility to solvent of the fluorotyrosyl residues of M13 coat protein reconstituted in synthetic phospholipid vesicles, in S. Opella and P. Lu (Editors), *NMR and Biochemistry*, Marcel Dekker, New York, 1979, p. 51.
33 J.J. Dechter, R.A. Komorowski, D.E. Axelson and L. Mandelkern, The carbon-13 NMR linewidths of polyethylenes and related polymers, *Journal of Polymer Science, Polymer Physics Edition*, 19 (1980) 631–651.
34 D.P. Lin, H. Grates and F. Valeriote, Spatial distribution of water molecules in normal and leukemia cells as studied by NMR–DISPA, *Society of Magnetic Resonance in Medicine*, 3 (1984) 474–475.
35 F.G. Herring, Fig. 50 in ref. 2.
36 E.C. Craig, I. Santos and A.G. Marshall, Dispersion vs. absorption (DISPA) method for automatic phase correction of Fourier transform ion cyclotron resonance mass spectra, *Rapid Communications in Mass Spectrometry*, 1 (1987) 33–37.
37 L. Chen, C.E. Cottrell and A.G. Marshall, Effect of signal-to-noise ratio and number of data points upon precision in measurement of peak amplitude, position, and width in Fourier transform spectrometry, *Chemometrics and Intelligent Laboratory Systems*, 1 (1986) 51–58.
38 M.R. Philpott and J.D. Swalen, Exciton surface polaritons on organic crystals, *Journal of Chemical Physics*, 69 (1978) 2912–2921.
39 J.D. Swalen, personal communication.
40 R.A. Creswell and R.H. Schwendeman, Microwave dispersion and absorption for a rotational transition in methanol, *Chemical Physics Letters*, 38 (1976) 297–299.
41 W.H. Flygare, Pulsed Fourier transform microwave spectroscopy, in A.G. Marshall (Editor), *Fourier, Hadamard, and Hilbert Transforms in Chemistry*, Plenum, New York, 1982, pp. 207–270.
42 J.M. Ramsey and W.B. Whitten, Fourier transform microwave spectrometer using an electric-field cross-correlation technique, *Review of Scientific Instruments*, 57 (1986) 1329–1337.
43 H. Dreizler, Experiences with microwave Fourier transform spectroscopy of molecular gases, *Molecular Physics*, 59 (1986) 1–28.

Chapter 14

Sampling Theory

G. KATEMAN

Laboratory for Analytical Chemistry, Catholic University Nijmegen, Toernooiveld, 6525 ED Nijmegen (The Netherlands)

(Received 23 November 1987; accepted 11 March 1988)

CONTENTS

1 Introduction	196
2 Classification	197
2.1 Objectives	198
2.1.1 Description	198
2.1.2 Monitoring	198
2.1.3 Control	198
2.2 Objects	198
2.3 Analysts	199
3 Descriptors	199
3.1 Autocorrelation	199
3.2 Semivariance	200
4 Sampling schemes	202
4.1 Random particulate objects	202
4.2 (Pseudo)continuous objects	202
4.2.1 Sampling for gross description	202
4.2.2 Sampling for detailed description and control	204
4.2.3 Sampling for monitoring	206
5 Terms used in sampling	207
References	208

1 INTRODUCTION

A widely accepted definition of analytical chemistry states: "Analytical chemistry produces (or studies, or uses) strategies for obtaining and validation of relevant information by the optimal use of available procedures in order to characterize matter or systems." This means that analytical chemists try to tell us something new about objects, something that can be used to solve problems. The objects can be goods, bulk material or material systems e.g. chemical processes or biological systems.

The way this information can be obtained changes with the problem. In many cases this information depends on the qualitative or quantitative composition of the material studied. A vast array of instruments and methods is available to the analytical chemist, but most have one thing in common: their size is limited and the way they

0169-7439/88/$03.50   © 1988 Elsevier Science Publishers B.V.

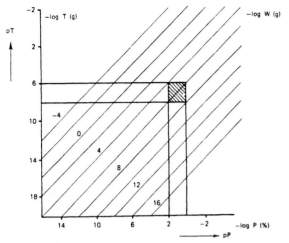

Fig. 1. Nomogram for the interdependence of sample size W, composition P and total amount of component T. (Reprinted by permission of Springer Verlag from Arbeitskreis "Automation in der Analyse". *Fresenius Z. Anal. Chem.*, 261 (1972) 7.)

obtain information is destructive. That means that, as a rule, the analytical chemist cannot or will not use the whole object in his analytical machine, but that he or she uses only a small part of the object under investigation. In practice this fraction can be very small: the amount of material introduced in the analytical method rarely exceeds 1 g, but as a rule is not more than 10^{-2} to 10^{-1} g of objects as large as 10^{-1}–10^{15} g. In many instances this fraction is larger, but a fraction of the object to be analysed of 10^{-4}–10^{-6} is common practice (Fig. 1).

The implications of such a small fraction of the object to be investigated are enormous. This fraction must fulfill a series of requirements before it may be called a sample and used as such. A sample must:
— represent the properties under investigation faithfully: e.g. composition, color, crystal type;
— be of a size that can be handled by the sampler (1–10^4 g);
— be of a size that can be handled by the analyst (10^{-3}–1 g);
— retain the properties the object had at the time of sampling, or change its properties in the same way as the object;
— be suitable to furnish the information required e.g. mean composition, composition as a function of time or place;
— maintain its identity throughout the whole procedure of sampling, transport and analysis.

Assuming that a strategy is the best way to achieve a known goal, sampling strategies depend on the type of object to be sampled, the objective of the principal and the means that are available for analysis.

The quality of a sampling method is given by the similarity between the reconstruction of the composition of an object and the object itself, insofar as it is influenced by the sampling strategy. It depends on the characteristics of the object and the purpose of the reconstruction. For a mere description of the object, the quality can be expressed in terms of the sample quality, i.e. in uncertainties concerning the means of results of sample and total object.

To satisfy these demands, the analyst can use the results of much theoretical and practical work from other disciplines. Statistics and probability theory provide the analyst with the theoretical framework that predicts the uncertainties in estimating properties of populations when only a part of the population is available for investigation. Unfortunately this theory is not well suited to analytical sampling. Analytical samples do not always belong to populations with a nicely modelled composition, e.g. a Gaussian distribution of independent items. In practice the analyst does not know the type of distribution of the composition, he usually has to deal with correlations within the object, and the sample or the number of samples must be small, since obtaining a sample is expensive. Moreover, the art of constructing instruments for sampling and sample handling developed in advance of the theory, so many sampling instruments do not give good samples.

2 CLASSIFICATION

A rough division of the different viewpoints of sampling can be made. Three points of view can be distinguished:
— the principal and his objectives;
— the object and its properties;
— the analyst and his restrictions.

2.1 Objectives

The information that is required by the principal should comply with his objectives. Most information will be used for:
— describing the object globally or in detail;
— monitoring an object or system;
— controlling a system or process.

2.1.1 Description

The description of an object can be the determination of the gross composition of e.g. lots of a manufactured product, lots of raw material, single objects, the mean state of a process. Regarding sampling effort it is desirable to collect a sample that has the minimum size set by the condition of representativeness or that is demanded by handling.

When the objective is to provide a global description of a lot then it should be born in mind that a lot is a finite part of the whole object, which means that the composition of the lot depends not only on the object, but also on the lot size. The sample, however, must represent the lot, not the object as a whole.

Another goal can be the description of the object in detail, e.g. the composition of a metal part as a function of distance from the surface, or the composition of various particles in a mixed particulate product such as a bulk blended fertilizer. Here it is necessary to know the size and number of sample increments, the distance between sample increments or the sampling frequency.

2.1.2 Monitoring

The next objective can be monitoring or threshold control of an object or system as a function of time. Here it often suffices to know that a certain value of the property under investigation has been reached, or will be reached with a certain probability. If action is required, it can be approximate.

As a rule monitoring requires that the sampling rate is as low as possible, and that one can predict with a known probability that, between the sampling times, no fatal concentration change will occur. If it is expected that the monitored value will exceed a preset value a simple action can prevent this: administer some drug, open or close a valve, etc.

2.1.3 Control

The purpose of control is to keep a process property, e.g. the composition, as close to a preset value as is technically possible and economically desirable. The deviation from the set-point is caused by intentional or random fluctuations of the process condition. In order to control the fluctuating process, samples must be taken with such frequency and analyzed with such reproducibility and speed that the process condition can be reconstructed. From this reconstruction predictions can be made for the near future and control action can be optimal. Another goal can be the detection of nonrandom deviations, such as drift or cyclic variations. This also sets the conditions for sampling frequency and sample size.

2.2 Objects

In general only two types of objects can be distinguished: homogeneous and heterogeneous. Truly homogeneous objects are rare and homogeneity may be assumed only after verification. Heterogeneous objects can be divided into two subsets: those with discrete changes of the properties and those with continuous changes. Examples of the first type are ore pellets, tablets, bulk blended fertilizer and coarse crystallized chemicals. Examples of the other type are larger quantities of fluids and gases, including air, mixtures of reacting components and finely divided granular material. This type has a correlated distribution of random properties. The behaviour of the properties can be understood by viewing the object or the process under consideration as a series of mixing tanks that connect input and output. The degree of correlation can be described by Fourier techniques or, better, by autocorrelation or semivariance. The results of Fourier techniques are given in the frequency domain, a quite unfamiliar domain when dealing with material objects. Autocorrelation and semivariance deal with the time or distance domain.

A special type of heterogeneous objects exhibits cyclic changes in its properties. Frequency of cyclic variations can be a sign of daily influences such as temperature of the environment or shift-to-shift variations. Seasonal frequencies are common in environmental objects like air or surface water.

2.3 Analysts

The analyst and his instruments suffer many restrictions. The (limited) size of the sample, the accuracy of the method of analysis, the limit of detection and the sensitivity influence the sampling strategy. These analytical parameters set an upper limit on the information that can be obtained.

Another important analytical parameter is speed. The time span between sampling and availability of the result affects the usable information for monitoring and control. The time required for analysis constrains the time available for sampling. But there is another interrelation between sampler and analyst. The frequency of sampling causes a workload for the analyst and affects the analysis time by queueing of the samples on the laboratory bench, thus diminishing the information output. Obviously an optimal sampling frequency exists for a given size of the laboratory and its organization.

The theory of queueing can be applied to study the effects on information yield of such limited facilities. Basically, limited facilities result in a limited number of practicable analytical measurements. If this number is occasionally exceeded, waiting times occur, and a situation may arise in which more samples enter the system than leave it. Besides this quantity the ways in which a certain number of samples is offered and processed are important. In this respect, organizational quality aspects can be quantified. Such systems have been extensively studied using queueing theory.

3 DESCRIPTORS

3.1 Autocorrelation

A thorough treatment of this subject can be found in ref. 1. The object can be described as being more or less internally correlated in time or space. Factors causing the internal correlation of objects include diffusion or mixing within the object, for example in tanks, compressors and pipelines, sediments or rocks; or the varying properties of the producer of the object. In both situations samples are mutually dependent. The difference between two samples increases with greater distance, however. To estimate the number of increments that have to be taken to obtain a "representative" sample or a gross sample, the correlation of the object must be known. An efficient and much used method is to calculate the autocorrelation.

The autocorrelation can be obtained from the time series that represents the object, in this case a collection of measurements on the same variable quantity measured as a function of time. An assumption is that the object is stationary; that is, the physical variable as a function of time does not depend on the starting time of the observation period.

The mean of this stationary object is given by:

$$\mu = \frac{1}{n} \sum_{t=1}^{n} x_t \quad \text{for } n \to \infty \tag{1}$$

or

$$\bar{x} = \frac{1}{n} \sum_{t=1}^{n} x_t \tag{1a}$$

The variance is also a constant:

$$\sigma^2 = \frac{1}{n} \sum_{t=1}^{n} (x_t - \mu)^2 \quad \text{for } n \to \infty \tag{2}$$

or

$$s^2 = \frac{1}{n-1} \sum_{t=1}^{n} (x_t - \bar{x})^2 \tag{2a}$$

The covariance between x_t and $x_{t+\tau}$ (τ being a number smaller than n) can be calculated and is called the autocovariance σ_τ:

$$\sigma_\tau = \frac{1}{n} \sum_{t=1}^{n} ((x_t - \mu)(x_{t+\tau} - \mu)) \tag{3}$$

or

$$s_\tau = \frac{1}{n-\tau} \sum_{t=1}^{n-\tau} ((x_t - \bar{x})(x_{t+\tau} - \bar{x})) \tag{3a}$$

By normalizing the autocovariance the autocorrelation function can be obtained:

$$\rho_\tau = \sigma_\tau / \sigma^2 \qquad (4)$$

or

$$r_\tau = s_\tau / s^2 \qquad (4a)$$

Because τ is between $-\infty$ and $+\infty$ and the function is symmetric around $\tau = 0$, in general only that part of the function between 0 and $+\tau$ is represented. The autocorrelation function for $\tau = 0$ equals 1; a higher value of τ usually results in a lower value of ρ_τ. For noncorrelated measurements the autocorrelation function sharply drops to 0 for all $\tau > 0$. In practical situations correlated time series frequently produce an autocorrelation that may be represented by an exponential function:

$$r_\tau \cong e^{-\tau/T_x} \qquad (5)$$

Here T_x is called the time constant of the process. It can be found from the τ-value at $\rho_\tau = 0.37$, because $\ln 0.37 = -1 = -\tau/T_x$, or by one of the methods indicated in Fig. 2.

A non-stationary process yields r_τ values that do not tend to zero for high τ-values. A measured signal that includes drift is always autocorrelated. A periodic function such as a sine function yields an autocorrelation function with the same periodicity. A series of measurements suffering from many random high frequent disturbances can produce a very statisfactory autocorrelation function. The calculation of r_τ may be useful in situations where a series of process values is hidden by sources of noise. Because white noise produces a zero contribution to the autocorrelation function for $\tau \neq 0$, the influence of random disturbances on the autocorrelation function is eliminated and the calculation of r_τ offers a method of reducing noise from a time series of measurements (Fig. 3).

In most practical situations the autocorrelation function can be represented by an exponential function. This may be used to predict the probability of occurrence of a process value for time t, based on the process value x_0 at time 0:

$$x_t = x_0 \, e^{-\tau/T_x} \qquad (6)$$

The accuracy of the prediction x_t decreases with an increased t-value. According to Bartlett [2] for $\tau \gg T_x$:

$$\mathrm{var}(r_\tau) = \frac{1}{n-\tau} \left[\frac{\{1+\exp(-2/T_x)\}}{\{1-\exp(-2/T_x)\}} \right] \frac{T_x}{n-\tau} \qquad (7)$$

This implies that mostly the confidence interval is large and the precision is low. For limiting t-values toward ∞, a prediction of all values is possible within the limit of standard deviation and a probability corresponding to the Gaussian distribution.

3.2 Semivariance

Unmixed, stratified objects such as soil and rock, cannot usually be represented by a simple correlation constant derived from the autocorrelation model. Here another measure is introduced, the semivariogram, γ_τ:

$$\gamma_\tau = \frac{\sum_{i=1}^{n-\tau} (x_i - x_{i+\tau})^2}{2n} \qquad (8)$$

where x_i = measurement at location i, $x_{i+\tau}$ = measurement taken τ intervals Δ away, and n = number of sampling points.

The distance at which γ_τ approaches σ_0^2 is called the range, a. σ_0^2 is the variance of the observations, or the autocovariance at lag 0. There is a mathematical relationship between autocorre-

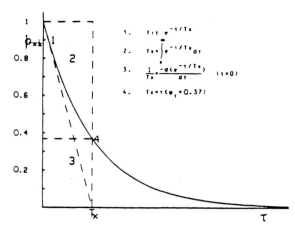

Fig. 2. Methods for estimation of the time constant T_x from the autocorrelogram of a first-order autoregressive time series.

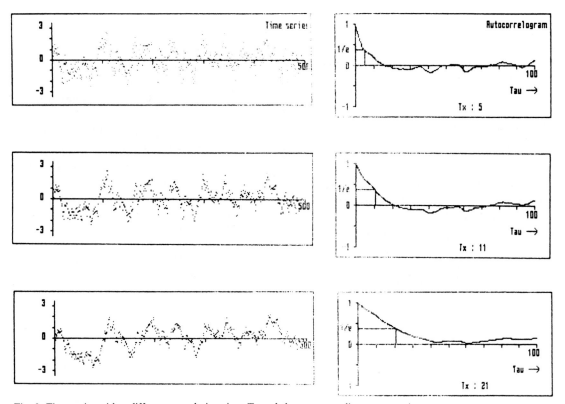

Fig. 3. Time series with a different correlation time T_x and the corresponding autocorrelograms.

lation and semivariance. If the variable x is stationary, the semivariance for a distance Δ is equal to the difference between the variance and the autocovariance for the same distance (Fig. 4).

As in the model for the autocorrelation we neglect drift and deterministic disturbances and

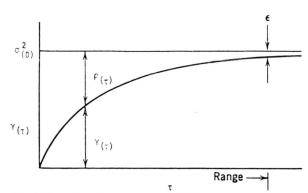

Fig. 4. The relation between autocorrelation and semivariance.

assume a stationary, stochastic object (constant mean and variance).

There are many ways to model the semivariogram. A model that resembles the exponential autocorrelation model is:

$$\gamma_\tau = \sigma_0^2 (1 - \exp(-\tau/a)) \qquad (9)$$

where $a =$ the range, the distance at which $\gamma_\tau \cong \sigma_0^2$.

An often-used model is the spherical model described by:

$$\gamma_\tau = \sigma^2 \left(\frac{3\tau}{2a} - \frac{\tau^3}{2a^3} \right) \qquad (10)$$

Care must be taken when interpreting autocorrelograms or semivariograms in which $\rho_1 \ll 1$ or $\gamma_1 \gg 0$. Statistically this means that the correlation between adjacent points is low. The usual interpretation for autocorrelation is that the measuring system has a high noise level. When using semivariograms the same interpretation could be given but the phenomenon is often explained by assuming that the object had different statistical properties: the so-called "nugget effect". Of course, both interpretations can be valid but the interpretation is influenced by the subculture from which it comes.

4 SAMPLING SCHEMES

4.1 Random particulate objects

The starting point for sampling theory was the object that consists of two or more particulate parts. The particulate composition is responsible for a discontinuous change in composition. A sample must be of sufficient size to represent the average composition of the object. The cause of any difference between an aliquot of insufficient size and the object itself is the inevitable statistical error. The composition of the sample, the collection of all increments, is given by the mean composition of the object and the standard deviation of this mean. As the standard deviation depends on the number of particles, and the size and the composition of these particles, an equation can be derived that gives the minimum number of particles.

Visman et al. [3–5] described the sampling variance as the sum of random and segregation components according to:

$$S^2 = (A/wn) + (B/n) \quad (11)$$

Here A and B are constants determined by preliminary measurements on the system. The A term is the random component; its magnitude depends on the weight w and number of sample increments collected, as well as the size and variability of the composition of the units in the population. The B term is the segregation component; its magnitude depends only on the number of aliquots collected and on the heterogeneity of the population. Different ways to estimate the values of A and B are possible. It has been shown [5–7] that the relation

$$WR^2 = K_s \quad (12)$$

is valid in many situations. Here W represents the weight of the sample, R is the relative standard deviation (in percent) of the sample composition and K_s is the sampling constant, the weight of sample required to limit the sampling uncertainty to 1% with 68% confidence (assuming a normal distribution). The magnitude of K_s may be determined by estimating the standard deviation from a series of measurements of samples of weight W.

An alternative way [6] to calculate the sample size is by defining a shape factor f as the ratio of the average volume of all particles having a maximum linear dimension equal to the mesh size of a screen to that of a cube which will just pass the same screen. $f = 1.00$ for cubes and 0.524 for spheres. For most materials $f \approx 0.5$.

The particle size distribution factor g is the ratio of the upper size limit (95% pass screen) to the lower size limit (5% pass screen). For homogeneous particle sizes $g = 1.00$. The composition factor c is:

$$c = (1-x)[(1-x)d_x + xd_g]/x \quad (13)$$

where x = the overall concentration of the component of interest, d_x = the density of this component, and d_g = the density of the matrix.

c is in the range 5×10^{-5} kg/m^3 for high concentrations of c to 10^3 for trace concentrations.

The liberation factor l is:

$$l = (d_l/d)^{1/2} \quad (14)$$

where d_l = mean diameter of component of interest, and d = diameter of largest particles.

The standard deviation of the sample s is estimated by:

$$s^2 = fgcld^3/w \quad (15)$$

Ingamells [8] has derived a relation between eq. 12 and eq. 15:

$$K_s = fgcl(d^3 \cdot 10^4) \quad (16)$$

A number of other sample constants exists, each for some specific goal.

4.2 (Pseudo)continuous objects

4.2.1 Sampling for gross description

When the object to be analyzed is a process, defined as a stream of material of infinite length with properties varying in time, the sampling parameters can be derived from the process parameters. When the object is a finite part of a process, however, which is usually called a lot, the description of the real composition of the lot depends not only on the parameters of the process from which the lot is derived, but also on the size of the lot. The sample must now represent the lot, not the process.

Lots derived from stationary, stochastic processes with normally distributed properties allow us to adopt a theoretical approach. In practice most lots seem to fulfill the above requirements with sufficient accuracy to justify the following equations [10,11]. An estimate of the mean m of a lot with size P can be obtained by taking n samples of size G, equally spaced with a distance A. In this case $P = nA$ and the size of the gross sample $S = nG$. Here it is assumed that it is not permitted to have overlapping samples and that $A > G$.

The relevant parameter is the variance in the composition of the gross sample compared to the variance in the lot. This variance σ_e^2 is thought of being composed of the variance in the composition of the gross sample itself σ_m^2, the variance in the composition of the whole lot σ_μ^2, and the covariance between m and μ: $\sigma_{m\mu}$.

$$\sigma_e^2 = \sigma_m^2 + \sigma_\mu^2 - 2\sigma_{m\mu} \tag{17}$$

This leads to:

$$\sigma_m^2 = \frac{2\sigma_x^2}{ng^2}\left(g - 1 + \exp(-g)\right.$$
$$+ [\exp(-g) + \exp(g) - 2]$$
$$\times \left(\frac{\exp(-a)}{1 - \exp(-a)}\right.$$
$$\left.\left.- \frac{\exp(-a)[1 - \exp(-p)]}{n[1 - \exp(-a)]^2}\right)\right) \tag{18}$$

$$\sigma_\mu^2 = \frac{2\sigma^2}{p^2}[p - 1 + \exp(-p)] \tag{19}$$

$$\sigma_{m\mu} = \frac{\sigma_x^2}{npg}\left(2ng + [1 - \exp(-p)]\right.$$
$$\left.\times \left(\frac{\exp(-g) - 1}{1 - \exp(-a)} + \frac{\exp(g) - 1}{1 - \exp(a)}\right)\right) \tag{20}$$

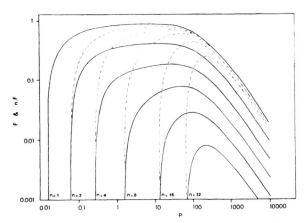

Fig. 5. The relative increment size $F = G/P$ (———) and the relative gross sample size nF (— — —) as a function of the lot size p for a number of samples n. $(\sigma_e/\sigma_x) = 0.1$ [10,11].

where σ_x^2 = variance of process and $p = P/T_x$, $g = G/T_x$, $a = A/T_x$, where P = lot size, G = increment size, A = distance between increments, and T_x = time constant of the process.

As can be seen in these equations, σ_e^2 depends on a number of factors.

The properties of the process from which the lot stems are described by σ_x, the standard deviation of the process, and T_x, the correlation factor of the process.

The relevant property of the lot is its size p, expressed in the same units as T_x. These units may be time units, such as when T_x is measured in hours. The lot size is expressed in hours as well. When describing a river, the lot size can be the mass of water that flows by in one day or one year. The constant T_x can be called a space constant when the unit used is length, however. Here the lot size is expressed in length units. This is the case, for example, when a lot of manufactured products contained in a conveyor belt or stored in a pile of material in a warehouse is considered.

The correlation factor can also be expressed in dimensionless units, as when bags of products are produced. In this case the lot size is also expressed in terms of items (number of bags, drums, tablets). The properties of the sample are the increment size G, expressed in the same units as T_x and P, the distance between the centre of adjacent increments A (also in units of T_x, P and G), and the number of increments n that form the sample. If the sample size is chosen such that the subsequent samples are taken without interruption, it can be shown that $\sigma_e^2 = 0$. This result is not surprising because here the whole lot has been sampled: $A = P/n$, and the sample size $nG = P$. For an uncorrelated lot, with $T_x = 0$, it can be shown that σ_m, σ_μ, and $\sigma_{m\mu}$ and consequently σ_e are all 0. If, however, the size of the sample increments $G = 0$, $\sigma_m^2 = \sigma_x^2/n$ and $\sigma_e^2 = \sigma_x^2/n$.

In other words, if the lot size is very large in time constant units, one sample of finite size suffices; but if the increment size is near zero, a number of increments must be taken to form a sample. Note that here the sample can be obtained from one part of the object (e.g. when $G \neq 0$) or from several, equally spaced sampling sites (e.g. when $G = 0$).

When $T_x \to \infty$ the situation of an homogeneous object is approached. In this case it can be shown that σ_m, σ_μ, and $\sigma_{m\mu}$ all approach the value of σ_x, therefore σ_e will be 0 and one sample of any desired size suffices. In highly autocorrelated lots only one small sample is needed to describe the whole lot with sufficient accuracy. For medium lot sizes P, the smallest sample needed to obtain a certain value is one made up of many small increments. When fewer but larger increments are taken, the size of the sample increases. For large values of P and with σ_e/σ_x constant, $\log F$ is proportional to $\log P$.

In soil science one usually has to deal with two- or three-dimensional objects that cannot be represented by correlation constants. In this case an intermediate step is kriging, i.e. the mapping of lines or planes of equal composition. For the simplest case, punctual kriging, the composition of points p is estimated by solving W from the matrix equation

$$[A] \cdot [W] = [B] \qquad (21)$$

with

$$A = \begin{bmatrix} \gamma(\tau_{11}) & \gamma(\tau_{12}) & \cdots & 1 \\ \gamma(\tau_{12}) & \gamma(\tau_{22}) & \cdots & 1 \\ \gamma(\tau_{13}) & \gamma(\tau_{23}) & \cdots & 1 \\ 1 & 1 & 1 & 0 \end{bmatrix} \qquad (22)$$

with $\gamma(\tau_{ij})$ being the semi-variance over a distance τ corresponding to the control points i and j.

$$W = \begin{bmatrix} W_1 \\ W_2 \\ \cdot \\ \cdot \\ \cdot \\ \gamma \end{bmatrix} \qquad B = \begin{bmatrix} \gamma(\tau_{1p}) \\ \gamma(\tau_{2p}) \\ \cdot \\ \cdot \\ \cdot \\ 1 \end{bmatrix} \qquad (23)$$

with $\gamma(\tau_{1p})$ being the semi-variance over a distance τ equal to that between known point i and the location p where the estimate is to be made.

Now

$$x_p = W_1 x_1 + W_2 x_2 + W_3 x_3 \qquad (24)$$

with a variance

$$\sigma_e^2 = W_1 \gamma(\tau_{1p}) + W_2 \gamma(\tau_{2p}) + W_3 \gamma(\tau 3p) + \lambda \qquad (25)$$

By estimating "compartments" of compositions that do not vary more than a given amount, sampling can be restricted to one sample per compartment.

4.2.2 Sampling for detailed description and control

When the objective of sampling is the estimation of the composition of the object in detail, the sampling strategy will be different from the strategy aimed at the estimation of the mean composition of the lot. Use of a model of the distribution of the composition allows us to reconstruct the value of the composition.

Application of these reconstruction methods can be described as the interpolation of object composition between sample points by means of an exponential function, characterised by the correlation constant. The estimation algorithm can be used for real time process control when the time between sampling and availability of the analytical result is introduced. A measure of reconstruction efficiency is the measurability factor [13,14]:

$$m = \left((\sigma_x^2 - \sigma_e^2)/\sigma_x^2 \right)^{1/2} \qquad (26)$$

where σ_x^2 = variance of composition of object, and σ_e^2 = residual variance after reconstruction.

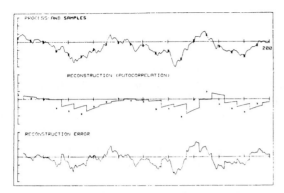

Fig. 6. Reconstruction and reconstruction error [13,14].

When the object is a process, an object whose composition changes with time, this factor m can be estimated by:

$$m = [\exp -(d + a/2 + g/3)](1 - s_a t^{1/2}) \quad (27)$$

where $d = D/T_x$, $a = A/T_x$, $g = G/T_x$, $t_e = T_e/T_x$ and $s_a = \sigma_a/\sigma_x$ with D = analysis time, A = (sampling frequency)$^{-1}$, G = sample size, T_x = time constant of process, T_e = time constant of measuring device, σ_a^2 = variance of method of analysis, and σ_x^2 = variance of process, or

$$m = m_D \cdot m_A \cdot m_G \cdot m_N \quad (28)$$

where $m_D = \exp(-d)$, $m_A = \exp(-a/2)$, $m_G = \exp(-g/3)$, and $m_N = 1 - s_a t_e^{1/2}$ (Fig. 6).

The maximum obtainable measurability factor will never exceed the smallest of the composing factors. This implies that all factors should be considered, in order to eliminate the restricting one. It also means that a trade-off is possible between high and low values of the various factors. If, for example, m_N, the factor influenced by the reproducibility of the measuring device or the method of analysis is high and m_A is low, the frequency of sampling can be increased, causing m_A to increase.

From the fact that $m_A = \exp(-a/2)$ it follows that decreasing a, the distance between samples or the reciprocal of the sampling frequency, causes m_A to increase. The higher the sampling rate, the better the possibility of controlling the process. This does not mean, however, that the highest obtainable frequency is the best, for sampling costs rise roughly linearly with $1/a$. Moreover the optimal value of a depends on balancing the costs of analysis against costs of a possible process failure.

From the fact that $m_G = \exp(-g/3)$ it is clear that decreasing g, the sample size, increases m_G. The conclusion is that, when sampling for control, it is always better to use samples that have the minimum size required by analysis. This is in contrast to the situation encountered when describing a lot. A special case is encountered when the time constant of the instrument, an automatic analyzer, for example, sets the sampling rate. Here $T_e = a$ so

$$m_N = 1 - s \cdot a^{1/2} \quad (29)$$

The sampling rate is not only set by the sampling scheme, but also by the laboratory. If, for instance, the frequency of sampling is too high, queueing of samples will occur, which implies a loss of information if the results are used for process control [15]. The effect can be calculated for simple systems with one service point (one analyst or one instrument). For more complicated systems simulation should be applied. For an $M/M/1$ system (random sampling/random analysis time/1 server), which is, in fact, the least organized system, the measurability factor is:

$$m^2 = [T_x/(T_x + 2\overline{A})][(1-\rho)T_x + \overline{D}] \quad (30)$$

where T_x = time constant, $\overline{A}$ = mean interarrival time of samples, $\overline{D}$ = mean interanalysis time, and ρ = utilization factor of the lab.

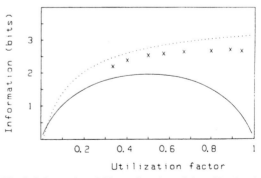

Fig. 7. Information yield as a function of the utilization factor. (———) $M/M/1$ system FIFO, ($\times\times\times$) $M/M/1$ system LIFO, ($\cdots\cdots$) $D/D/1$ system [15].

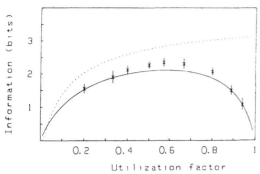

Fig. 8. Information yield as a function of the utilization factor. $M/D/1$ system FIFO, ($\times \times \times$) simulation results with 95% probability intervals [15].

For a $D/D/1$ system (fixed sampling rate/fixed analysis time) there are no waiting times (Fig. 7).

The situations $D/M/1$ (fixed sampling rate) and $M/D/1$ (fixed analysis time) are intermediate between the two extremes described above (Figs. 8 and 9).

$D/M/1$ system:

$$m^2 = \exp(-2\bar{A}/T_x)((1-\rho)T_x \\ /[(1-\rho)T_x + 2\bar{D}]) \tag{31}$$

$M/D/1$ system:

$$m^2 = [T_x/(T_x + 2\bar{A})](2(1-\rho)\bar{A} \\ /[(2\bar{A} - T_x)\exp(2\bar{D}/T_x) + T_x]) \tag{32}$$

There is an optimal mean sampling frequency: however, at a utilization factor of 0.5 this is

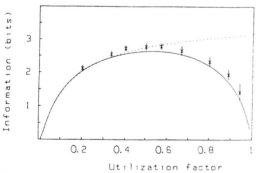

Fig. 9. Information yield as a function of the utilization factor. $D/M/1$ system FIFO, ($\times \times \times$) simulation results with 95% probability intervals [15].

surprisingly low. Only half of the analyst's capacity is used in obtaining the maximal information yield, or in gaining the best process control when using such system. At lower utilization factors, the information decreases because of the increasing waiting time in the system: the information-decreasing effect of the waiting times overrides the information-increasing effect of the higher sampling frequency. In a $D/D/1$ system there are no waiting times. Here, it is obvious that a utilization factor of 1 gives maximal information.

With the theory outlined above, it is possible to develop an analytical planning scheme. First it is important to distinguish between the influence of a stochastic input ($M/D/1$ system) and a stochastic processing time ($D/M/1$ system). The latter is clearly more favourable.

Various plans to improve the performance of the laboratory or to increase the information yield are possible. One conclusion drawn from the above is that smoothing the input is advantageous (from $M/M/1$ to $D/M/1$).

A totally different kind of plan is to fix priorities for the various samples entering the laboratory. Two possibilities are considered. In one, the priorities influence the sequence in which the samples are processed. With respect to process control, it can be shown by simulation that a Last-In-First-Out (LIFO) scheme gives a higher information yield than the standard First-In-First-Out (FIFO) scheme. This is hardly surprising since, for an auto-regressive process, only the most recent sample gives relevant information. In fact all the samples that queue could be discarded.

4.2.3 Sampling for monitoring

Monitoring is the measurement of the changing properties of an object in order to detect too large deviations from a preset value. If a (first order auto-regressive) object is analyzed and the result is immediately available then, at that moment, there is no uncertainty about the object.

During the next period, however, no new information is obtained, and only old predictions are available. When the value of these predictions (their probability for instance) decreases too much, a new sample should be taken and analyzed. As

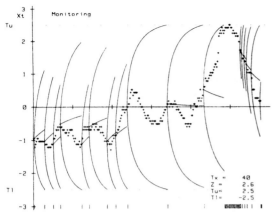

Fig. 10. Graphic representation of the operation of the monitoring system, showing the predicted process values and the 95% reliability interval of the prediction. (+ + +) actual process values, |(bottom line) sampling points with intervals τ_L. T_u and T_l are upper and lower threshold values T_r [16].

long as the predicted value, including the prediction error, does not exceed the warning threshold, no new sample is required [16]. The next sample should be taken at a time τ_L after an analytical result x_t:

$$\tau_L = -T_x \times \ln\left(T_r x_t + z\left[x_t^2 - q(T_r^2 - z^2)\right]^{1/2} \right. $$
$$\left. /(x_t^2 + qz^2)\right) - T_d \qquad (33)$$

where T_r = threshold value, normalized to zero mean and unit standard deviation, x_t = last measured value (also normalized), z = reliability factor, $q = (\sigma_x^2 + \sigma_a^2)/\sigma_x^2$, T_d = time between sampling and availability of result, σ_x^2 = variance of object, and σ_a^2 = variance of method of analysis (Fig. 10).

5 TERMS USED IN SAMPLING [17]

Bulk sampling. Sampling of a material that does not consist of discrete, identifiable, constant units, but rather of arbitrary, irregular units.
Gross sample. (Also called bulk sample, lot sample.) One or more increments of material taken from a larger quantity (lot) of material for assay or record purposes.

Homogeneity. The degree to which a property or substance is randomly distributed throughout a material. Homogeneity depends on the size of the units under consideration. Thus a mixture of two minerals may be inhomogeneneous at the molecular or atomic level but homogeneous at the particulate level.
Increment. An individual portion of material collected by a single operation of a sampling device, from parts of a lot separated in time or space. Increments may be either tested individually or combined (composited) and tested as a unit.
Individuals. Conceivable constituent parts of the population.
Laboratory sample. A sample, intended for testing or analysis, prepared from a gross sample or otherwise obtained. The laboratory sample must retain the composition of the gross sample. Often reduction in particle size is necessary in the course of reducing the quantity.
Lot. A quantity of bulk material of similar composition whose properties are under study.
Population. A generic term denoting any finite or infinite collection of individual things, objects, or events in the broadest concept; an aggregate determined by some property that distinguishes things that do and do not belong.
Reduction. The process of preparing one or more subsamples from a sample.
Sample. A portion of a population or lot. It may consist of an individual or groups of individuals.
Segment. A specifically demarked portion of a lot, either actual or hypothetical.
Strata. Segments of a lot that may vary with respect to the property under study.
Subsample. A portion taken from a sample. A laboratory sample may be a subsample of a gross sample; similarly, a test portion may be a subsample of a laboratory sample.
Test portion. (Also called specimen, test specimen, test unit, aliquot.) That quantity of material of proper size for measurement of the property of interest. Test portions may be taken from the gross sample directly, but often preliminary operations such as mixing or further reduction in particle size are necessary.

REFERENCES

1. G.E.P. Box and G.M. Jenkins, *Time Series Analysis*, Holden-Day, Oakland, CA, 1976.
2. M.S. Bartlett, On the theoretical specification and sampling properties of autocorrelated time series, *Journal of the Royal Statistical Society*, B8 (1946) 27–41.
3. J. Visman, A general sampling theory, *Materials Research and Standards*, 9 (11) (1969) 8–64.
4. J. Visman, A.J. Duncan and M. Lerner, General theory of sampling, *Materials Research and Standards*, 11 (8) (1971) 32.
5. J. Visman, General theory of sampling, *Journal of Materials*, 7 (1972) 345–350.
6. M. Gy, *Sampling of Particulate Materials*, Elsevier, Amsterdam, 1979.
7. C.O. Ingamells and P. Switzer, A proposed sampling constant for use in geochemical analysis, *Talanta*, 20 (1973) 547–568.
8. C.O. Ingamells, New approaches to geochemical analysis and sampling, *Talanta*, 21 (1974) 141–155.
9. C.O. Ingamells, Derivation of the sampling constant equation, *Talanta*, 23 (1976) 263–264.
10. P.J.W.M. Muskens and G. Kateman, Sampling of internally correlated lots, the reproducibility of gross samples as a function of sample size, lot size and number of samples. Part I: Theory, *Analytica Chimica Acta*, 103 (1978) 1–9.
11. G. Kateman and P.W.J.M. Muskens, Sampling of internally correlated lots, the reproducibility of gross samples as a function of sample size, lot size and number of samples. Part II: Implication for practical sampling and analysis, *Analytica Chimica Acta*, 103 (1978) 11–20.
12. R.A. Olea, *Measuring Spatial Dependence with Variograms*, Kansas Geological Survey Series on Spatial Analysis, no. 3, University of Kansas, Lawrence, 1976.
13. P.M.E.M. van der Grinten, Control effects of instrument accuracy and measuring speed, *Journal of the Instrument Society of America*, 12 (1) (1965) 48–50.
14. P.M.E.M. van der Grinten, Control effects of instrument accuracy and measuring speed, *Journal of the Instrument Society of America*, 13 (2) (1966) 58–61.
15. T.A.H.M. Janse and G. Kateman, Enhancement performance of analytical laboratories; a theoretical approach to analytical planning, *Analytica Chimica Acta*, 150 (1983) 219–231.
16. P.J.W.M. Muskens, The use of autocorrelation techniques for selecting optimal sampling frequency. Application to surveillance of surface water quality, *Analytica Chimica Acta*, 103 (1978) 445–457.
17. B. Kratochvil, D. Wallace and J.K. Taylor, Sampling for chemical analysis, *Analytical Chemistry*, 56 (1984) 113R–129R.

Principal Component Analysis

SVANTE WOLD *

Research Group for Chemometrics, Institute of Chemistry, Umeå University, S 901 87 Umeå (Sweden)

KIM ESBENSEN and PAUL GELADI

Norwegian Computing Center, P.B. 335 Blindern, N 0314 Oslo 3 (Norway) and Research Group for Chemometrics, Institute of Chemistry, Umeå University, S 901 87 Umeå (Sweden)

CONTENTS

1 Introduction: history of principal component analysis 209
2 Problem definition for multivariate data 210
3 A chemical example 212
4 Geometric interpretation of principal component analysis 213
5 Mathematical definition of principal component analysis 213
6 Statistics; how to use the residuals 215
7 Plots 216
8 Applications of principal component analysis 218
   8.1 Overview (plots) of any data table 218
   8.2 Dimensionality reduction 218
   8.3 Similarity models 219
9 Data pre-treatment 219
10 Rank, or dimensionality, of a principal components model 220
11 Extensions; two-block regression and many-way tables 221
12 Summary 222
   Appendix 222
   References 223

1 INTRODUCTION: HISTORY OF PRINCIPAL COMPONENT ANALYSIS

Principal component analysis (PCA) in many ways forms the basis for multivariate data analysis. PCA provides an approximation of a data table, a data matrix, X, in terms of the product of two small matrices T and P'. These matrices, T and P', capture the essential data patterns of X.

Plotting the columns of T gives a picture of the dominant "object patterns" of X and, analogously, plotting the rows of P' shows the complementary "variable patterns".

Consider, as an example, a data matrix containing absorbances at $K = 100$ frequencies measured on $N = 10$ mixtures of two chemical constituents. This matrix is well approximated by a (10×2) matrix T times a (2×100) matrix P', where T describes the concentrations of the constituents and P describes their spectra.

PCA was first formulated in statistics by Pearson [1], who formulated the analysis as finding "lines and planes of closest fit to systems of points in space". This geometric interpretation will be further discussed in Section 4. PCA was briefly mentioned by Fisher and MacKenzie [2] as more suitable than analysis of variance for the modelling of response data. Fisher and MacKenzie also outlined the NIPALS algorithm, later rediscovered by Wold [3]. Hotelling [4] further developed PCA to its present stage. In the 1930s, the development of factor analysis (FA) was started by Thurstone and other psychologists. This needs mentioning here because FA is closely related to PCA and often the two methods are confused and the two names are incorrectly used interchangeably.

Since then, the utility of PCA has been rediscovered in many diverse scientific fields, resulting in, amongst other things, an abundance of redundant terminology. PCA now goes under many names. Apart from those already mentioned, singular value decomposition (SVD) is used in numerical analysis [5,6] and Karhunen-Loéve expansion [7,8] in electrical engineering. Eigenvector analysis and characteristic vector analysis are often used in the physical sciences. In image analysis, the term Hotelling transformation is often used for a principal component projection. Correspondence analysis is a special double-scaled variant of PCA that is much favoured in French-speaking countries and Canada and in some scientific fields.

Many good statistical textbooks that include this subject have been published, e.g., by Gnanadesikan [9], Mardia et al. [10], Johnson and Wichern [11] and Joliffe [12]. The latter is devoted solely to PCA and is strongly recommended for reading.

In chemistry, PCA was introduced by Malinowski around 1960 under the name principal factor analysis, and after 1970 a large number of chemical applications have been published (see Malinowski and Howery [13]), and Kowalski et al. [14]).

In geology, PCA has lived a more secluded life, partly overshadowed by its twin brother factor analysis (FA), which has seen ups and downs in the past 15-20 years. The one eminent textbook in this field of geological factor analysis is that by Jöreskog, Klovan and Reyment [15]. Davis [16], who has set the standards for statistics and data analysis in geology for more than a decade, also included a lucid introduction to PCA.

2 PROBLEM DEFINITION FOR MULTIVARIATE DATA

The starting point in all multivariate data analysis is a data matrix (a data table) denoted by **X**. The N rows in the table are termed "objects". These often correspond to chemical or geological samples. The K columns are termed "variables" and comprise the measurements made on the objects. Fig. 1 gives an overview of the different goals one can have for analysing a data matrix. These are defined by the problem at hand and not all of them have to be considered at the same time.

Fig. 2 gives a graphical overview of the matrices and vectors used in PCA. Many of the goals of PCA are concerned with finding relationships between objects. One may be interested, for example, in finding classes of similar objects. The class membership may be known in advance, but it may also be found by exploration of the available data. Associated with this is the detection of outliers, since outliers do not belong to known classes.

AVAILABLE

A matrix of data, measured for N objects, K variables per object

GOALS OF PCA

SIMPLIFICATION
DATA REDUCTION
MODELING
OUTLIER DETECTION
VARIABLE SELECTION
CLASSIFICATION
PREDICTION
UNMIXING

Fig. 1. Principal component analysis on a data matrix can have many goals.

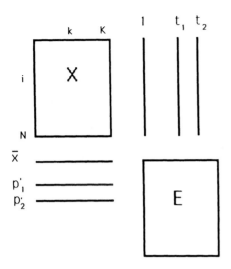

$$X = 1\bar{x} + TP' + E$$

Fig. 2. A data matrix **X** with its first two principal components. Index i is used for objects (rows) and index k for variables (columns). There are N objects and K variables. The matrix **E** contains the residuals, the part of the data not "explained" by the PC model.

Another goal could be data reduction. This is useful when large amounts of data may be approximated by a moderately complex model structure.

In general, almost any data matrix can be simplified by PCA. A large table of numbers is one of the more difficult things for the human mind to comprehend. PCA can be used together with a well selected set of objects and variables to build a model of how a physical or chemical system behaves, and this model can be used for prediction when new data are measured for the same system. PCA has also been used for unmixing constant sum mixtures. This branch is usually called curve resolution [17,18].

PCA estimates the correlation structure of the variables. The importance of a variable in a PC model is indicated by the size of its residual variance. This is often used for variable selection.

Fig. 3 gives a graphical explanation of PCA as a tool for separating an underlying systematic

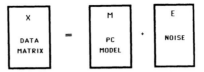

Fig. 3. The data matrix **X** can be regarded as a combination of an underlying structure (PC model) **M** and noise **E**. The underlying structure can be known in advance or one may have to estimate it from **X**.

data structure from noise. Fig. 4a and b indicate the projection properties of PCA. With adequate interpretation, such projections reveal the dominating characteristics of a given multivariate data set.

Fig. 5 contains a small (3 × 4) numerical illustration that will be used as an example.

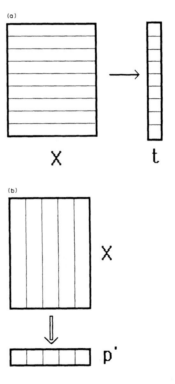

Fig. 4. (a) Projecting the matrix **X** into a vector t is the same as assigning a scalar to every object (row). The projection is chosen such that the values in t have desirable properties and that the noise contributes as little as possible. (b) projecting the matrix **X** into a vector p' is the same as assigning a scalar to every variable (column). The projection is chosen such that the values in p' have desirable properties and that the noise contributed as little as possible.

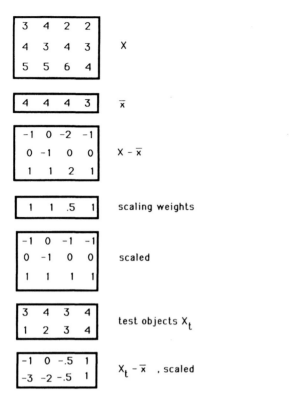

Fig. 5. A training data matrix used as an illustration. Two extra objects are included as a test set. The actions of mean-centring and variance-scaling are illustrated.

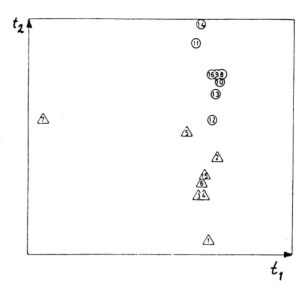

Fig. 6. Plot of the first two PC score vectors (t_1 and t_2) of the swede data of Cole and Phelps [19]. The data were logarithmed and centred but not scaled before the analysis. Objects 1-7 and 15 are fresh samples, whereas 8-14 and 16 are stored samples.

3 A CHEMICAL EXAMPLE

Cole and Phelps [19] presented data relating to a classical multivariate discrimination problem. On 16 samples of fresh and stored swedes (vegetables), they measured 8 chromatographic peaks. Two data classes are present: fresh swedes and swedes that have been stored for some months. The food administration problem is clear: can we distinguish between these two categories on the basis of the chromatographic data alone?

Fig. 6 shows the PC score plot (explained later) for all 16 samples. Two strong features stand out: sample 7 is an outlier and the two classes, fresh and stored, are indeed separated from each other. Fig. 7 shows the corresponding plot calculated on the reduced data set where object 7 has been deleted. Here the separation between these two data classes is even better. Further details of the analysis of this particular data set will be given below.

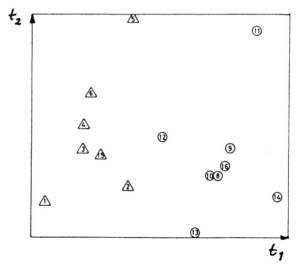

Fig. 7. PC plot of the same data as in Fig. 6 but calculated after object 7 was deleted.

4 GEOMETRIC INTERPRETATION OF PRINCIPAL COMPONENT ANALYSIS

A data matrix **X** with N objects and K variables can be represented as an ensemble of N points in a K-dimensional space. This space may be termed M-space for measurement space or multivariate space or K-space to indicate its dimensionality. An M-space is difficult to visualize when $K > 3$. However, mathematically, such a space is similar to a space with only two or three dimensions. Geometrical concepts such as points, lines, planes, distances and angles all have the same properties in M-space as in 3-space. As a demonstration, consider the following BASIC program, which calculates the distance between the two points I and J in 3-space:

100 KDIM = 3
110 DIST = 0
120 FOR L = 1 TO KDIM
130 DIST = DIST + (X(I, L) − X(J, L)) ∗ ∗2
140 NEXT L
150 DIST = SQR(DIST)

How can be change this program to calculate the distance between two points in a space with, say, 7 or 156 dimensions? Simply change statement 100 to KDIM = 7 or KDIM = 156.

A straight line with direction coefficients $P(K)$ passing through a point with coordinates $C(K)$ has the same equation in any linear space. All points (I) on the line have coordinates $X(I, K)$ obeying the relationship (again in BASIC notation)

X(I, K) = C(K) + T(I) ∗ P(K)

Hence, one can use 2-spaces and 3-spaces as illustrations for what happens in any space we discuss henceforth. Fig. 8 shows a 3-space with a point swarm approximated by a one-component PC model: a straight line. A two-component PC model is a plane — defined by two orthogonal lines — and an A-components PC model is an A-dimensional hyperplane. From Fig. 8 it may be realized that the fitting of a principal component line to a number of data points is a least squares process.

Lines, planes and hyperplanes can be seen as

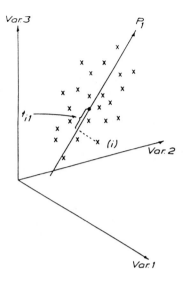

Fig. 8. A data matrix **X** is represented as a swarm with N points in a K-dimensional space. This figure shows a 3-space with a straight line fitted to the points: a one-component PC model. The PC score of an object (t_i) is its orthogonal projection on the PC line. The direction coefficients of the line from the loading vector p_k.

spaces with one, two and more dimensions. Hence, we can see PCA also as the projection of a point swarm in M-space down on a lower-dimensional subspace with A dimensions.

Another way to think about PCA is to regard the subspaces as a windows into M-space. The data are projected on to the window, which gives a picture of their configuration in M-space.

5 MATHEMATICAL DEFINITION OF PRINCIPAL COMPONENT ANALYSIS

The projection of X down on an A-dimensional subspace by means of the projection matrix **P**′ gives the object coordinates in this plane, **T**. The columns in **T**, t_a, are called score vectors and the rows in **P**′, p'_a, are called loading vectors. The latter comprise the direction coefficients of the PC (hyper) plane. The vectors t_a and p_a are orthogonal, i.e., $p'_i p_j = 0$ and $t'_i t_j = 0$, for $i \neq j$.

The deviations between projections and the original coordinates are termed the residuals. These are collected in the matrix **E**. PCA in matrix form is the least squares model:

$$X = 1\bar{x} + TP' + E$$

Here the mean vector $\bar{x}$ is explicitly included in the model formulation, but this is not mandatory. The data may be projected on a hyperplane passing through the origin. Fig. 9 gives a graphical representation of this formula.

The sizes of the vectors t_a and p_a in a PC dimension are undefined with respect to a multiplicative constant, c, as $tp = (tc)(p/c)$. Hence it is necessary to anchor the solution in some way. This is usually done by normalizing the vectors p_a to length 1.0. In addition, it is useful to constrain its largest element to be positive. In this way, the ambiguity for $c = -1$ is removed.

An anchoring often used in FA is to have the length of p_a be the square root of the corresponding eigenvalue l_a. This makes the elements in p_a correspond directly to correlation coefficients and the score vectors t_a be standardized to length 1.0.

It is instructive to make a comparison with the singular value decomposition (SVD) formulation:

$$X = 1\bar{x} + UDV' + E$$

In this instance, **V'** is identical with **P'**. **U** contains the same column vectors as does **T**, but normalized to length one. **D** is a diagonal matrix containing the lengths of the column vectors of **T**. These diagonal elements of **D** are the square roots of the eigenvalues of **X'X**.

In the statistical literature, PCA has two slightly different meanings. Traditionally, PCA has been viewed as an expansion of **X** in as many components as $\min(N, K)$. This corresponds to expressing **X** in new orthogonal variables, i.e., a transformation to a new coordinate system. The one which is discussed here refers to PCA as the approximation of the matrix **X** by a model with a relatively small number of columns in **T** and **P**. The possibility of deciding on a specific cut-off of the number of components gives a flexible tool for problem-dependent data analysis: several contributions to this issue make ample use of these PCA facilities.

A basic assumption in the use of PCA is that the score and loading vectors corresponding to the largest eigenvalues contain the most useful information relating to the specific problem, and that the remaining ones mainly comprise noise. Therefore, these vectors are usually written in order of descending eigenvalues.

Often the obtained PC model is rotated by the rotation matrix **R** to make the scores and loading easier to interpret. This is possible because of the equivalence

$$TP' = TRR^{-1}P' = SQ'$$

The FA literature contains numerous discussions about various rotation schemes, which we refrain from adding to here because the choice of rotation is very problem specific and often problematic.

Once the PC model has been developed for a "training matrix", new objects or variables may be fitted to the model giving scores, t, for the new objects, or loadings, p, for the new variables, respectively. In addition, the variance of the residuals, e, is obtained for each fitted item, providing a measure of similarity between the item and the "training data". If this residual variance is larger than that found in the training stage, it can be concluded that the new object (or variable) does not belong to the training population. Hypothesis tests can be applied to this situation. The residuals may alternatively be interpreted as residual distances with respect to a pertinent PC model.

The formulae for a new object x are as follows: multiply by the loadings from the training stage of obtain the estimated scores t:

$$t = xP$$

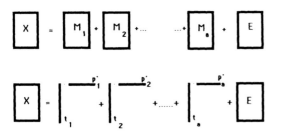

Fig. 9. A data matrix **X** can be decomposed as a sum of matrices M_i and a residual **E**. The M_i can be seen as consisting of outer products of a score vector t_i and a loading vector p_i'.

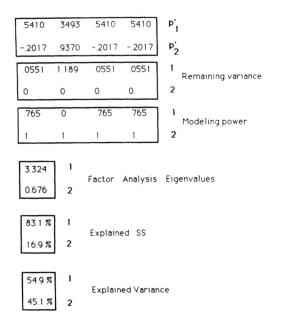

Fig. 10. Results for the PC model built from the training set in Fig. 5.

x is projected into the A-dimensional space that was developed in the training stage.

Calculated the residuals vector e:

$$e = x - tP' \text{ or } e = x(I - PP')$$

Here I is the identity matrix of size K. This calculation of the new scores t_a or loadings p_a is equivalent to linear regression because of the orthogonality of the vectors.

Figs. 10 and 11 show the results of a PCA of the (3×4) matrix in Fig. 5.

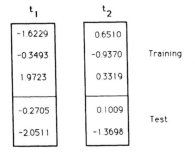

Fig. 11. Scores obtained for training and test set in Fig. 5.

6 STATISTICS; HOW TO USE THE RESIDUALS

This section describes two different, but related, statistics of a data matrix and its residuals, the usual variance statistics and influence statistics (leverage).

The amount of explained variance can be expressed in different ways. In FA the data are always scaled to unit variance and the loadings are traditionally calculated by diagonalization of the correlation matrix $X'X$. To obtain a measure corresponding to the factor analytical eigenvalue, in PCA one may calculate the fraction of the explained sum of squares (SS) multiplied by the number of variables, K. Thus, in Fig. 13 the first PC explains 83.1% of the SS. Hence the first eigenvalue is $0.831 \times 4 = 3.324$.

For a centred matrix (column means subtracted), the variance is the sum of squares (SS) divided by the number of degrees of freedom. Sums of squares can be calculated for the matrix X and the residuals E. The number of degrees of freedom depends on the number of PC dimensions calculated. It is $(N - A - 1)(K - A)$ for the Ath dimension when the data have been centred (column averages subtracted), otherwise $(N - A)(K - A)$. It is also practical to list the residual variance and modelling power for each variable (see Fig. 10).

The modeling power is defined as explained standard deviation per variable $(1 - s_k/s_{0k})$. A variable is completely relevant when its modeling power is 1. Variables with a low modeling power, below ca. (A/K), are of little relevance. This follows from the eigenvalue larger than one rule of significance (see below).

The total sum of squares and the sums of squares over rows and columns can all be used to calculate variance statistics. These can be shown as histograms and allow one to follow the evolution of the PCA model as more dimensions are calculated. Fig. 12 gives an idea of how this would look in the case of variable statistics.

The topic of influential data has been introduced recently, mainly in multiple regression analysis [20–23]. A measure of influence that can be visualized geometrically is leverage. The term leverage is based on the Archimedian idea that

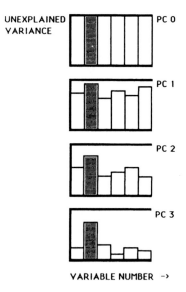

Fig. 12. Statistics for the variables. The data are shown as histogram bars representing the variance per variable for mean-centred and variance-scaled data. With 0 PCs, all variables have the same variance. For PC models with increasing dimensionality, the variance of the variables is used up. The hatched bar shows a variable that contributes little to the PC model. A similar reasoning can be used for object variances.

anything can be lifted out of balance if the lifter has a long enough lever.

The least squares method used for fitting principal components to object points in M-space makes leverage useful for PCA. Fig. 13 shows an

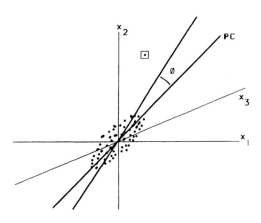

Fig. 13. Leverage. A point of high leverage (indicated by a square) can rotate the principal component axis over an angle ϕ.

illustration of the effect of a high leverage point on a principal component. It can be seen that high leverage is not necessarily bad. A high leverage observation falling near a PC axis only reinforces the PC model. A high leverage observation lying far away from a PC line causes a rotation of the PC.

Leverage is calculated as follows [24,25]:

$$H_0 = T(T'T)^{-1}T'$$

The diagonal element h_{ii} of H_0 is the leverage for the ith object. The h_i values are between 0 and 1.

For the variables, leverage is calculated as

$$H_v = PP'$$

The diagonal elements h_{kk} of H_v are the leverages for the variables.

The interpretation of leverage is closely related to the concepts of outliers and of construction of a set of samples, i.e., the experimental (or sampling) design.

7 PLOTS

Perhaps the most common use of PCA is in the conversion of a data matrix to a few informative plots. By plotting the columns t_a in the score matrix T against each other, one obtains a picture of the objects and their configuration in M-space. The first few component plots, the t_1–t_2 or t_1–t_3, etc., display the most dominant patterns in X. As was commented upon above, this tacitly assumes that the directions of maximum variance represent the directions of maximum information. This need not apply to all types of data sets, but it is a well substantiated empirical finding.

Fig. 14 shows the loading plot corresponding to fig. 6 for the swedes example. In this plot one can directly identify which variables cause No. 7 to be an outlier and which variables are responsible for the separation of the two classes, fresh and stored. The directions in Fig. 6 correspond directly to the directions in Fig. 14. The horizontal direction separates No. 7 from the others in Fig. 6. Hence, variables far from zero in the horizontal direction in Fig. 14 (Nos. 5 and 6) are those responsible for this. Analogously, the vertical direction in Fig. 6

Chapter 15

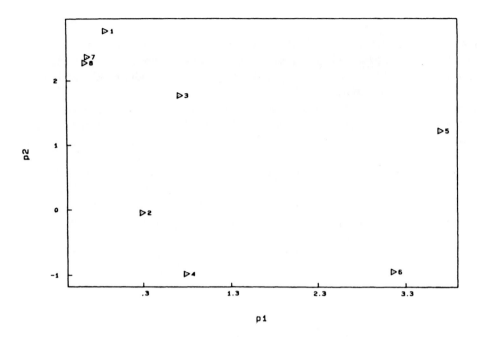

Fig. 14. Plot of the first two loading vectors (p_1 and p_2) corresponding to Fig. 6.

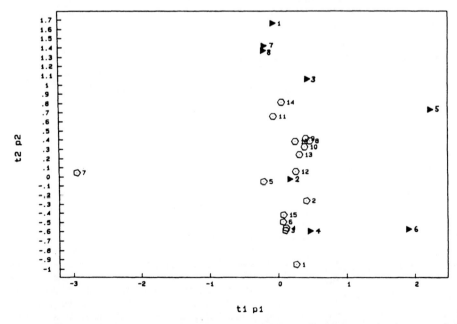

Fig. 15. Figs. 6 and 14 superimposed (origin in the same place). The loadings (triangles) are scaled up by a factor of 3.

separates the two classes. Hence, variables vertically far from zero in Fig. 7 (all except No. 2, but mainly 1, 3, 7 and 8) contribute to this separation.

One may also superimpose these two plots (Figs. 6 and 14) to obtain Fig. 15, which simultaneously displays both the objects and the variables. This type of plot is common in correspondence analysis, but can also be accomplished by ordinary PCA. It is largely a matter of choice whether one wishes to inspect these two complementary plots in one or two separate figures. For larger data sets it is probably best to keep the two plots separate to improve clarity.

8 APPLICATIONS OF PRINCIPAL COMPONENT ANALYSIS

PCA can be applied to any data matrix (properly transformed and scaled, see Section 9). This is also recommended as an initial step of any multivariate analysis to obtain a first look at the structure of the data, to help identify outliers, delineate classes, etc. However, when the objective is classification (pattern recognition levels 1 or 2 [26]) or relating one set of variables to another (e.g., calibration), there are extensions of PCA that are more efficient for these problems.

Here a rather partisan view of PCA is presented. It reflects the experience that well considered projections encompass a surprisingly large range of typical scientific problem formulations [27,28].

8.1 Overview (plots) of any data table

The score plot of the first two or three score vectors, t_a, shows groupings, outliers and other strong patterns in the data. This can be seen in the score plot of the swedes data. Many other applications can be found in this issue.

8.2 Dimensionality reduction

As pointed out by Frank and Kowalski [29], two main groups of FA and PCA applications can be seen in analytical chemistry, namely the extraction of the underlying factors — the latent variables — and the resolution of spectra of multi-component mixtures. In two-dimensional high-performance liquid chromatographic analysis, for instance, this is a way to find the smallest number of species in a sample. Reviews of the numerous applications of these types can be found in Kowalski et al. [14] and Malinowski and Howery [13].

The first few score vectors, t_a, may be seen as latent variables that express most of the information in the data. This has recently been used to find "principal properties" of amino acids, solvents and catalysts, which then later find use in the quantitative description of the selection of these entities [30,31]. See Fig. 16 for an example.

PCA and FA are used in many other types of applications. As PC models can be also calculated for matrices with incomplete data, PCA may be used to predict the values for the "holes" in a data matrix. This, however, is more efficiently done with partial least quares analysis, mentioned in Section 11.

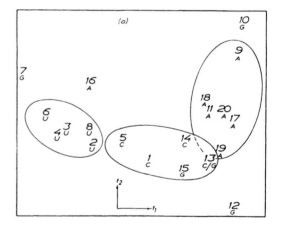

Fig. 16. Plot of the first and second PC score vectors of a table with 20 properties for the 20 common "natural" amino acids. The grouping indicates a relationship between the physical-chemical properties of the amino acids and which nucleotide is used in the second codon in the amino acid synthesis (adenosine, uracil and cytosine). The amino acids coded for by guanine (G) do not seem to participate in this relationship [46]. These PC score vectors numerically describe the structural change within families of peptides and thus constitute a basis for structure–activity relationships for peptides [30].

8.3 Similarity models

Wold [32] showed that a PC model has the same approximation property for a data table of similar objects as does a polynomial for bivariate data in a limited interval; the PC model can be seen as a Taylor expansion of a data table. The closer the similarity between the objects, the fewer terms are needed in the expansion to achieve a certain approximation goodness. The prerequisites for this interpretation of PCA are a few assumptions about differentiability and continuity of the data generating process, which have been found to hold in many instances. This explains, at least partly, the practical utility of PCA.

Hence, a pattern recognition method may be based on separate PC models, one for each class of objects. New objects are classified according to their fit or lack of fit to the class models, which gives a probabilistic classification. This is the basis for the SIMCA method (Soft Independent Modeling of Class Analogies) [33]. If one wishes to distinguish optimally between the two classes of swedes, a PC model can be fitted to each of the classes. The results show that indeed there is similarity within the classes (two- or three-component models adequately describe the data) and that the classes are well separated (fresh samples fit the "stored class" badly and vice versa); see Wold et al. [34].

9 DATA PRE-TREATMENT

The principal components model parameters depend on the data matrix, its transformation and its scaling. Hence, these must be explicitly defined.

First, the data matrix: In statistics it is usually customary to put all available data into the matrix and analyse the lot. This conforms with the convention that *all* data reflect legitimate phenomena that have been sampled. When one has full control over the sampling, or over the experimental design, this approach may be correct.

Practical experience from many sciences often outlines a less ideal reality, however. One is frequently faced with outlying data in real data sets. One also often knows certain pertinent external facts (such that cannot be coded into the data matrix itself) relating to the problem formulation. In chemistry, for example, one often has one set of objects (e.g., analytical samples) about which certain essential properties are known. In the svedes example, for instance, there were 16 samples that were known or assumed to be either fresh or stored. The analysis confirmed this, but the PCA recognized that one sample did not comply. This type of external information can be used to compute a polished PC model. On may subsequently investigate new objects and project them on to the same scores plot without letting the new samples influence the PC model as such.

Hence, it is practical to distinguish between a training set that is used for calculating problem-dependent PC models and a test set of objects that are later subjected to the same projection as that developed in the training phase.

From the least squares formulation of the PC models above it is seen that the scores, t, can be viewed as linear combinations of the data with the coefficients p'. Conversely, the loadings, p', can also to be understood as linear combinations of the data with the coefficients t. This duality has resulted in the coining of the term bilinear modeling (BLM) [35]. From this follows that if one wishes to have precise t-values, i.e., precise information about the objects, one should have many variables per object and vice versa. The rule for multiple regression that the number of variables must be much smaller that the number of objects does not apply to PCA.

As PCA is a least squares method, outlier severely influence the model. Hence it is essential to find and correct or eliminate outliers before the final PC model is developed. This is easily done by means of a few initial PC plots; outliers that have a critical influence on the model will reveal themselves clearly. Irrelevant variables can also be identified by an initial PC analysis (after taking care of outliers). Variables with little explained variance may be removed without changing the PC model. However, usually it does not matter if they are left in. The reduction of the data set should be made only if there is a cost (computational or experimental) connected with keeping the variables in the model.

Second, the data matrix can be subjected to transformations that make the data more symmetrically distributed, such as a logarithmic transformation, which was used in the swedes example above. Taking the logarithm of positively skewed data makes the tail of the data distribution shrink, often making the data more centrally distributed. This is commonly done with chromatographic data and trace element concentration data. Autocorrelation or Fourier transforms are recommended if mass spectrometric data are used for classification purposes [36]. There exist a great many potential transformations that may be useful in specialized contexts. Generally, one should exercise some discipline in applying univariate transformations with different parameters for different variables, lest a "shear" be introduced in the transformed covariance (correlation) matrix relative to that pertaining to the original data.

Centring the data by subtracting the column averages corresponds to moving the coordinate system to the centre of the data.

Third, the scaling of the data matrix must be specified. Geometrically, this corresponds to changing the length of the coordinate axes. The scaling is essential because PCA is a least squares method, which makes variables with large variance have large loadings. To avoid this bias, it is customary to standardize the data matrix so that each column has a variance 1.0. This variance scaling makes all coordinate axes have the same length, giving each variable the same influence on the PC model. This is reasonable in the first stages of a multivariate data analysis. This scaling makes the PC loadings be eigenvectors of the correlation matrix.

Variance scaling can be recommended in most instances, but some care must be taken if variables that are almost constant have been included in the data set. The scaling will then scale up these variables substantially. If the variation of these variables is merely noise, this noise will be scaled up to be more influential in the analysis. *Rule*: If the standard deviation of a variable over the data set is smaller than about four times its error of measurement, leave that variable unscaled. The other variables may still be variance scaled if so desired.

When different types of variables are present, say 6 infrared absorbances and 55 gas chromatographic peak sizes, a blockwise scaling may be performed so that the total variance is the same for each type of variables with autoscaling within each block. This is accomplished by dividing each variable by is standard deviation times the square root of the number of variables of that type in the block.

When both objects and variables are centred and normalized to unit variance, the PCA of this derived matrix is called correspondence analysis. In correspondence analysis the "sizes" of the objects are removed from the model. This may be desirable in some applications, e.g., contingency tables and the like, but certainly not in general. This double-scaling/centring in effect brings about a complete equivalence between variables and objects, which can confuse the interpretation of the results of the data analysis. The loading and score plots can now be superimposed. However, with proper scaling this can also be done with ordinary PCA, as shown in Fig. 15.

The treatment of missing data forms a special topic here. There may often be "holes" in the data matrix. The treatment of these can proceed mainly in two ways. One is to "guess" a value from knowledge of the population of objects or of the properties of the measuring instrument. A simple guess that does not affect the PCA result too much is to replace the missing value with the average for the rest of the column. Secondly, the PC algorithm may be able to cope with missing values (see Appendix). This allows the calculation of a model without having to fill in the missing value.

No matter how missing values are dealt with, an evaluation of the residuals can give extra hints. Objects (or variables) with missing values that show up as outliers are to be treated with suspicion. As a rule of thumb, one can state that each object and variable should have more than five defined values per PC dimension.

10 RANK, OR DIMENSIONALITY, OF A PRINCIPAL COMPONENTS MODEL

When PCA is used as an exploratory took, the first two or three components are always extracted. These are used for studying the data structure in terms of plots. In many instances this serves well to clean the data of typing errors, sampling errors, etc. This process sometimes must be carried out iteratively in several rounds in order to pick out successively less outlying data.

When the purpose is to have a model of **X**, however, the correct number of components, A, is essential. Several criteria may be used to determine A.

It should be pointed out here that rank and dimensionality are not used in the strict mathematical sense. The ideal case is to have a number A of eigenvalues different from zero [A smaller or equal to $\min(K, N)$] and all the others being zero. Measurement noise, sampling noise or other irregularities cause almost all data matrices to be of full rank [$A = \min(K, N)$]. This is the most difficult aspect of using PC models on "noisy" data sets.

Often, as many components are extracted as are needed to make the variance of the residual of the same size as the error of measurement of the data in X. However, this is based on the assumption that all systematic chemical and physical variations in **X** can be explained by a PC model, an assumption that is often dubious. Variables can be very precise and still contain very little chemical information. Therefore, some statistical criterion is needed to estimate A.

A criterion that is popular, especially in FA, is to use factors with eigenvalues larger than one. This corresponds to using PCs explaining at least one Kth of the total sum of squares, where K is the number of variables. This, in turn, ensures that the PCs used in the model have contributions from at least two variables. Jöreskog et al. [15] discussed aspects in more detail.

Malinowski and Howery [13] and others have proposed several criteria based on the rate of decrease of the remaining residual sum of squares, but these criteria are not well understood theoretically and should be used with caution [10,15].

Criteria based on bootstrapping and cross-validation (CV) have been developed for statistical model testing [37]. Bootstrapping uses the residuals to simulate a large number of data sets similar to the original and thereafter to study the distribution of the model parameters over these data.

With CV the idea is to keep parts of the data out of the model development, then predict the kept out data by the model, and finally compare the predicted values with the actual values. The squared differences between predicted and observed values are summed to form the prediction sum of squares (PRESS).

This procedure is repeated several times, keeping out different parts of the data until each data element has been kept out once and only once and thus PRESS has one contribution from each data element. PRESS then is a measure of the predictive power of the tested model.

In PCA, CV is made for consecutive model dimensions starting with $A = 0$. For each additional dimension, CV gives a PRESS, which is compared with the error one would obtain by just guessing the values of the data elements, namely the residual sum of squares (RSS) of the previous dimension. When PRESS is not significantly smaller than RSS, the tested dimension is considered insignificant and the model building is stopped.

CV for PCA was first developed by Wold [38], using the NIPALS algorithm (see Appendix) and later by Eastment and Krzanowski [39], using SVD for the computations. Experience shows that this method works well and that with the proper algorithms it is not too computationally demanding. CV is slightly conservative, i.e., leads to too few components rather than too many. In practical work, this is an advantage in that the data are not overinterpreted and false leads are not created.

11 EXTENSIONS; TWO-BLOCK REGRESSION AND MANY-WAY TABLES

PCA can be extended to data matrices divided into two or more blocks of variables and is then called partial least squares (PLS) analysis (partial least squares projection to latent structures). The two-block PLS regression is related to multiple linear regression, but PLS applies also to the case with several Y-variables. See, e.g., Geladi and Kowalski [40], for a tutorial on PLS. Wold et al. [41] detailed the theoretical background for PLS regression. Several contributions to the present issue illustrate the diverse application potential for PLS modeling and prediction.

Multivariate image analysis can use PCA beneficially, as explained by Esbensen et al. [42]. Multivariate image analysis is important in the chemical laboratory [43] and Geographical Information Systems (GIS).

PCA and PLS have been extended to three-way and higher order matrices by Lohmöller and Wold [44], and others. See the expository article by Wold et al. [45] for the theoretical background and application possibilities for these new bilinear concepts.

12 SUMMARY

Principal component analysis of a data matrix extracts the dominant patterns in the matrix in terms of a complementary set of score and loading plots. It is the responsibility of the data analyst to formulate the scientific issue at hand in terms of PC projections, PLS regressions, etc. Ask yourself, or the investigator, why the data matrix was collected, and for what purpose the experiments and measurements were made. Specify before the analysis what kinds of patterns you would expect and what you would find exciting.

The results of the analysis depend on the scaling of the matrix, which therefore must be specified. Variance scaling, where each variable is scaled to unit variance, can be recommended for general use, provided that almost constant variables are left unscaled. Combining different types of variables warrants blockscaling.

In the initial analysis, look for outliers and strong groupings in the plots, indicating that the data matrix perhaps should be "polished" or whether disjoint modeling is the proper course.

For plotting purposes, two or three principal components are usually sufficient, but for modeling purposes the number of significant components should be properly determined, e.g. by cross-validation.

Use the resulting principal components to guide your continued investigation or chemical experimentation, not as an end in itself.

APPENDIX

Denoting the centred and scaled data matrix by U, the loading vectors p_a are eigenvectors to the covariance matrix $U'U$ and the score vectors t_a are eigenvectors to the association matrix UU'. Therefore, principal components have earlier been computed by diagonalizing $U'U$ or UU'. However, singular value decomposition [5] is more efficient when all PCs are desired, while the NIPALS method [3] is faster if just the first few PCs are to be computed. The NIPALS algorithm is so simple that it can be formulated in a few lines of programming and also gives an interesting interpretation of vector–matrix multiplication as the partial least squares estimation of a slope. The NIPALS algorithm also has the advantage of working for matrices with moderate amounts of randomly distributed missing observations.

The algorithm is as follows. First, scale the data matrix X and subtract the column averages if desired. Then, for each dimension, a:

(i) From a start for the score vector t, e.g., the column in X with the largest variance.

(ii) Calculate a loading vector as $p' = t'X/t't$. The elements in p can be interpreted as the slopes in the linear regressions (without intercept) of t on the corresponding column in X.

(iii) Normalize p to length one by multiplying by $c = 1/\sqrt{p'p}$ (or anchor it otherwise).

(iv) Calculate a new score vector $t = Xp/p'p$. The ith element in t can be interpreted as the slope in the linear regression of p' on the ith row in X.

(v) Check the convergence, for instance using the sum of squared differences between all elements in two consecutive score vectors. If convergence, continue with step vi, otherwise return to step ii. If convergence has not been reached in, say, 25 iterations, break anyway. The data are then almost (hyper)spherical, with no strongly preferred direction of maximum variance.

(vi) Form the residual $E = X - tp'$. Use E as X in the next dimension.

Inserting the expression for t in step iv into step ii gives $p = X'Xp * c/t't$ (c is the normalization constant in step iii). Hence p is an eigenvector to $X'X$ with the eigenvalue $t't/c$ and we see that the NIPALS algorithm is a variant of the power method used for matrix diagonalization (see, e.g., Golub and VanLoan [5]). As indicated previously, the eigenvalue is the amount of variance explained by the corresponding component multiplied by the number of variables, K.

REFERENCES

1 K. Pearson, On lines and planes of closest fit to systems of points in space, *Philosophical Magazine*, (6) 2 (1901) 559-572.
2 R. Fisher and W. MacKenzie, Studies in crop variation. II. The manurial response of different potato varieties, *Journal of Agricultural Science*, 13 (1923) 311-320
3 H. Wold, Nonlinear estimation by iterative least squares procedures, in F. David (Editor), *Research Papers in Statistics*, Wiley, New York, 1966, pp. 411-444.
4 H. Hotelling, Analysis of a complex of statistical variables into principal components, *Journal of Educational Psychology*, 24 (1933) 417-441 and 498-520.
5 G. Golub and C. VanLoan, *Matrix Computations*, The Johns Hopkins University Press, Oxford, 1983.
6 J. Mandel, Use of the singular value decomposition in regression analysis, *American Statistician*, 36 (1982) 15-24.
7 K. Karhunen, Über lineare Methoden in der Wahrscheinlichkeitsrechnung, *Annales Academiae Scientiarum Fennicae*, Series A, 137 (1947).
8 M. Loéve, Fonctions aleatoires de seconde ordre, in P. Levy (Editor), *Processes Stochastiques et Mouvement Brownien*, Hermann, Paris, 1948.
9 R. Gnanadesikan, *Methods for Statistical Data Analysis of Multivariate Observations*, Wiley, New York, 1977.
10 K. Mardia, J. Kent and J. Bibby, *Multivariate Analysis*, Academic Press, London, 1979.
11 R. Johnson and D. Wichern, *Applied Multivariate Statistical Analysis*, Prentice-Hall, Englewood Cliffs, NJ, 1982.
12 J. Joliffe, *Principal Component Analysis*, Springer, Berlin, 1986.
13 F. Malinowski and D. Howery, *Factor Analysis in Chemistry*, Wiley, New York, 1980.
14 L.S. Ramos, K.R. Beebe, W.P. Carey, E. Sanchez, B.C. Erickson, B.E. Wilson, L.E. Wangen and B.R. Kowalski, Chemometrics, *Analytical Chemistry*, 58 (1986) 294R-315R.
15 K. Jöreskog, J. Klovan and R. Reyment, *Geological Factor Analysis*, Elsevier, Amsterdam, 1976.
16 J. Davis, *Statistics and Data Analysis in Geology*, Wiley, New York, 1973 and 1986.
17 W. Lawton and E. Sylvestre, Self modeling curve resolution, *Technometrics*, 13 (1971) 617-633.
18 W. Full, R. Ehrlich and J. Klovan, Extended Q model — Objective definition of external end members in the analysis of mixtures, *Mathematical Geology*, 13 (1981) 331-334.
19 R. Cole and K. Phelps, Use of canonical variate analysis in the differentiation of swede cultivars by gas-liquid chromatography of volatile hydrolysis products, *Journal of the Science of Food and Agriculture*, 30 (1979) 669-676.
20 R. Hocking, Developments in linear regression methodology: 1959-1982, *Technometrics*, 25 (1983) 219-230.
21 R. Beckman and R. Cook, Outlier...s, *Technometrics*, 25 (1983) 119-149.
22 D. Belsley, E. Kuh and R. Welsch, *Identifying Influential Data and Sources of Collinearity*, Wiley, new York, 1980.
23 D. Cook and S. Weisber, *Residuals and Influence in Regression*, Chapman and Hall, New York, 1982.
24 P. Velleman and R. Welsch, Efficient computing a regression diagnostics, *American Statistician*, 35 (1981) 234-242.
25 H. Martens, *Multivariate Calibration*, Thesis, University of Trondheim, 1985.
26 C. Albano, W. Dunn, U. Edlund, E. Johansson, B. Horden, M. Sjöström and S. Wold, Four levels of pattern recognition, *Analytica Chimica Acta*, 103 (1978) 429-443.
27 S. Wold, C. Albano, W. Dunn, U. Edlund, K. Esbensen, P. Geladi, S. Hellberg, E. Johansson, W. Lindberg and M. Sjöström, Multivariate data analysis in chemistry, in B.R. Kowalski (Editor), *Chemometrics: Mathematics and Statistics in Chemistry*, Reidel, Dordrecht, 1984, pp. 17-95.
28 S. Wold, C. Albano, W.J. Dunn III, K. Esbensen, P. Geladi, S. Hellberg, E. Johansson, W. Lindberg, M. Sjöström, B. Skagerberg, C. Wikström and J. Öhman, Multivariate data analysis: Converting chemical data tables to plots, in I. Ugi and J. Brandt (Editors), *Proceeding of the 7th International Conference on Computers in Chemical Research and Education (ICCCRE), held in Garmisch-Partenkirchen, BDR, June 10-14, 1985*, in press.
29 I. Frank and B.R. Kowalski, Chemometrics, *Analytical Chemistry*, 54 (1982) 232R-243R.
30 S. Hellberg, M. Sjöström and S. Wold, The prediction of bradykinin potentiating potency of pentapeptides. An example of a peptide quantitative structure-activity relationship, *Acta Chemica Scandinavica*, Series B, 40 (1986) 135-140.
31 R. Carlson, T. Lundstedt and C. Albano, Screening of suitable solvents in organic synthesis. Strategies for solvent selection, *Acta Chemica Scandinavica*, Series B, 39 (1985) 79-84.
32 S. Wold, A theoretical foundation of extrathermodynamic relationships (linear free energy relationships), *Chemica Scripta*, 5 (1974) 97-106.
33 S. Wold, Pattern recognition by means of disjoint principal components models, *Pattern Recognition*, 8 (1976) 127-139.
34 S. Wold, C. Albano, W. Dunn, K. Esbensen, S. Hellberg, E. Johansson and M. Sjöström, Pattern recognition; finding and using regularities in multivariate data, in H. Martens and H. Russwurm (Editors), *Food Research and Data Analysis*, Applied Science Publishers, London, 1983, pp. 147-188.
35 J. Kruskal, Bilinar methods, in W. Kruskal and J. Tanur (Editors), *International Encyclopedia of Statistics*, Vol. I, The Free Press, New York, 1978.
36 S. Wold and O. Christie, Extraction of mass spectral information by a combination of autocorrelation and principal

components models. *Analytica Chimica Acta*, 165 (1984) 51–59.
37 P. Diaconis and B. Efron, Computer-intensive methods in statistics, *Scientific American*, May (1983) 96–108.
38 S. Wold, Cross validatory estimation of the number of components in factor and principal components models, *Technometrics*, 20 (1978) 397–406.
39 H. Eastment and W. Krzanowski, Crossvalidatory choice of the number of components from a principal component analysis, *Technometrics*, 24 (1982) 73–77.
40 P. Geladi and B.R. Kowalski, Partial least squares regression (PLS): a tutorial, *Analytica Chimica Acta*, 185 (1986) 1–17.
41 S. Wold, A. Ruhe, H. Wold and W.J. Dunn III, The collinearity problem in linear regression. The partial least squares approach to generalized inverses, *SIAM Journal of Scientific and Statistical Computations*, 5 (1984) 735–743.
42 K. Esbensen, P. Geladi and S. Wold, Bilinear analysis of multivariate images (BAMID), in N. Raun (Editor), *Proceedings of Nordisk Symposium i Anvendt Statistik*, Danmarks EDB-center for Forskning og Uddannelse, Copenhagen, 1986, pp. 279–297.
43 P. Geladi, K. Esbensen and S. Wold, Image analysis, chemical information in images and chemometrics, *Analytica Chimica Acta*, (1987) in press.
44 J. Lohmöller and H. Wold, Three-mode path models with latent variables and partial least squares (PLS) parameter estimation, *Paper presented at the European Meeting of the psychometrics Society, Groningen, Holland, 1980*, Forschungsbericht 80.03 Fachbereich Pedagogik, Hochschule der Bundeswehr, Munich.
45 S. Wold, P. Geladi, K. Esbensen and J. Öhman, Multi-way principal components and PLS-analysis, *Journal of Chemometrics*, 1 (1987) 41–56.
46 M. Sjöström and S. Wold, A multivariate study of the relationship between the genetic code and the physical-chemical properties of amino acids, *Journal of Molecular Evolution*, 22 (1985) 272–277.

Chapter 16

Multivariate Data Analysis: Its Methods *

MICHEL MELLINGER

Saskatchewan Research Council, 15 Innovation Blvd., Saskatoon, Saskatchewan S7N 2X8 (Canada)

CONTENTS

1 Introduction 225
   1.1 Data analysis 225
   1.2 Multivariate data analysis 226
2 Multivariate data analysis methods 227
   2.1 Factor analysis methods 227
   2.2 Classification methods 229
   2.3 Comments on vocabulary and usage 230
3 Acknowledgements 231
References 231

1 INTRODUCTION

In this paper, multivariate data analysis methods are presented from the point of view of users asking themselves the following questions: what do multivariate data analysis techniques do? What kind of method is this author using? Multivariate data analysis methods are powerful tools for the investigation of large and complex data sets such as those generated today at a large rate by users working in various domains. Unfortunately, many users cannot apply these tools to their advantage, because they face a great difficulty when trying to first understand the methods and then apply them to their own data. It is hoped that this paper will help clarify the often-confusing field of multivariate data analysis.

1.1 Data analysis

The process of data analysis may be explained by considering the three following concepts: facts, data, and information. A fact is 'something that has actual existence' such as a rock formation, a drill core sample, or a human being; facts make up reality. Data (singular: datum) are 'something given, some measurements used as a basis for reasoning, discussion, or calculation', such as the thickness and porosity of a rock formation, the

Fig. 1. Fundamental concepts related to data analysis, and their relationships.

---

* SRC Publication No. R-851-1-A-87.

0169-7439/87/$03.50 © 1987 Elsevier Science Publishers B.V.

mineralogical composition of a rock sample, or the age, sex, and weight of a human being. Information is 'knowledge obtained from the investigation or study of facts and data', it is 'something that justifies change, that is a prerequisite to decision-making', such as the oil-bearing potential of a rock formation, the metamorphic grade of a rock, or the health of a human being.

The relationships between these three concepts are illustrated in Fig. 1. Reality cannot be accessed directly nor fully, and one must carry out measurements in order to derive data from facts. Next, data must be investigated before information is obtained ('extracted') from the data: this is the process of data analysis. Finally, decisions are made on the basis of the information obtained: decision-making can thus be seen as the ultimate purpose of data analysis.

Data analysis is carried out using various methods and techniques which have either a fairly wide or a narrow applicability, and are based either on few or on many assumptions about the mathematical properties of the data under study. For example, one field of statistics, inferential statistics, is concerned with building specific statistical models for the purpose of extrapolating (predicting) unknown properties of a population from known properties of a sample which represents this population. Another major field of statistics is descriptive statistics (see ref. 1), concerned with the detection and interpretation of data patterns, usually within large data sets, not only for the purpose of data reduction, but also for the purpose of analyzing, verifying, testing, and proving hypotheses. As stated by Benzécri [2]: "A model must be derived from the data, not the opposite. ... What is required is an accurate method which permits us to extract structures from data." Multivariate data analysis methods, reviewed in the next section, have given much power to the field of descriptive statistics, for the reasons described below.

1.2 Multivariate data analysis

For a long time, statisticians have been concerned mainly with the availability of numerous observations for a limited number of variables, aiming at validating a given dependency model or at testing specific hypotheses. The major obstacle to studying simultaneously many variables had of course been the practical limitations of computational procedures. With the advent of the electronic computer, however, this computational obstacle has disappeared, with the result that the implementation of multivariate data analysis has progressed very rapidly. It has become possible to investigate the "variable" dimension, that is, the complex relationships between many variables considered together. This is not to say that inferential statistics has lost its usefulness; but many situations ('systems') which could not be investigated successfully with inferential methods can now be studied using descriptive multivariate methods. In some fields of study, such as physics, the reductionist approach, which limits the representation of a system to some of its presumed components and then studies each component separately, is valid and produces very useful results. In other fields of study, however, such as economics or geology, one deals with a complex multi-dimensional system which must be studied as a whole, and whose components do not make much sense when they are isolated from each other; the multivariate extraction of information in such cases becomes necessary.

One intriguing aspect of the multivariate nature of some systems is that of the a priori dimensionality of the system: how many variables should be measured on a given system in order to describe its properties or behaviour adequately? There is no practical answer to that question: the dimensionality of a problem is usually limited by our limited capability in selecting and measuring all variables relevant to the description of the system and by the somewhat arbitrary nature of this 'relevancy'. Then, can a variable that was not measured, an absent variable, still influence the data patterns obtained from those variables that were measured? It may, and this will depend upon whether the absent variable is correlated (in a generic sense) with the measured variables: if it is, then the observed data patterns will contain features that cannot be explained fully without knowledge of the absent variable; if it is not correlated, then the data patterns will not contain features resulting from that absent variable. This fact is sometimes unfortunate — when absent variables could, and should, have been measured in order to increase information contained in the data set —, but is also fortunate — when the absent variables are precisely those whose influence we are trying to discover through the study of the data patterns.

2 MULTIVARIATE DATA ANALYSIS METHODS

Many multivariate data analysis methods exist, and users who are approaching the field of multivariate statistics commonly find it quite confusing, faced as they are with a prolific vocabulary which developed in a generally uncoordinated manner. Further comments about vocabulary and usage of multivariate statistics will be given in Section 2.3; let us first try to put some order in the apparent multitude of methods we must deal with.

Multivariate data analysis methods fall into only two broad categories: factor analysis methods and classification methods (Fig. 2). The purpose of methods from the former category is to calculate new variables from the initial variables and their approach is geometrical in nature; often (but not necessarily), these new variables are used to examine the data using some kind of geometrical projection or representation. The purpose of classification methods, on the other hand, is to allocate variables (or cases) to classes in order to obtain generally homogeneous subgroups of variables (or cases), and their approach is essentially algebraic in nature. Each category of methods will now be examined in more detail.

2.1 Factor analysis methods

Factor analysis methods calculate, from the initial variables, new variables called factors, which are linear combinations of the initial variables.

Why calculate factors? The initial data table is somewhat redundant because it contains various correlations (in a generic sense) between rows and between columns. This results from the fact that each of the measured variables may not explain alone a particular aspect of the phenomenon we are trying to investigate; it may also be that, among all the measurements that were made, several cover a same range of values for some of the variables. Factors are calculated in such a way that they take into account the correlations (in a generic sense) present in the data table, and that they are uncorrelated (mathematically-speaking: orthogonal to one another). In this way, data structures become apparent in the 'factor space' (as opposed to the 'data space'); these can be interpreted more readily by the user because such data patterns are usually more directly related to the phenomena under study than the somewhat redundant measured variables.

How are factors calculated? Firstly, the data table is transformed in a certain fashion to produce a matrix. Secondly, the eigenvalues and eigenvectors of this matrix are calculated following standard numerical procedures; each eigenvalue and its related eigenvector define a factor, and each eigenvalue measures the amount of variability in the data (in a generic sense) that is accounted

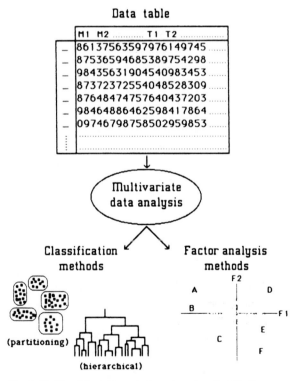

Fig. 2. Multivariate data analysis: categories of methods.

for by the factor. Thirdly, the data table is transformed into the calculated factor space. The key step in this procedure is the first step above, where the data table is transformed into a matrix assumed to contain enough information about the initial data that the factors calculated from it help describe the data patterns. The reason for the variety of available factor analysis methods is found in the many variations used in this data-to-matrix transformation.

For principal components analysis [1,3,4], the matrix used to calculate the factors is the correlation matrix (in the standard statistical sense) for the variables. The covariance–variance matrix may be used instead, but only when the variables are homogeneous, i.e. when they have essentially equal variances, otherwise a notable 'scale factor' disturbs the factor space. The factors obtained from this matrix define a factor space for the variables only. Another factor space exists for the cases which, although related mathematically to the variables factor space by means of the data table (the two spaces are said to be 'dual' spaces), cannot be directly superimposed onto it; nevertheless, this may be accomplished indirectly using one or another convention, to produce what is often referred to as bi-plots. For historical reasons, the correlation coefficient has been noted r (as in regression) since its invention, and the calculation of the variables factor space is thus known as R-mode factor analysis; when the duality relationship with the cases factor space was applied to the calculation of this new space, this new procedure was named Q-mode factor analysis.

For correspondence analysis [1,2,5,6], the matrix used to calculate the factors is equivalent to the table of weighted data profiles derived directly from the data table. As the profiles can be calculated for the rows or for the columns of the data table, one might expect that two factor spaces will be obtained: one for the rows and one for the columns. Fortunately, this is not the case: only one factor space is obtained, which is simultaneously the factor space for the rows and the columns. Actually, as described in ref. 1, correspondence analysis can be viewed as finding the best simultaneous representation of two data sets that comprise the rows and columns of a data matrix.

Another factor analysis method is discriminant analysis [1,7,8] often (wrongly) perceived by users as being a classification method. The purpose of discriminant analysis is to find factors, although in a somewhat constrained context: the user tells the method beforehand which cases are believed to belong to which group among a set of pre-defined groups, and the method then finds those factors which discriminate best between the groups of cases. For N groups of cases defined by the user, discriminant analysis finds $N-1$ factors, and derives from them N classification functions which permit the eventual allocation of new cases to one or the other of the pre-defined groups. It is this latter application of discriminant analysis that creates confusion in the mind of users who see discriminant analysis as a classification method.

Two closely related factor analysis methods have proved very useful in understanding the relationships between the various factor analysis methods that were developed over time: canonical correlation analysis, developed by Hotelling [9], and multiple canonical correlation analysis, an extension of the previous method developed by Carroll [10]. Canonical correlation analysis considers two sets of variables, sets $\{A\}$ and $\{B\}$ measured for the same cases — thus producing the data table $(\{A\},\{B\})$ — and finds two factor spaces — one for each of $\{A\}$ and $\{B\}$ — so that each factor from $\{A\}$ is correlated as well as possible with one factor from $\{B\}$. Although this method is very powerful, its results are difficult to interpret and use, which explains the limited practical success of canonical correlation analysis. Multiple canonical correlation analysis is an extension of the previous method which correlates more than two sets of variables, for example the four sets of variables in the data table $(\{A\},\{B\},\{C\},\{D\})$. This method is of great theoretical value and provides an excellent framework for discussing several other multivariate methods [1,11]:

— for a data table $(\{A\},\{B\})$, one has Hotelling's canonical correlation analysis;
— for a data table $(\{x\},\{Y\})$ where x is a single variable and $\{Y\}$ a set of variables, one has

multiple regression with x the dependent variable and $\{Y\}$ the independent (explicative) variables;

— for a data table ($\{A\}, \{I\}$), where $\{A\}$ is a set of quantitative variables and $\{I\}$ is a set of indicator variables (Boolean vectors), one has discriminant analysis, with $\{I\}$ being the description of group memberships as pre-defined by the user;

— for a data table ($\{A\}, \{B\}$) where each of $\{A\}$ and $\{B\}$ describes a partition of a population as tabulated in a contingency table, one has ('simple') correspondence analysis, which can thus be considered as being essentially a double discriminant analysis;

— for a data table ($\{I\}, \{J\}, \{K\},...$) where each set is made of one indicator variable (Boolean vector), one has correspondence analysis of complete disjunctive variables, also called by some authors 'multiple' correspondence analysis;

— finally, for a data table ($\{a\}, \{b\}, \{c\},...$) where only one quantitative variable is present in each set, one has principal components analysis.

2.2 Classification methods

Classification methods, also known as clustering or cluster analysis methods, analyze a data table by considering only one entry at a time and do not relate directly the two entries to each other like factor analysis methods can do. In this section, only the classification of the columns of a data table is mentioned; classification of the rows of a data table follows exactly the same procedure. In this respect, it is obvious that if one were to classify first the columns and then the rows of a data table, the two results would be related in some fashion because rows and columns are related through the contents of the data table; this is where the combination of classification and factor analysis methods proves to be very powerful, as explained in refs. 12 and 13.

Classification methods, when applied to the columns of a data table, will group these columns into classes which are non-empty, usually disjoint, sets of columns found to be similar enough to be clustered together. Two key steps are involved: (1) as for any method, the definition of a distance criterion (or similarity criterion, which has opposite properties), which will permit us to calculate how dissimilar (or similar) two columns are, and (2) the definition of an aggregation criterion, which is an extension of the concept of distance criterion to the measuring of a distance (or similarity) between one column and an existing class of columns (i.e. several columns and not only one, as with the normal distance criterion), and which permits the classification to proceed. Many distance criteria exist which are adapted to various types of data and various types of problems. Several aggregation criteria have also been developed, the most commonly found in classification software being (under these or fairly similar names): average linkage, single linkage, complete linkage, and Ward's inertia criterion. The reader is referred to refs. 1 and 12–14 and various statistical software manuals for more details. This should suffice for the reader to realize that several dozens of classification techniques exist. In practice, you only need to remember that understanding the two criteria, explained above, in a particular application will allow you to understand each particular technique. In addition, classification methods belong to one of two types: partitioning methods and hierarchical methods.

A partition of a set of elements is defined as a collection of non-empty and disjoint subsets whose union is equal to the initial set. Partitioning methods, also know as K-means methods, will thus allocate the columns of a data table each to one of a pre-defined number of classes until all columns have been classified. The user must help such a method at its start by first specifying how many classes are required, and then giving some initial definition of each class (e.g. by providing locations of centres of gravity, or by providing a set of class nuclei) even though the algorithm may modify the latter as it proceeds. The algorithms of partitioning methods are usually convergent (i.e. usually find a solution), but may not be optimal because their results depend upon the user's initial choices. Their relatively easy implementation, even when very large data sets are involved (see further), is their main strength.

Hierarchical methods will aggregate the columns of a data table, starting with the two most similar (less distant) columns and continuing with the leftover columns in accordance with the distance and aggregation criteria selected, until only one class is obtained which is the set of all columns. These methods thus produce a series of nested classes which form a binary tree or dendrogram. One may note that when this tree is 'cut' at any level between the initial state (at the 'leaves': all columns separate) and the final state (at the 'root': all columns together), it yields a partition of the columns into as many classes as there are 'branches' at that level. The result of a hierarchical classification method is thus a collection of nested partitions, and the user can select the level at which a partition may be obtained, which is clearly more flexible than calculating partitions for an a priori number of classes. There are cases when partitioning methods are still of interest, however, essentially when very large data sets (i.e. larger than 5000×100?) are involved and must be reduced to a more manageable size, before eventual further processing by hierarchical classification or other methods.

The domain of classification methods is actually somewhat more complex than this summary indicates, although the two types recognized here — partitioning and hierarchical — are fundamental and are the only ones generally encountered in statistical software. Other methods that can be mentioned are: classification involving overlapping clusters [15,16], and classification involving fuzzy sets [17].

2.3 Comments on vocabulary and usage

As stated earlier, a prolific vocabulary has emerged during the development of the field of multivariate statistics, which is both confusing and discouraging for new users. It is hoped that the few comments that follow will not add to that confusion.

The most common source of confusion can be traced to unclear statements made by users about the method or technique they employ and the usage they make of it. For example, when an author states that principal components analysis was used to 'classify' samples, doubts are bound to appear in a beginner's mind. It is an unfortunate fact that the human vocabulary is very limited considering the variety of concepts that human beings can generate; the additional fact that discipline in vocabulary usage is a difficult task does not help the situation. To come back to the example above, a factor analysis method (principal components analysis) is used to project sample distribution patterns onto factorial planes: this is the technical aspect. Then, because groups of samples appear on such projections, the user decides, legitimately enough, to consider such groupings as significant for the purpose of the project under study: this is the usage aspect. Maybe the user should next state that 'sample groups were identified' or 'new samples were assigned to recognized groups', rather than 'samples were classified': this is the aspect of vocabulary discipline. I have only one advice for users:

(a) try to distinguish a technique (data → information, see in Fig. 1) from its usage (information → decisions, see in Fig. 1), then

(b) develop a clear understanding of the type of technique used and its specifications, and

(c) appreciate the usage of this technique.

Here are some other examples of general problem areas related to vocabulary in multivariate statistics that are clarified when one uses the simple typology of methods explained in the two previous sections. 'Mapping techniques', whether linear or non-linear, are projection methods and are thus akin to the factor analysis type. 'Discriminant analysis', as explained in Section 2.1, is a factor analysis method and not a classification method. 'Supervised' classification methods assume that the user has information about the classes before the algorithm is applied, as is the case with partitioning methods for which class nuclei are given as input; unfortunately, this expression is often used, for example in chemometrics, to designate discriminant analysis or another form of sample identification in a factor space! Conversely, 'unsupervised' classification methods do not assume anything about the elements to be classified, as is the case for hierarchical classification. It is suggested that the expressions 'unsupervised classification' and 'supervised classification' be used and interpreted with extreme care, because they relate to usage more than to specific techniques. 'Modelling' consists in using mathematical formulae to describe a phenomenon, or observed correlations (in the generic sense), or any feature that can be described quantitatively: it is thus not related more to multivariate statistics than it is to other statistical techniques. Similarly with 'prediction', which consists essentially in extrapolating results from known cases to unknown cases. Finally, a commonly used post-processing technique in factor analysis is 'factor rotations', which can be done following various criteria: users frequently ask whether it is an appropriate procedure. Without going into the details of the battle waged by statisticians around this question, well presented in ref. 18, it is worth noting that the initial purpose of factor rotations, which may even involve oblique (i.e. correlated) factors, was to avoid data structures of general interest in favour of those of specific a priori interest to the user. In the general context of data analysis, it is safer to accept the data structures obtained without rotations and to try to interpret them for what they are; if some geometrical transformation of the factor coordinates is useful for some practical reason, such as measuring a distance along an oblique trend of cases, users can take the responsibility of making such simple calculations on their own.

A few last comments may be made which are more specifically related to the field of chemometrics as it is exposed in this volume. The SIMCA method (Soft Independent Modelling of Class Analogy [19]) is used following the identification of subsets of cases in a principal components factor space, and consists in the local modelling of each of these subsets separately by local principal components analyses. The result is a series of local principal components models in a common initial principal components space, described by a series of parameters. The PLS method (Partial Least Squares [20]) is based on a double principal components analysis, and finds factors from one variable space which are correlated as well as possible to factors from another variable space, both being 'measured' on the same set of samples. As a method, PLS has thus a purpose similar to that of canonical correlation analysis, discussed briefly in Section 2.1.

3 ACKNOWLEDGEMENTS

I wish to thank the organizers and sponsors of the workshop "Multivariate Statistics for Geochemists and Geologists" for inviting me to participate. I also wish to thank the participants of this workshop for the enthusiasm they showed during the course of the workshop, which encouraged me to write this contribution; I hope it will in turn encourage them to use multivariate data analysis methods in their work. R.G. Brereton, G. Nickless, and an anonymous reviewer are gratefully acknowledged for their constructive criticisms and their suggestions for improving the manuscript.

REFERENCES

1 L. Lebart, A. Morineau and K.M. Warwick, *Multivariate Descriptive Statistical Analysis: correspondence analysis and related techniques for large matrices*, Wiley, New York, 1984, 231 pp.
2 J.P. Benzécri, *L'Analyse des Données: 2. L'Analyse des Correspondances*, Dunod, Paris, 1st ed. 1973, 2nd ed. 1980, 632 pp.
3 K. Pearson, On lines and planes of closest fit to systems of points in space, *Philosophical Magazine*, 2, No. 11 (1901) 559–572.
4 C.R. Rao, The use and interpretation of principal component analysis in applied research, *Sankhya, Series A*, 26 (1964) 329–357.
5 M.O. Hill, Correspondence analysis: a neglected multivariate method, *Applied Statistics*, 3 (1974) 340–354.
6 M.J. Greenacre, *Theory and Applications of Correspondence Analysis*, Academic Press, London, 1984, 364 pp.
7 R.A. Fisher, The use of multiple measurements in taxonomic problems, *Annals of Eugenics*, 7 (1936) 179–188.
8 C.R. Rao, *Advanced Statistical Methods in Biometric Research*, Wiley, New York, 1952.
9 H. Hotelling, Relations between two sets of variables, *Biometrika*, 28 (1936) 129–149.
10 J.D. Carroll, A generalization of canonical correlation analysis to three or more sets of variables, *Proceedings of the American Psychological Association*, 1968, pp. 227–228.
11 J.M. Bouroche and G. Saporta, *L'Analyse des Données*, Presses Universitaires de France, Paris, 1980, 127 pp.
12 M. Jambu and M.O. Lebeaux, *Cluster Analysis and Data Analysis*, North Holland, New York, 1983, 898 pp.
13 J.P. Fenelon, *Qu'Est-Ce Que l'Analyse des Données?*, Lefonen, Paris, 1981, 311 pp.
14 J.P. Benzécri, *L'Analyse des Données: 1. La Taxinomie*, Dunod, Paris, 1st ed. 1973, 2nd ed. 1980, 625 pp.

15 N. Jardine and R. Sibson, A model for taxonomy, *Mathematical Biosciences*, 2 (1968) 465–482.
16 N. Jardine and R. Sibson, *Mathematical Taxonomy*, Wiley, New York, 1971.
17 J.C. Bezdek, *Pattern Recognition with Fuzzy Objective Functions*, Plenum Press, New York, 1981.
18 J.P. Benzécri, *Histoire et Préhistoire de l'Analyse des Données*, Dunod, Paris, 1982, 159 pp.
19 S. Wold, Pattern recognition by means of disjoint principal components models, *Pattern Recognition*, 8 (1976) 127–139.
20 H. Wold, Non-linear estimation by iterative least squares procedures, in *Research Papers in Statistics: Festschrift for Neyman*, Wiley, New York, 1966, pp. 411–444.

Chapter 17

Reprinted from:
Chemometrics and Intelligent Laboratory Systems, 2 (1987) 61-77
Elsevier Science Publishers B.V., Amsterdam - Printed in The Netherlands

Correspondence Analysis: The Method and Its Application *

MICHEL MELLINGER

Saskatchewan Research Council, 15 Innovation Blvd., Saskatoon, Saskatchewan S7N 2X8 (Canada)

CONTENTS

1 Introduction 233
2 Correspondence analysis: the method 234
   2.1 Short history 234
   2.2 Symmetry of the data treatment 234
   2.3 Wide-ranging applicability 235
3 Correspondence analysis: data analysis strategies and application 237
   3.1 Usage 237
   3.2 Principal and supplementary elements 237
   3.3 Interpretation 238
4 An example of application: the Fox Wells data set 239
   4.1 The data set 239
   4.2 Preliminary examination of the data and data analysis strategy 241
   4.3 Eigenvalues 243
   4.4 Factor characteristics for the rows 243
   4.5 Factor characteristics for the columns 244
   4.6 Factor plots and conclusions 246
5 Conclusion 248
6 Acknowledgements 248
Appendix: Computer programs for correspondence analysis 248
References 249

1 INTRODUCTION

Correspondence analysis is a multivariate data analysis method which is part of the family of factor analysis methods [1]. Although the paternity of correspondence analysis can be the source of some debate [2,3], it is clear that its usefulness and wide-ranging applicability has been demonstrated extensively under the guidance of Benzécri [4] and other statisticians directly or indirectly linked to his group of researchers, such as Lebart et al. [5], Fenelon [6] and Greenacre [7].

Correspondence analysis has had some difficulty entering the English-speaking scene, although articles about the method were published some time ago [2,8]. One major reason for this appears to be — naturally enough — resistance to change, whereby one treats a new method by first forcing it into the mold of well-established methods, and then by disclaiming its merits and origi-

---

* SRC Publication No. R-851-2-A-87.

nality. Correspondence analysis has had its share of such treatment, particularly in the field of mathematical geology, essentially as a result of its resemblance with principal components analysis [9–14]. Now that textbooks covering 'French-School' statistics are available in English [5,7,15], it is hoped that the controversy will reach a more constructive level.

In this paper, the method of correspondence analysis is presented, with an emphasis on its unique properties. Next, the most important topic of data analysis strategy is discussed in the context of correspondence analysis, and finally, a short example of its application is given which illustrates the salient points of the discussion.

2 CORRESPONDENCE ANALYSIS: THE METHOD

The main interest of correspondence analysis lies in its properties and flexibility. These are explained below, after a short reminder of the origin of the method.

2.1 Short history

Although statisticians had for some time been interested in the analysis of tables of positive numbers such as contingency tables [2,4], the expression 'correspondence analysis' was first coined by Benzécri in 1962. The type of data table investigated then was a contingency table of the type used in linguistics to describe the occurrence of nouns (the rows) as subjects of verbs (the columns). Such a table describes in a very simple form the concept of linguistic context of a noun: the verb; and vice versa. The nature of the problem under study was such that the following decisions were made (see ref. 3, Chapter 5 for a detailed account): (a) use row and column profiles as descriptive entities; (b) use a geometrical approach (that is, calculate factors in a Euclidian metric); (c) apply the principle of distributional equivalence (that is, if two rows/columns have the same profile, they can be merged without producing any change in the distance between any two columns/rows), which lead to using the χ^2 distance between profiles.

After the first trial calculations were made, one for the rows factor space and one for the columns factor space, it was realized that these spaces were not only dual but also isomorphous, that is, they were the same.

From then on, correspondence analysis was applied to various situations and proved to be very powerful indeed.

2.2 Symmetry of the data treatment

A unique property of correspondence analysis is that it produces a factor space that simultaneously represents rows and columns, as a result of the symmetrical treatment of the data table, as shown in Fig. 1. Actually, the formulae of correspondence analysis can also be found by searching for the best simultaneous representation of the rows and column of a data table, as demonstrated in ref. 5, Section II.2.5.

The initial data table is transformed into a table of row profiles (Fig. 1, top) by first calculating the sum of each row and then dividing each element in a row by this sum for the row. The sum of the elements in a row profile, which are the coordinates of the row profile, is now equal to 1.0, and the initial sum for the row is kept as the weight of the row; thus, no information is lost during this transformation. Also, an average row profile is calculated from the data table, which is the profile of the bottom margin of the table; it defines the coordinates of the centre of gravity of the row profiles space. The distance between rows is calculated using the formula of the distributional distance between row profiles, which is the χ^2 distance between row profiles using the average row profile as reference (or 'theoretical' profile in the χ^2 formula); the distance between a row and the centre of gravity of the row space is the χ^2 distance between the row profile and the average row profile for the table. The same data transformation can be applied to columns (Fig. 1, bottom) and all formulae are symmetrical. In order to calculate the factor space for the data table, a symmetric matrix is calculated from the frequency table deduced from the data table (sum of all elements equal to 1.0) and from the two diagonal matrices formed respectively by the row weights and the column weights. Eigenvalues and eigen-

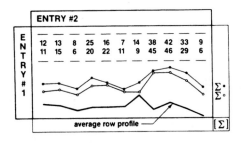

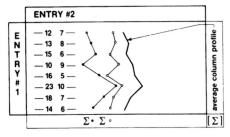

Fig. 1. Symmetrical treatment of the two entries of a data table in correspondence analysis (see text for discussion).

vectors are calculated by diagonalizing this symmetric matrix (see for example ref. 5, Section II.5 for details).

2.3 Wide-ranging applicability

The only prerequisite for correspondence analysis is that the data table contain positive numbers (including zero values), which is a very soft constraint on the type of data. The meaning of the results is the responsibility of the user, who will interpret them with the help of the various output parameters (see Section 3.3).

As a result, the types of data that can be treated by correspondence analysis can be either discrete or continuous. Among discrete data, that is, data which only have integer values, one may use:

(a) logical data (Table 1a), whose values can be 0 or 1 only, such as presence/absence data when each character can be either present (1) or absent (0) (Mr. A has a car and a dog but no bicycle and no cat; Mr. B has a car, a dog and a cat, but no bicycle; etc.), or such as disjunctive data when two or more characters are related but mutually exclusive (Mr. A has a car of type S; Mr. B has a car of type F; etc.);

(b) nominal data (Table 1b), whose values are purely conventional (Mr. A has a car of type 1 and Mr. B of type 3; etc.; but one may have changed this order before writing the data in the table, as shown on the two examples to the right of Table 1b);

(c) ordinal data (Table 1c), for which the order of the values does have a meaning (Mr. A has 3 bedrooms in his house, which is less than Mr. B — who has 4 — but more than Mr. C — who has only 2.

Continuous data will typically be measurements (Table 1d).

Such data can also be recoded from one type to another following certain rules:

(a) logical data types are treated as they are;

(b) nominal data must be recoded as disjunctive data, because statistics such as an average do not make sense for them, (in Table 1a, the type of car has been recoded this way from Table 1b, leftmost case); the only case when a nominal variable would not be recoded is when it is used as an illustrative variable and its values are plotted as case symbols on factorial plots;

(c) ordinal data can be treated as they are, or recoded as disjunctive data, in which case the order of the logical characters still has a meaning;

(d) continuous data can be treated as they are,

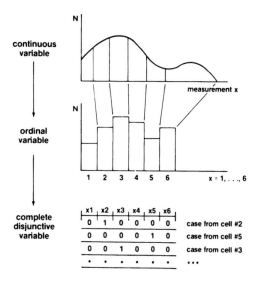

Fig. 2. Data recoding schemes (see text for discussion).

TABLE 1

Examples of data types treated by correspondence analysis

DISCRETE DATA

(a) logical data:

	Car	Bic.	Dog	Cat	S. Car	C. Car	F. Car
Mr. A	1	0	1	0	1	0	0
Mr. B	1	0	1	1	0	0	1
Mr. C	1	1	0	1	0	1	0

(Bic. = bicycle; S./C./F. Car = sports/compact/family car)

(b) nominal data:

	Type of car		Type of car		Type of car	
Mr. A	1		1		3	
Mr. B	3	or:	2	or:	1	or etc.
Mr. C	2		3		2	
	1 = sports		1 = sports		1 = family	
	2 = compact		2 = family		2 = compact	
	3 = family		3 = compact		3 = sports	

(c) ordinal data:

House of	Number of bedrooms	Number of stories	Number of bathrooms
Mr. A	3	1	2
Mr. B	4	2	2
Mr. C	2	1	1

CONTINUOUS DATA

(d) measurements:

	revenue (arbitrary)	height (cm)	weight (kg)	age (years)
Mr. A	35,000	175	80	38
Mr. B	45,000	180	76	43
Mr. C	30,000	178	72	27

or recoded as ordinal data ('reduced' coding), or recoded further as disjunctive data ('complete disjunctive' coding) (Fig. 2).

Recoding may also involve other schemes (see other examples in ref. 6), but those of Fig. 2 will suffice for illustrating this particular aspect of data analysis which is rich in possibilities.

Correspondence analysis can also treat a mixture of data types, for example when a nominal variable is recoded as a disjunctive variable, and then used as a supplementary variable to illustrate the relationship between the values of the nominal variable and the data patterns calculated from a set of continuous variables. Such an example is discussed in Section 3.4, and the concept of supplementary variables is discussed in Section 3.2.

As a result of its flexibility with respect to data types, correspondence analysis can treat various types of data tables, i.e. frequency, contingency, measurements, intensity levels, preference, ranks, qualitative, logical, presence/absence, Burt, n-dimensional, and other tables.

3 CORRESPONDENCE ANALYSIS: DATA ANALYSIS STRATEGIES AND APPLICATION

Making use of a statistical method is more complex than just processing data. The statistical method is only a tool that carries out calculations for the user; of more significance are the pre-processing decisions that must be made by the user, as well as the interpretation that follows the processing of the data. This whole package is the concern of a data analysis strategy, and Fenelon [6] showed how any data analysis project can be described by 10 key issues. In this section, only those concepts which are important for our purpose are discussed; they are applicable to any data analysis project, whether correspondence analysis is involved as a data reduction technique or not.

3.1 Usage

There is a fundamental difference between a technique and the usage one makes of it. Correspondence analysis is a data reduction technique of the factor analysis (geometrical) type; its main usage is the description of a data table into the more synthetic form of data patterns projected into a calculated factor space. Three major types of usage can be further identified [16].

Correspondence analysis can be used as an orientation tool: one has a large data table to examine and would like to gain some general knowledge about the relationships between the rows and columns of this table. The data patterns that are observed will help grasp the general nature of the data and help set a strategy for further analysis of the data using correspondence analysis or other methods: for instance, one may notice data groupings which justify the use of discriminant analysis, or variables groupings which justify the creation of independent subsets of variables.

Correspondence analysis can also be used as a direct synthesis tool, which is the most common usage: the calculated factors are used as new variables to describe the data. For example in the study of the organic geochemistry of oil-bearing sediments, one factor may be used as a measurement of paleo-environment characteristics; or in a socio-economic study, one factor may be found to represent the degree of urbanization of various areas surveyed as related to the age of the residence owners.

Finally, correspondence analysis can be used as a multistage synthesis tool whereby factor characteristics are used as a basis for further multivariate statistical analysis. For example in ref. 17, correspondence analysis of rock geochemical data from drill core samples produced a factor which characterizes hydrothermal alteration of interest to the exploration geologist, and the coordinate of a sample along that factor is a measure of the degree of alteration reached in the sample. Each of a series of exploration drill holes was then characterized by its down-hole profile in that factor coordinate, and correspondence analysis of all down-hole profiles produced a typology map of the alteration profiles observed on the property. This information could be applied directly in the field for selecting subareas of high priority for further drilling.

3.2 Principal and supplementary elements

The concept of principal and supplementary elements (rows and columns), extensively used by French data analysts, seems to have been essentially overlooked or excluded from data analysis strategies elsewhere. The need for this concept becomes evident when dealing with multivariate data about complex systems, because many variables are measured which do not all address the same features of the system under study. Why then try to include them all together into one analysis? Lebart et al. [5] recommend that the data analyzed shall be homogeneous in its substance and content, following the principle of relevance put forward by linguists: 'out of the heterogeneous mass of facts, retain only those facts that are related to one point of view'. Thus, if some variables represent a set of (e.g. demographic) characteristics different from that represented by other variables (e.g. people's attitude toward work), one should not mix them in the analysis but should instead calculate a factor space related to the first set of (demographic) variables and then project into that space the second set of (behavioural) variables. The variables of the first set are said to be principal or active variables, while those from the second set are said to be supplementary or illustrative variables. The same concept can of course be applied to cases; it is often done in multiple regression analysis when a set of test cases is used after calculations to validate the regression model. Also, in the analysis of one data set, variables may in turn be principal and supplementary, thus providing a more complete and richer analysis, each successive analysis being complementary to the others. This approach has been recommended for example in lithogeochemistry [18].

The use of principal and supplementary variables and/or cases is also important in the context of factor stability. It may happen that one or more of the most important factors calculated is created by only one variable, or one or two cases (also commonly called 'outliers' in statistical analysis), which indicates that these elements have profiles that are notably different from all other profiles in the data table. While this in itself constitutes information of interest, the results of this particular analysis may not represent reasonably well the whole data table, especially considering that subsequent factors have directions orthogonal to, and thus influenced by, the most important factors. In such a situation, it is wise to put the peculiar variables and/or cases as supplementary elements, to re-run the calculations, and then to compare the newly calculated factor space to the ones obtained previously (supplementary elements also have coordinates in the new factor space). The stability of the initial factors can thus be evaluated. As a more general test, after a satisfactory factor space is obtained, one may consider putting as supplementary elements a number of cases chosen at random and then evaluate the stability of the factor space by recalculating it; this is the basis of validation procedures in correspondence analysis and other statistical methods.

3.3 Interpretation

It is worth explaining here some aspects of the interpretation of an output from correspondence analysis, when the user must decide whether the observed data patterns provide any useful information. The factor space presents data patterns in such a way that the redundancy of the data table is greatly reduced. In this sense, factors and data structures are usually related to some natural phenomenon that affects the data but could not be measured directly.

Another general observation is that, for correspondence analysis as well as for other factor analysis methods, the first factors extracted contain much synthetic information (i.e. mix well several 'correlated' variables), whereas smaller factors become less synthetic and contain more specific information related to one or another variable or to few cases, and to variability in the data not yet explained by the most important (synthetic) factors.

The output from correspondence analysis comprises several sets of parameters which help the user judge the nature of the factors and the quality of the analysis as a whole (Table 2):

(a) the first eigenvalue in correspondence analysis is trivial and equal to 1.0 (a consequence of this method being a particular case of canonical correlation analysis). Other eigenvalues have values between 0.0 and 1.0 and are usually presented in a table in order of decreasing value;

(b) each row and column of the data table has certain characteristics along each factor: a coordi-

TABLE 2

Output parameters for correspondence analysis

Eigenvalues (non-trivial)
    number
    value ($0.0 < v < 1.0$)
    percent of total inertia
    histogram
Characteristics along each factor (each row and column)
    coordinate
    proximity ($\cos^2\theta$)
    contribution to factor inertia*
Characteristics in factor space (each row and column)
    quality (or communality: sum of proximities)
    weight (or mass: m)
    inertia ($m \times d^2$, with d = distance to centre of gravity)*
Factor plots
    using only coordinates...
    for principal/supplementary/all rows and/or columns

---

* These parameters do not have a meaning for supplementary elements.

nate (the centre of gravity of the data cloud is at coordinate 0.0 on all factors); a proximity which indicates how close that row/column is to the factor (θ = angle between the vector origin-element and the factor; $\cos \theta$ is their correlation coefficient; $\cos^2 \theta$ is their proximity), or how well the element is represented by that factor; and a contribution to the factor's inertia which measures to what extent (proportion) the inertia of the row/column is responsible for the appearance of the factor (*Warning*: that latter parameter has no meaning for a supplementary row/column, while the two former parameters do);

(c) each row and column of the data table also has characteristics in the factor space: a quality (or communality: the sum of proximities for a number of factors, equal to 1.0 for all factors); a weight (or mass: see Section 2.2 and Fig. 1); and an inertia (in the physical sense: equal to its mass times the squared distance to the centre of gravity of the data cloud);

(d) factor plots are used to display data patterns, and use only coordinates as information.

Several guidelines for interpretation are worth mentioning, because the method only carries out calculations without attaching any meaning to the numbers obtained, and because it is the user who is responsible for the interpretation.

There is a natural tendency on the part of users to consider that factors associated with larger eigenvalues are more 'significant' or 'important' than others. This may not always be the case and depends largely upon the purpose of the project. It is usually true for projects where even major data patterns are unknown. It may be wrong for projects where major data patterns are known and the user seeks patterns hidden by the more intense patterns. Examples of the latter case abound in lithogeochemistry where the important patterns are usually related to the presence of various rock types which were already identified in the field, and are expressed along the first two or three factors; in this case, the geologist may be more interested in alteration effects which are expressed in smaller factors. These considerations are related to the difference between purely quantitative information (of the entropy type) and the 'quantitative–qualitative information' (non-thermodynamic in nature) as defined in ref. 19, where the concept of usefulness is combined with that of intensity.

A second important guideline is to verify that the factor space obtained is not unduly influenced by a limited number of rows and/or columns: one wishes to obtain a synthetic picture of the whole data cloud, not only of some of its peculiar elements. The examination of contributions to the factors' inertias will help detect such problems, and the concept of supplementary elements permits the verification of the stability of the obtained factor space as discussed in the previous section.

A third warning concerns the interpretation of factor plots: they will display, by projection, all selected elements of the data cloud, whether close or not to the factorial plane considered. It is very important to take into account, in the interpretation of such plots, only those variables which either contribute significantly to at least one of the two factors, or have a reasonable proximity to the factorial plane (which is equal to the sum of the proximities to each of the two factors considered). Similarly for cases, although phenomena are usually interpreted in terms of the variables.

Finally, one should remember that the data table is the reason for the factor space, so that the user should always refer back to the data in order to understand data patterns.

4 AN EXAMPLE OF APPLICATION: THE FOX WELLS DATA SET

This short example illustrates many of the points made in the previous sections.

4.1 The data set

The data table chosen is one first used as a test data set by Krumbein [20], and which was later used for illustrating correspondence analysis and also for arguing about its presumed merits and statistical originality (see ref. 14 for the latest episode of this adventure). The data (Table 3) are formation thicknesses measured (in feet) for 31 drill holes (the rows of the table) and four types of formations (the columns of the table) encountered

TABLE 3

The Fox Wells data set [20]

Well No. *	X **	Y **	Total ***	Sand	Shale	Carb.	Evap.	Carb. + Evap.
01FW1014	2.60	1.85	608	365	148	20	75	95
02FW1015	2.85	2.35	640	224	304	14	98	112
03FW1023	2.30	2.60	464	104	242	18	100	118
04FW2019	2.20	4.50	532	157	238	0	137	137
05FW2021	2.30	5.50	562	120	316	0	126	126
06FW2034	1.40	5.55	530	30	461	0	39	39
07FW1007	2.95	0.20	447	293	116	12	26	38
08FW1006	3.30	1.15	844	451	311	42	40	82
09FW1004	3.40	2.30	906	337	432	60	77	137
10FW1001	3.55	3.10	845	266	350	24	205	229
11FW1019	3.80	2.90	915	295	355	43	222	265
12FW1020	4.00	3.60	1139	179	643	20	297	317
13FW2002	3.65	3.70	1118	180	568	0	370	370
14FW2003	4.20	3.85	1224	207	758	11	248	259
15FW2008	3.45	4.80	1162	130	659	13	360	373
16FW2009	3.30	5.10	1003	224	542	21	216	237
17FW2011	3.10	5.55	721	229	400	12	80	92
18FW2012	3.00	6.20	775	223	477	28	47	75
19FW1012	4.35	0.60	374	240	110	24	0	24
20FW1017	4.30	1.15	614	255	272	28	59	87
21FW1021	4.95	2.25	702	237	341	39	85	124
22FW1010	5.00	2.60	933	275	435	41	182	223
23FW1009	4.85	3.10	1001	348	450	17	186	203
24FW2004	4.40	4.25	1204	277	610	10	307	317
25FW2005	5.10	4.10	1144	310	520	12	302	314
26FW2006	5.50	3.80	1048	362	510	12	164	176
27FW2016	5.30	4.30	1114	246	528	32	308	340
28FW2015	5.50	4.20	1023	295	501	18	209	227
29FW2017	4.60	5.70	955	267	502	24	162	186
30FW2031	5.10	5.75	1005	271	637	8	89	97
31FW8001	5.80	3.40	1126	270	558	68	230	298

* Well No. = ⟨Control Point⟩ FW ⟨Code⟩, with ⟨ ⟩ as in Krumbein's paper [20].
** X and Y are the well coordinates as in ref. 20.
*** Formation thicknesses are in feet, with: Total = Sand + Shale + Carbonate + Evaporite.

in the Upper Permian of western Kansas and eastern Colorado: sand, shale, evaporite, and carbonate. Both sand and shale formations are of detrital origin, i.e. they are comprised of transported sediment particles, with shale being finer-grained than sand. Evaporite and carbonate formations are of chemical origin, i.e. they were precipitated from saline waters: evaporite refers to salts such as chlorides and sulfates, whereas carbonate refers to various carbonate minerals. The abundance of each formation in one drill hole provides information on the type of environment that was prevalent at the time of deposition (detrital environment: linked to erosion elsewhere; chemical environment: linked to evaporation and lack of erosional activity in the surroundings). Two new variables are calculated from the four measured thicknesses for each drill hole: the sum evaporate plus carbonate (i.e. total thickness of chemical sediments), and the total thickness intersected in each well. Finally, each well has two geographical coordinates, X and Y.

4.2 Preliminary examination of the data and data analysis strategy

The first step in the analysis of this data table is to examine the data. This was done plotting the data as a series (Fig. 3). It is apparent that the shale and evaporite thicknesses appear to be generally correlated with one another and with total thickness. Sand thickness, however, appears to follow a different pattern. Carbonate thickness (not shown) is about an order of magnitude smaller than the three other variables, and it can thus be expected to be of small significance in the data patterns.

Next, one has to decide upon a strategy for correspondence analysis. Firstly, the variables 'carbonate plus evaporite' and 'total thickness' are linear combinations of the other four thickness variables, and thus do not add any information to the data table: they are treated here as supplementary variables. Secondly, the coordinates X and Y are included in the analysis as supplementary variables in order to examine the geographical character (if it exists at all) of the observed data patterns. Because we are interested in each geographical location separately, and not in average geographical locations, both X and Y were recoded first as ordinal variables ($X = 1-5$, $Y = 1-7$), and then as complete disjunctive variables ($X1-X5$; $Y1-Y7$), which produced a total of 12 location variables. The four thickness variables: sand, shale, evaporite, and carbonate are thus the only principal variables, and the factors are calculated from the 31×4 portion of the $31 \times (4 + 2 + 12)$ data table.

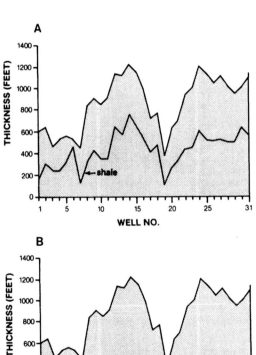
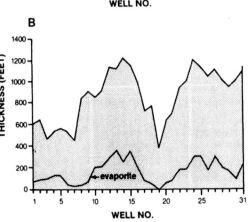
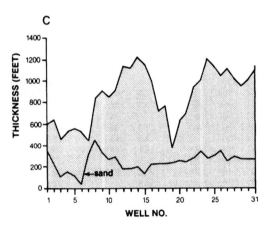

Fig. 3. The Fox Wells data set: profiled formation thickness data compared to total thickness intersected.

TABLE 4

The Fox Wells data set: output of correspondence analysis of the 31 wells and four thickness variables

(QLT = quality or communality, WGHT = weight, INR = inertia, $i\#F$ = factor coordinate for factor $\#i$, COR = correlation with or proximity to the factor, CTR = contribution to the factor's inertia; note that all COR values add up to 1000 horizontally for all factors calculated (= QLT), and that INR and CTR values add up to 1000 vertically in the I1 and J1 tables. See text for discussion.

```
EIGENVALUES:      VAL(1)=   1.00000
-----------------------------------------------------------------
|VAL()|FACTOR#| EIGENVALUE | PERCENT |  CUMUL. |*|>>HISTOGRAM>>>>
-----------------------------------------------------------------
|  2  |   1   |  0.09853   |  73.681 |  73.681 |*|*****|*****|*****
|  3  |   2   |  0.02455   |  18.358 |  92.040 |*|****
|  4  |   3   |  0.01064   |   7.960 | 100.000 |*|**
```

FACTOR CHARACTERISTICS FOR THE ROWS (WELLS):

```
  | I1   | QLT WGHT INR|  1#F  COR CTR|  2#F  COR CTR|  3#F  COR CTR|
  1|1014| 1000   23  77|  647  934  98|  170   64  27|   28    2   2|
  2|1015| 1000   24   3|  112  750   3|  -35   74   1|   54  176   7|
  3|1023| 1000   18   4| -139  602   3|  -34   36   1| -108  362  19|
  4|2019| 1000   20   8|  -79  122   1|  186  667  28|  104  211  21|
  5|2021| 1000   21  11| -230  775  11|    6    1   0|  124  225  31|
  6|2034| 1000   20  90| -484  390  48| -560  523 256|  229   87  98|
  7|1007| 1000   17  78|  775  969 103|   71    8   3|  119   23  23|
  8|1006| 1000   32  82|  574  961 107| -114   38  17|  -23    1   2|
  9|1004| 1000   34  36|  259  473  23| -218  335  67| -165  192  88|
 10|1001| 1000   32   6|   15    8   0|  148  860  29|  -58  132  10|
 11|1019| 1000   35  14|   62   71   1|  142  378  29| -172  551  96|
 12|1020| 1000   43  35| -329  993  47|   21    4   1|  -17    3   1|
 13|2002| 1000   42  61| -380  746  62|  219  247  82|   38    7   6|
 14|2003| 1000   46  36| -295  827  41| -107  109  22|   82   64  29|
 15|2008| 1000   44  70| -451  953  91|   99   46  18|  -19    2   1|
 16|2009| 1000   38   8| -171  984  11|  -21   15   1|    2    0   0|
 17|2011| 1000   27  10|   55   63   1| -176  652  34|  116  285  35|
 18|2012| 1000   29  28|   49   18   1| -355  973 150|   33    8   3|
 19|1012| 1000   14  76|  835  973 100| -123   21   9|  -67    6   6|
 20|1017| 1000   23  19|  307  846  22| -124  138  15|  -43   17   4|
 21|1021| 1000   27  13|  157  372   7| -157  374  27| -130  254  42|
 22|1010| 1000   35   4|   20   28   0|   -9    5   0| -119  966  47|
 23|1009| 1000   38   4|   85  549   3|   50  188   4|   59  263  12|
 24|2004| 1000   46  17| -195  760  18|   98  190  18|   50   50  11|
 25|2005| 1000   43  13| -113  308   6|  166  666  49|   33   26   4|
 26|2006| 1000   40   6|   83  333   3|  -24   27   1|  115  640  50|
 27|2016| 1000   42  19| -188  591  15|  126  265  27|  -93  144  34|
 28|2015| 1000   39   1|  -41  445   1|   28  207   1|   36  348   5|
 29|2017| 1000   36   2|  -32  141   0|  -77  824   9|   16   36   1|
 30|2031| 1000   30 147|  754  858 171|  242   88  70|  190   54 100|
 31|8001| 1000   43  20|  -66   71   2|  -62   61   7| -231  867 214|
  Σ |    |           1000|          1000|          1000|          1000|
```

FACTOR CHARACTERISTICS FOR THE PRINCIPAL COLUMNS (VARIABLES):

```
  | J1   | QLT WGHT INR|  1#F  COR CTR|  2#F  COR CTR|  3#F  COR CTR|
  1|SAND| 1000  299 472|  454  980 628|   54   14  35|   37    6  38|
  2|SHAL| 1000  485 177| -178  650 156| -128  333 321|   29   17  38|
  3|CARB| 1000   25 106|  355  226  32| -250  112  64| -607  662 878|
  4|EVAP| 1000  191 245| -308  551 184|  273  434 580|  -51   15  46|
  Σ |    |          1000|          1000|          1000|          1000|
```

TABLE 4 (continued)

FACTOR CHARACTERISTICS FOR THE SUPPLEMENTARY COLUMNS (VARIABLES):

JSUP		QLT	WGHT	INR	1#F	COR	CTR	2#F	COR	CTR	3#F	COR	CTR
5	TOTL	352	974	168	-64	177	-	-41	73	-	-49	103	-
6	CAEV	1000	216	179	-228	472	-	213	409	-	-115	119	-
7	X1	410	0	14	-1542	49	-	-3576	261	-	2215	100	-
8	X2	103	0	12	576	46	-	387	21	-	519	37	-
9	X3	48	0	6	4	0	-	-196	17	-	-267	31	-
10	X4	49	0	7	212	14	-	-335	34	-	-61	1	-
11	X5	82	0	5	-162	8	-	241	18	-	-417	55	-
12	Y1	212	0	18	2564	209	-	-163	1	-	254	2	-
13	Y2	221	0	10	1623	218	-	-146	2	-	-124	1	-
14	Y3	239	0	9	250	12	-	-331	21	-	-1034	206	-
15	Y4	83	0	5	-404	64	-	223	19	-	-17	0	-
16	Y5	250	0	6	-567	88	-	747	153	-	181	9	-
17	Y6	355	0	9	-549	46	-	-1057	172	-	944	137	-
18	Y7	183	0	9	1279	102	-	-361	8	-	1078	73	-

4.3 Eigenvalues

Correspondence analysis of this data table produces three factors. Let us examine the output from this analysis (Table 4).

The first non-trivial eigenvalue, which defines the first factor, accounts for about 74% of the total inertia of the data cloud, and is thus much larger than the next two eigenvalues, which account respectively for about 18% and 8% of total inertia (Table 4, top).

4.4 Factor characteristics for the rows

The factor characteristics for the 31 wells (Table 4, part I1) show that if some wells have relatively high contributions to the factors' inertias (CTR = about 100 or more), no single outlier appears to be present. Some amount of judgment is involved, but one must also remember that only 31 rows are involved here and that some heterogeneity in the data cloud can be expected as a result. One could rerun the analysis after designating well No. 6 (identifier 2034) as a supplementary row after noticing its larger contribution to the second factor (CTR = 256 out of 1000), or well No. 31 (identifier 8001) due to its larger contribution to the third factor (CTR = 214 our of 1000). For our purpose, however, such experimenting will not be carried out, and one will investigate the 31 wells as per this analysis. Several other factor characteristics can be mentioned:

(a) all wells are fully projected into the factor space defined by the three factors (communality QLT = 1000 out of 1000), as is expected for the full factor space; however, when proximities are examined separately for the three factors (columns COR in Table 4, part I1, given as a portion of 1000), one can notice that some wells are well projected along one factor only (e.g. well No. 1, 1014, along F1: COR = 934), or along two factors (e.g. well No. 3, 1023, along F1 and F3: COR = 602 + 362 = 964) or more seldom along all three factors (e.g. well No. 21, 1021, COR = 372 + 374 + 254 = QLT); this means that, with respect to the centre of gravity of the data cloud, the particular character of a well is more or less complex, while its exact nature is explained by the factor characteristics of the variables (see next section); thus, proximities are examined for one well at a time, across factors;

(b) the distribution of the contributions to each factor (columns CTR in Table 4, part I1) shows that several wells create each factor, some having positive coordinates, some negative coordinates (columns i#F); thus, contributions are examined for one factor at a time, across wells;

(c) the weight of each well (column WGHT in

Table 4, part I1) is proportional to total thickness intersected (see data in Table 3); the inertia, however, varies over a much larger range (column INR) and increases for wells which are more dissimilar to the 'average well' than others (e.g. well No. 2, 1015, is very similar to the 'average well' or centre of gravity: WGHT = 24 and INR = 3; whereas well No. 30, 2031, is very dissimilar to it: WGHT = 30 and INR = 147); the larger the inertia, the larger the influence on one or more factors.

4.5 Factors characteristics for the columns

The factor characteristics for the four active variables (Table 4, part J1) are examined next and are interpreted following the same principle as above. This shows that:

(1) 'sand thickness' has the largest inertia, while 'carbonate' thickness has the smallest. Even though 'shale thickness' has the largest weight (WGHT = 485), its inertia is third in intensity (INR = 177): this means that even though shale thicknesses are relatively large (see data in Table 3), their variability is relatively small, and variability in sand thicknesses are much larger;

(b) the first factor (F1), which accounts for about 74% of total inertia, is due to variability in 'sand thickness' (CTR = 628, positive coordinate) and also — but to a smaller degree — in 'shale thickness' (CTR = 156, negative coordinate) and in 'evaporite thickness' (CTR = 184, negative co-

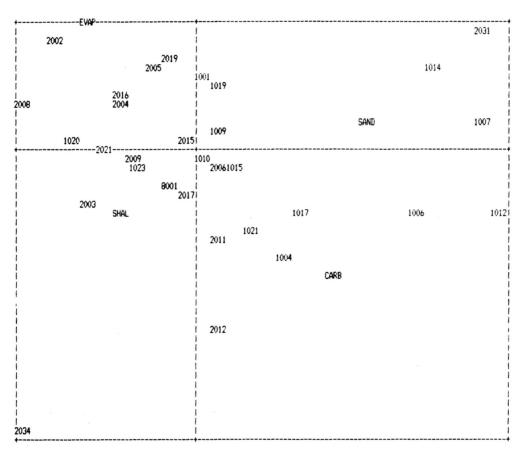

Fig. 4. The Fox Wells data set: factor plot F1–F2 showing the four principal variables and all wells (SAND = sand thickness, SHAL = shale thickness, CARB = carbonate thickness, EVAP = evaporite thickness; each well is identified by its four digit code as in Table 3; F1 is horizontal, F2 is vertical). See text for discussion.

ordinate); the signs of the coordinates indicate that along factor F1, shale and evaporite are positively correlated, but that both are negatively correlated with sand — this is consistent with our preliminary observations on the data profiles (Fig. 3); thus, factor F1 is a factor 'sand versus shale-evaporite';

(c) the second factor (F2), which accounts for about 18% of total variability, expresses a negative correlation between shale (CTR = 321, negative coordinate) and evaporite (CTR = 580, positive coordinate); this negative correlation between these two variables may be considered as a residual correlation notwithstanding their common larger negative correlation with the 'sand thickness' variable as expressed along F1;

(d) the third factor (F3), which accounts for about 8% of total inertia, is essentially a 'carbonate thickness factor' (CTR = 878, COR = 662); as a relatively small proximity between 'carbonate thickness' and factors F1 (COR = 226) and F2 (COR = 112) is noted, this would indicate that variability in carbonate thickness is not much correlated with the variability in the thickness of the three other formations.

The last factor characteristics to be examined are those of the supplementary variables (no supplementary well/row was selected), in Table 4, part JSUP:

(a) 'total thickness' has a poor communality in this factor space (QLT = 352 out of 1000), although it is calculated as the sum of the four active variables; this indicates that total thickness variability contains additional information which cannot be expressed in terms of this factor space;

(b) 'carbonate + evaporite thickness' is well represented (QLT = 1000), in particular by the first two factors (COR = 472 + 409 = 881) and with coordinates similar to those of 'evaporite thickness' (Table 4, part J1); this is not surprising,

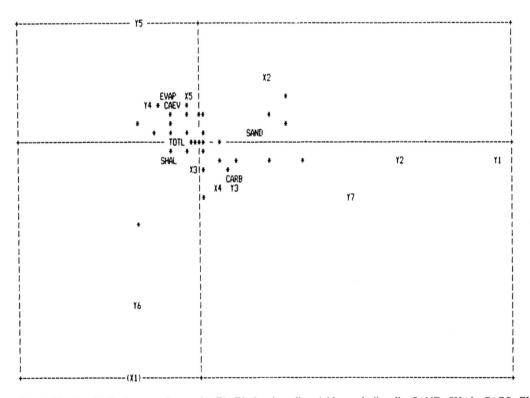

Fig. 5. The Fox Wells data set: factor plot F1–F2 showing all variables and all wells (SAND, SHAL, CARB, EVAP as in Fig. 4; CAEV = carbonate + evaporite thickness, TOTL = total thickness; $X1-X4$ and $Y1-Y7$ are the geographical coordinate intervals, with $X1$ located outside of the plot frame in the direction shown; each * represents a well; same factors as in Fig. 4). See text for discussion.

because the sum 'carbonate + evaporite' is essential equal to 'evaporite' (Table 3);

(c) X and Y coordinate intervals have reasonable proximities considering their disjunctive nature and indicate some consistent geographical patterns; central intervals appear not as well defined, however, (see QLT for X3, X4, and Y4, relative to the others); these will be examined in more detail in the factor plots in the next section.

4.6 Factor plots and conclusions

Finally, factor plots are used to visualize and summarize the various features noticed in the output tables. Two factor plots are examined respectively: F1–F2, which shows sand–shale–evaporite patterns; and F2–F3, which shows shale–evaporite–carbonate patterns. In addition, each plot is examined from two viewpoints: firstly by considering only well patterns and the four active variables; secondly, by projecting in addition to the preceding elements the geographical coordinate intervals, in order to examine the geographical distribution of lithological patterns.

Factor plot F1–F2 shows very well the relationships sand–shale–evaporite (Fig. 4).Ns 2031, 1014, 1007, 1006, and 1012 typically have high sand thicknesses, at least in terms of their formation profile: they may not have the highest absolute sand thickness (see data in Table 3), but relatively to total thickness and to shale and evaporite thicknesses, they do stand out as having

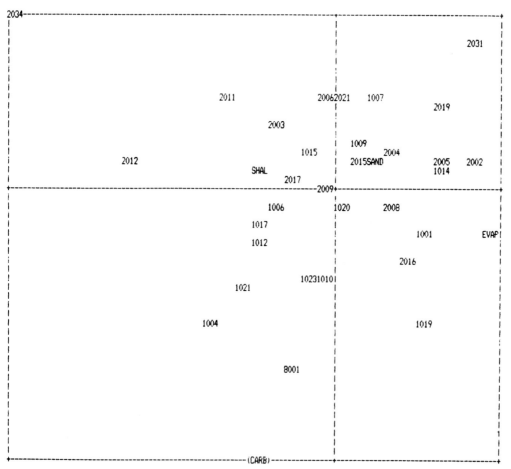

Fig. 6. The Fox Wells data set: factor plot F2–F3 showing the four principal variables and all wells (same notations as in Fig. 4; CARB is located outside of the plot frame in the direction shown; F2 is horizontal, F3 is vertical). See text for discussion.

a sandy character. Well 2034 is obviously shaly, and the data explain why it is so (Table 3): it intersected almost only shale formations; well 2012 is also relatively shaly, but here, its lack of evaporite also explains its location in the factor space away from this variable. On the upper left, well 2002 is of an evaporitic character compared to the other wells. Thus, the character of each well with respect to intersected sand, shale, and evaporite thicknesses can be quickly evaluated on this plot. When geographical coordinate intervals are projected onto the same factorial plane F1–F2 (Fig. 5), one notices that X coordinates do not display any obvious patterns (e.g. $X1 \rightarrow X2 \rightarrow \ldots \rightarrow X5$), except for the outlier character of $X1$, which is due entirely to well 2034, the shaly well (see Table 3). Conversely, Y coordinates seem to display a clear pattern; the strong sandy character in $Y1$ decreases progressively to $Y2$ and $Y3$, then $Y5$ is evaporitic, $Y6$ is shaly, and one goes back to a sandy character in $Y7$; this indicates a zonation in an east–west direction across the sedimentary basin.

On factor plot F2–F3 (Fig. 6), the carbonate

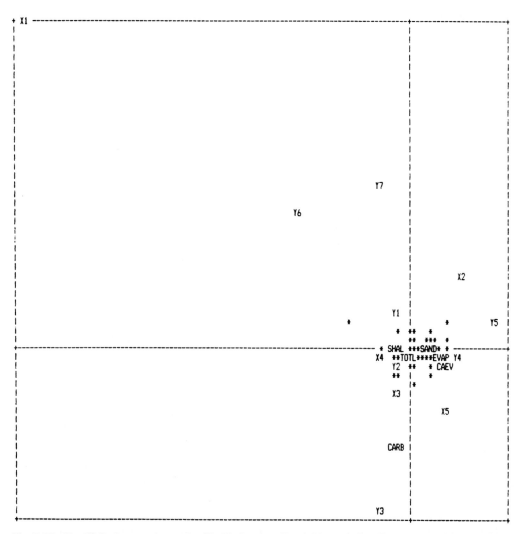

Fig. 7. The Fox Wells data set: factor plot F2–F3 showing all variables and all wells (same notations as in Fig. 5; same factors as in Fig. 6). See text for discussion.

character of the wells is displayed along factor F3 (with a strong negative coordinate); however, only a few wells have a notable carbonate intersection (relate wells with relatively large negative F3 coordinate to relative carbonate intersections in Table 3). Geographical distribution of the carbonate character (Fig. 7) shows that it is present in $Y3$ and possibly $X5$, while in $Y6$ and $Y7$ it is its absence which is notable.

Due to the limited scope of this example, a more detailed examination of the geographical distribution of the rock formations cannot be carried out. A logical step would be to produce maps of factor coordinates F1, F2 and F3 respectively, which would permit a detailed study of the geographical distribution of, respectively, sandy versus shaly-evaporitic character, shaly versus evaporitic character, and carbonate character. What has been shown here is that such patterns do exist and that they are dominated by an east–west component.

This brief example illustrates the flexibility of correspondence analysis as a data analysis tool, as well as many of the points made in the previous sections on the necessity for establishing a data analysis strategy which is adapted to the data and the problem at hand. In particular, the choice of principal and supplementary variables and the recoding of the well coordinates have yielded very useful results (for a comparison see ref. 14, where no data analysis strategy relevant to the data is even mentioned).

5 CONCLUSION

Correspondence analysis is a powerful and flexible factor analysis method which has had a profound impact on the data analysis philosophy of French statisticians and users of multivariate statistics. In conclusion, let us quote Benzécri (ref. 3, p. 101): "Correspondence analysis (as it is used today) does not just extract factors from any table containing positive numbers. It also provides rules for the preparation of the data, such as complete disjunctive coding; helps examine critically the results obtained, essentially using the calculated contributions; provides effective means of implementing discrimination and regression procedures; and blends itself harmoniously with hierarchical classification. Thus, a unique method with a simple mathematical basis was able to unify numerous ideas and difficulties which had developed separately at first, some of them several decades ago. We will attribute this success to two reasons: firstly, the initial formula of the distributional distance alone gives a table of positive numbers a mathematical structure which compensates as much as is possible for the arbitrary nature of the selection of weights and of data subsets; secondly, the numerous researchers involved [see the 70 co-authors of ref. 4] have set as a goal for themselves not to invent each a new variation of an existing statistical method, but to reduce the variety of problems defined by many different types of data to a unified data processing approach."

6 ACKNOWLEDGEMENTS

I wish to thank the organizers and sponsors of the workshop "Multivariate Statistics for Geochemists and Geologists" for inviting me to participate and submit this tutorial paper. R.G. Brereton, G. Nickless, and an anonymous reviewer are gratefully acknowledged for their constructive criticisms and their suggestions for improving the manuscript.

APPENDIX

Computer programs for correspondence analysis

Two large software libraries are distributed by non-profit French organizations, and contain programs for data recoding, correspondence analysis, and classification techniques. For further information, contact: (1) ADDAD (Association pour le Développement et la Diffusion de l'Analyse des Données), 22, rue Charcot, 75013 Paris, France, or (2) CESIA (Centre de Statistique et d'Informatique Appliquées), 82, rue de Sèvres, 75007 Paris, France.

To this date it is not sure whether either organization is offering an English version of their software on computer compatible medium; however, an earlier version of the ADDAD software library is fully published in print and in English as the second part of the book by Jambu and Lebeaux [15], and the book by Lebart et al. [5] contains the listing of a program for correspondence analysis which is taken from the CESIA software library. Both software libraries are written in FORTRAN (1966 or later) and were meant for use on mainframe computers or minicomputers.

Greenacre [7,21] has written a program for correspondence analysis which runs on IBM-PC and compatible microcomputers and which if offered to interested persons on a non-profit basis. Individual programs from the two software libraries mentioned above can also be downloaded to a microcomputer and compiled there using an appropriate FORTRAN compiler: the author has used for the Fox Wells example of this paper the last version of the ANCORR program from the ADDAD software library, downloaded to and compiled on an Apple MacIntosh Plus microcomputer.

REFERENCES

1 M. Mellinger, Multivariate data analysis: its methods, *Chemometrics and Intelligent Laboratory Systems*, 2 (1987) 29-36.
2 M.O. Hill, Correspondence analysis: a neglected multivariate method, *Applied Statistics*, 3 (1974) 340-354.
3 J.P. Benzécri, *Histoire et Préhistoire de l'Analyse des Données*, Dunod, Paris, 1982, 159 pp.
4 J.P. Benzécri, *L'Analyse des Données: 2. L'Analyse des Correspondances*, Dunod, Paris, 1st ed. 1973, 2nd ed. 1980, 632 pp.
5 L. Lebart, A. Morineau and K.M. Warwick, *Multivariate Descriptive Statistical Analysis: Correspondence Analysis and Related Techniques for Large Matrices*, Wiley, New York, 1984, 231 pp.
6 J.P. Fenelon, *Qu'Est-Ce Que l'Analyse des Données?*, Lefonen, Paris, 1981, 311 pp.
7 M.J. Greenacre, *Theory and Applications of Correspondence Analysis*, Academic Press, London, 1984, 364 pp.
8 H. Teil, Correspondence factor analysis: an outline of its method, *Mathematical Geology*, 7 (1975) 3-12.
9 R.W. May, Progresses in R- and Q-mode analysis: correspondence analysis and its application to the study of geological processes: discussion, *Canadian Journal of Earth Sciences*, 11 (1974) 1494-1497.
10 M. David, C. Campiglio and R. Darling, Progresses in R- and Q-mode analysis: correspondence analysis and its application to the study of geological processes: reply, *Canadian Journal of Earth Sciences*, 11 (1974) 1497-1499.
11 F. Valenchon, The use of correspondence analysis in geochemistry, *Mathematical Geology*, 14 (1982) 331-342.
12 A.T. Miesch, Correspondence analysis in geochemistry, *Mathematical Geology*, 15 (1983) 501-504.
13 F. Valenchon, Correspondence analysis in geochemistry: reply, *Mathematical Geology*, 15 (1983) 505-509.
14 Di Zhou, T. Chang and J.C. Davis, Dual extraction of R-mode and Q-mode factor solutions, *Mathematical Geology*, 15 (1983) 581-606.
15 M. Jambu and M.O. Lebeaux, *Cluster Analysis and Data Analysis*, North-Holland, Amsterdam, 1983, 898 pp.
16 M. Mellinger, Statistical analysis and modelling of geochemical data: what can data analysis do for you? in L. Lindquist, I. Lundholm and T. Stark (Editors), *Geochemistry and Data Analysis*, Swedish Geological Co., Luleå, Sweden, 1986, pp. 61-73.
17 M. Mellinger, Evaluation of lithogeochemical data by use of multivariate analysis: an application to the exploration for uranium deposits in the Athabasca Basin of Saskatchewan (Canada), *Proceedings of the APCOM '84 conference, held in London, U.K., March, 1984*, Institute of Mining and Metallurgy, London, 1984, pp. 21-27.
18 M. Mellinger, Correspondence analysis in the study of lithogeochemical data: general strategy and the usefulness of various data-coding schemes, *Journal of Geochemical Exploration*, 21 (1984) 455-469.
19 M. Belis and S. Giuasu, A quantitative-qualitative measure of information in cybernetic systems, *IEEE Trans. Inform. Theory*, 14 (1968) 593-594.
20 W.C. Krumbein, Open and closed number systems in stratigraphic mapping, *Bulletin of the American Association of Petroleum Geologists*, 46 (1962) 2229-2245.
21 M.J. Greenacre, Correspondence analysis on a personal computer, *Chemometrics and Intelligent Laboratory Systems*, 2 (1987) 233-234.

Chapter 18

Spectral Map Analysis: Factorial Analysis of Contrasts, Especially from Log Ratios

PAUL J. LEWI

Janssen Pharmaceutica NV, B-2340 Beerse (Belgium)

(Received 14 May 1987; accepted 3 May 1988)

CONTENTS

1 Introduction 250
2 Data table 251
3 Method of SMA 253
4 Interpretation of the spectral map 256
5 Results of SMA 257
6 Comparison of SMA with PCA and CFA 258
7 Conclusion 258
8 Glossary 259
9 Appendix 260
10 Acknowledgement 260
References 261

1 INTRODUCTION

Spectral map analysis (SMA) has been designed for the graphical analysis of contrasts. Analysis of contrasts can be likened to analysis of shapes. By way of illustration, let us consider a table of measurements of widths and lenghts in a collection of cats, tigers and rhinoceroses. With respect to size, tigers compare with rhinoceroses. But, according to shape, tigers are classified with cats. A classification of various animals by shape results from a comparison of individual width/length ratios with their geometric mean ratio. (In what follows, the term "mean" will refer to geometric mean, unless specified otherwise.)

For our purpose, we define contrast as the logarithm of an individual width/length ratio divided by the mean width/length ratio.

If an animal is found with a large positive contrast, we determine a rhinoceros. If the contrast is largely negative we have to decide between a tiger or a cat. Otherwise, if the contrast is near zero, this may indicate an overfed cat, a starving rhinoceros, or some error in the data. We will show that SMA produces a classification by contrasts from all possible log ratios in a table.

The name "spectral map" is tied to drug research, i.e. the screening of chemical compounds for biological activity. (Activity is related to the amount of substance that is required to produce a

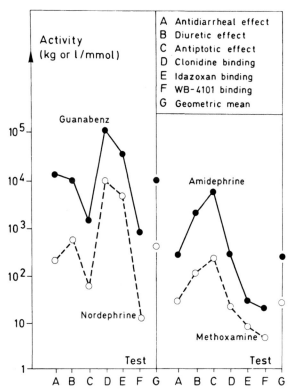

Fig. 1. Activity spectra as defined by Janssen et al. [1] of four selected alpha-agonists in six tests on rats. Data are from Table 1. The vertical scale represents the logarithm of biological activity. The two contrasting pairs of compounds (guanabenz, nordephrine and amidephrine, methoxamine) belong to distinct subclasses of alpha-agonists.

given effect.) In drug research, the distinction between size and shape appears as one of activity and specificity. (Specificity is defined as the ratio of activities in a pair of tests.) On the one hand, highly active compounds are sought which require the least amount of substance to produce effects. On the other hand, the researchers seek highly specific compounds, with high activity in one test and as little as possible in another.

In our laboratory it is common practice to construct so-called "activity spectra" in the form of histograms, showing on logarithmic scales the activities of the compounds in the various tests. (See Fig. 1 for an example.) Activity spectra were introduced for the classification of drugs (in a classic paper) by Janssen et al. [1]. In this early study, the tests revealed one major contrast which allowed to classify the compounds visually. In a later study [2] activity spectra were drawn on pieces of paper and spread out on a table. Several people tried in vain to arrange the spectra into a satisfactory classification according to specificity. The reason for the failure, as appeared afterwards, was that more than one major contrast accounted for the differences in the spectra.

Principal components analysis (PCA) [3] does not explicitly separate the spectra into activity and specificity. Correspondence factor analysis (CFA) [4] detects contrasts, but not from log ratios which are basic tools in drug research.

At this point, the author devised a method which would map the activity spectra according to specificity of the compounds [5]. SMA is a factorial method of data analysis. It involves logarithmic transformation and double-centering, followed by factorization and biplot. The biplot in SMA is special in that it represents multiple ratios between elements of the data table. The properties of the biplot are also invariant under a transposition of rows and columns of the table. SMA originated from the analysis of pharmacological data. The method is described from this perspective, using the results from a published study of alpha-adrenergic compounds. The method is of general use and has been applied subsequently to competitive positioning and financial analysis, all of which make routine use of characteristic ratios [6]. In general, SMA is applicable to tabulated data that are strictly positive and defined on ratio scales (not necessarily with the same units of measurement).

2 DATA TABLE

The 18 compounds listed in Table 1 have been studied pharmacologically by Megens et al. [7]. They belong to the class of alpha-adrenergic agonists (alpha-agonists for short). Alpha-agonists stimulate alpha-adrenergic receptors (alpha-receptors for short) and produce an increase in blood pressure and constriction of blood vessels, among other effects. Alpha-receptors are further divided into alpha-1 and alpha-2 subclasses. Some compounds in the table are specific alpha-1 agonists

TABLE 1

Biological activities of 18 alpha-agonist compounds in six tests on rats observed by Megens et al [7]

Tests A, B and C are performed on live animals (in vivo). Tests D, E and F are performed on isolated tissues (in vitro). Biological activity is defined as the reciprocal of the amount of substance required to produce an effect. Mean activity of all in vitro data is made equal to mean activity of all in vivo data for the purpose of comparison of compounds and tests. See text for further explanation.

Compound	In-vivo (kg/mmol) test			In-vitro (l/mmol) test			Geometric mean
	A	B	C	D	E	F	
	Antidiarrheal effect	Diuretic effect	Antiptotic effect	Clonidine binding	Idazoxan binding	WB-4101 binding	
1. Amidephrine	240	1995	6026	244	24	17.3	258
2. Azepoxole	81	214	12	405	102	0.5	40
3. B-HT920	1479	3802	60	3866	1539	10.9	529
4. Clonidine	16596	33113	3891	50959	5096	255.4	7221
5. Guanabenz	12023	9120	1318	114083	64154	719.8	9556
6. Guanfacine	7586	17783	5012	32153	12800	67.2	5152
7. Lidamidine	96	96	21	57	64	5.1	39
8. Methoxamine	26	105	219	19	7	4.0	26
9. Naphazoline	2884	6607	1445	57177	40478	1611.5	6844
10. Nordephrine	214	562	62	9062	4542	11.4	389
11. Oxymetazoline	21380	16596	21380	71982	20287	17267.2	24004
12. ST587	32	7	1549	454	534	101.7	142
13. Tetrahydrozoline	1259	1072	1259	10168	7198	509.6	1997
14. Tiamenidine	871	1148	977	18081	8077	360.8	1929
15. Tramazoline	2570	4467	1950	32153	11408	807.6	4335
16. UK-14304	12023	10471	1122	106469	8077	57.2	4368
17. Xylazine	501	759	63	994	287	9.1	199
18. Xylometazoline	1862	2138	1072	25540	11408	4047.8	4140
Geometric mean	1021	1569	633	5502	1993	92.5	1005

(e.g. amidephrine and methoxamine) some are specific alpha-2 agonists (e.g. guanabenz and nordephrine), and others are mixed-type alpha-agonists [7]. In drug research, the screening for active and specific compounds is performed both in vivo and in vitro.

Columns labeled A to C in Table 1 describe the effects of compounds when studied in vivo, i.e. in live rats. Activity is defined here as the reciprocal of the amount of substance per unit body weight that produces a given effect in half of the animals on test (kg body weight/mmol compound). An antidiarrheal effect is marked by the absence of diarrhea after the ingestion of castor oil (which reputedly produces diarrhea). Diuresis is defined as an excess production of urine. The antiptotic effect is measured by the degree of opening of the eyelids after the administration of prazosin (which produces eye closure).

Columns D to F in Table 1 represent the abilities by which the compounds bind to alpha-receptors, as measured in vitro, i.e. in test tubes. Samples of tissue are saturated with a radioactively labeled compound (or marker) which binds specifically to alpha-receptors. When a test compound also binds to the same alpha-receptors, then it will compete with the marker and tend to displace it. Activity is defined here as the reciprocal of the amount of test substance per unit volume of sample that displaces half of the marker (liter sample/mmol compound). Clonidine is a specific marker for alpha-2 receptors, and WB-4101 is a marker that binds specifically to alpha-1 receptors.

For the sake of comparison we made the mean activity of all in vitro data equal to the mean activity of all in vivo data (by multiplication with a proportionality factor of 0.255). We readily see from the last column of Table 1 that oxymetazoline is on average about 900 times more active than methoxamine. It is also observed from the bottom row in the table that the mean activity in clonidine binding is some 60 times larger than that in WB-4101 binding.

As for the study of specificity between compounds and tests, this requires a method, such as SMA, which is sensitive to contrasts, especially from log ratios. Let it be reiterated that specificity of a compound means the ratio of activities in two tests, and that contrast is equal to the logarithm of specificity in proportion to its mean. SMA is described hereafter in the context of drug research. A more formal definition is presented in the Appendix.

3 METHOD OF SMA

Fig. 1 shows "activity spectra" for selected alpha-agonists. (Note that the vertical scales of the spectra are divided logarithmically.) The spectra of guanabenz and nordephrine are seen to be similar in shape, although the mean activities of the two compounds differ by a factor of 25. Likewise, the spectra of amidephrine and methoxamine are also similar in shape, although mean activities vary by a factor of 10. Clearly, the shapes of the two types of spectra appear to be dissimilar. If all 18 compounds were to resemble one of these two types of spectra (perhaps with some mixed types in between) the problem would be that of a straightforward classification of the compounds according to a single contrast. Such a contrast can be made visible by means of a special logarithmic plot which was introduced in drug research by Janssen et al. [1]. Fig. 2 shows the activities (on logarithmic scales) in two strongly contrasting tests, clonidine binding (a specific alpha-2 test) and WB-4101 binding (a specific alpha-1 test). A series of dashed lines has been added at regular intervals and parallel to the line of identity (or bisector) of the plot. In this way, a

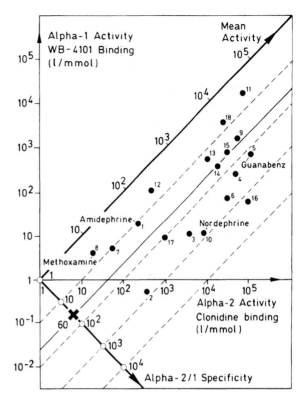

Fig. 2. Logarithmic plot of activity and specificity as defined by Janssen et al. [1] for 18 alpha-agonists in two selected in vitro tests in rats. Vertical projection of compounds upon the line of identity (bisector) produces mean activities. Vertical projection upon the line drawn perpendicularly to the line of identity defines the specificity with respect to the ratio of alpha-2 and alpha-1 activities. The mean specificity of the 18 compounds is 60. Compounds with specificities in excess of the mean possess a positive contrast (guanabenz and nordephrine). Those below the mean obtain a negative contrast (amidephrine and methoxamine).

visual classification of the compounds according to the alpha-2/1 activity ratio (or specificity) is obtained. The range of this ratio is from 4 to about 2000 with a mean of 60. The solid dividing line separates compounds with ratios larger than the mean (positive alpha-2/1 contrast) such as guanabenz and nordephrine, from compounds with ratios smaller than the mean (negative alpha-2/1 contrast) such as amidephrine and methoxamine. In this context of drug research, we define contrast of a compound as the logarithm of the specificity of the compound divided by the mean specificity, where the mean is computed over all compounds in the data table.

From this plot one can at once determine the mean activity and the specificity of each compound. Vertical projection of a point, representing a compound, upon the line of identity produces its mean activity in the two tests. Vertical projection upon a line which is perpendicular to the line of identity produces the specificity of the compound in the two tests. (The mean specificity is indicated by a cross.) These two lines represent the axis of mean activity and the axis of contrast of the two-dimensional spectra.

From the data in Table 1 one calculates that nordephrine (compound 10) possesses a mean activity in clonidine and WB-4101 binding tests of $(9062 \times 11.4)^{1/2} = 321$ and a specificity in these two tests of $9062/11.4 = 795$. These results can also be estimated from Fig. 2 by means of perpendicular projection of the point representing nordephrine on the axes representing mean activity and specificity. Since the mean specificity in the two tests equals 60, the contrast of the compound amounts to $\log(795/60) = 1.12$. This highly positive contrast indicates high specificity of nordephrine for alpha-2 activity (clonidine binding) in comparison to its alpha-1 activity (WB-4101 binding).

Fig. 3 represents the compounds by means of an ellipsoid in a space spanned by three tests (clonidine, WB-4101 and idazoxan binding). The barycenter of the ellipsoid is represented by a cross. Mean activity of the compounds is defined by the line of identity which makes equal angles with all three coordinate axes. (For convenience, we have represented the barycenter on the line of identity.) Compounds possessing high activity in all three tests, such as guanabenz, are projected high-up on this line. Compounds with low overall activity, such as methoxamine, are projected at the low end. A plane has been constructed perpendicularly to the line of identity. This is the plane of contrasts. After perpendicular projection the clus-

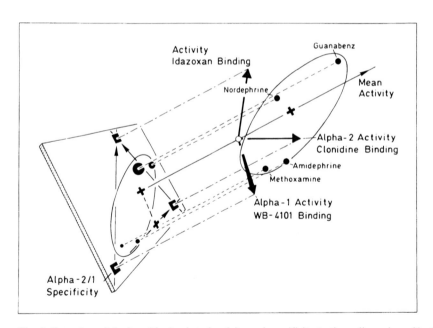

Fig. 3. Extension of the logarithmic plot of activity and specificity to three dimensions. Vertical projection of compounds upon the line of identity produces mean activities. Vertical projections upon the plane, drawn perpendicularly to the line of identity, define the specificities with respect to the three tests. The center of the triangular plot is the projection of the center of gravity of the compounds (indicated by a cross). The horizontal side of the triangle defines the specificity ratio of alpha-2 to alpha-1 activities. Mean specificity is defined by the projection of the center of the triangle upon the horizontal side of the triangle. Positive contrasts are obtained by compounds with specificities higher than the mean (guanabenz and nordephrine). Negative contrasts are carried by compounds with specificities less than the mean (amidephrine and methoxamine).

ter of compounds is represented as an ellipse in the plane of contrasts. Areas of the circles, which represent individual compounds in the plane of contrasts, are made proportional to the corresponding mean activities of the compounds. The projections of the end points of the three coordinate axes represent the vertices of a triangle in the plane of contrasts. Three axes of contrast can be constructed in this plane, but only two of them are independent. The base of the triangle represents the alpha-2/1 contrast, the other two sides define a contrast of the idazoxan binding test with the specific alpha-2 and alpha-1 binding tests. From the shape and orientation of the projected pattern, one deduces that idazoxan binding correlates with clonidine binding. Note also the characteristic positions in the plane of contrasts of guanabenz and nordephrine (with high alpha-2/1 contrast) and of amidephrine and methoxamine (with low alpha-2/1 contrast).

In the three-dimensional case we have decomposed the activity spectra into a component of mean activity and two components of contrast. In the general n-dimensional case, the spectra can be decomposed into a component of mean activity and $(n-1)$ components of contrast [5].

It can be shown mathematically that projection of the compounds from the original space of data into the space of contrasts is equivalent to row-centering of the data [8]. By row-centering we mean the subtraction of the mean activity of a compound from its activities in the various tests. It is somewhat perplexing to find that row-centering collapses the original data space into a space with one dimension less than the original one. Usually, data are also centered column-wise which results, geometrically, in a translation of the center of gravity of the points toward the origin of space. In double-centering the order of centering — rows first or columns first — is irrelevant.

The spectra of the 18 compounds in the six tests can thus be decomposed into one component of mean activity and a five-dimensional space of contrasts. It is natural to attempt a reduction of this high-dimensional space into a lower-dimensional space of orthogonal factors. (Orthogonal factors are mutually independent or uncorrelated.) For this purpose, we apply factorial analysis to double-centered data, in the same way as is done in principal components analysis to column-centered data. This procedure yields factor scores and factor loadings which define the coordinates of the compounds and the tests along axes representing orthogonal factors. Since these factors are meant to reproduce contrasts rather than original data, they will be called factors of contrast. Each factor explains a fraction of the total variance of the double-centered data (called variance of contrasts for short).

The importance of a factor is determined by its contribution to the variance of contrasts, and is called factor variance. In the case of the 18 components and six tests we can find at most five factors of contrast (in addition to the component of mean activity).

The three most important factors account for 48, 40 and 8% of the variance of contrasts, respectively. The remaining two factors carry the residual 4% of the variance of contrasts, which will not be considered further.

An important step in the method is the scaling of factor scores and factor loadings by means of multiplication by suitable constants. Scaling causes an expansion or contraction of the pattern of points in the directions of the factor axes. It greatly influences the way in which the result can be interpreted. In PCA factor scores and factor loadings are scaled such that their variances are equal to the corresponding factor variances [9]. In biplot graphic display [10] factor loadings are often scaled to factor variances, and factor scores are scaled to unit variances. Although this is an asymmetrical form of scaling, the advantage is that one can meaningfully display compounds and tests in one and the same plot (hence the name biplot). In spectral map analysis we opted for an intermediate form of scaling in which the variances of factor scores and factor loadings are equal to the square roots of the factor variances. This way we preserve the symmetrical form of scaling, while retaining the properties of a biplot. The mathematical details of the procedure are defined in the Appendix. The result of applying SMA to Table 1 is displayed in Fig. 4.

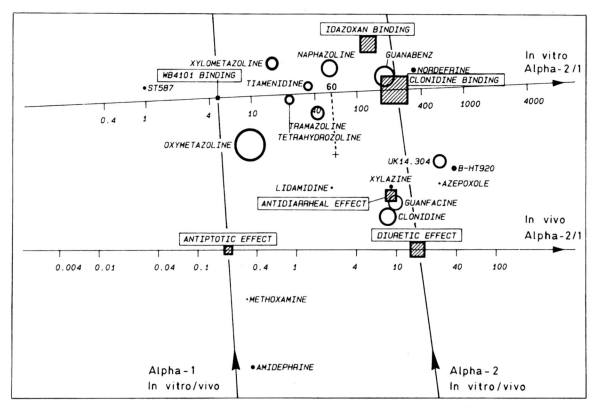

Fig. 4. Spectral map or multi-dimensional extension of the logarithmic plot of activity and specificity. Vertical projection of compounds upon the horizontal axes defines their specificity with respect to alpha-2 and alpha-1 activities. Mean specificity is defined by projection of the center of the plot (indicated by a cross). Compounds that project to the right of the mean, have positive contrast (guanabenz and nordephrine in the upper right part of the plot). Compounds that project to the left possess negative contrast (amidephrine and methoxamine in the lower left part). Areas of circles and squares are related to mean activity of compounds and tests. The vertically oriented axes represent an in vitro/vivo contrast. See text for further explanation. The horizontal and vertical directions of the biplot represent the first and second factors of contrast, respectively.

4 INTERPRETATION OF THE SPECTRAL MAP

In order to read the spectral map of Fig. 4 correctly the following rules must be considered.

Rule 1. Circles represent compounds, squares denote tests.

Rule 2. Areas of circles and squares are proportional to the mean activities of compounds and tests. Oxymetazoline possesses the largest mean activity, methoxamine the smallest. Specific alpha-2 tests (on the right side of the plot) show a larger mean activity than specific alpha-1 tests (on the left side).

Rule 3. The positions of compounds and tests are defined by scores and loadings on the first two dominant factors of contrast. A compound is attracted by a test for which it is highly specific. A compound is repelled by a test for which it has low specificity. Symmetrically, a test is attracted or repelled by compounds according to its specificity for the compounds. The plane of the two dominant factors renders 88% of the variance of the contrasts. Compounds and tests that possess little contrast are displayed near to the center of the map (which is indicated by a cross).

Rule 4. A third dominant factor is encoded in the variation of the thickness of the contours of the symbols. A thick contour indicates a point that is above the plane of the plot (e.g. xylometazoline). A thin contour means that the point lies below the plane (e.g. guanfacine). The third factor contributes 8% to the variance of contrasts.

Rule 5. Compounds and tests that cannot be represented in the space spanned by the three dominant factors, appear with a broken contour. All compounds and tests are well-represented, and the residual variance is only 4% of the variance of contrasts.

Rule 6. An axis of contrast can be defined through any two squares representing tests. Vertical projection of the centers of the circles, representing compounds, upon this axis determines their specificity (which is defined as the ratio of activities in the two tests). Calibration of an axis of contrast is obtained by means of linear regression between the ratios obtained from the data in Table 1 and from projection. Projection of the center of the plot (indicated by a cross) determines the mean specificity. High or low contrast of a compound depends on its specificity with respect to the mean specificity. (Symmetrically, an axis can also be defined through two circles representing compounds, and the contrasts of tests can be determined by vertical projection of the corresponding squares.)

5 RESULT OF SMA

Clonidine and WB-4101 are known to be specific markers for alpha-2 and alpha-1 receptors, respectively [7]. Hence the upper horizontal axis explicitly defines alpha-2/1 specificities. Mean specificity is 60, as can be determined by vertical projection of the center of the map (marked by a cross) upon the axis. Compounds that project on the right of the mean possess a positive alpha-2/1 contrast. These possess high specificity for alpha-2 receptors (e.g. guanabenz, nordephrine). Those that project on the left side obtain a negative alpha-2/1 contrast. These interact specifically with alpha-1 receptors (e.g. methoxamine and amidephrine). Mixed alpha-2/1 agonists possess near-zero contrasts (e.g. naphazoline and lidamidine). It appears from the plot that idazoxan is also a specific marker for alpha-2 receptors, possibly the same as those marked by clonidine.

The parallelism of the lower horizontal axis with the upper one indicates a strong correlation ($r = 0.81$) between diuretic/antiptotic specificity obtained from live animals with alpha-2/1 specificity observed from binding to receptors in isolated tissues. This leads to the conclusion that diuresis and antidiarrheal effects are produced by stimulation of alpha-2 receptors, and that antiptotic effect is caused by stimulation of alpha-1 receptors [7]. The vertically oriented axes represent the in vitro/vivo contrast. Note that the contrast obtained with alpha-2 specific tests (on the right) correlates ($r = 0.46$) with the contrast from the alpha-1 specific tests (on the left), as can be deduced from the parallelism of the lines. Classification of the compounds according to the two types of contrast (alpha-2/1 and in vitro/vivo) follows immediately from a visual inspection of the spectral map. This classification is of particular interest as the two types of contrast appear to be orthogonal (i.e. independent or uncorrelated).

The spectrum of compounds ranges from highly specific alpha-1 compounds (ST587) to highly specific alpha-2 compounds (B-HT920), from compounds that are specifically active in vitro (naphazoline) to those that are specifically active in vivo (amidephrine), and with many shades of specificity in between. The question is to determine how reliably contrasts are reproduced by the spectral map. Considering the variance of the third factor (8%) and the residual variance (4%) we assume that the degree of reproduction of the contrasts by the first two factors is fair. There are no compounds or tests that are particularly ill-represented. An indication of reliability is also given by the rank order correlations between ratios obtained from the data in Table 1 and projections upon the axes of contrast in the spectral map of Fig. 4. As shown in Table 2 these correlations are excellent for the alpha-2/1 contrasts ($r = 0.94$ and 0.92) and fair for the in vitro/vivo contrasts ($r = 0.86$ and 0.77).

TABLE 2

Rank order correlations of contrasts computed from the data in Table 1 and determined by projection upon axes of contrast in spectral map analysis (SMA), principal components analysis (PCA, factors 2 and 3) and correspondence factor analysis (CFA)

The four axes of contrast are exemplified in Fig. 4.

Contrast	SMA	PCA	CFA
Alpha-2/1, in vivo	0.94	0.94	0.98
Alpha-2/1, in vitro	0.92	0.92	0.87
In vitro/vivo, alpha-2	0.86	0.78	0.85
In vitro/vivo, alpha-1	0.77	0.61	0.57

6 COMPARISON OF SMA WITH PCA AND CFA

Another question is whether the results of SMA can be obtained from other methods such as PCA and CFA.

We have applied PCA to the data in Table 1, using logarithmic re-expression and column-centering. In this analysis, the first factor comes out strongly as the component of mean activity of the compounds, representing 78% of the total variance of the data. The second and third factors reveal two contrasts that are similar to the ones obtained by SMA in Fig. 4.

From the rank order correlations in Table 2 we observe that the alpha-2/1 contrasts are reproduced as well in PCA as in SMA ($r = 0.94$ and 0.92). But the in vitro/vivo contrasts come out less well ($r = 0.78$ and 0.61). The reason for this lies in the degree of correlation of the mean activity with the two types of contrast. Mean activity is almost independent of the alpha-2/1 contrast ($r = 0.19$), while being correlated with the in-vitro/vivo contrast ($r = 0.40$). Hence the third factor of PCA is not a pure factor of contrast as it is partly masked by the mean activity. In other words, PCA produced a dominant factor of size, a factor of shape, and a factor which expresses a mixture of size and shape.

As a general rule, in order to find pure contrasts in the second, third and following factors of PCA, it is required (a) that the variance of mean activity exceeds the variance of the largest contrast and (b) that mean activity is approximately orthogonal to the second, third and following factors. In practice, these conditions are not always satisfied, as has been shown by Kvalheim [11].

CFA also produced a pattern of contrasts which is similar to that of SMA. The major difference lies in the bad representation in CFA of the WB-4101 binding test. This is due to the weighting which is applied in CFA to compounds and tests in proportion to their arithmetic mean activity.

The weight assigned to WB-4101 is particularly small, such that it plays a lesser role in the analysis, which will be dominated by the other five tests. This explains the rank order correlations of CFA contrasts with contrasts computed from the data in Table 2. The in vivo alpha-2/1 contrast is reproduced in CFA even better than in SMA ($r = 0.98$). But the in vitro contrast of alpha-2/1 is less ($r = 0.87$), because the WB-4101 test for alpha-1 activity is not well-represented. Likewise the in vitro/vivo contrast is comparable to that of SMA for alpha-2 activity ($r = 0.85$), but is worse for alpha-1 activity ($r = 0.57$) again because of the bad fit of the WB-4101 test.

Contrasts are related in CFA to distances of chi-square, a concept which applies to contingency data (i.e. counts). In practice, the method is extended to data defined on ratio scales, provided that they are defined with the same units and that they can be added to meaningful totals both row- and column-wise (e.g. weight, currency, length, time, etc.). Data that are defined with different units can be analyzed by means of SMA, as this method produces a multivariate analysis of (log) ratios.

7 CONCLUSION

Spectral map analysis (SMA) is a method of choice for the analysis of contrasts, when these can be related to log ratios. We have defined contrast in the narrow sense of the log of a ratio in proportion to its mean.

In such cases as occur in the screening of drugs, when two independent contrasts are present in the data, SMA allows the identification of the contrasts. These can also be determined quantitatively in terms of log ratios by vertical projection upon calibrated axes, in the same way as is customary with rectangular coordinate axes.

In addition to the well-recognized synthetic properties of multivariate diagrams (classification and reduction of dimensions), SMA also adds analytic capabilities (identification and quantification or contrasts).

8 GLOSSARY

Receptor: Macromolecule, often within the membrane of cells, that mediates in the expression of a biological effect.

Agonist: Chemical compound that binds to receptors of a particular kind and that hereby elicits a biological effect.

In vivo test: Observation, in live animals, of biological effects produced by compounds.

In vitro test: Determination, in isolated tissues, of the degree of binding of a compound to receptors.

Activity: Reciprocal of the amount of compound required to produce a well-defined effect (in vivo) or type of binding (in vitro).

Activity spectrum: Graphic representation, in the form of a histogram, of the various activities of a compound in a battery of tests. Activities are defined on a logarithmic scale.

Specificity: Ratio of activities of a compound in two tests.

Contrast: Logarithm of the specificity of a compound divided by the mean specificity computed over all compounds. A compound possesses positive contrast when its specificity is higher than the mean specificity. A negative contrast results when its specificity is less than the mean specificity.

Line of identity: The line through the origin of space and which makes equal angles with all coordinate axes. Bisector line in two-dimensional space.

Barycenter: Center of gravity of a data structure. In a space spanned by the columns of a table the center of gravity is defined by the mean values computed column-wise.

Column-centering: Calculation of the mean values column-by-column, followed by subtraction of these means from all elements in the corresponding columns. Column-centering produces a translation of the origin of column-space toward the barycenter of the structure that represents the rows in the table.

Row-centering: Calculation of the mean values row-by-row, followed by subtraction of these means from all elements in the corresponding rows. Row-centering produces a projection of the structure, representing rows in the table, upon a subspace which is orthogonal to the line of identity of column-space. The subspace possesses one dimension less than the original space. Row-centering eliminates differences in elevation of the data points, were elevation is defined by the projection of a point upon the line of identity.

Double-centering: Centering by rows, followed by centering by columns (or vice versa). The effect of double-centering is a partition of the variance of the data into a component of mean activity and one or more components of contrast.

Variance–covariance matrix: Square matrix obtained by multiplication, rows-by-columns, of a data table with its transpose, followed by division by the number of rows in the data table.

Total variance: Sum of the elements on the principal diagonal (trace) of the variance–covariance matrix, divided by the number of columns in the data table.

Variance of contrasts: Total variance of the double-centered data.

Factor: Orthonormal vectors extracted from the variance–covariance matrix. The first factor amounts for a maximum of the total variance. The second factor explains a maximum of the residual variance, i.e. the variance not accounted for by the first one, etc. Eigenvector. Principal component.

Factor axes: Set of factors that span the data space.

Factor variance: Part of the total variance accounted for by a factor. The sum of the factor variances is equal to the total variance. Eigenvalue.

Factor of contrast: Factor extracted from the variance–covariance matrix of double-centered data.

Factor scores and factor loadings: coordinates of

rows and columns, respectively, on the factor axes.

Scaling of factors: Multiplication of factor scores and factor loadings by a set of constants, one for each factor. Scaling of factors produces expansion or contraction of the data pattern along the factor axes.

Contribution: Part of the total variance explained by a factor.

Dominant factors: Factors with the largest contribution to the total variance.

Residual factors: Factors that account for only a small part of the total variance.

Biplot: Joint graphical representation of rows and columns by means of scores and loadings on two dominant factors.

9 APPENDIX

Spectral map analysis is defined formally by the following steps.

1. Logarithmic re-expression of the original data in table X:

$$Y_{ij} = \log X_{ij}$$

where the row-index i ranges from i to I, and the column-index j ranges from j to J.

It is assumed that all elements of X are strictly positive and defined on ratio scales (not necessarily with the same units), and that I is larger than or equal to J.

2. Double-centering:

$$Z_{ij} = Y_{ij} - M_i^* - M_j + MG$$

with

$$M_i^* = \sum_{j=1}^{J} Y_{ij}/J, \quad \text{row-mean}$$

$$M_j = \sum_{i=1}^{I} Y_{ij}/I, \quad \text{column-mean}$$

$$MG = \sum_{i=1}^{I}\sum_{j=1}^{J} Y_{ij}/IJ, \quad \text{grand mean}$$

3. Calculation of variance–covariance matrix of double-centered data:

$$V = Z^t \cdot Z/I$$

where the dot indicates matrix multiplication.

4. Calculation of the orthonormal factor loadings F from the variance–covariance matrix V, satisfying the relations:

$$F^t \cdot V \cdot F = \Lambda$$

$$F^t \cdot F = U$$

where Λ represents the diagonalized variance-covariance matrix, and U denotes the unit matrix of factor space.

5. Calculation of factor scores S^* and factor loadings S:

$$S^* = X \cdot F \cdot \Lambda^{-1/4}$$

$$S = J^{1/2} F \cdot \Lambda^{1/4}$$

The special scaling of factor scores and factor loadings produces a biplot which is symmetrical with respect to rows and columns of the table.

6. The contribution c to the total variance by the first L factors is defined by means of:

$$c = \sum_{k=1}^{L} \Lambda_{kk} \bigg/ \sum_{k=1}^{K} \Lambda_{kk}$$

where K is the number of elements on the principal diagonal of Λ.

Factorial data analysis can be provided with variable weighting of rows and columns [12]. In such a case, the weight coefficients assigned to rows and columns determine their effect on the computed factors. In the case of a zero weight, a row or column is merely fitted to the factors that have been computed from the remaining rows and columns.

10 ACKNOWLEDGEMENT

The author is grateful to A. Megens for a critical reading of the manuscript. Mrs A. Biermans, Mrs A. Vermeiren and Mr B. Joossen are thanked for their help with the preparation of the manuscript. Thanks are due to reviewers for valuable comments and suggestions.

REFERENCES

1 P.A.J. Janssen, C.J.E. Niemegeers and K.H.L. Schellekens, Is it possible to predict the clinical effects of neuroleptic drugs (major tranquillizers) from animal data? Part I: 'Neuroleptic activity spectra' of rats, *Arzneimittel Forschung*, 15 (1965) 104–117.
2 L.K.C. Desmedt, C.J.E. Niemegeers and P.A.J. Janssen, Antagonism of maximal Metrazol seizures in rats and its relevance to an experimental classification of antiepileptic drugs, *Arzneimittel Forschung*, 26 (1976) 1592–1603.
3 H. Hotelling, Analysis of a complex of statistical variables into principal components, *Journal of Educational Psychology*, 24 (1933) 417–441.
4 J.-P. Benzécri, *L'Analyse des Données. Vol. II, L'Analyse des Correspondences*, Dunod, Paris, 1973.
5 P.J. Lewi, Spectral mapping, a technique for classifying biological activity profiles of chemical compounds, *Arzneimittel Forschung*, 26 (1976) 1295–1300.
6 P.J. Lewi, *Multivariate Data Analysis in Industrial Practice*, Wiley, Chichester, 1982.
7 A.H.P. Megens, J.E. Leysen, F.H.L. Awouters and C.J.E. Niemegeers, Further validation of 'in vivo' and 'in vitro' pharmacological procedures for assessing alpha-2/alpha-1-selectivity of tests: (2) alpha-adrenoceptor agonists, *European Journal of Pharmacology*, 129 (1986) 57–64.
8 L.J. Cronbach and G.C. Gleser, Assessing similarities between profiles, *Psychological Bulletin*, 50 (1953) 456–473.
9 W.W. Cooley and P.R. Lohnes, *Multivariate Procedures For the Behavioral Sciences*, Wiley, New York, 1962.
10 K.R. Gabriel, The biplot graphic display of matrices with applications to principal components analyis, *Biometrika*, 58 (1971) 453–467.
11 O.M. Kvalheim, Scaling of analytical data, *Analytica Chimica Acta*, 177 (1985) 71–79.
12 P.J. Lewi, Multivariate data representation in medicinal chemistry, in B.R. Kowalski (Editor), *Chemometrics: Mathematics and Statistics in Chemistry*, Reidel, Dordrecht, 1984, pp. 351–376.

Chapter 19

Similarities and Differences among Multivariate Display Techniques Illustrated by Belgian Cancer Mortality Distribution Data

A. THIELEMANS

Farmaceutisch Instituut, Vrije Universiteit Brussel, Laarbeeklaan 103, B-1090 Brussels (Belgium)

P.J. LEWI

Information Science Department, Janssen Research Foundation, B-2340 Beerse (Belgium)

D.L. MASSART*

Farmaceutisch Instituut, Vrije Universiteit Brussel, Laarbeeklaan 103, B-1090 Brussels (Belgium)

(Received 20 May 1987; accepted 23 December 1987)

CONTENTS

1 Introduction	263
2 Objectives of the three techniques	263
2.1 Analysis procedure	263
2.2 Characteristics of each method	264
2.2.1 Principal component analysis	264
2.2.2 Correspondence factor analysis	265
2.2.3 Spectral map analysis	266
3 Application	267
3.1 Data	267
3.2 Results	267
3.2.1 Interpretation of the figures	267
3.2.2 PCA	271
3.2.3 CFA	271
3.2.4 SMA	272
3.3 Discussion	273
4 Conclusion	274
Appendix: Analysis procedures	274
A.I PCA	274
A.II CFA	278
A.III SMA	281
Acknowledgement	284
References	284

> Experience of delivering tutorial material at recent courses and conferences has suggested that the audience is of two types. Some want a very basic treatment of topics, whereas others are familiar with most elementary concepts of chemometrics and want an in-depth explanation and illustration of advanced methods. This tutorial is written as an Advanced tutorial aimed at the latter readership.

1 INTRODUCTION

Multivariate numerical information resulting from chemical analysis, research projects, surveys and so on, is usually collected in a table which is called a data matrix. The rows of such a data matrix represent the different objects of the study, while the columns of the data matrix are the variables which describe the objects. Relationships between objects could possibly be derived from their point-to-point distances in the 'variable-space', i.e. a hyperspace in which the p variables represent p axes. Conversely, resemblances between variables could be evaluated from their position in 'object-space', i.e. a hyperspace in which the n objects define n axes. Since the data matrix usually consists of more than three objects and more than three variables, these graphical representations are difficult to realize in practice. Multivariate display techniques aim at reproducing objects and variables in a low-dimensional common 'factor-space' so that (dis)similarities between objects and/or variables can be interpreted in the more familiar two-dimensional (or eventually three-dimensional) space. The popularity of display techniques proceeds from the fact that the results are reproduced in graphical format. Pictures are usually more useful than numerical data in elucidating the most important trends that reside in the multivariate data table.

Different display techniques exist [1,2] and the fact that they are used on so many different data types and, furthermore, that they are described in so many different terms may sometimes lead to confusion about the correctness of their application and interpretation. In this article a comparison will be made between principal component analysis (PCA), correspondence factor analysis (CFA) and spectral map analysis (SMA). Other techniques such as, for instance, factor analysis (in the strict sense) [3,4] and nonlinear mapping [5] are also rather frequently applied in analytical chemistry but are, however, not discussed in this article. The three techniques mentioned above are all based on the same fundamental principle, namely 'data space simplification'. This goal is achieved by factorial decomposition of the covariance matrix and by projecting the data onto the axes defined by the extracted factors. The process can also be regarded as loading the original information described by the p original variables onto a set of uncorrelated (i.e. orthogonal) factors. One tries to do this in such a way that the total variance between the objects and variables in the original p-dimensional hyperspace is preserved as well as possible in a reduced space spanned by the first two or three (i.e. the most dominant) factors. Therefore, the principle of factorization is often considered to be a dimensionality reduction process. The techniques differ from each other in their way of preprocessing of the data i.e. in the operations that are carried out prior to factorization. Depending upon the kind of transformation, specific characteristics or structures may come to the front or remain in the background. Hence, the projections need to be interpreted somewhat differently.

Similarities and differences concerning the objectives, the mathematical background and the interpretation of the three techniques will be discussed and illustrated in greater detail in the following sections of this article.

2 OBJECTIVES OF THE THREE TECHNIQUES

2.1 Analysis procedure

The data analysis procedure is analogous for all three display techniques. The different steps included in the procedure can be enumerated as follows: (1) re-expression of the original data, (2) definition of row- and column-weights, (3) transformation of the data, (4) (column-)weighting of the transformed matrix, (5) calculation of the covariance matrix as a (row-)weighted product of the transformed matrix with its transpose, (6) diagonalization of the covariance matrix or in other words computation of the eigenvectors (factors) and eigenvalues (factor variances) of this matrix and, finally, (7) calculation of coordinates of objects and coordinates of variables on the new axes or factors. The techniques mainly differ in the first four steps (the 'preprocessing' part) whereas steps 5 to 7 (factorization part) are exactly the same. The Appendix describes the mathematical equations of these different steps in more detail and illustrates them with a numerical example.

2.2 Characteristics of each method

2.2.1 Principal component analysis

Principal component analysis (PCA) [6–10] is probably the best known and most commonly used display method. Other names for the same techniques are principal factor analysis, eigenvector analysis, eigenvector decomposition, Karhunen–Loewe expansion, etc. [2,3]. The technique is often applied in the analysis of tables where the columns are expressed in different units.

In PCA, row- and column-weights are usually constant (see Appendix). The transformation procedure usually consists of column-centering (i.e. subtracting the column-mean from each corresponding element) followed by column-standardization (i.e. dividing the result by the corresponding column-standard deviation). This transformation corrects for differences in size (i.e. importance) and differences in range between the variables. Since row-centering and row-standardization are usually not included in this process, differences in size and range between the objects are not eliminated. As a result of the asymmetry of the transformation procedure with regard to objects and variables, objects are represented as points and variables are interpreted as vectors in the display. Therefore, although objects and variables can be plotted simultaneously on the same graph, their position on the projection should be interpreted differently.

A PCA projection describes the information in a data table as expressed by correlation coefficients between variables. When working with absolute data and when large differences in size between the objects exist, correlation coefficients between variables are all positive and consequently all variables are located at the same side of the first (and most important) factor. In that case, the first factor mostly expresses size information. It can be regarded as a global size component reproducing a classification of the objects according to their size. 'Large' objects, i.e. objects with high values for all the variables will appear together at the one end of the size component while small-sized objects, i.e. objects that have low values for the variables. Will be located at the opposite end. The remaining part of the variation in the data set — reproduced by the second and following axes — is due to relative differences between the objects. Therefore, it can be advisable to plot the second against the third factor in order to observe these relative differences [11]. When all objects are of the same size (e.g. closed data, that is, data summing row-wise to 100%), variables will be located on either side of the first factor (positive as well as negative correlations occur) and the first factor will then show relative differences between the objects instead of absolute differences. In what follows, we shall refer to these relative differences as contrasts.

As a result of column-centering and column-standardization, variables are all situated at the same distance from the barycentre (which functions as a new origin) in real multidimensional data space. When the percentage of variance explained by the two-factorial display is 100%, this leads to a circular arrangement of the variables around the origin of the plot (correlation circle). Since this is rarely the case, the position of the variables on the two-dimensional plot will always deviate to some extent from this circular projection. Variables that are well-represented by two factors are projected close to the periphery of this circle, whereas variables that are less well reproduced are situated closer to the origin.

The direction and angular separation of the

variables in the two-dimensional plot is important in the interpretation of their relationship towards each other and towards the objects taken as a whole. In PCA, the angular separation provides an estimate of the correlation between the two corresponding variables since the correlation coefficient between two variables is defined as the cosine of the angle between the two corresponding variable vectors. When the percentage of variance explained by the two-dimensional plot is 'sufficiently' large (i.e. when the PCA plot reproduces a large fraction of the variation in the data matrix), it is appropriate to draw axes from the origin of the plot towards the different variables. In the biplot variant of PCA [12], these axes can then be viewed as coordinate axes. Projecting the objects on such a variable axis results in a classification of the objects with respect to their value for that particular variable. This means that objects that lie in the same direction as a variable have rather high values for this variable compared to objects that lie in the opposite direction to this same variable.

2.2.2 Correspondence factor analysis

Correspondence factor analysis (CFA) [9,10, 13–15] is a display method that was originally designed for the analysis of contingency tables. The objects and variables of such a table can be described as different categories or contingencies of a certain characteristic or feature. Its elements are expressed as counts or frequencies and hence they can be added, in a meaningful sense, row-wise as well as column-wise. However, provided that the elements are positive and that the different variables are expressed in the same units, the application of CFA can be extended to other types of data tables (e.g. ranks, presence-absence, etc.) [15]. Depending upon the field in which the technique is used, it is also given such different names as, for instance, dual scaling, reciprocal averaging, canonical analysis of contingency tables and so on [14].

Initially, the data are re-expressed by dividing each element by the matrix-total (the sum over all elements). Weight coefficients of the objects and variables are then defined as the corresponding marginal row- and colum-totals (see Appendix).

The transformation procedure consists of subtracting and subsequently dividing each element in the matrix by its 'expected' value (or theoretical frequency in the chi-square sense). The 'expected' value of an element can be defined as the product of the corresponding marginal row-total with the marginal column-total. It equals the value that would be obtained if no interaction existed between the corresponding row and column. The procedure of subtraction and division by expectation can be seen as both a correction for differences in size (importance) between variables and for differences in size between the objects. Consequently, in CFA, the roles of objects and variables are symmetric. Exactly the same results are obtained if the analysis is carried out starting from the transposed data table. Geometrically, this kind of transformation results in the evaluation of distances between objects or variables in a chi-square metric (i.e. a 'weighted' Euclidean space where the weights of the dimensions are the inverses of the theoretical frequencies) [14]. Consequently, point-to-point distances on a CFA plot are not related to the more common Euclidean distances and therefore they are perhaps less easily interpreted from a purely intuitive point of view.

After factorization of the covariance matrix and the computation of factor coordinates of objects and variables, it is possible to project objects and variables simultaneously in the reduced space spanned by the most dominant factors. Instead of reproducing correlations, as in PCA, results from CFA are related to contrasts (i.e. differences in relative frequencies), irrespective of the size of objects and variables. More precisely, CFA displays interactions. In our terminology, interaction is defined as the variance of contrasts among objects and among variables. On the corresponding maps, the actual positions of objects and variables are meaningfully interpreted in terms of interaction between rows and column. Proximities between objects indicate that their variable-profiles (i.e. their relative frequencies for the different variables) are similar. Objects with high relative frequencies for one or more variables are attracted by those variables. Consequently they are close together on the CFA map. On the other hand, large distances between objects and variables can be considered as being due to repulsive forces caused by low relative frequencies of the object for the variable in question or vice versa. The final dispersion of objects and variables on the correspondence factor analysis plot can be regarded as the result of interacting forces of attraction and repulsion that are related to the relative specificities of objects and variables towards each other.

The further the objects and variables are removed from the centre of the plot (i.e. the centre of gravity of the data), the more specific or characteristic they are for that particular plot and the more reliable the interpretation of their position becomes. Objects or variables located close to the origin of the map can be regarded as being neutral or aspecific; that is, such points do not show any particular contrast (i.e. variability). Their profile or pattern is very similar to the mean profile.

2.2.3 Spectral map analysis

Spectral map analysis (SMA) [10,16,17] is closely related to CFA. In practice, the method is also applicable to data that are expressed in different units. The approach in SMA is to carry out a logarithmic transformation prior to all other computations, thus making it possible to interpret the final plots in terms of log ratios (vide infra). In the case of additive data, weight coefficients are generally defined in the same way as in CFA. They are derived from the marginal totals. However, in the case of non-additive data (when row- or column-totals have no significant meaning), weight coefficients can be kept constant or they can be defined externally; for instance, as being proportional to a particular row or column.

In analogy with CFA, the transformation procedure is symmetrical with respect to both objects and variables. The process is defined as double-centering and consists of subtracting consecutively the corresponding column-mean and the corresponding row-mean from each (logarithmic) element. This procedure corrects for differences in size between the objects as well as for differences in size (importance) between the variables. From this point of view, the result of preprocessing in CFA and SMA is similar. The difference lies in the fact that double-centering includes only linear operations (subtractions) whereas division by expectation in CFA is a non-linear operation. As a result of the linear operations in SMA (carried out after logarithmic transformation), the statistical principles can be thought of as being derived from a two-way analysis of variance since the total variance in the logarithmic data set can be described in terms of a linear decomposition of different sources of variation:

$$V_{tot} = V_0 + V_m + V_i$$

The terms V_0 and V_m stand for the variances of row-means and column-means, respectively. Those are the main effects that are eliminated by double-centering. The source of variation that remains in the centered data set is the so-called interaction term (V_i) which is reproduced by the contrasts of objects and variables on the graphical display.

The information reproduced by an SMA map is two-fold. On the one hand, the dispersion of objects and variables on the plot translates the existing relationships between them, irrespective of their size. As in CFA, the plot shows relative specificities of the objects towards the different variables (or vice versa, since objects and variables play symmetrical roles). Point-to-point distances are meaningful in estimating resemblances between objects or between variables. An object is attracted by those variables for which its relative frequency is higher than the mean relative frequency and vice versa. By contrast, objects are repelled by variables for which they have only low specificity (lower than average). The position of an object and, consequently, the final appearance of the map is determined by the net result of these interaction forces. The fact that the more specific objects are situated near the border of the map also applies here. This kind of information extracted from an SMA plot is thus quite analogous to the information obtained by CFA. However, as a consequence of a different transformation procedure, the subspace in which this information is displayed is also somewhat different (see Fig. 4 below). Note that in CFA a point in two-dimensional data space is projected from the origin radially upon an axis of contrast which is orthogonal to the bisector. In SMA the point is projected perpendicularly upon an axis of contrast which is also perpendicular to the bisector. The projection of the point upon the bisector represents the size of the object.

Moreover, SMA also gives size-information. Since in SMA the total variation in a data table can be described as the sum of three individual variation terms, the total information can also be displayed as such. As has already been mentioned, the interaction term is reproduced in the dispersion of objects and variables on the plot. Differences in size of the objects (V_0) and differences in size of the variables (V_m) are expressed by varying the areas of the corresponding symbols on the plot. In CFA, differences in size of objects and variables are straightened out by a non-linear operation (division by expectation). As a result, the size variation cannot be expressed as individual terms of a linear decomposition of the total variance. Consequently, those terms are not represented on the corresponding plot and all symbols appear with equal size.

A particular way of scaling factor coordinates in SMA makes it possible to draw axes in the projected factor space by connecting any two variables (or any two objects) with each other. The values read off along these axes are immediately associated with the contrasts among the numbers in the data table. More precisely, if an axis is formed by the connection of two variables and if objects are projected upon it, calibrations along the axis allow estimation of the ratios of the corresponding two variables. This is due to the fact that the difference between the logarithms of two values is equal to the logarithm of their ratio. It explains why SMA is also applicable to data that are defined with different units of measurement (one evaluates ratios instead of algebraic differences).

The logarithmic re-expression has still further consequences. It corrects for the so-called positive skewness of a distribution (i.e. it makes the data more uniformly distributed). Extremely large values are substituted by more 'moderate' values while, on the other hand, it extends the range of values bounded by the value of zero towards minus infinity. Values that originally fall in the range between 0 and 1 result in negative values after logarithmic re-expression. This explains why small values have a much more pronounced repulsive effect in SMA than in CFA. Values near zero may exert a leverage effect in SMA whereas in CFA they may pass unnoticed (see the application to cancer data described below).

3 APPLICATION

3.1 Data

An epidemiologic data set was analyzed. It describes the cancer mortality in the Belgian male population. Standardized mortality rates (i.e. expressed as a number of deaths per 100 000 males per year) were extracted from *The Atlas of Cancer Mortality in Belgium* [18] and multiplied by the size of the male population in the different Belgian districts (population figures were made available by the National Institute of Statistics). The final data set (Table 1) gives for each of the 43 Belgian districts (objects), the approximate number of deaths due to 12 different sites of cancer (variables). This 43×12 data matrix was then submitted to the three display techniques under consideration. As a complement to this application, the reader may be referred to a study describing the results of an investigation of the corresponding Belgian female data by the use of pattern recognition techniques [19].

3.2 Results

3.2.1 Interpretation of the figures

The maps are reproduced with the help of the SPECTRAMAP program [17], which allows evaluation by SMA as well as by PCA and CFA. Objects are represented as circles and variables are represented as squares. The centre of the projection is indicated by means of a cross. The first and second factors represent the abscissa and ordinate of the map (prior to rotation). The importance of the third factor ('depth' factor) is introduced by varying the contour of the corresponding circles and squares. A thick contour means that the object or variable lies above the two-dimensional plane while objects and variables that lie below the surface are represented with a thin contour.

TABLE 1

Tumor mortality in males, Belgium: 1969–1976

Mean yearly number of deaths over the period 1969–1976. A = esophagus, B = stomach, C = colon, D = rectum, E = gall-bladder, F = pancreas, G = larynx, H = trachea-lung, I = prostate, J = bladder, K = brain, L = leukemia. Source: table extracted from the *Atlas of Cancer Mortality in Belgium (1969–1976)* [18] and multiplied by the male population figures.

	A	B	C	D	E	F	G	H	I	J	K	L	TT
1: AAL	8	47	26	18	4	10	11	107	34	15	10	8	300
2: ARL	1	8	4	2	0	3	1	25	5	3	1	1	54
3: ATH	2	8	5	6	1	4	2	30	10	3	1	4	76
4: ANT	21	155	82	59	17	48	26	477	102	66	34	34	1122
5: BXL	29	115	100	56	18	54	36	459	129	66	17	49	1120
6: BRU	6	38	21	19	1	12	8	101	29	11	8	9	263
7: BAS	1	4	2	1	0	3	2	18	5	1	0	1	39
8: CHA	10	68	36	22	2	23	21	214	61	30	14	19	521
9: DIK	1	10	5	3	0	3	1	14	4	3	1	2	46
10: DEN	7	39	17	12	3	9	8	77	20	8	6	5	211
11: DIN	3	8	9	5	2	3	2	50	10	4	3	3	102
12: EEK	3	16	10	6	1	3	2	40	10	4	3	3	99
13: GEN	14	99	51	33	2	25	13	204	61	19	16	14	551
14: HAS	7	66	30	20	6	16	10	161	32	19	10	11	388
15: HAL	14	77	51	32	8	22	15	200	52	23	9	21	524
16: HUY	2	11	8	4	2	5	3	53	13	6	3	4	113
17: IEP	2	25	10	11	1	4	2	35	12	6	3	4	114
18: KOR	7	50	23	18	2	14	7	97	36	11	7	10	283
19: LIE	15	74	47	28	11	33	26	336	84	41	19	22	737
20: LEU	8	70	37	25	5	19	12	168	41	20	9	15	429

Chapter 19

21:MOU	3	15	6	8	1	3	3	30	10	5	1	3	87
22:MEC	5	50	32	23	5	12	7	136	31	16	11	11	339
23:MON	7	25	24	14	3	15	8	105	33	16	3	10	263
24:MAR	1	4	4	2	1	2	1	21	5	2	0	2	46
25:MAA	1	39	17	10	3	8	3	108	18	10	4	5	226
26:NEU	1	7	3	3	2	3	2	32	6	1	1	3	64
27:NAM	4	29	22	9	7	14	8	123	32	10	7	10	277
28:NIV	4	32	25	15	2	13	9	112	31	16	5	12	275
29:OUD	6	21	11	9	1	5	3	46	13	6	3	4	128
30:OOS	4	23	11	7	0	7	6	65	12	5	3	6	150
31:PHI	2	7	6	1	1	4	2	27	9	2	2	2	65
32:ROE	3	31	14	13	1	7	4	51	14	7	4	5	152
33:SOI	2	20	16	10	2	8	5	67	22	9	4	6	172
34:STN	6	45	19	14	1	11	5	87	25	13	6	7	240
35:TON	3	30	13	13	2	5	4	79	17	13	4	6	190
36:TUR	7	58	30	22	6	20	7	160	32	13	13	12	380
37:TIE	1	15	6	5	0	2	1	27	9	3	2	3	75
38:THU	3	18	10	6	1	8	5	62	20	8	5	6	153
39:TOU	6	18	14	9	1	6	5	52	21	8	2	8	151
40:VEU	2	10	4	3	0	2	2	20	8	2	2	2	59
41:VIR	1	4	4	2	1	2	1	25	4	2	0	2	47
42:VER	5	29	22	10	5	9	6	123	33	14	7	10	273
43:WAR	0	7	6	2	1	2	2	33	7	2	2	2	66
TT: Table total	240	1527	896	593	132	482	308	4454	1160	544	266	376	10977

Mortality by units

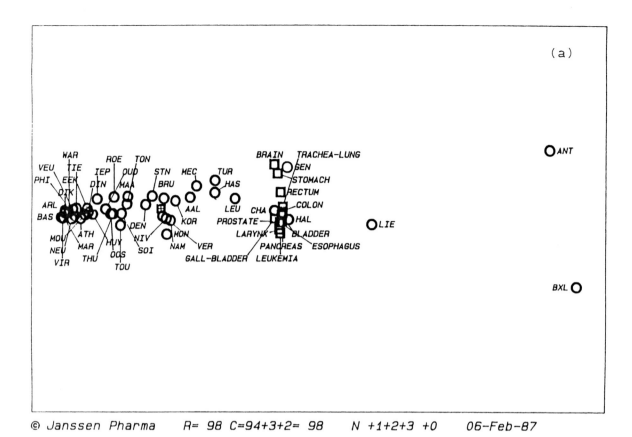

© Janssen Pharma   R= 98 C=94+3+2= 98   N +1+2+3 +0   06-Feb-87

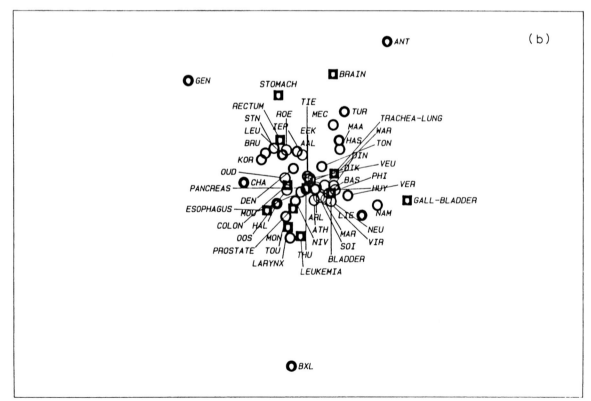

© Janssen Pharma   R= 98 C=3+2+94= 98   N -2+3+1 +90   15-Dec-87

Objects and variables that cannot be reproduced in a satisfactory way by the three factors (i.e. objects or variables for which the residual with respect to the three-dimensional factorial space is larger than a predefined threshold), are represented with a dotted contour. Elements that originally are not included in the calculations of the factors (because of their divergent behavior) but which are positioned afterwards on the computed factors, are indicated with an asterisk.

3.2.2 PCA

The first three factors explain about 99% of the information contained in the data set (see Fig. 1a and b). Clearly, the first factor, accounting for 94% of the variation in the data set, shows global differences in size between the districts. As can be derived from the deviating position of the three most densely populated districts (i.e. Brussels, Antwerp, Liège), districts are arranged according to their total cancer mortality rate, which is proportional to the population size of the district.

Information about relative specificities in cancer mortality between the different districts remains masked by the size component in the first factor. The map (Fig. 1a) shows only a slight separation of the Flemish districts from the Walloon districts along the second axis. Since all cancer sites are located rather close to each other, it is difficult to determine which cancer sites are predominant in the respective regions. Brain and stomach cancer seem to diverge towards the 'Flemish' side whereas leukemia, pancreas and larynx cancer seem to be directed more towards the 'Walloon' side or the negative side of the second axis. These specificities are not pronounced, however.

The angular separations between all cancer sites are small, which means that they all are positively correlated with each other. The positive correlations are due to the strongly pronounced size factor in the original data. Districts that are densely populated show high absolute mortality figures for all cancer sites, whereas sparsely populated districts exhibit low absolute mortality figures for all cancer sites.

Because of the presence of a large size-component, relative differences between cancer sites would appear more distinctly in a PCA plot of the second and third factors (see Fig. 1b). In order to study these more closely we resort to methods that have been developed specifically for the analysis of contrasts, such as in CFA and SMA.

3.2.3 CFA

The data variability expressed by the map (Fig. 2) — which is equal to the variance accounted for by the first three factors — amounts to 73%. This percentage refers to the fraction of the global variance in the data set *after* subtraction and division by expectation (i.e. after removing the size information).

In contrast to the first factor in PCA, CFA does not show differences in size but rather stresses differences in relative specificities between the different districts and/or cancer sites. It is possible to identify the tumor types that are typical for certain regions or, alternatively, to detect groups of districts characterized by high or low relative cancer mortality for certain tumor types.

Districts and cancer sites are more or less uniformly spread over the whole area of the plot. The disposition of the districts on the map coincides very closely with their real geographical location, indicating that cancer patterns are strongly geographically determined. The north–south division (i.e. Flemish–Walloon), as well as the west-to-east arrangement of the districts, is preserved on the plot.

The more the objects and variables deviate from the centre of the map, the more reliable the interpretation of their position becomes. Stomach, rectum and brain cancer diverge towards the upper (or Flemish) side of the plot. Stomach and rectum cancer lie more in the direction of the western Flemish districts while brain cancer is positioned slightly towards the eastern side. The triplet formed by leukemia, larynx and prostate cancer, on the other hand, lies more in the

---

Fig. 1. SPECTRAMAP — Tumor mortality in males, Belgium: 1969–1976. (a) PCA plot of the first against the second factor; (b) PCA plot of the second against the third factor.

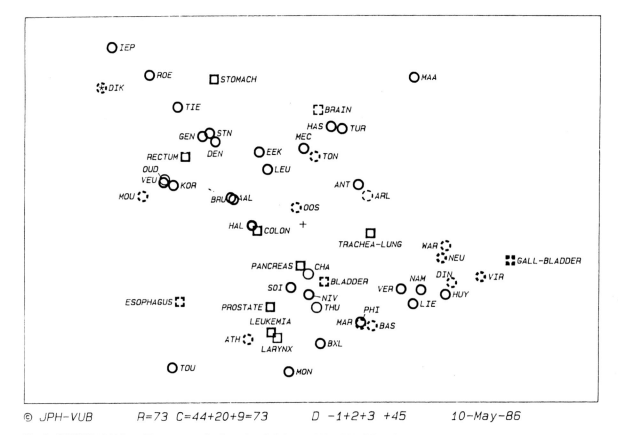

Fig. 2. SPECTRAMAP — Tumor mortality in males, Belgium: 1969–1976. CFA plot.

neighbourhood of the Walloon districts. From its rather extreme position along the east–west axis, gall-bladder cancer seems to be characteristic for the eastern side of the country, while the opposite is true for esophagus cancer, which is especially close to the western Walloon districts.

3.2.4 SMA

In order to be able to carry out a logarithmic transformation, values equal to zero were substituted by small positive numbers by means of extrapolation (Fig. 3).

The contribution of the first three dominant factors in explaining the interaction variance amounts to 67%. The degree of reproduction refers to the fraction of the total variance in the data set *before* double-centering (i.e. including the variances due to differences in size) and amounts to 99%.

Unlike the two previous techniques, the SMA map reveals information about the size-component as well as information about contrasts between objects and variables.

Most of the elements fall within the area of a triangle formed by gall-bladder, brain and esophagus cancer that occupy the most extreme positions. Other outlying positions are observed for the districts of Maaseik (MAA), Virton (VIR) and Neufchâteau (NEU). The districts and cancer sites that are furthest removed from the centre are all of small absolute size and exhibit one or more extremely small values in the absolute mortality rates (Table 1). This result illustrates the strong influence of small values — and thus of repulsive forces — on the final dispersion of an SMA plot.

As in CFA, Walloon and Flemish districts can be distinguished according to their vertical position while the horizontal position of the districts

Chapter 19

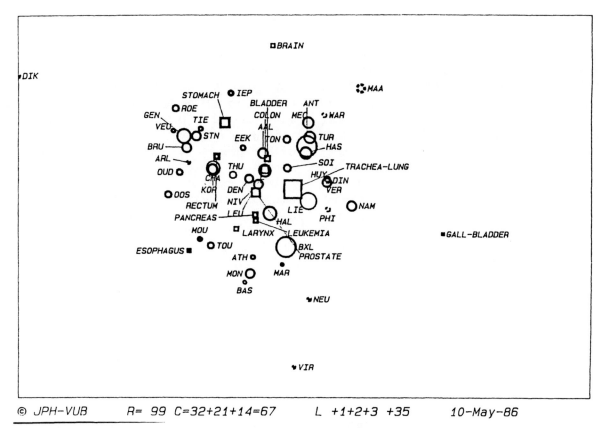

Fig. 3. SPECTRAMAP — Tumor mortality in males, Belgium: 1969–1976. SMA plot.

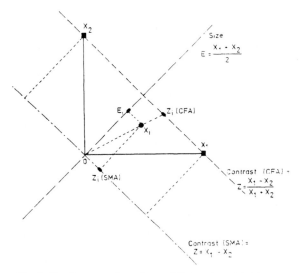

Fig. 4. X_1, X_2: two column-items defining the original two-dimensional data space. One-dimensional subspaces orthogonal to the bisector line of the data space as defined by CFA (- - - - -) and SMA (· – · – ·). X_i: representation of i th row-item

coincides well with their west-to-east situation. The cancer sites that are typical for the observed contrasts (i.e. gall-gladder, esophagus, brain, etc.) are also more or less the same. However, differences between the two techniques also occur. They are observed for those elements that show one or more zero values (i.e. Virton, Neufchâteau, Arlon, Waremme, brain cancer etc.).

3.3 Discussion

Although the data that were used in the three analyses are exactly the same, the dispersion of objects and variables on the resulting plots are

(i.e. i th object) in the original data space. Z_i: one-dimensional representation of i th row-item in the subspaces respectively defined by CFA and SMA. E_i: size of i th row-item along the bisector line of the two-dimensional space.

different. The PCA results are the most different from the others. The plot of the first against the second factor is totally determined by the large differences in absolute cancer mortality between the different districts. The total cancer mortality rate in the districts varies from ±1000 per year (Brussels, Antwerp, Liège) to ±50 per year (Bastogne, Diksmuide, Marche, Virton, etc.). These differences result in a strongly pronounced size effect translated by the first axis. In contrast to CFA and SMA, the first vs. second factor plot in PCA expresses information which is obvious from the row-sums in the data table whereas information about relative specificities of cancer mortality for the various districts remains hidden. In order to extract information about relative specificities, it is advisable to plot the second against the third factor.

On the CFA plot, districts and cancer sites are uniformly spread over the whole area while, on the SMA plot, districts with very small cancer frequencies and cancer sites with low mortality rates are strongly repelled towards the border of the plot. Nevertheless, both plots present more or less the same information by showing contrasts (i.e. information which is not immediately obvious from the data table) rather than size-differences. A north–south and an east–west tendency can be observed on both plots. The cancer sites that seem to be typical for these geographical regions can also be identified in both plots, although the strength of their discriminating power is somewhat different in the two analyses. Abnormally small values have a stronger influence in SMA compared to CFA.

4 CONCLUSION

The differences and similarities in the algorithms of the three methods are obvious from the results. In short, PCA can be described as an analysis of correlation and size, whereas correspondence factor analysis and spectral map analysis can be seen as analyses of interaction and of contrasts. The advantage of CFA and SMA over PCA is that their result is the same whether it is computed from rows or from columns. All factors in CFA are pure components of contrast, but one loses the size-component. Furthermore, contrasts in CFA are defined in a so-called chi-square metric, which is difficult to interpret intuitively. In SMA all factors are also pure components of contrast. In this method, one recovers the size-component, which is reintroduced graphically by variation of the areas of the symbols. Furthermore, contrasts in SMA can be interpreted as log ratios.

Which method is to be preferred depends mainly upon the data one wishes to analyze and, consequently, on the facts and features one is interested in. Overall, the main result of multivariate display techniques is a visual and synthetic description of the the major tendencies and main interacting forces in the observed data. These techniques are therefore more appropriate in the formulation of hypotheses and suggestions than in providing precise answers to predefined questions.

APPENDIX: ANALYSIS PROCEDURES

Table A.1 gives a schematic overview of the analysis procedure in PCA, CFA and SMA. The mathematical equations corresponding to the different steps included in the analysis procedure are explained separately for each of the three techniques (Sections A.I, A.II, A.III). To illustrate and compare the way in which the different analyses proceed, they were carried out on the same small data table consisting of 10 objects characterized by 6 variables (Table A.0). The intermediate results obtained during each of the consecutive steps are tabulated for each of the three techniques (Tables A.I.1 to A.III.7).

A.I PCA

*A.I.0 Original data matrix (**A**)*

The original data matrix which is used as a starting point in the analysis will be denoted as:

$$\mathbf{A}_{(n \times p)}$$

with: n = total number of rows (objects); p = total number of columns (variables); and A_{ij} being the

Chapter 19

TABLE A.1

Analysis procedure		PCA	CFA	SMA
PART I.	Pre-processing of the data			
Step 1.	Re-expression of the original data	(Usually) none	Division by matrix-total	Logarithmic re-expression and division by matrix-total
Step 2.	Definition of diagonal weighting matrices	(Usually) constant	Marginal totals	Marginal totals, constant or proportional to a row or column
Step 3.	Transformation of the data	Column-centering Column-standardisation	Subtraction and division by expectation	Double-centering
PART II.	Dimension reduction			
Step 4.	Weighting of the transformed matrix	Weighting according to previously defined column weights		
Step 5.	Factorization or calculation of the	Calculation of the covariance matrix		
Step 6.	new factor space	Diagonalization of the covariance matrix		
Step 7.	Representation of objects and measurements in the new factor space	Calculation of factor coordinates		

TABLE A.0

Original data matrix (A)

The objects are:
1: Antwerp (ANT)
2: Roeselare (ROE)
3: Sint-Niklaas (STN)
4: Hasselt (HAS)
5: Leuven (LEU)
6: Liège (LIE)
7: Tournai (TOU)
8: Ath (ATH)
9: Namur (NAM)
10: Waremme (WAR)

The variables are:
1: esophagus cancer
2: stomach cancer
3: colon cancer
4: gall-bladder cancer
5: lung cancer
6: prostate cancer

	1	2	3	4	5	6	Tot
1	21	155	82	17	477	102	854
2	3	31	14	1	51	14	114
3	6	45	19	1	87	25	183
4	7	66	30	6	161	32	302
5	8	70	37	5	168	41	329
6	15	74	46	11	336	84	566
7	6	18	14	1	52	21	112
8	2	8	6	1	30	10	57
9	4	29	22	7	123	32	217
10	0	7	6	1	33	7	54
Tot	72	503	276	51	1518	368	2788

value of object i for variable j, where i ranges from 1 to n and where j ranges from 1 to p (see Table A.0).

A.I.1 Re-expression of the original data (A → X)
The re-expressed data-matrix will be denoted as:

$$\mathbf{X} \atop (n \times p)$$

In PCA, usually no re-expression is carried out, therefore:

$$X_{ij} = A_{ij}$$

A.I.2 Definition of diagonal weighting matrices (W* and W)

Because PCA is usually applied to data where the columns are expressed in different units, it is not indicated to assign different weights to the different column- and row-items. Therefore, weight coefficients are kept constant and diagonal weighting matrices for row- and column-items are

TABLE A.I.2.1

Row-weights (**W***)

	1	2	3	4	5	6	7	8	9	10
1	0.1									0
2		0.1								
3			0.1							
4				0.1						
5					0.1					
6						0.1				
7							0.1			
8								0.1		
9									0.1	
10	0									0.1

TABLE A.I.2.2

Column-weights (**W**)

	1	2	3	4	5	6
1	0.167					0
2		0.167				
3			0.167			
4				0.167		
5					0.167	
6	0					0.167

defined as respectively:

$\mathbf{W}^*$ with $W_{ii}^* = 1/n$
$(n \times n)$

(weight coefficient for row i)
(see Table A.I.2.1)

$\mathbf{W}$ with $W_{jj} = 1/p$
$(p \times p)$

(weight coefficient for column j)
(see Table A.I.2.2)

A.I.3 Transformation of the (re-expressed) data matrix ($\mathbf{X} \rightarrow \mathbf{Y}$)

The transformation procedure consists of column-centering (i.e. subtracting the corresponding column-mean from each value in the data table) followed by column-standardization (i.e. divide each resulting value by the corresponding column-standard deviation):

$\mathbf{Y}$ with $Y_{ij} = \dfrac{X_{ij} - M_j}{SD_j}$ (see Table A.I.3)
$(n \times p)$

TABLE A.I.3

Means, standard deviations and transformed matrix (**Y**)

	1	2	3	4	5	6
M	7.2	50.3	27.6	5.1	151.8	36.8
SD	6.0	42.1	21.9	5.1	139.3	30.1

Y	1	2	3	4	5	6
1	2.29	2.49	2.48	2.31	2.33	2.17
2	−0.70	−0.46	−0.62	−0.80	−0.72	−0.76
3	−0.20	−0.13	−0.39	−0.80	−0.47	−0.39
4	−0.03	0.37	0.11	0.17	0.07	−0.16
5	0.13	0.47	0.43	−0.02	0.12	0.14
6	1.30	0.56	0.84	1.15	1.32	1.57
7	−0.20	−0.77	−0.62	−0.80	−0.72	−0.52
8	−0.86	−1.00	−0.99	−0.80	−0.87	−0.89
9	−0.53	−0.51	−0.26	0.37	−0.21	−0.16
10	−1.20	−1.03	−0.99	−0.80	−0.85	−0.99

where

$\mathbf{M}(p)$ = vector containing the p column-means

with

$M_j = \sum_i X_{ij}/n$ (the jth column-mean)

and

$\mathbf{SD}(p)$ = vector containing the p column-standard deviations

with

$SD_j = \sqrt{\sum_i (X_{ij} - M_j)^2/n}$

(the jth column-standard deviation)

Column-centering eliminates the differences in importance between the different variables. Column-standardization neutralizes the differences in range between the variables. The global effect of this transformation is thus that every variable possesses the same contribution in the analysis.

A.I.4 Weighting of the transformed matrix ($\mathbf{Y} \rightarrow \mathbf{Z}$)

The transformed matrix is multiplied by the square root of the diagonal matrix of column weights:

$\mathbf{Z} = \mathbf{Y} \cdot \mathbf{W}^{1/2}$ (see Table A.I.4)
$(n \times p)$ $(n \times p)$ $(p \times p)$

TABLE A.I.4

Weighted matrix (**Z**)

	1	2	3	4	5	6
1	0.94	1.02	1.01	0.94	0.95	0.88
2	−0.29	−0.19	−0.25	−0.33	−0.30	−0.31
3	−0.08	−0.05	−0.16	−0.33	−0.19	−0.16
4	−0.01	0.15	0.04	0.07	0.03	−0.07
5	0.05	0.19	0.18	−0.008	0.05	0.06
6	0.53	0.23	0.34	0.47	0.54	0.64
7	−0.08	−0.31	−0.25	−0.33	−0.29	−0.21
8	−0.35	−0.41	−0.40	−0.33	−0.36	−0.36
9	−0.22	−0.21	−0.10	0.15	−0.08	−0.07
10	−0.49	−0.42	−0.40	−0.33	−0.35	−0.40

where the dot indicates matrix multiplication.

Since, in PCA, column-weights are kept constant (see A.I.2), this operation is the same as multiplying each value by the same coefficient, namely $\sqrt{1/p}$. This operation can thus be considered as an unnecessary step since it changes nothing in the mutual relationships between the data in the data table. Final results will be exactly the same with or without inclusion of the 'weighting' step in the PCA-analysis. In this appendix, the 'weighting' step was taken up in the PCA analysis procedure only to be able to show the close analogy of this procedure with the other two techniques, SMA and CFA.

A.I.5 Calculation of the matrix **V**

The matrix **V** is calculated as a weighted product of the matrix **Z** with its transpose **Z**':

$$\underset{(p \times p)}{\mathbf{V}} = \underset{(p \times n)}{\mathbf{Z}^t} \cdot \underset{(n \times n)}{\mathbf{W}^*} \cdot \underset{(n \times p)}{\mathbf{Z}}$$

(see Table A.I.5.1)

TABLE A.I.5.1

Matrix **V**

	1	2	3	4	5	6
1	0.167	0.156	0.161	0.152	0.162	0.163
2	0.156	0.167	0.164	0.149	0.157	0.151
3	0.161	0.164	0.167	0.158	0.163	0.160
4	0.152	0.149	0.158	0.167	0.162	0.159
5	0.162	0.157	0.163	0.162	0.167	0.165
6	0.163	0.151	0.160	0.159	0.165	0.167

TABLE A.I.5.2

Correlation matrix (**R**)

	1	2	3	4	5	6
1	1.000	0.936	0.965	0.911	0.971	0.978
2	0.936	1.000	0.983	0.896	0.943	0.907
3	0.965	0.983	1.000	0.949	0.979	0.958
4	0.911	0.896	0.949	1.000	0.974	0.955
5	0.971	0.943	0.979	0.974	1.000	0.990
6	0.978	0.907	0.958	0.955	0.990	1.000

The resulting matrix coincides with the correlation matrix (i.e. containing the correlation coefficients between each pair of variables) multiplied by a constant factor $1/p$:

$$V_{jj'} = R_{jj'} \times 1/p$$

with $R_{jj'}$ being the correlation coefficient between variables j and j'.

The constant factor $1/p$ is due to the inclusion of the 'weighting' step in the PCA analysis procedure. If this step were to be left out of the analysis, the matrix **V** would contain the correlation coefficients themselves (see Table A.I.5.2).

PCA can therefore be carried out directly on a correlation matrix.

A.I.6 Diagonalization of the matrix **V**

Diagonalization or calculation of the eigenvectors (factors) and corresponding eigenvalues (factor variances) of the symmetric matrix **V**:

$$\underset{(q \times q)}{\Lambda} = \underset{(q \times p)}{\mathbf{F}^t} \cdot \underset{(p \times p)}{\mathbf{V}} \cdot \underset{(p \times q)}{\mathbf{F}}$$

with $\mathbf{F}^t \cdot \mathbf{F} = \mathbf{U}$.

Λ is a diagonal matrix containing the q eigenvalues of matrix **V** in descending order with Λ_{kk} being the kth eigenvalue corresponding to eigenvector k. The number of eigenvalues q is at most equal to the number of objects (n) or the number of variables (p), whichever is the smallest.

F is an orthonormal matrix containing the q column eigenvectors of matrix **V** with F_{jk} denoting the jth term of eigenvector k.

U is the unit matrix.

(See Tables A.I.6.1 and A.I.6.2 for the first three eigenvalues and eigenvectors ($k = 1, 2, 3$) of matrix **V**.)

TABLE A.I.6.1

Matrix of eigenvalues (Λ) (first three eigenvalues only)

	1	2	3	4	5	6
1	0.961	0	0			
2	0	0.022	0			
3	0.	0	0.014			
4				.		
5					.	
6						.

TABLE A.I.6.2

Matrix of eigenvectors (**F**) (first three eigenvectors only)

	1	2	3	4	5	6
1	0.408	0.100	−0.638	.	.	.
2	0.401	0.691	0.328			
3	0.413	0.271	0.201			
4	0.403	−0.532	0.532			
5	0.415	−0.193	−0.015			
6	0.410	−0.346	−0.402			

Remark: In our example eigenvectors and eigenvalues were calculated making use of the JK method (= modification of the Jacobi method) [20].

*A.I.7 Representation of objects and variables in the new factor space (**S*** and **S**)*

(a) Calculation of the coordinates of the objects in the factor space by projecting the matrix **Z** onto the factor matrix **F**:

$$\underset{(n \times q)}{\mathbf{S}^*} = \underset{(n \times p)}{\mathbf{Z}} \cdot \underset{(p \times q)}{\mathbf{F}} \quad \text{(see Table A.I.7.1)}$$

S* is the matrix of factor coordinates of the objects with S_{ik}^* being the coordinate of object i on factor k.

(b) Calculation of the coordinates of the variables in the factor space by re-scaling the factor matrix:

$$\underset{(p \times q)}{\mathbf{S}} = \underset{(p \times p)}{\mathbf{W}^{-1/2}} \cdot \underset{(p \times q)}{\mathbf{F}} \cdot \underset{(q \times q)}{\Lambda^{1/2}}$$

(see Table A.I.7.2)

S is the matrix of factor coordinates of the variables with S_{jk} being the coordinate of variable j on factor k.

The re-scaling is such that the coordinates of

TABLE A.I.7.1

Matrix of factor coordinates of the objects (**S***) (first three factors only)

	1	2	3	4	5	6
1	2.346	0.077	0.072	.	.	.
2	−0.676	0.111	0.025			
3	−0.395	0.178	−0.103			
4	0.087	0.095	0.131			
5	0.211	0.160	0.035			
6	1.125	−0.270	−0.210			
7	−0.604	0.010	−0.184			
8	−0.903	−0.060	−0.012			
9	−0.216	−0.234	0.158			
10	−0.975	−0.068	0.088			

TABLE A.I.7.2

Matrix of factor coordinates of the variables (**S**) (first three factors only)

	1	2	3	4	5	6
1	0.979	0.036	−0.188	.	.	.
2	0.963	0.249	0.097			
3	0.992	0.098	0.059			
4	0.967	−0.192	0.157			
5	0.996	−0.069	−0.004			
6	0.984	−0.125	−0.119			

objects and the coordinates of variables possess equal factor variances so that they can be projected simultaneously on the same plot.

Fig. A.1 is a representation of objects and variables in a plane defined by the first two factors as defined by principal component analysis.

A.II CFA

*A.II.0 Original data matrix (**A**)*

The original data matrix used as a starting point in the analysis is denoted as:

$$\underset{(n \times p)}{\mathbf{A}} \quad \text{(see Table A.0)}$$

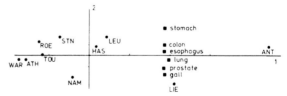

Fig. A.1. PCA plot (first against second factor) of the objects (●) and variables (■) of Table A.0.

TABLE A.II.1

Re-expressed data matrix (**X**) (in 10^2)

	1	2	3	4	5	6	Tot
1	0.75	5.56	2.94	0.61	17.11	3.66	30.63
2	0.11	1.11	0.50	0.04	1.83	0.50	4.09
3	0.22	1.61	0.68	0.04	3.12	0.90	6.56
4	0.25	2.37	1.08	0.22	5.77	1.15	10.83
5	0.29	2.51	1.33	0.18	6.03	1.47	11.80
6	0.54	2.65	1.65	0.39	12.05	3.01	20.30
7	0.22	0.65	0.50	0.04	1.87	0.75	4.02
8	0.07	0.29	0.22	0.04	1.08	0.36	2.04
9	0.14	1.04	0.79	0.25	4.41	1.15	7.78
10	0.00	0.25	0.22	0.04	1.18	0.25	1.94
Tot	2.58	18.04	9.90	1.83	54.45	13.20	100.00

A.II.1 Re-expression of the original data matrix $(\mathbf{A} \to \mathbf{X})$

Division of each value in the data matrix by the matrix total:

$$\mathbf{X}_{(n \times p)} \quad \text{with} \quad X_{ij} = \frac{A_{ij}}{\sum_i \sum_j A_{ij}} \quad \text{(see Table A.II.1)}$$

A.II.2 Definition of diagonal weighting matrices (**W*** *and* **W**)

Weight coefficients for objects and variables are defined according to respectively the marginal row-totals and marginal column-totals. Diagonal weighting matrices for row- and column-items are denoted as:

$$\mathbf{W}^*_{(n \times n)} \quad \text{with} \quad W_{ii}^* = \sum_j X_{ij}$$

(weight coefficient for row i)
(see Table A.II.2.1)

TABLE A.II.2.1

Row-weights (**W***) (in 10^2)

	1	2	3	4	5	6	7	8	9	10
1	30.63									0
2		4.09								
3			6.56							
4				10.83						
5					11.80					
6						20.30				
7							4.02			
8								2.04		
9									7.78	
10	0									1.94

TABLE A.II.2.2

Column-weights (**W**) (in 10^2)

	1	2	3	4	5	6
1	2.58					0
2		18.04				
3			9.90			
4				1.83		
5					54.45	
6	0					13.20

$$\mathbf{W}_{(p \times p)} \quad \text{with} \quad W_{jj} = \sum_i X_{ij}$$

(weight coefficient for column j)
(see Table A.II.2.2.)

A.II.3 Transformation of the re-expressed data $(\mathbf{X} \to \mathbf{Y})$

The transformation procedure consists of subtraction and division by the 'expected value':

$$\mathbf{Y}_{(n \times p)} \quad \text{with} \quad Y_{ij} = \frac{X_{ij} - W_{ii}^* \times W_{jj}}{W_{ii}^* \times W_{jj}}$$

with: $W_{ii}^* \times W_{jj}$ being the expected value for element X_{ij} (X_{ij} being the observed value).
(See Tables A.II.3.1 and A.II.3.2.)

The expected value or theoretical frequency of a particular element is thus equal to the product of the corresponding marginal row- and column-total. It is the value that would be obtained if no interaction existed between the corresponding row and column.

TABLE A.II.3.1

Expected values (in 10^2)

	1	2	3	4	5	6
1	0.79	5.53	3.03	0.56	16.68	4.04
2	0.11	0.74	0.40	0.07	2.23	0.54
3	0.17	1.18	0.65	0.12	3.57	0.87
4	0.28	1.95	1.07	0.20	5.90	1.43
5	0.30	2.13	1.17	0.22	6.43	1.56
6	0.52	3.66	2.01	0.37	11.05	2.68
7	0.10	0.72	0.40	0.07	2.19	0.53
8	0.05	0.37	0.20	0.04	1.11	0.27
9	0.20	1.40	0.77	0.14	4.24	1.03
10	0.05	0.35	0.19	0.04	1.05	0.26

TABLE A.II.3.2

Transformed data matrix (**Y**)

	1	2	3	4	5	6
1	−0.049	0.006	−0.030	0.088	0.026	−0.095
2	0.019	0.507	0.241	−0.520	−0.178	−0.070
3	0.270	0.363	0.049	−0.701	−0.127	0.035
4	−0.102	0.211	0.003	0.086	−0.021	−0.197
5	−0.058	0.179	0.136	−0.169	−0.062	−0.056
6	0.026	−0.275	−0.179	0.062	0.090	0.124
7	1.074	−0.109	0.263	−0.512	−0.147	0.421
8	0.359	−0.222	0.063	−0.041	−0.033	0.329
9	−0.286	−0.259	0.024	0.763	0.041	0.117
10	−1.000	−0.281	0.122	0.012	0.122	−0.018

The transformation procedure corrects for differences in size between the objects as well as for differences in importance between the variables and is symmetrical with respect to both objects and variables.

*A.II.4 Weighting of the transformed matrix (**Y** → **Z**)*

The transformed matrix is multiplied by the square root of the diagonal matrix of column weights:

$$\underset{(n \times p)}{\mathbf{Z}} = \underset{(n \times p)}{\mathbf{Y}} \cdot \underset{(p \times p)}{\mathbf{W}^{1/2}} \quad \text{(see Table A.II.4)}$$

with

$$Z_{ij} = \frac{X_{ij} - W_{ii}^* \times W_{jj}}{W_{ii}^* \times \sqrt{W_{jj}}}$$

The result of this weighting step is that the variables with larger weight (i.e. larger absolute

TABLE A.II.4

Weighted matrix (**Z**)

	1	2	3	4	5	6
1	−0.008	0.003	−0.009	0.012	0.019	−0.035
2	0.003	0.215	0.076	−0.070	−0.132	−0.025
3	0.043	0.154	0.015	−0.095	−0.094	0.013
4	−0.016	0.090	0.001	0.012	−0.015	−0.072
5	−0.009	0.076	0.043	−0.023	−0.046	−0.020
6	0.004	−0.117	−0.056	0.008	0.067	0.045
7	0.173	−0.046	0.083	−0.069	−0.109	0.153
8	0.058	−0.094	0.020	−0.006	−0.025	0.120
9	−0.046	−0.110	0.008	0.103	0.030	0.043
10	−0.161	−0.120	0.039	0.002	0.090	−0.007

TABLE A.II.5

Covariance matrix (**V**) (in 10^{-3})

	1	2	3	4	5	6
1	2.12	0.45	0.43	−1.16	−1.36	1.37
2	0.45	9.28	2.20	−2.61	−4.45	−2.94
3	0.43	2.20	1.45	−0.73	−1.84	−0.02
4	−1.16	−2.61	−0.73	1.95	1.80	−0.19
5	−1.36	−4.45	−1.84	1.80	3.29	0.06
6	1.37	−2.94	−0.02	−0.19	0.06	2.79

mortality) will have a more important contribution in the analysis.

*A.II.5 Calculation of the matrix **V** (covariance matrix)*

The matrix **V** is calculated as a weighted product of the matrix **Z** with its transpose **Z**':

$$\underset{(p \times p)}{\mathbf{V}} = \underset{(p \times n)}{\mathbf{Z}^t} \cdot \underset{(n \times n)}{\mathbf{W}^*} \cdot \underset{(n \times p)}{\mathbf{Z}}$$

(see Table A.II.5)

The result of the 'object-weighting' during this multiplication is that objects with greater weight (i.e. higher total mortality) will have a more important contribution in the analysis.

*A.II.6 Diagonalization of the matrix **V** (**V** → **F**)*

Diagonalization or calculation of the eigenvectors (factors) and corresponding eigenvalues (factor variances) of the symmetric matrix **V**. (See A.I.6 for the mathematical equations.)

*A.II.7 Representation of objects and variables in the new factor space (**S*** and **S**)*

(a) Calculation of the factor coordinates of the objects (**S***). (See A.I.7.a for the mathematical equations.)

TABLE A.II.6.1

Matrix of eigenvalues (Λ) (first three eigenvalues only)

	1	2	3	4	5	6
1	0.014	0	0			
2	0	0.005	0			
3	0	0	0.001			
4						
5						
6						

Chapter 19

TABLE A.II.6.2

Matrix of eigenvectors (**F**) (first three eigenvectors only)

	1	2	3	4	5	6
1	0.089	−0.566	−0.328	·	·	·
2	0.805	0.199	−0.171			
3	0.224	−0.160	0.674			
4	−0.259	0.301	0.503			
5	−0.432	0.337	−0.322			
6	−0.201	−0.641	0.228			

TABLE A.II.7.1

Matrix of factor coordinates of the objects (**S***) (first three factors only)

	1	2	3	4	5	6
1	−0.005	0.039	−0.012	·	·	·
2	0.271	−0.020	0.014			
3	0.194	−0.064	−0.045			
4	0.089	0.071	−0.015			
5	0.100	0.004	0.017			
6	−0.146	−0.021	−0.026			
7	0.031	−0.276	0.042			
8	−0.078	−0.141	0.043			
9	−0.139	0.017	0.091			
10	−0.140	0.096	0.069			

TABLE A.II.7.2

Matrix of factor coordinates of the variables (**S**) (first three factors only)

	1	2	3	4	5	6
1	0.065	−0.250	−0.070	·	·	·
2	0.224	0.033	−0.014			
3	0.084	−0.036	0.074			
4	−0.226	0.158	0.128			
5	−0.069	0.032	−0.015			
6	−0.065	−0.125	0.022			

(b) Calculation of factor coordinates of the variables (**S**). (See A.I.7.b for the mathematical equations.)

Fig. A.2 is a representation of objects and variables in a plane defined by the first two factors as defined by correspondence factor analysis.

A.III SMA

A.III.0 Original data matrix (**A**)

The original data matrix used as a starting point in the analysis is denoted as:

A   (see Table A.0)
$(n \times p)$

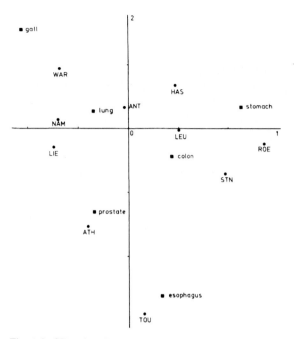

Fig. A.2. CFA plot (first against second factor) of the objects (●) and variables (■) of Table A.0.

Remark: To be able to carry out a logarithmic re-expression (see A.III.2.a), the zero value for object 10, variable 1 was initially substituted by the value of 0.1.

A.III.1 Re-expression of the original data matrix (**A** → **X**)

(a) Logarithmic re-expression:

TABLE A.III.1

Re-expressed data matrix (**X**) (in 10^2)

	1	2	3	4	5	6	Tot
1	1.74	2.39	2.18	1.67	2.75	2.25	12.99
2	1.11	1.87	1.61	0.75	2.03	1.61	8.96
3	1.33	1.99	1.71	0.75	2.20	1.80	9.77
4	1.38	2.11	1.85	1.33	2.40	1.88	10.96
5	1.42	2.13	1.92	1.27	2.41	1.96	11.12
6	1.63	2.15	1.99	1.53	2.64	2.19	12.13
7	1.33	1.69	1.61	0.75	2.03	1.74	9.15
8	0.97	1.42	1.33	0.75	1.85	1.50	7.83
9	1.20	1.84	1.75	1.38	2.31	1.88	10.37
10	3E−10	1.38	1.33	0.75	1.89	1.38	6.73
Tot	12.12	18.97	17.29	10.93	22.53	18.17	100.00

$$\underset{(n \times p)}{\mathbf{L}} = \log \underset{(n \times p)}{(\mathbf{A})}$$

(b) If the logarithmic data table contains negative values, it is necessary to subtract the largest negative value from each value in the data table to ensure that all data are non-negative:

$$L_{ij} = L_{ij} - \min(L_{ij})$$

(c) Division of each value by the matrix total (i.e. the sum over all values in the data table):

$$\underset{(n \times p)}{\mathbf{Z}} \quad \text{with} \quad X_{ij} = \frac{L_{ij}}{\sum_i \sum_j L_{ij}}$$

A.III.2 Definition of diagonal weighting matrices ($\mathbf{W}^*$ and $\mathbf{W}$)

In the case of additive data, weight coefficients for objects and variables are defined according to respectively the marginal row-totals and marginal column-totals. (See A.II.2 for the mathematical equations.)

In the case of non-additive data, weight coefficients can be kept constant or they can be defined as being proportional to a particular row or column.

A.III.3 Transformation of the re-expressed data ($\mathbf{X} \to \mathbf{Y}$)

The transformation procedure is defined as double centering and consists of successively subtracting the corresponding row-mean and corresponding column-mean from each value in the data table:

$$\underset{(n \times p)}{\mathbf{Y}} \quad \text{with} \quad Y_{ij} = X_{ij} - (M_i^* + M_j - MG)$$

where

$M^*(n)$ = vector containing the n row-means

with

$$M_i^* = \sum_j X_{ij} \times W_{jj}$$

(the 'weighted' mean for row i)

and

$M(p)$ = vector containing the p column-means

with

$$M_j = \sum_i W_{ii}^* \times X_{ij}$$

(the 'weighted' mean for column j)

and

TABLE A.III.2.1

Row-weights ($\mathbf{W}^*$) (in 10^2)

	1	2	3	4	5	6	7	8	9	10
1	12.99									0
2		8.96								
3			9.77							
4				10.96						
5					11.12					
6						12.13				
7							9.15			
8								7.83		
9									10.37	
10	0									6.73

TABLE A.III.2.2

Column-weights ($\mathbf{W}$) (in 10^2)

	1	2	3	4	5	6
1	12.12					0
2		18.97				
3			17.29			
4				10.93		
5					22.53	
6	0					18.17

TABLE A.III.3

Transformed data matrix ($\mathbf{Y}$) (in 10^3)

	1	2	3	4	5	6
1	0.168	0.013	−0.309	0.824	0.129	−0.487
2	0.412	1.350	0.514	−1.818	−0.570	−0.374
3	1.257	1.154	0.098	−3.226	−0.241	0.103
4	−0.029	0.612	−0.204	0.813	−0.027	−0.881
5	0.086	0.484	0.159	−0.100	−0.208	−0.395
6	0.514	−0.951	−0.749	0.848	0.430	0.321
7	2.474	−0.609	0.322	−2.010	−0.698	0.753
8	1.152	−0.995	−0.182	0.241	−0.236	0.591
9	−0.804	−1.017	−0.168	2.359	0.142	0.163
10	−7.185	−0.026	1.221	1.645	1.478	0.835

$$MG = \sum_i \sum_j W_{ii}^* \times X_{ij} \times W_{jj}$$

(the 'weighted' global mean)

The transformation procedure corrects for differences in size between the objects as well as for differences in importance between the variables and is symmetrical with respect to both objects and variables. The order of centering (row before column or column before row) is of no importance.

A.III.4 Weighting of the transformed matrix (Y → Z)

The transformed matrix is multiplied by the square root of the diagonal matrix of column-weights:

$$\underset{(n \times p)}{Z} = \underset{(n \times p)}{Y} \cdot \underset{(p \times p)}{W^{1/2}} \quad \text{(see Table A.III.4)}$$

The result of this weighting step is that the variables with greater weight (i.e. higher absolute mortality) will have a more important contribution to the analysis.

TABLE A.III.4

Weighted matrix (**Z**) (in 10^3)

	1	2	3	4	5	6
1	0.059	0.006	-0.129	0.272	0.061	-0.208
2	0.143	0.588	0.214	-0.601	-0.270	-0.159
3	0.438	0.503	0.041	-1.067	-0.115	0.044
4	-0.010	0.267	-0.085	0.269	-0.013	-0.376
5	0.030	0.211	0.066	-0.033	-0.099	-0.168
6	0.179	-0.414	-0.312	0.280	0.204	0.137
7	0.861	-0.265	0.134	-0.664	-0.331	0.321
8	0.401	-0.433	-0.076	0.080	-0.112	0.252
9	-0.280	-0.443	-0.070	0.780	0.067	0.070
10	-2.501	-0.011	0.508	0.544	0.701	0.356

*A.III.5 Calculation of the matrix **V** (covariance matrix)*

The matrix **V** is calculated as a weighted product of the matrix **Z** with its transpose Z^t:

$$\underset{(p \times p)}{V} = \underset{(p \times n)}{Z^t} \cdot \underset{(n \times n)}{W^*} \cdot \underset{(n \times p)}{Z}$$

(see Table A.III.5)

TABLE A.III.5

Covariance matrix (**V**) (in 10^7)

	1	2	3	4	5	6
1	5.345	0.007	-0.782	-2.095	-1.534	-0.276
2	0.007	1.307	0.301	-1.137	-0.246	-0.480
3	-0.782	0.301	0.394	-0.293	0.051	0.118
4	-2.095	-1.136	-0.239	2.945	0.861	-0.084
5	-1.534	-0.246	0.051	0.861	0.591	0.128
6	-0.276	-0.480	0.118	-0.084	0.128	0.524

The result of the 'object-weighting' during this multiplication is that objects with greater weight (i.e. higher total mortality) will have a more important contribution in the analysis.

*A.III.6 Diagonalization of the matrix **V** (**V** → **F**)*

Diagonalization or calculation of the eigenvectors (factors) and corresponding eigenvalues (factor variances) of the symmetric matrix **V**. (See A.I.6 for the mathematical equations.)

*A.III.7 Representation of objects and variables in the new factor space (**S*** and **S**)*

(a) Calculation of factor coordinates of the objects (**S***)

TABLE A.III.6.1

Matrix of eigenvalues (Λ) (in 10^7) (first three eigenvalues only)

	1	2	3	4	5	6
1	7.15	0	0			
2	0	2.73	0			
3	0	0	1.03			
4						
5						
6						

TABLE A.III.6.2

Matrix of eigenvectors (**F**) (first three eigenvectors only)

	1	2	3	4	5	6
1	0.828	-0.357	-0.203	.	.	.
2	0.098	0.589	-0.538			
3	-0.075	0.286	0.191			
4	-0.480	-0.664	-0.364			
5	-0.261	-0.030	0.101			
6	-0.040	-0.041	0.700			

TABLE A.III.7.1

Matrix of factor coordinates of the objects (**S***) (first three factors only)

	1	2	3	4	5	6
1	−0.003	−0.010	−0.015	.	.	.
2	0.018	0.034	−0.013			
3	0.033	0.038	0.003			
4	−0.003	−0.001	−0.029			
5	0.003	0.007	−0.012			
6	−0.002	0.026	0.008			
7	0.037	0.0006	0.024			
8	0.009	−0.021	0.015			
9	−0.023	−0.031	0.003			
10	−0.088	0.028	0.041			

$$\underset{(n \times q)}{\mathbf{S}^*} = \underset{(n \times p)}{\mathbf{Z}} \cdot \underset{(p \times q)}{\mathbf{F}} \cdot \underset{(q \times q)}{\Lambda^{-1/4}}$$

(see Table A.III.7.1)

The factor coordinates of the objects are scaled in such a way (according to a coefficient $\Lambda^{-1/4}$) that axes can be drawn in the new factor space by connecting any two objects. Vertical projections of the variables on the calibrated axes allow to estimate their ratio for the two corresponding objects.

(b) Calculation of factor coordinates of the variables (**S**)

$$\underset{(p \times q)}{\mathbf{S}} = \underset{(p \times p)}{\mathbf{W}^{-1/2}} \cdot \underset{(p \times q)}{\mathbf{F}} \cdot \underset{(q \times q)}{\Lambda^{1/4}}$$

(see Table A.III.7.2)

The factor coordinates of the variables are scaled in such a way (according to a coefficient $\Lambda^{1/4}$) that axes can be drawn in the new factor space by connecting any two variables. Vertical projections of the objects on the calibrated axes

TABLE A.III.7.2

Matrix of factor coordinates of the variables (**S**) (first three factors only)

	1	2	3	4	5	6
1	0.069	−0.023	−0.010	.	.	.
2	0.007	0.031	−0.022			
3	−0.005	0.016	0.008			
4	−0.042	−0.046	−0.020			
5	−0.016	−0.001	0.004			
6	−0.003	−0.002	0.029			

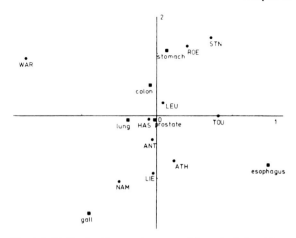

Fig. A.3. SMA plot (first against second factor) of the objects (●) and variables (■) of Table A.0.

allow to estimate their ratio for the two corresponding variables.

Fig. A.3 is a representation of objects and variables in a plane defined by the first two factors as defined by spectral map analysis.

ACKNOWLEDGEMENT

Anne Thielemans is research assistant of the National Fund for Scientific Research.

REFERENCES

1 B.S. Everitt, *Graphical Techniques for Multivariate Data*, Heinemann, 1978.
2 D.L. Massart and L. Kaufman, *The Interpretation of Analytical Chemical Data by the Use of Cluster Analysis*, Wiley, New York, 1983, Ch. 2.
3 E.R. Malinowski and D.G. Howery, *Factor Analysis in Chemistry*, Wiley, New York, 1980.
4 K.G. Joreskog, J.E. Klovan and R.A. Reyment, *Geological Factor Analysis*, Elsevier, Amsterdam, Oxford, New York, 1976.
5 B.R. Kowalski and C.F. Bender, Nonlinear mapping, *Journal of the American Chemical Society*, 94 (1972) 5632–5639.
6 S. Wold, K. Esbensen and P. Geladi, *Chemometrics and Intelligent Laboratory Systems*, 2 (1987) 37–52.
7 M. Kendall, *Multivariate Analysis*, Griffin & Co., London, 1975, Ch. 2.
8 W.W. Cooley and P.R. Lohnes, *Multivariate Data Analysis*, Wiley, New York, 1971, Ch. 4.

9 L. Lebart, A. Morineau and J.-P. Fénelon, *Traitement des Données Statistiques. Méthodes et Programmes*, Dunod, Paris, 1982, Part IV.
10 P.J. Lewi, Multivariate data representation in medicinal chemistry, in B.R. Kowalski (Editor), *Chemometrics — Mathematics and Statistics in Chemistry*, Reidel, Dordrecht, 1984, pp. 351–376.
11 M.O. Eide, O.M. Kvalheim and N. Telnaes, Routine analyses of crude oil fractions by principal component modelling of gas chromatographic profiles, *Analytica Chimica Acta*, 191 (1986) 433–437.
12 K.R. Gabriel, The biplot-graphic display of matrices with application to principal component analysis, *Biometrika*, 58 (1971) 453–467.
13 J.P. Benzécri, *L'Analyse des Données. Tome II: L'Analyse des Correspondences*, Dunod, Paris, 1973.
14 M.J. Greenacre, *Theory and Applications of Correspondence Analysis*, Academic Press, London, 1984.
15 M. Mellinger, Correspondence analysis: the method and its application, *Chemometrics and Intelligent Laboratory Systems*, 2 (1987) 61–77.
16 P.J. Lewi, *Multivariate Data Analysis in Industrial Practice*, Research Studies Press (Wiley), Chichester, 1982.
17 *SPECTRAMAP — V1.2*, Information Science Dept., Janssen Research Foundation NV, B-2340 Beerse, Belgium, 1985.
18 R. Ryckeboer, G. Janssens and G. Thiers, *Atlas of Cancer Mortality in Belgium (1969–1976)*, Institute of Hygiene and Epidemiology, Brussels, 1983.
19 A. Thielemans, P.K. Hopke, P. De Quint, A.M. Depoorter, G. Thiers and D.L. Massart, Investigation of the geographical distribution of female cancer patterns in Belgium using pattern recognition techniques, *International Journal of Epidemiology*, (1988) in press.
20 H.F. Kaiser, The JK method: a procedure for finding the eigenvectors and eigenvalues of a real symmetric matrix, *Computer Journal*, 15 (1972) 271.

Some Fundamental Criteria for Multivariate Correlation Methodologies

OLAV H.J. CHRISTIE

Rogaland Research Institute, P.O.B. 2503 Ullandhaug, N-4001 Stavanger (Norway)

CONTENTS

1 Computation force fosters validity 286
2 The Cinderella glass shoe versus the adjustable track shoe 287
3 From numbers to graphics 287
4 Poor definition results in crash-landing 289
5 Soft is wonderful 289
6 Models and nature 290
7 Acknowledgements 291
References 291

1 COMPUTATION FORCE FOSTERS VALIDITY

Before the computer age correlation analysis was, to many of us, the art of finding the correlation coefficient, the arithmetic mean, the standard deviation and a few other parameters typical of the Gaussian distribution. Some tests of similarity were based on the same parameters, and the whole art of correlation ended in a test of the hypothesis that two populations were different. Most often, the art of probability testing never offered a definitive and undoubted "yes" to any question; there was always a chance that the answer was "no". This poor situation was a matter of necessity, simply because one did not have the computing force at hand to go any further.

The limitations were severe, and many methods required special data characteristics, such as independence of variables impossible with chemical data. The battery of available methods could only be used on the inflexible, hard models of Nature, and statistics was but an exotic piece of training in fields such as chemistry and geology.

With the development of computers, the calculation capacity increased astronomically. This opened up much better solutions to data analytical problems and, as a first rule of thumb, one may state that the more generally valid a methodology is, the more computing capacity is needed.

Therefore, the development of computers and data analytical methodologies took place more or less hand-in-hand. The early minicomputers allowed the fast calculation of large numbers of correlation coefficients and in 1963 the famous book *Principles of Numerical Taxonomy* by R.R. Sokal and P.H.A. Sneath offered an example of a particularly successful application of correlation coefficient-based pair-group cluster analysis to the classification of plants. Soon thereafter, classical cluster analysis became a routine method in many fields of science.

In earlier terminology, which is still in use in parts of the modern literature, Q-mode analysis refers to correlation between objects, such as rock samples, plant individuals or human patients, and R-mode analysis refers to correlation between variables, such as chemical species, pertinent parts of flowers or other measurements. Classical cluster analysis can be carried out with the aid of a pocket calculator and, successful as they may have been in several early applications, they are risky methods that may lead us to draw conclusions in conflict with the information inherent in the data. Fortunately, we may abandon such outdated methods, as modern microcomputers have placed at our disposal the large computing force that is needed for methodologies with much broader ranges of validity [1–10].

2 THE CINDERELLA GLASS SHOE VERSUS THE ADJUSTABLE TRACK SHOE

The fact that Cinderella wore glass shoes at the Royal Ball made anxious-to-be-married girls cut their heels and toes the next day to fit the shoe. This, in fact, is what we did, consciously or not, for many years in data analysis. The old methods of hard modelling required specific data properties, and the restrictions were just as solid as those of a glass shoe. If we had a skew distribution, we just chopped off the ends in order to make it symmetrical. We had to, in order to fit the data to the method.

Fortunately, the present situation is completely different. We do have soft models at hand which are shaped from the properties of the data. We do not have to cut skew ends, and we do have trustworthy methods to weed out the outliers which may be a result of outlier samples or of measurement errors. However, all this requires computing force and speed. Therefore, to make a statement in the old Mao Tsetung style: the development from the glass shoe to the adjustable track shoe went hand-in-hand with the development of the modern computer.

3 FROM NUMBERS TO GRAPHICS

The choice of methods in data analysis must be based on a general knowledge of the peculiarities of numerical information. In this paper classification methods will be touched upon, leaving out other parts of statistics, such as time-series analysis. This means that the statements below are especially addressed to correlation work.

Let us, for the sake of the argument, consider the numbers in Table 1. They represent two peak heights of nine different chromatograms, and the only thing we know is that there are two groups of samples. Just by inspecting the numbers it appears that, for instance, samples 2 and 9 are different and that samples 8 and 9 are similar.

Tentatively we sort the samples according to the height of the first peak. A tentative sorting would be samples 1, 5, 8 and 9 in the first class and the other five samples in the second class. It is equally possible to sort the samples according to the height of the second-peak, and one would perhaps group samples 1, 6, 7, 8 and 9 in the first class and the other four samples in the second class, and this is a different result. We therefore realize that sorting according to peak height depends on which peak is used as the sorting criterion, and that at least one of the groupings must be wrong.

A poor solution to a problem adds new ones without really solving it. We started with the information that there are two different groups, but we have so far been unable to find a method that adequately extracts that information from the data.

TABLE 1

Peak heights of a set of nine samples

The samples belong to two different groups. A ranked according to sample number: B, ranked according to intensity of peak a: C, ranked according to intensity of peak b

A			B			C		
1	140	92.0	7	110	92.3	2	111	53.3
2	111	53.3	2	111	53.3	3	133	65.8
3	133	65.8	4	125	78.1	5	165	72.0
4	125	78.1	3	133	65.8	4	125	78.1
5	165	72.0	6	135	93.4	9	182	90.8
6	135	93.4	1	140	92.0	1	140	92.0
7	110	92.3	5	165	72.0	7	110	92.3
8	197	94.1	9	182	90.8	6	135	93.4
9	182	90.8	8	197	94.1	8	197	94.1

In Fig. 1 the data have been plotted in a two-dimensional system of orthogonal axes, each axis representing a peak. We see at a glance that the following information can be extracted from the numbers in Table 1: there are two clusters of points (i.e., samples) and one cluster is round and the other one elongated. This is the point of departure for explaining the term "class", so frequently used in the multivariate statistical literature. A class means an entity of objects (samples, individuals, etc.) having a specific combination of common features that we can measure in some way or another. Often, the more features there are, the better is the class description, provided that the features are relevant. For instance, there is an obvious difference between green and brown frogs, but the "green frog" and the "brown frog" classes would possibly be more adequately described if other measurements were included, such as length of legs, weight at a given age or specific metabolic products, supplementing the skin colour.

Returning to Fig. 1, we realize that the information that there were two clusters was given in advance, but the second important information about the shape and location of the classes had to be extracted without prior knowledge. This is an important point: by using an adequate methodology, we have extracted the ultimate information from the few data in Table 1. How could this be?

Table 1 contains only two variables per sample. Therefore, a two-dimensional diagram displays the complete numerical information. The sample point representation in this diagram is, therefore, unique: the location of a given sample point represents the complete numerical information. Because, there are only two variables, the two-dimensional diagram in Fig. 1 is an exhaustive display of the numerical information in Table 1.

If there had been data for a third peak, the point location in Fig. 1 would not have been unique. We would need a third axis to obtain a unique point representation of the numerical information. This leads us to the statement that, "In order to transfer numerical information to graphical information completely and without distortion, one needs a diagram of as many axes as there are variables."

There are some other important aspects that we have not yet touched upon. One aspect is the transformation of information from a numerical to a geometric world. The geometric world is well known to us because we interpret our surroundings visually in terms of geometric patterns. This is a process that goes on in our brain as long as we are awake and the human geometric pattern recognition capability is unsurpassed by any computer so far.

On the other hand, we were unsuccessful in making a proper classification of the samples in Table 1 just by inspecting the numbers, because we do not have any intuitive perception of numerical information. Therefore, the transformation of numerical information into graphical patterns is a key to the efficient extraction of numerical information. We have already stressed that the dimensionality of the graphical space is important to the complete transformation of numerical information into graphical information.

There is, finally, a fundamental statement that we have tacitly taken for granted, concerning the shape of the clusters in Fig. 1. It is the shape and the point configurations that allow us to make a proper classification of the samples. This is so self-evident that we accept it implicitly. It still deserves to be stated explicitly: "Similarity and difference are characterized in terms of point configurations." This fundamental statement opens up the discussion of criteria for the choice of methods in correlation studies.

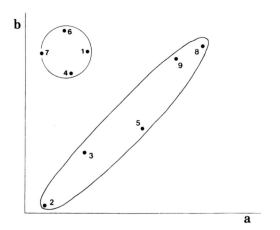

Fig. 1. Data in Table 1 plotted to illustrate the two classes of samples and that their shapes are different.

4 POOR DEFINITION RESULTS IN CRASH-LANDING

From Fig. 1, it can be seen that configuration is a matter of combined distance and direction. Neither of them is sufficient when standing alone. The distance between points 3 and 4 is much smaller than that between 3 and 9. However, 3 and 9 still belong to the same group and 3 and 4 to different groups.

We shall exemplify some earlier statistical routines with every-day situations. Consider that we travel by air from London to Paris, and we are told that the distance between the cities is 320 km. After flying for that distance we run short of gasoline and crash-land on the Dogger Bank in the North Sea because we did not go in the right direction. Next time, with another pilot, we are told to go South-East, and because we have sufficient gasoline this time, we crash-land in the Mediterranean off St. Tropez because we missed Paris.

We recognize the stupidity of this situation immediately, because going by air is a familiar situation. In the world of numbers, however, we do not have adequate reference frames, and for decades data analysts have, so to say, crash-landed in the North Sea or the Mediterranean without knowing it. They have disregarded the simple fact that clustering depends on point configuration, and that configuration is a matter of combined direction and distance. Neither of them work satisfactorily alone.

Let us first look at correlation coefficients and distance measures. The correlation coefficient has been used extensively for correlation work. Points of high correlation fall along a straight line and, therefore, correlation is a direction measure independent of configuration. Distance is, in a similar fashion, a configuration-independent measure. We therefore make the following statements: "Distance measure-based classification methods are dangerous, and they should be avoided" (Dogger Bank crash); "correlation coefficient (viz., direction)-based classification methods are dangerous, and they should be avoided" (St. Tropez crash).

A few remarks about these statements are needed. There are techniques of extraction of eigenvectors from a data matrix which lean on covariance matrices. They are correlation coefficient-based steps in a methodology, and do not serve as the fundamental basis for a classification method. A second remark is also important: we often speak about distances taking a direction measure as implicit. The statement "distance between the two cities" is direction non-specific, but the statement "distance between London and Paris" is direction-specific, insofar that London and Paris are geographically defined. We shall meet successful applications of distance measures in other papers in this issue, and they are direction-specific.

In the world of numbers, the most convenient direction and distance measure is the vector. Therefore, we can make a final statement: "Vector-based classification methods are configuration consistent, and are to be preferred." This is a fundamental statement of major importance in multivariate statistics.

Having reached an important milestone on our way to a better understanding of the choice of good methodologies for classification purposes, we still need a few additional observations in order to construct a good decision device for classification methodologies.

5 SOFT IS WONDERFUL

The most prominent feature of the long cluster in Fig. 1 is its elongation. This leads us to the statement that "Point spread means systematic information." Hence the direction of largest point spread is also the direction of high information density, it also corresponds to the direction of the first eigenvector of the data matrix of that class.

The lengths of the eigenvectors diminish with order. For instance, the first eigenvector is longer than the second, the second is longer than the third, and so forth. Because point spread means systematic information, this means that the most prominent information is found along the direction of the first eigenvector, and that less and less information is found along the directions of the second, third, etc., eigenvectors.

The length of the eigenvector is a measure of

the length of the cluster — in statistical terminology equal to the square of sum of variances of the samples in the class.

The variance along one direction can easily be compared with the variance in all directions and, again referring to the first statement above, we can easily see how much of the total systematic information (point spread) is found along one eigenvector.

In Fig. 1 the first eigenvector runs in the direction of the elongation of the long cluster. It is as long as the cluster itself and it has a zero point in the middle. Fig. 2 illustrates the loadings (a_e and b_e), i.e., how the variables a and b are used to describe the direction of the eigenvector. Because direction is consistent with correlation, the loadings also express the correlation between the variables along the eigenvector. This is why the loading plot is a common illustration of variable correlation in multivariate data analysis.

One of the important virtues of multivariate data analysis is the soft modelling principle. Referring to Fig. 1, one model is established for the circular class and another for the elongated class. The relationship between the samples of a given class is then described in terms of its location relative to the eigenvectors of the class. The important thing is that the eigenvectors are calculated from the class samples, so that the model is shaped according to their properties. Talking in

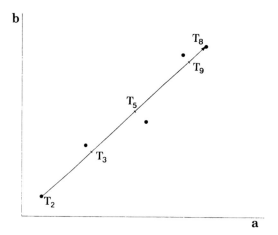

Fig. 3. Location of the samples along the eigenvector given by the score T_i. Using the eigenvector method we also describe the correlation between the variables and between the samples in terms of loadings and scores, respectively.

the language of fairy-tales, the glass shoe of Cinderella corresponds to statistical hard modelling, whereas the modern track shoe is constructed to adapt the shape to the soft model of the foot.

Fig. 3 illustrates the relationship between the samples and the eigenvector of the elongated class. Each sample is described in terms of its score, i.e., its location along the eigenvector measured as distance from its middle point.

6 MODELS AND NATURE

If I have a pond in the garden and I transfer a number of genetically identical frog eggs of identical weight to it, after a few weeks I shall have frogs of different weights, because the frogs will have been influenced by a number of different factors, such as aggression, temperature distribution in the pond and access to feedstuffs. Influx of sunlight influences the metabolic processes of the frogs and, consequently, imprints a systematic data structure in the weight record. Generally, we can state that "Whenever a process acts on a chemical system, it imprints a systematic structure on it." Obviously, this structure may, at least partly, be extracted in terms of an eigenvector, and we proceed with the next, important statement: "Eigen-

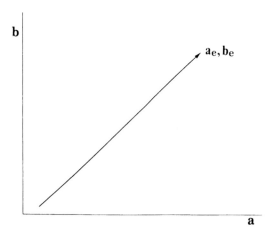

Fig. 2. Description of the direction of the first eigenvector of the elongated class with the aid of its loadings, a_e and b_e.

vectors may be the heralds of processes."

Orthogonal eigenvectors may be referred to independent processes, the independence being stated by the orthogonality. There are, however, processes of more complex interrelationships that call for more subtle treatment. In cases where a given eigenvector is attributed to a given process and the effect of this process is known, it may be desirable to rotate the eigenvector in such a fashion that the loadings of the actual variables are optimized. This belongs to advanced levels of multivariate data analysis, which are treated in other papers in this issue.

Let us, then, return to the construction of class models. The models are constructed in a very simple fashion by describing the location of each sample, y_i. Let the first data set consist of samples of no systematic configuration, i.e., the samples form a balloon-shaped cluster. Then we describe their location by referring to the middle of the balloon, m, and its radius, e_i, which is also called the residual

$$y_i = m + e_i; \quad e_i < y_i \tag{1}$$

This model described the circular class in Fig. 1.

If a process imprints a structure on the data set, the balloon will be stretched in one direction. We describe this structure by the first eigenvector of the data matrix, b, and the location of the sample y_i is described by the model

$$y_i = m + b \cdot t_i + e_i \tag{2}$$

where t_i is the sample score. This model described the elongated class in Fig. 2. The product $b \cdot t$ is often referred to in the current literature as the principal component, or the factor.

If another process is active, the model will have two principal components, and the most important process may appear as the first principal component:

$$y_i = m + b_1 t_{i1} + b_2 t_{i2} + e_i \tag{3}$$

Looking at these three models we see that, in fact, we retrieve the systematic structure, i.e., the numerical information, and push the rest of the total variation into e_i. Going from model 2 to 3, we extract more systematic information from the residuals, e_i in model 2, and describe it as the second principal component, $b_2 t_{i2}$ in model 3. We can extend the model as long as there is systematic information in e_i, and there is a good test (cross-validation) for the presence or absence of such information.

This is the underlying philopsophy of the disjoint principal components modelling of the program package SIMCA, which will be referred to several times in this issue.

There are a large number of successful application of principal components models and, for instance, in organic geochemistry one speaks about a maturation eigenvector, a depositional environment eigenvector and an organic input eigenvector (which in the current literature are referred to as principal components).

The development of vector-based classification methodologies is a formidable contribution to the progress of statistics. It has freed us from many of the intolerable restrictions of earlier methods, and it opens up a much more detailed insight into the variety of active processes. I am inclined to say to students of modern classification strategies that soft modelling is wonderful, and multivariate is a must, because Nature is soft and multivariate.

7 ACKNOWLEDGEMENTS

An earlier version was critically read by Dr. Olav Kvalheim and Professor Svante Wold. Their kind suggestions improved the paper, the content of which is the full responsibility of the author.

REFERENCES

1 O.H.J. Christie, Eigenvector and principal components analysis of granitic rocks from the Bergell Alps, *Schw. mineral. petrogr. Mitt.*, 58 (1978) 303–313.
2 K.H. Esbensen and S. Wold, SIMCA, MACUP, SELPLS, GDAM, SPACE & UNFOLD: The way towards regionalized principal components analysis and subconstrained N-way decomposition — with geological illustrations, in O.H.J. Christie (Editor), *Proceedings, Nordic Symposium on Applied Statistics, 1983, Stavanger*, in commission by Stokkand Forlag Publishers, Stavanger, 1983, pp. 11–36.
3 K.G. Jöreskog, J.E. Klovan and R.A. Reyment, *Geological Factor Analysis*, Elsevier, Amsterdam, 1976, 178 pp.
4 O.M. Kvalheim and N. Telnaes, Visualizing in formation in

multivariate data: Applications to petroleum geochemistry. Part 1. Projection methods. Part 2. Interpretation and correlation of North Sea oils using three different biomarker fractions, *Analytica Chimica Acta*, (1987) in press.
5 H.A. Martens, *Multivariate Calibration*, Dr. techn. Thesis, Technical University of Norway, Trondheim, 1985, 415 pp.
6 D.L. Massart and L. Kaufman, *The Interpretation of Analytical Chemical Data by the Use of Cluster Analysis*, J. Wiley, New York, 1983, 596 pp.
7 S. Wold, Pattern recognition by means of disjoint principal components models, *Pattern Recognition*, 8 (1976) 127–139.
8 S. Wold, Crossvalidity estimation of the number of components in factor and principal components models, *Technometrics*, 20 (1978) 397–405.
9 S. Wold, C. Albano, W.J. Dunn, U. Edlund, K. Esbensen, P. Geladi, S. Hellberg, E. Johansson, W. Lindberg and M. Sjöström, Multivariate data analysis in chemistry, in B.R. Kowalski (Editor), *Chemometrics, Mathematics and Statistics in Chemistry*, Reidel, Dordrecht, NATO ASI C 138, 1984, pp. 17–98.
10 S. Wold, C. Albano, W.J. Dunn, K. Esbensen, S. Hellberg, E. Johansson and M. Sjöström, Pattern recognition: finding and using regularities in multi-variate data, in H. Martens and H. Russwurm, Jr. (Editors), *Food Research and Data Analysis*. Applied Science Publishers, London, 1983, pp. 147–188.

Mixture Analysis of Spectral Data by Multivariate Methods

WILLEM WINDIG *

GEO CENTERS, Inc., c/o U.S. Army Chemical Research, Development and Engineering Center, SMCCR-RSL, Aberdeen Proving Ground, MD 21010-5423 (U.S.A.)

(Received 4 January 1988; accepted 28 March 1988)

CONTENTS

1 Introduction	293
2 The ternary diagram	294
3 Multivariate analysis	295
3.1 Factor analysis	295
3.2 Discriminant analysis	296
4 Mixture analysis	296
4.1 Score based methods	296
4.1.1 Principal components regression	298
4.1.2 Target factor analysis	299
4.2 Loading based methods	300
4.2.1 Graphical rotation	300
4.2.2 Pure mass method	301
4.2.3 Variance diagram	303
5 Concluding remarks	304
6 Acknowledgements	305
References	305

1 INTRODUCTION

This tutorial will present some of the principles of mixture analysis by multivariate data analysis methods. In order to explain the approaches, geometrical representations will be used as much as possible. The tutorial will start with representing a small data set of a three component mixture system in a form most scientists are familiar with: the ternary diagram. Using this ternary diagram, the application of factor analysis to mixture analysis will be introduced. From then on, several approaches to mixture analysis will be introduced by using a simulated data set of mass spectra. The use of mass spectral data does not imply that the techniques cannot be used for other spectral data. The order of the presentation of the several techniques does not necessarily reflect the historical

---

* Present address: Chemometrics Laboratory, B49 F2 ATD KP, Eastman Kodak Company, Rochester, NY 14650, U.S.A.

0169-7439/88/$03.50 © 1988 Elsevier Science Publishers B.V.

development of mixture analysis methods and the highly related self-modeling curve-resolution methods. Furthermore, this tutorial is not meant to give an up-to-date literature review. Its aim is to give some basic ideas about multivariate mixture analysis, which should be enough to serve as a basis for further study.

2 THE TERNARY DIAGRAM

As a model for mixture analysis, the data in Table 1 are used. An important feature of this model is that the concentrations are expressed as fractions; in other words, the relative amounts are given. Plotting the values of this Table results in Fig. 1. Two important properties of this type of data can be observed in this plot. First, all the data-points, representing the mixtures, lie in a plane. This is because the concentrations are expressed as fractions. In this case, one degree of freedom is lost: if two of the fractional concentrations are known, the third one can be derived easily by simply subtracting the sum of the two known fractional concentrations from the total concentration (i.e. 100).

Second, all the mixtures lie within a triangle spanned by the data-points representing the pure compounds. Since the dimensionality of this data set is only two, it can just as well be plotted in a two dimensional representation rather than in the

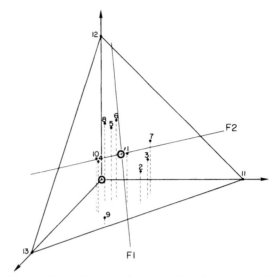

Fig. 1. Three-dimensional plot of the mixture compositions given in Table 1. The mixture points lie in a plane due to the fact that the sum of the three components in the mixtures is 100%. The vectors labeled $F1$ and $F2$ are the first two factors as found by principal component analysis, after subtracting the mean (or standardizing) the data set.

three dimensional representation as given in Fig. 1. The result of using a two dimensional plot is the well known ternary diagram, given in Fig. 2.

In this plot the so called 'component axes' are indicated. These axes can be considered projections of the three axes given in Fig. 1 into the plane in which the mixture points lie. The con-

TABLE 1

The composition of the mixtures in relative concentrations (percentages)

Mixture No.	Relative component concentrations		
	A	B	C
1	34	34	32
2	44	23	33
3	48	25	27
4	20	35	45
5	21	51	28
6	22	54	24
7	44	37	19
8	17	54	29
9	33	5	62
10	19	37	44
11	100	0	0
12	0	100	0
13	0	0	100

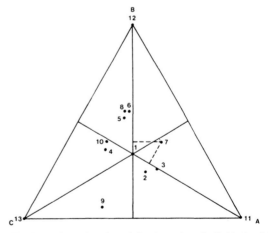

Fig. 2. A triangular plot of the data given in Table 1, with the component axes. The coordinates of one of the spectral points have been indicated.

centrations of the three components in a certain mixture can be derived from projecting a spectral point onto these three component axes. As can easily be understood, every component axis is perpendicular on the line connecting the two other pure components; only in this case the other components project onto the same point on that axis, i.e. 0% concentration.

The next step is to introduce the use of factor analysis as a tool for mixture analysis on this model data set. Although factor analysis is the subject of another tutorial [1], a mathematical rationalization will be given here, since some of the mathematical descriptions will be needed.

3 MULTIVARIATE ANALYSIS

3.1 Factor analysis

Prior to factor analysis, the data set is often transformed. When the original data are represented by the data matrix $\mathbf{V}^*$, size $c \times v$, c is the number of cases and v is the number of variables, the transformations are as follows:

$$v_{i,j} = (v^*_{i,j} - a_j)/b \tag{1}$$

where $v^*_{i,j}$ is an element of $\mathbf{V}^*$, $v_{i,j}$ is an element of the matrix with the transformed values, and a and b are substitutions which can have the following values:

(1) $a_j = 0,$ $b_j = 1$ (2a)

(2) $a_j = u_j,$ $b_j = 1$ (2b)

(3) $a_j = 0,$ $b = w_j$ (2c)

(4) $a_j = u_j,$ $b_j = s_j$ (2d)

s_j is the standard deviation of variable j; u_j is the mean of variable j; w_j is the length of the mass vector:

$$w_j = \left[\sum_{i=1}^{c} (v_{i,j})^2 \right]^{1/2} \tag{3}$$

The transformation by substituting the constants given in eq. 2b in eq. 1 is called centering (the mean of the transformed variables equals zero); substituting the constants given in eq. 2d in eq. 1 is called standardizing or autoscaling (the mean of the transformed variables is zero, the standard deviation equals one).

Factor analysis is a data reduction method. Its aim is to express a (transformed) data matrix more efficiently. The general formula is:

$$\mathbf{V} = \mathbf{S} * \mathbf{F}^T \tag{4}$$

In this formula the data matrix $\mathbf{V}$, size $c \times v$, is expressed as the product of the score matrix $\mathbf{S}$, size $c \times f$, and (the transpose of) the loading matrix $\mathbf{F}$, size $v \times f$, and f represents the number of factors. In order to be a data reduction method, f should be smaller than the smaller of c and v.

An efficient way to obtain data reduction is by calculating the eigenvectors of the dispersion matrix $\mathbf{D}$. The matrix $\mathbf{D}$ is calculated from the transformed data in the following way:

$$\mathbf{D} = (1/c)\mathbf{V}^T\mathbf{V} \tag{5}$$

The dispersion matrices resulting from the transformations described in eqs. 1 and 2 are, respectively:
(1) Covariance around the origin;
(2) Covariance around the mean (the regular variance–covariance matrix);
(3) Correlation around the origin;
(4) Correlation around the mean (the regular correlation matrix).

The eigenvectors of the dispersion matrix are independent linear combinations of the original variables that describe the maximum variance in the data set. Each eigenvector has an eigenvalue that describes in principle the amount of variance explained by that eigenvector. It is generally possible to represent a data set with a few factors, i.e. a few factors describe most of the variance in a data set. The remaining factors, each describing a low variance, represent the noise in the data set [1,2]. It needs to be mentioned here that determining the right number of factors is far from a trivial procedure. Most of the work in the field of error analysis has been done by Malinowski and Howery [2]; a comparison of several methods has been published recently in this journal [3].

The eigenvectors are often scaled by their eigenvalues in the following way:

$$\mathbf{F} = \mathbf{E}\Lambda^{1/2} \tag{6}$$

where **F**, size $v \times f$, contains the factor loadings, v being the number of variables and f the number of significant factors. The matrix **E**, size $v \times f$, contains the eigenvectors. The matrix Λ is a diagonal matrix, size $f \times f$, containing the eigenvalues.

The scores are calculated in the following way:

$$\mathbf{S} = \mathbf{V}\mathbf{E}\Lambda^{-1/2} \qquad (7)$$

This scaling results in a convenient format, e.g. when correlation around the mean is used, the loadings represent the correlation coefficients of the variables with the factor scores, and the factor scores are standardized. This mathematical technique described above is known as principal component analysis.

The relation between principal component analysis and mixture analysis will be clear after considering that a mixture is expressed in the same way (see eq. 4). In the case of mixture analysis, the matrix **S** contains the fractions of the f mixture constituents, each row containing the fractional concentrations of a mixture. The data in Table 1 in fact represent the matrix **S** for the mixture data set that will be used throughout this tutorial. The matrix **F** (for mixture analysis) contains the spectra of the pure components, each column representing a pure component spectrum.

Since the matrices **F** and **S** resulting from principal component analysis have been determined by mathematical criteria, they do not generally represent meaningful information from a chemical point of view. The goal of factor analysis is to obtain more meaningful 'basic' factors out of the above described 'abstract' factors [2]. In our case the basic factors should represent pure chemical components. This is possible, since the results of principal component analysis and of mixture analysis are related; they both describe the same data set. Therefore, the principal component results can be transformed into the results of mixture analysis. Once a matrix **T** has been found that transforms **S** to a matrix with the proper chemical information, the matrix **F** needs to be multiplied by $\mathbf{T}^{-1}$ in order to preserve the relation described by eq. 4:

$$\mathbf{D} = (\mathbf{ST})(\mathbf{T}^{-1}\mathbf{F}^\mathrm{T}) = \mathbf{S}\mathbf{F}^\mathrm{T} \qquad (8)$$

3.2 Discriminant analysis

A technique related to principal component analysis is discriminant analysis. The same technique is also known as canonical variate analysis. When replicate analyses of samples are present, discriminant analysis is a much better tool for mixture analysis than principal component analysis, since discriminant analysis is able to find a space in which the noise in the replicate analysis is minimized. Several studies show the superior performance of discriminant analyses when replicate analyses are present [4–6]. In the field of pyrolysis mass spectrometry, discriminant analysis is used as a more or less routine method. Since the results of discriminant analysis are just a rotation of the principal component analysis results, the representation is exactly the same. The spectral data points are represented in the form of (discriminant) scores and the contributions of the variables to the discriminant functions (the rotated factors) in the form of (discriminant) loadings. Discussing discriminant analysis in more detail falls out of the scope of this tutorial, but I strongly suggest consideration of discriminant analysis for data in which replicate samples are present.

4 MIXTURE ANALYSIS

4.1 Score based methods

In a practical situation, a mixture data set (for demonstration purposes limited to three components) will have more than three variables, and often no (known) pure masses. A pure mass is a mass that is characteristic for one of the components in the mixtures under consideration [7] and can be used to retrieve (relative) concentrations, which will be discussed later in this tutorial.

As has been demonstrated above, the fractional concentrations of the mixtures can be derived from the triangle. Therefore, when the pure components are available, it is useful to retrieve the plane in which the triangle lies. Although the data set from Table 1 can easily be plotted manually in the form of the triangular diagram, it will be employed to introduce the use of principal components analysis for this type of problems.

In order to retrieve the triangular diagram, the covariance around the mean dispersion matrix is an appropriate one for factor analysis: by subtracting the mean from the data set the origin (O in Fig. 1) is shifted towards the center of the triangle (O' in Fig. 1). The same effect is achieved by standardizing the data set. The latter transformation has the advantage that all variables have equal weight in the data analysis. After the origin of the data set has been shifted (translated) to the plane in which the triangle lies, principal component analysis describes the variance of the data set as efficiently as possible, and it will find two axes describing the plane in which the triangle lies. The first factor, $F1$, lies in the direction of the most 'spread' of the data points, similar to a least square approach. The next factor, $F2$, describes the maximum of the residual variance and is thus orthogonal (perpendicular) to $F1$. After determining these two factors components, the space in which the mixture data points lie has been described completely. The result of principal component analysis on the standardized data set from Table 1 is given in Table 2. In this table we see the 'eigenvalues', which are a direct measure of the variance in the directions of the principal components. Since all variables have the same variance, i.e. one, due to the standardization, the total eigenvalue equals the number of variables, i.e. three. Calculation of the relative eigenvalues results in the values under the header relative variance. After standardizing the data, the loadings equal the correlation coefficients of the original variables with the principal components. The last part of Table 2 lists the scores, which represent the coordinates of the mixtures in this new coordinate system. Just as the original variables, these scores are standardized. A visual representation of the results of Table 2 is given in Fig. 3.

At this point a few important observations can be made. First, the scores of the pure components don't lie on the component axes. The reason is that the origin of the data set is the mean, due to the transformation applied. In this case an origin which results in component axes that go through the corner points of the triangle seems more appropriate. One has to keep in mind that the mass variables, as given by the loadings, only give their directions, and can be shifted if we want. It does not make any difference to the concentrations that can be derived from these results. Second, the

TABLE 2

The results of factor analysis on the data set of Table 1

	Eigenvalues	Relative variance
Factor 1	1.55	51.6%
Factor 2	1.45	48.4%
Sum	3.00	100.0%
Variable	Loadings	
	Factor 1	Factor 2
A	0.455	0.890
B	−0.998	−0.070
C	0.588	−0.809
Mixture No.	Scores	
	Factor 1	Factor 2
1	0.029	0.120
2	0.419	0.355
3	0.329	0.567
4	0.035	−0.494
5	−0.583	−0.135
6	−0.701	−0.031
7	−0.119	0.631
8	−0.685	−0.257
9	1.145	−0.495
10	−0.039	−0.500
11	1.123	2.440
12	−2.397	−0.122
13	1.440	−2.090

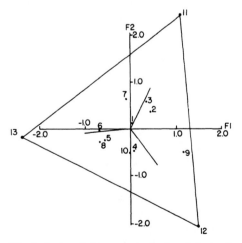

Fig. 3. A plot of the scores (points) and the loadings (vectors) as given in Table 2.

component axes, i.e. the plotted loadings, are perpendicular to the sides of the triangle, which was to be expected. This property of the component axes (i.e. the perpendicularity to the sides of the triangle) gives us an important tool for mixture analysis. If no variable (loading) is known that represents the component axis (in case of more complex data than this), the component axis can easily be constructed because of its known perpendicularity property. Third, the triangle as given by principal component analysis results, is not necessarily equilateral. This is due to the scaling procedures applied, but the composition of the mixtures can just as well be read from this triangle.

The example described above was very simple and principal component analysis did not give us anything that could not be done by simple means. It is a different story, however, if the data set is more complex, e.g. when the mixture data exists of mass spectra with a mass range of 100 or more variables. In case of complex spectra, i.e. spectra which do not have masses that are 100% characteristic for each of the components (pure masses), it is not possible to pick three variables to determine the composition of the mixtures. Even in case there are more or less pure variables, it is hard to decide whether one should use just one variable per component, or combine two, or use even more complex 'linear combinations' of variables. And how can one find these variables? At this point principal component analysis comes in as a useful tool, since principal component analysis applied on a data set like this will still yield the ternary diagram. The reason is that, whatever the number of variables in this data set is, the dimensionality is based on the three components. In other words: every variable is a linear combination of the component axes and, as a consequence, lies in exactly the same space as the component axes. In order to demonstrate this, the data of Table 1 were used to create mixtures of pyrolysis mass spectra given in Fig. 4. The scores of this data set have been plotted in Fig. 5. In this figure the scores again represent the mixture data-points, but now they are based on spectral data. In contrast to the results presented in Fig. 3, this representation of the spectral mixture data cannot be obtained by

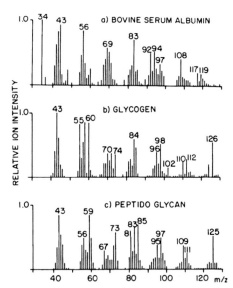

Fig. 4. Low voltage pyrolysis mass spectra of (a) bovine serum albumin (BSA), (b) glycogen (GLY) and (c) peptidoglycan (PG).

simple plotting of intensities of three variables from the raw data. Although no variable (loading) was known to represent the component axes, they could be constructed by using the perpendicularity property. In practice, the component axes will be calculated.

4.1.1 Principal components regression

As will be clear from the material presented above, a component axis is a linear combination

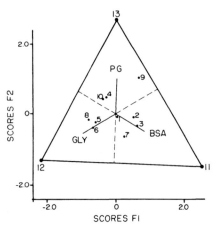

Fig. 5. The results of factor analysis on the data set of simulated mixtures.

of the principal components, that is orthogonal on a side of the triangle, formed by the scores of the three components. The scores of a component axis will give the fractional concentrations of a certain component. The concentration of a component ranges from 0 to 100% Since the original scores are standardized, an offset needs to be added to the scores in order to obtain positive concentrations. Thus the general equation for the transformation of the principal component scores into fractional concentrations is:

$$aF1 + bF2 + c \qquad (9)$$

When we focus on the component axis of bovine serum albumin (BSA), the following set of requirements needs to be fulfilled:
(a) Substituting the scores of BSA for $F1$ and $F2$, eq. 9 has to equal 100;
(b) Substituting the scores of glycogen for $F1$ and $F2$, eq. 9 has to equal 0;
(c) Substituting the scores of peptidoglycan for $F1$ and $F2$, eq. 9 has to equal 0.

This results in the following set of linear equations, respectively:

$$2.373a - 1.473b + c = 100 \qquad (10a)$$

$$-2.152a - 1.268b + c = 0 \qquad (10b)$$

$$0.057a + 2.639b + c = 0 \qquad (10c)$$

The solutions for a, b and c are: $a = 21.5$, $b = -12.2$ and $c = 30.9$.

The values of a and b amount to an angle of 30° clockwise between the positive part of the first principal components ($F1+$) and the component axis of BSA. This is in agreement with the results presented above (Fig. 5). Calculating the concentrations with eq. 9, using the values for a, b and c given above and the scores of the mixtures on the first two principal components, results in the same concentrations as given in Table 1.

In order to perform the calculations described above, calibration standards have been used (the spectral points representing the pure components). This type of calculation is related to what is known as principal component regression. Similar calculations can be made if more than two principal components are involved and/or if other calibration points than the pure spectra are available.

The question may arise if similar results can be obtained by expressing the mixture spectra in a least squares approximation as linear combinations of the pure components. If data are almost without noise, least squares approximations are the right approach. However, when the data are of a noisy character, least squares approximations may give wrong results [8]. Due to the noise reduction capabilities of principal component analysis, this method is the one to use. At this point it needs to be stressed again that discriminant analysis will be even better than factor analysis.

4.1.2 Target factor analysis

In the examples given above, the spectra of the pure components were included in the data set. This is not always a realistic approach. If it is just a hypothesis that a mixture consists of certain components, the inclusion of spectra of these components in the data set might give biased results. In this case a test should first be carried out to see if a spectrum of a certain candidate of the model compound lies in the space described by the data of the mixtures. If a component is a true part of the mixtures, it will coincide with the space. If it is not present at all, it will be orthogonal to the mixture space. This procedure, i.e. testing if a certain spectrum is part of the mixture space, is the principle of target factor analysis, as developed by Malinowski and Howery [2]. The principle is that a spectrum, called the target, is projected in the mixture space, as described by factor analysis. Projection is mathematically simply the inner product of the target spectrum and the factor analysis space, which is exactly the way factor scores are calculated (see eq. 7). In case of target rotation, however, the matrix **V** is replaced by a matrix **U** containing the targets. In case the matrix **V** was transformed prior to factor analysis, the data in matrix **U** have to be transformed the same way (N.B. the transformation variables a and b (see eq. 2) are based on the data in **V**):

$$\mathbf{Z} = \mathbf{U}\mathbf{E}\Lambda^{-1/2} \qquad (11)$$

The scores of the target, in matrix **Z**, can be used to reconstruct the projected target spectra using eq. 4, in which the matrix **S** is replace by a matrix **Z** containing the scores of the targets.

$$W = Z * F^T \qquad (12)$$

A visual inspection of the fit of the targets is possible by comparing the original targets in **U** and the projection of the targets in the factor space in **W**. There are several ways to calculate the fit: for more information see ref. 2. If the fit is acceptable, the composition of the mixtures in terms of the targets can be calculated as described above.

Summarizing, we have seen that it is possible to perform mixture analysis on data sets using a library of reference spectra. The question that arises now, is how to perform mixture analysis without the use of reference spectra or (prior knowledge of) pure masses.

4.2 Loading based methods

4.2.1 Graphical rotation

In a lot of practical cases there are no calibration points in a mixture data set in the form of pure components or mixtures with a known composition. For this type of data so called self-modeling methods have been developed. This type of method is based on the work of Lawton and Sylvestre [9], Knorr and Futrell [7], and Malinowski [2,10].

Since the scores-approaches discussed above cannot be used, the self-modeling techniques are based on the use of the loadings. Most of these methods are based on the presence of a pure mass. Pure masses have a typical behavior in the loading pattern, due to which they can be tracked down. This type of approach will be discussed later. For complex data the presence of a pure mass is not always a realistic assumption; therefore, other methods have been developed to extract the chemical information from this type of data. One of these methods is a visually assisted method, the so-called graphical rotation. In order to present the loadings visually a bar plot is appropriate, since the original spectral data are normally presented in this form. When standardized data are used, however, a direct plot of the variables shows the correlation coefficients. In order to obtain

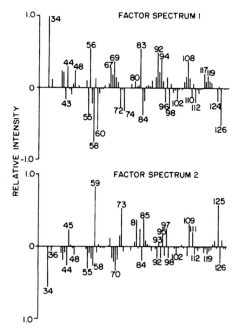

Fig. 6. The factor spectra of the biopolymer mixture data.

'factor spectra' which are comparable to the original data, one simply multiplies the variables with their respective standard deviations. A plot of the transfomed loadings results in the factor spectra given in Fig. 6. In contrast to original spectra, the factor spectra have positive and negative intensities. The reason is that by subtracting the mean of a data set, the differences between spectra are studied. In the most simple case, this is the difference between two spectra. Since both spectra have the same total intensity after normalizing, the difference spectrum will have a total intensity of 0, which is the cause of the positive and the negative intensities.

Since the position of the component axes is known (perpendicular to the sides of the triangle, see Fig. 5) it is possible to predict what the factor spectra should look like. The positive part of the first factor ($F1+$) has a projection of BSA. The negative part of the first factor ($F1-$), shows a projection on GLY and PG. The second factor shows PG on the positive part ($F2+$) and BSA and GLY on the negative part. This is confirmed by the factor spectra in Fig. 6; it is obvious that the component axis of BSA lies somewhere be-

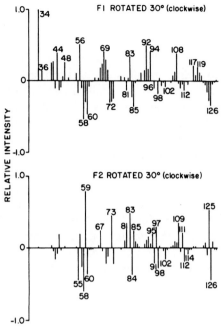

Fig. 7. Rotated (30° clockwise) factor spectra of the biopolymer mixture data set; compare with the spectra in Figs. 3 and 4.

tween $F1+$ and $F2-$. This gives us a tool to determine the component axes without using the spectra of the pure components. By rotating the $F1$ and $F2$ axes clockwise, a rotation can be reached where the BSA pattern is projected only of $F1+$, i.e. in the case that $F1$ coincided with the BSA component axis.

The formulae for rotations of factors are:

$$F1_r = F1 \cos \alpha + F2 \sin \alpha \quad (13a)$$

$$F2_r = -F1 \sin \alpha + F2 \cos \alpha \quad (13b)$$

In these equations α is the angle between $F1$ and $F1_r$. This rotational procedure is in practice done in steps of $10°$. For the data set under consideration the value of $30°$ clockwise (i.e. α equals $-30°$) gives us the direction of the BSA component axis (see Fig. 7). When more than two factors are present, a third factor can be included in the rotation by rotating the linear combination of the first and second factor versus the third factor. This visually assisted rotation is called graphical rotation. It can be applied on data sets in which the reference spectra are not included. In graphical rotation, the operator uses his knowledge of reference spectra, in order to determine the correct rotations. In cases of complex data this visually assisted method can have advantages over mathematical procedures, because the human being has pattern recognition capabilities which can recognize structures in the factor spectra that are hard to describe mathematically [11–13]. Despite the subjectivity of the method, high precisions have been obtained [14].

4.2.2 Pure mass method

The method described above is simple, but subjective. Another widely used method for mixture analysis is based on the presence of a so-called pure mass. A pure mass approach, however, requires factor analysis applied on a matrix around the origin. The principle of this type of factor analysis can be seen in Fig. 8 (compare with Fig. 1). The reason why factor analysis needs to be performed on a non-centered matrix will be explained below with a simple two-mixture system. When correlation or covariance around the origin is applied on a data set like this, two factors will result. Suppose we know the component axes, as given in Fig. 9, a pure mass will coincide with its component axes. All masses that are not pure lie

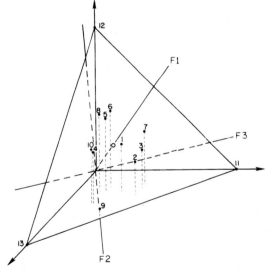

Fig. 8. The visualization of the factors found, when the mean has not been subtracted from the data set, see also Fig. 2.

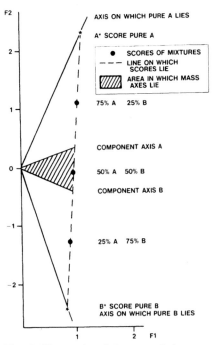

Fig. 9. The results of factor analysis on a two-component mixture system, when the mean has not been subtracted.

in between the component axes, since they are linear combinations of the component axes. Therefore, if there are reasons to assume the presence of pure masses, they can be found by finding those variables that bracket all other variables. As can be seen in Fig. 9, the first pure mass can be found by taking the one with the lowest loading on the first factor. The second one can be found by the mass that makes the largest angle with the first pure mass, mathematically expressed: the mass vector that has the lowest inner product with the first pure mass. The third one (for more complex mixtures than in the example discussed here) must have the lowest inner product with the previously found pure masses.

The scores of the mixtures have also been plotted in Fig. 9. It may look confusing that the scores do not lie in between the component axes. This is due to the scaling procedure applied in order to get loadings from which the dispersion matrix can be reconstructed and to get standardized scores. One has to keep in mind that one can multiply (one of) the rows of S by a certain number as long as the corresponding column of F is multiplied by the inverse of that number, as has been expressed by eq. 8.

As can be seen in Fig. 9, all the spectral points (scores) lie on a line, which is due to the normalization of the spectra. A mathematical way to find the pure mass variables has been described by Malinowski. This is especially useful when more than two factors are involved [10].

After the pure masses have been determined, the axes on which the scores of the pure spectra lie, can be easily determined: the spectrum of pure A, i.e. 100% A, is bound to project into the origin of the pure mass of B, since 100% A is 0% B. This means that pure A lies on an axis perpendicular on the pure mass of B. For the same reasons, pure B lies on a line perpendicular on A (see Fig. 9). In order to see the mathematical link between the lines indicated by A and B (the pure masses) and the lines A* and B* (which give the pure spectra), one has to realize that a matrix containing A* and B* is the graphical representation of the matrix containing the inverse of A and B, see eq. 8.

The pure mass is not sufficient for calculating the spectra of the pure components and their fractional concentrations. At this point the axis on which the spectrum lies is known; its length, however, is not known. In Fig. 9, it is clear that the length of the pure mass axis can easily be derived, since all the spectral points lie on a line. Mathematical details about how to derive the length in this way can be found elsewhere [15]. It is also possible to derive the length of this vector by calculating the associated spectrum and use total intensity to scale it with respect to normalized spectra [2,9]. Although the pure mass method has a history in the field of analysis of time-resolved data, such as GC–MS data [2,7,10], it will be demonstrated on the data set used throughout this tutorial, since the principles are exactly the same. The pure masses of the data set under consideration are: m/z 114, 34 and 125, respectively. These masses are typical for glycogen, bovine serum albumin and peptidoglycan, respectively (also see Fig. 4). The directions of the pure masses in the factor space are exactly in agreement with the directions of the component axes, as determined by graphical rotation and the variance diagram, a technique that will be discussed below. After the proper scaling of the length of the component axes found, the resulting concentrations and spectra are exact [15].

In the sections above, it was argued that the pure mass method was not suited to pyrolysis data. The approach works for this simulated data set, since the data set is without noise and pure masses were put in for demonstration purposes.

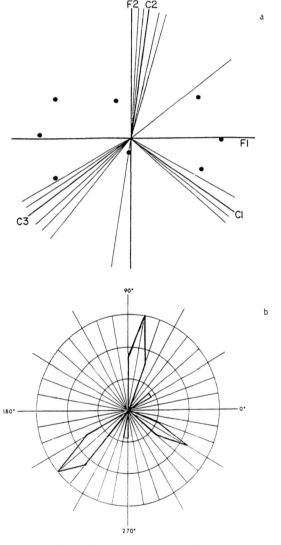

Fig. 10. (a) Plot of the scores and the loadings of a fictitious data set. Loadings tend to cluster around the component axes. (b) The Variance Diagram (VARDIA) of the data presented in (a).

4.2.3 Variance diagram

As has been mentioned under graphical rotation, the pure mass approach has its limitations for complex spectra. Although the graphical rotation technique has proved to be a useful tool for this type of data, it has some disadvantages. It is a subjective method and basic factors not recognized by the researcher may be missed. Another method has therefore been developed for self-modeling mixture analysis, which uses a mathematical approach in combination with a visually assisted rotation.

As has been shown above, factor analysis on the standardized data of a normalized three-component mixture system results in two factors. Just as with the simple data from Table 1, it is also possible for data with more variables to plot the loadings in the space described by the first two factors. This is demonstrated in Fig. 10a, where a plot of the loadings of a more complex (fictitious) data set is visualized. In this plot the directions of the component axes have been indicated by the thick lines. If a mass is more or less characteristic of a certain component, it will lie in the same direction as the component axes. As a result, a plot of all the loadings will show clusters of loadings in the directions of the component axes. It will be clear that this approach only works when many variables are present. Also, when no reference spectra are available, this clustering behavior of the loadings will generally indicate the presence of a component axis. In other words: a clustering of variables (as indicated by their loadings) indicates a common origin, which often is a chemical component. In order to find the clusters of loadings the following procedure is followed:

(a) In the direction of the positive part of $F1$, $F1+$, all loadings within a certain limit are taken, e.g. $\pm 5°$. The variance of the variables within that window is then calculated (for details see below). The variance is based on the number of variables within this window, so it is a measure for the clustering behavior of the loadings in that pie shaped window.

(b) Next, the window is rotated $10°$ counterclockwise with respect to $F1+$ and the variance is calculated again. In this way, the whole $F1-F2$ space is scanned.

The mathematical formulation is:

$$\text{Var}(W = \beta)_\gamma = \sum_{i=1}^{v} a_i^2$$

$$\text{for } a_i > \left(\alpha_{i,1}^2 + \alpha_{i,2}^2\right)^{1/2} \cos(\beta/2) \quad (14)$$

where $a_i = (\alpha_{i,1} \cos \gamma + \alpha_{i,2} \sin \gamma)$ and $\text{Var}(W = \beta)_\gamma$ is the variance in the factor space with an angle of γ degrees with respect to the first factor, using a window W of β degrees. Furthermore, a_i are the loadings of the mass variables on the rotated factor and $\alpha_{i,j}$ are the loadings of the mass variables in the unrotated factor j. Expressed in terms that are easier to visualize, eq. 14 accomplishes the following task: in a two-dimensional system, the sum of the squares of the lengths of all mass axes present in a pie-shaped window of degrees (generally 10 or 20°) is calculated while the window 'scans' the whole two-dimensional space in discrete steps (generally 10°).

The values obtained by the procedure described above are plotted in a polar plot, the so-called variance diagram (see Fig. 10b). As will be obvious, the directions of the clusters of the loadings are clearly indicated. This method can slso be used in factor spaces of more than two dimensions, similar to the graphical rotation procedure [16]. As an example, the results obtained on the data set under study in this tutorial will be used. The variance diagram (Fig. 11) indeed shows the directions of the component axes as found by graphical rotation and as given in Fig. 5. The technique has been applied to the results of different types of data, but its use has so far mainly been limited to pyrolysis data [16–18].

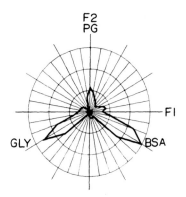

Fig. 11. VARDIA of the simulated data set of biopolymers.

A combination of the pure mass method is possible in the following way: by comparing Fig. 1 and Fig. 8 it will be clear that $F2$ and $F3$ of factor analysis around the origin are similar to $F1$ and $F2$ of factor analysis around the mean. As can be seen in Figs. 8 and 9, the pure mass has the highest loading in the second factor in the direction of the component axis. This means that the pure mass will be the mass with the highest loading in a maximum in the variance diagram of $F2$ and $F3$, when factor analysis has been applied on the correlation around the origin matrix. Since a noisy mass behaves the same way as a pure mass, this is an elegant way to verify the validity of a certain pure mass.

5 CONCLUDING REMARKS

The examples given above have shown the principles of self-modeling mixture analysis. The same techniques can be used for time resolved data, such as GC–MS data, in which case it is called self-modeling curve resolution. The techniques presented above all have their advantages and disadvantages and for an optimal approach the best technique is to combine several methods.

Recent developments in the area of self-modeling curve-resolution techniques are techniques that use the time-resolved information [19–21]. This in contrast to the methods presented above.

The most promising techniques are methods developed by Gemperline [20] and Vandeginste et al. [21]. The methods are designed for time-resolved data, such as GC–MS patterns. Both methods calculate a first approximation for the components in the curve. The plot of the scores of these components will show if the approximation is wrong, e.g. when the scores have negative intensities or when the curve has two maxima, while there are reasons to assume that there is only one. The next step (in case the approximation is not right) is to change the curve into one that is more realistic. This new curve is then used as a target for target factor analysis. This target will produce a new curve for a certain component. When it is close to the target, the process is completed. When the pattern is still wrong, it can be changed again and be used as a target.

These techniques do not assume the presence of a pure mass, although the change of a curve into a target is based on certain assumptions (e.g. only one maximum for a certain component) which may not be right. In an ideal approach the several techniques should be combined in one program, so that all the relative merits of the techniques can be combined into a powerful tool for self-modelling chemical evaluation.

6 ACKNOWLEDGEMENTS

The Biomaterials Profiling Center, University of Utah is acknowledged for providing the figures.

Tables 1 and 2 and Figs. 1, 3, 4, 5, 6, 7, 8, and 10 are (adaptations of) tables and figures from ref. 6 and are reproduced with permission of Plenum Publishing Corporation.

REFERENCES

1 S. Wold, K. Esbensen and P. Geladi, Principal component analysis, *Chemometrics and Intelligent Laboratory Systems*, 2 (1987) 37–52.
2 E.R. Malinowski and D.G. Howery, *Factor Analysis in Chemistry*, Wiley-Interscience, New York, 1980.
3 R.F. Hirsch, G.C. Wu and P.C. Tway, Reliability of factor analysis in the presence of random noise or outlying data, *Chemometrics and Intelligent Laboratory Systems*, 1 (1987) 265–272.
4 R. Hoogerbrugge, S.J. Willig and P.G. Kistemaker, Discriminant analysis by double stage principal component analysis, *Analytical Chemistry*, 55 (1983) 1710–1712.
5 L.V. Vallis, H.J.H. MacFie and C.S. Gutteridge, Comparison of canonical variates analysis with target rotation and least-squares regression as applied to pyrolysis mass spectra of simple biochemical mixtures, *Analytical Chemistry*, 57 (1985) 704–709.
6 W. Windig and H.L.C. Meuzelaar, Chemical component extraction by multivariate analysis, in H.L.C. Meuzelaar and T.L. Isenhour (Editors), *Computer-Enhanced Analytical Spectroscopy*, Plenum, New York, 1987, pp. 67–102.
7 F.H. Knorr and J.H. Futrell, Separation of mass spectra of mixtures by factor analysis, *Analytical Chemistry*, 51 (1979) 1236–1241.
8 K.R. Beebe and B.R. Kowalski, An introduction to multivariate calibration and analysis, *Analytical Chemistry*, 59 (1987) 1007A–1017A.
9 W.H. Lawton and A.S. Sylvestre, Self modeling curve resolution, *Technometrics*, 13 (1971) 617–633.
10 E.R. Malinowski, Obtaining the key set of typical vectors by factor analysis and subsequent isolation of component spectra, *Analytica Chimica Acta*, 145 (1982) 129–137.
11 W. Windig, J. Haverkamp and P.G. Kistemaker, Chemical interpretation of sets of pyrolysis mass spectra by discriminant analysis and graphical rotation, *Analytical Chemistry*, 55 (1983) 81–88.
12 K.J. Voorhees and R. Tsao, Smoke aerosol analysis by pyrolysis–mass spectrometry/pattern recognition for assessment of fuels involved in flaming combustion, *Analytical Chemistry*, 57 (1985) 1630–1636.
13 R.E. Aries, C.S. Gutteridge and R. Macrae, Pyrolysis–mass spectrometry investigations of reversed-phase high-performance liquid chromatography stationary phases, *Journal of Chromatography*, 319 (1985) 285–297.
14 D. van de Meent, J.W. de Leeuw, P.A. Schenck, W. Windig and J. Haverkamp, Quantitative analysis of biopolymer mixtures by pyrolysis–mass spectrometry, *Journal of Analytical and Applied Pyrolysis*, 4 (1982) 133–142.
15 W. Windig, W.H. McClennen and H.L.C. Meuzelaar, Determination of fractional concentrations and exact component spectra by factor analysis of pyrolysis mass spectra of mixtures, *Chemometrics and Intelligent Laboratory Systems*, 1 (1987) 151–165.
16 W. Windig and H.L.C. Meuzelaar, Nonsupervised numerical extraction from pyrolysis mass spectra of complex mixtures, *Analytical Chemistry*, 56 (1984) 2297–2303.
17 R.J. Evans and T.A. Milne, Molecular characterization of the pyrolysis of biomass. 1. Fundamentals, *Energy and Fuels*, 1 (1987) 123–138.
18 J.M. Bracewell and G.W. Robertson, Indications from analytical pyrolysis on the evolution of organic materials in the temperate environment, *Journal of Analytical and Applied Pyrolysis*, 11 (1987) 355–366.
19 H. Gampp, M. Maeder, C.J. Meyer and A.D. Zuberbuehler, Calculation of equilibrium constants from multiwavelength spectroscopic data — III, Model-free analysis of spectrophotometric and ESR titrations, *Talanta*, 32 (1985) 1133–1139.
20 P.J. Gemperline, A priori estimates of the elution profiles of the pure components in overlapped liquid chromatography peaks using target factor analysis, *Journal of Chemical Information and Computer Science*, 24 (1984) 206–212.
21 B.G.M. Vandeginste, W. Derks and G. Kateman, Multicomponent self-modeling curve resolution in high-performance liquid chromatography by iterative target transformation analysis, *Analytica Chimica Acta*, 173 (1985) 253–264.

Interpretation of Direct Latent-Variable Projection Methods and Their Aims and Use in the Analysis of Multicomponent Spectroscopic and Chromatographic Data

OLAV M. KVALHEIM

Department of Chemistry, University of Bergen, N-5007 Bergen (Norway)

(Received 17 September 1987; accepted 4 January 1988)

CONTENTS

1 Introduction	307
2 Data multivariability and collinearity in chemistry	307
2.1 Quantification and representation of chemical data	307
2.2 Chemical analysis and collinearity	308
2.3 Problem specification and aims	309
3 Latent-variable projections of collinear data	310
3.1 Historical background	310
3.2 Interpretation of variable and object space	311
3.3 Graphic representation of multivariate data on latent variables	312
3.4 Mathematical representation and interpretation of latent-variable projection methods	312
3.5 Aims and use of latent-variable projection methods	313
3.5.1 Principal-component projections	313
3.5.2 Partial-least-squares projections	314
3.5.3 Marker-variable projections	314
3.5.4 Marker-object projections	315
3.5.5 Choice of methods	316
3.6 Quantitative comparison of latent variables by means of congruence coefficients	317
4 Conclusion	319
5 Acknowledgements	319
References	319

0169-7439/88/$03.50 © 1988 Elsevier Science Publishers B.V.

Chapter 22

1 INTRODUCTION

Methods for projecting multivariate data onto latent variables can be divided into two broad classes: (i) Methods in which the projections are obtained directly from a data matrix (or covariance/correlation matrix), and (ii) methods in which the projections are obtained by linearly combining latent variables extracted by a direct approach. The latter methods are slightly more complex from a computational point of view than the direct methods [1,2], and often have other aims, too [3]. Singular-value decomposition (SVD) [2], partial-least-squares (PLS) decomposition [4], marker-variable projections (MVP) [5] and marker-object projections (MOP) [6] are direct methods. Target-transformation [7] represents an example of the combinational type of methods. This technique aims to provide interpretable latent variables by combining the orthogonal latent variables obtained by SVD [7] or PLS [8] using some external hypothetical structure or properties.

In this tutorial, the interpretative aspects of the above-mentioned direct methods, their aims, uses and relationships are discussed. Since these fundamental aspects of the analysis of multivariate data are based mainly on concepts from linear algebra, few statistical terms are needed for this tutorial. To illustrate major aspects of the methods, applications to chemical and geochemical analysis are referenced.

The outline of the paper is as follows: (1) Data multivariability and collinearity in chemical and geochemical analysis, (2) mathematical and geometrical representation of multivariate data, (3) categorization of problem areas where multivariate analysis is used, (4) historical background, mathematical and geometrical interpretation of latent-variable projection methods, (5) *comparison of aims and results of latent-variable projection methods*, and, finally, (6) interpretation of multiblock systems by use of congruence coefficients.

2 DATA MULTIVARIABILITY AND COLLINEARITY IN CHEMISTRY

2.1 *Quantification and representation of chemical data*

Instrumental analysis of oils, food samples, etc., generally creates huge amounts of data for each sample. In computerized data acquisition it is not unusual that a chromatographic or spectral profile is represented by several thousand data points. Thus an initial data reduction is usually performed. The approach most commonly used is to transform the point representation of the profiles to peak heights or peak areas. This peak representation of a profile is at the same time simple and conserves the main information. For many chemical methods, e.g. chromatography, the relative peak areas approximate fairly well to the molecular distribution in a sample. Thus the peak representation of the samples can provide information about distributional similarities and differences among samples and variables. Recently, maximum-entropy and similar criteria have been used to reduce the spectral or chromatographic profiles into a few hundred 'merged' variables which carry the same information as the raw point representation of the profiles [9,10]. Such approaches are useful for spectra with overlapping peaks since they avoid time-consuming, operator-intensive and, often, inaccurate procedures for peak identification and resolution.

After the initial reduction of the spectral or chromatographic profiles each sample (object) is described by M quantitative variables [11] which can be collected in a row vector x'_k, here called an object vector [6]. (Note that all vectors are given as columns in this work. Thus, the prime is used to imply transposition of column vectors into row vectors.) The object vectors $\{x'_k, k = 1, 2, \ldots, N\}$ can be arranged in a matrix $\mathbf{X}$, each row containing the measurements of one object. Alternatively, the matrix $\mathbf{X}$ can be viewed as being composed of column vectors, each column of the matrix $\mathbf{X}$ containing the values of one variable for all of the N objects. The term variable vector will be used here to imply a column in $\mathbf{X}$ representing a specific variable. To distinguish variable vectors from ob-

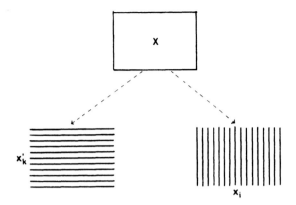

Fig. 1. The two ways of looking at a data matrix $\mathbf{X}$, either as composed of row object vectors $\{x'_k;\ k=1,2,\ldots,N\}$ or column variable vectors $\{x_i;\ i=1,2,\ldots,M\}$.

ject vectors, the indices i and j are reserved for variable vectors, i.e. x_i and x_j imply variable vectors i and j, while x'_k and x'_l represent object vectors. Fig. 1 summarizes these two ways of partitioning the data matrix $\mathbf{X}$ into vectors.

The vector space spanned by the matrix $\mathbf{X}$ can thus be pictured in two different ways depending on whether one wishes to focus on the relationships between objects (rows of $\mathbf{X}$) or the relationships between variables (columns of $\mathbf{X}$). In the former case, the object vectors are plotted on M orthogonal axes, each axis representing one variable. All the information in $\mathbf{X}$ about sample relations is quantitatively displayed in this variable space [2]. In the latter case, the variable values are plotted on N orthogonal axes, one axis for each object. This representation in object space [2] quantitatively displays the relationships between the variables in $\mathbf{X}$.

The two representations described above contain together all the information in the data matrix $\mathbf{X}$. A bivariate variable plot is a special projection of the M-dimensional variable space onto two selected variable axes. Similarly, a bivariate object plot [5] represents a projection of object space onto two selected object axes. These simple points appear to have been overlooked in most of the vast literature on multivariate data analysis.

All methods for extracting relevant information from multivariate data start from the configurations in variable or object space, thus, turning the problem of extracting relevant information from a data matrix $\mathbf{X}$ into a problem of finding useful projections of these two spaces. As will be discussed in later sections, the choice of criterion or method to be used for finding the 'best' projections depends on the purpose of the investigation, i.e. the information required [6].

If one has several sample sets, or the samples are analyzed using several types of independent measurement techniques, the matrix $\mathbf{X}$ may be partitioned into blocks giving a set of submatrices $\{\mathbf{X}_1, \mathbf{X}_2, \ldots, \}$. Each matrix then defines its own variable and object space. Such multi-block systems are discussed in Section 3.6.

2.2 Chemical analysis and collinearity

Spectroscopic and chromatographic data obtained for complex samples generally contain 'redundant' information. In the statistical literature the term collinearity is used to describe this phenomenon. Collinearity is inherent in all measurements of multi-component samples using spectroscopic, chromatographic or other chemical techniques. For many methods each constituent may give rise to many signals or resonances [nuclear magnetic resonance (NMR), gas chromatography–mass spectrometry (GC–MS), IR etc.]. On the other hand, a few underlying phenomena may be responsible for the variation in data. We can restate this in a slightly more precise way: collinearity arises because of the measurement technique, and because the number of underlying (chemical) phenomena giving rise to the variation is limited (Fig. 2).

Consider, for instance, the ^{13}C NMR spectra of a collection of crude oils. The composition of the oils differs because of differences in organic input, depositional environment, and maturity of the source, and because of differences in migration path [12]. However, these factors influence many of the hydrocarbons in similar ways, implying that the compositional variation can be explained by a few latent variables or factors [1,2] related to these geochemical factors (Fig. 2). Furthermore, as a consequence of the measurement technique used, each hydrocarbon is represented by a number of resonances (peaks) determined by the number of

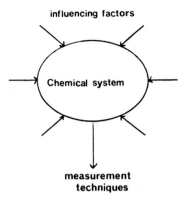

Fig. 2. Possible sources of collinearity in chemical data.

carbon atoms in different molecular surroundings. The ratios between the intensities of these peaks in different oils are constant, however, for a hydrocarbon producing non-overlapping resonances. Thus, the peaks originating from a particular hydrocarbon are covarying.

In the statistical literature data collinearity is often considered to be a problem due to computational complications, mainly stability problems connected with matrix-singularity, e.g., for matrix inversion in regression analysis. Such problems arise when the matrix is not of full rank [1,2], e.g., in the case where there are fewer objects than variables.

For chemical applications collinearity may often be an advantage. Two compounds with almost identical molecular structure may easily be recognized in a sample as long as only one resonance is non-overlapping in, e.g., the ^{13}C NMR spectrum. Thus, measurement techniques producing collinear data represent no drawback for qualitative analysis. For quantitative purposes, such as calibration, the inclusion of redundant information in the form of covarying or overlapping signals makes the solution more stable towards noise.

Thus, from the chemist's point of view, collinear data mean essentially that more information is available. In addition, chemical techniques producing collinear data may offer rapid and accurate alternatives to time-consuming and difficult experimental work-up of samples. The experimental aspects of chemical analysis are thus reduced in complexity. At the same time, familiarity with computational techniques for the analysis of collinear multivariate data becomes increasingly important.

Although the acquired data may possess a large degree of collinearity, low-dimensional projections of the variable and object spaces defined in Section 2.1 can usually be found which are representative for the major sources of variation (information) in the data. Sample score plots and variable loading plots represent commonly-used graphic displays of such projections. A score plot represents a projection of the variable space onto some latent variables, while a loading plot maps the object space onto the same latent variables. This will be discussed further in Section 3.3, and examples of such plots are provided in the applications referred to in this tutorial.

For some methods simultaneous display of sample scores and variable loadings is possible, i.e. in a so-called biplot representation [11,13]. Before discussing particular methods and their representations results, however, the aims of chemical analysis and the types of problems that can be analyzed by latent-variable projection methods require some further consideration.

2.3 Problem specification and aims

Multivariate analysis of multicomponent spectra or chromatogram by means of latent-variable projection methods aims at: (1) interpretation, i.e. identification of the major phenomena responsible for the variation in a data set, (ii) classification and correlation of samples, and (iii) calibration, i.e. establishing a predictive relation between two variable sets. These aims are not independent of each other. Their connections can be pictured geometrically by an equilateral triangle with classification and calibration at the two baseline vertices and interpretation at the apex (Fig. 3). Almost every problem using latent-variable projection methods can be defined within this closed simplex.

The extreme cases, represented by the vertices in Fig. 3, are normally not encountered in real situations. Consider, for instance, classification (termed discrimination in biometrics, see ref. 11): usually one wishes to check the chemical or geo-

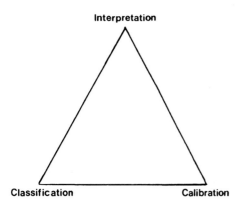

Fig. 3. Graphic representation of objectives of latent-variable projection methods.

chemical plausibility of a classification by relating the variables responsible for the classification to a priori knowledge, often of a qualitative nature. Thus, the classification problem is actually located somewhere between the classification corner and the interpretation corner along the side of the triangle connecting these two vertices. Similarly, a calibration problem always involves classification of the samples to check for multiple groups (classes) in the calibration step and to check for new unmodelled interferences in prediction samples. Furthermore, some level of interpretation may be required to check the plausibility of the model, especially in the presence of strong interferences [10]. It follows that the calibration problem may actually be located somewhere inside the triangle, the exact position being dependent on the weight of the prediction, classification and interpretation aspects.

In some cases, the interpretation of a data structure may be possible without any connection to classification and calibration. Such situations appear when one explores data structures to reveal correlation patterns among variables. Such patterns may be used for several purposes, e.g.: (i) to generate hypotheses about the number and characteristics of underlying causal chemical factors using external qualitative information [14], and (ii) for identification and assignment of chemical constituents in mixtures using loading plots to reveal variable relations [15].

3 LATENT-VARIABLE PROJECTIONS OF COLLINEAR DATA

3.1 Historical background

Despite the pioneering work by Spearman [16] and Pearson [17] at the very beginning of this century, the systematic development of methods for latent-variable modelling of collinear multivariate data started in 1931 with a paper on multiple factor analysis by the American psychologist Louis L. Thurstone [18]. Following the suggestions of the astronomer W. Barkley (see ref. 19, p. 474), Thurstone used a maximum-variance criterion to arrive at the principal-axes solution, usually called principal component analysis (PCA). This was soon replaced by the simple-structure criterion [19] invented by Thurstone to obtain so-called 'real' factors, i.e. factors amenable to interpretation in psychological terms. During the 1940s and 1950s graphic rotation to give simple structures developed into an art (see, e.g., Preface and Ch. 17 'The basic art of rotation by graphs' in ref. 20). Being both subjective and time-consuming, graphic rotation techniques were soon replaced by computer-implemented analytical procedures such as Varimax and Promax, having similar aims but based on more objective, mathematical defined criteria (ref. 2 and references therein).

At the end of the 1960s and the beginning of the 1970s the first applications of factor-analytical techniques appeared in chemistry (see ref. 7, Ch. 7). The major aims of the pioneering work of Malinowski, Howery, and others using target-factor analysis were interpretation of multi-component spectra and chromatograms and their correlation to physical properties of mixtures (ref. 7 and references therein). While Varimax and Promax (ref. 2 and references therein) are based on a simple-structure criterion with respect to variables, target-transformation of principal components aims at the production of chemically interpretable latent-variables by means of a least-squares fit to some hypothetical or independently determined structure.

In the beginning of the 1980s, Martens, Wold and coworkers (ref. 21 and references therein), applied the partial-least-squares (PLS) method [4] to the multivariate calibration of spectroscopic data. As was shown recently [6], the PLS method is only one of a large class of latent-variable projection methods of which some will be given special attention in later sections.

3.2 Interpretation of variable and object space

The interpretation of multivariate data by means of latent-variable projection methods can most easily be understood by reference to variable and object space (Section 2.1). A more detailed examination of these spaces is therefore necessary.

Consider a data set which contains only three samples k, l and m, each sample being characterized by the two variables i and j. This simple bivariate case is representative for all the properties of the two spaces also in the general multivariate case of N objects and M variables where both N and M are larger than three.

Fig. 4a is the geometrical representation of the relations between the three objects in the variable space of the two measured variables. Note that the data are assumed to be column-centered, i.e. the variable means are subtracted from the data matrix. From Fig. 4a we observe that the cosine of the angle ϕ_{kl} between samples k and l represents a quantitative measure of the similarity between the two samples. The angle ϕ_{kl} represents the cosine association measure introduced by Imbrie and Purdy (ref. 2, p. 124 and references therein). This similarity measure can be calculated directly from the scalar product of the two object vectors:

$$\cos \phi_{kl} = x'_k \cdot x_l / \| x'_k \| \, \| x'_l \| \quad (1)$$

To arrive at eq. 1 we have used the identity $(x'_l)' = x_l$ (see also Section 2.1). The $\| \, \|$ (norm) implies the length of a vector. The length of an object vector is equal to the 'size' of the object, e.g., its total chromatographic area. Often the objects are normalized to equal length, e.g., by representing each peak in a chromatogram as a percentage of the total area. In some cases, however, it may be desirable to retain the size factor. Distributional variations among the samples can also be revealed in the presence of the size factor [22]. In summary, the variable space contains all the information about the similarity or differences between objects, and its geometrical representation (Fig. 4a) is a graphic display of the sample relations.

Fig. 4b is the geometrical representation of the two measured variables in the object space spanned by the two objects. The angle θ_{ij} (Fig. 4b) between the two variable vectors represents a quantitative measure of the collinearity or correlation between the two measured variables:

$$\cos \theta_{ij} = x'_i \cdot x_j / \| x_i \| \, \| x_j \| \quad (2)$$

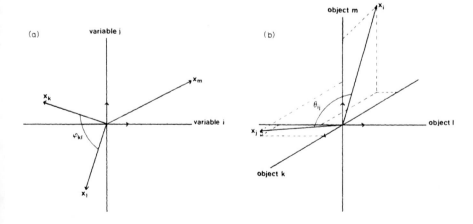

Fig. 4. (a) The configuration of three objects k, l and m in the variable space of the two variables i and j. The angle ϕ_{kl} between the object vectors x'_k and x'_l is calculated by use of eq. 1. (b) The configuration of the two variables i and j in the object space of objects k, l and m. The angle θ_{ij} between the variable vectors x_i and x_j is calculated by use of eq. 2.

Eq. 2 represents the definition of the correlation coefficient between two variables, and, thus, Fig. 4b can be viewed as a geometrical definition of correlation between variables.

The length of a variable vector i in object space is proportional to the square root of the variance accounted for by that variable,

$$s_i^2 = x_i' \cdot x_i / (N - 1) \qquad (3)$$

Thus, the projection of the variable vector x_i onto the variable vector x_j defines that part of the variance which variable j has in common with variable j, i.e. the covariance between i and j. Thus, the object space contains all the information about the relations between the measured variables for a set of samples.

3.3 Graphic representation of multivariate data on latent variables

When the number of variables is larger than three, it is no longer possible to find graphic representations that reproduce the sample and variable relations exactly as given by the data matrix **X**. The problem is thus to find low-dimensional projections of variable and object space which are as informative as possible in accordance with the aim of the investigation. One possibility is to select two or three measured variables and let the projections onto these represent the configuration in variable and object space. Projecting onto latent variables, which are obtained as linear combinations of measured variables, represents another approach. Several criteria are available for the determination of axes for such projections of which some will be closer examined in Section 3.5.

The coordinates or scores of the objects onto a set of latent variables define uniquely the sample relations in the reduced variable space spanned by the latent variables. The Euclidean distance between a pair of samples is a quantitative measure of the similarity between them in the reduced variable space [11,15]. These relations can be displayed in bivariate score plots. Such plots can be regarded as 'windows' into the multi-dimensional variable space. Examples illustrating the aims and use of such plots can be found in, e.g., refs. 15, 23, 24 and 25.

Similarly, the projections of the measured variables onto one or several latent variables can be displayed in so-called loading plots. The projections (scores) of the samples (in variable space) onto a latent variable uniquely define its position in object space and thus the axes of such plots [6]. The projection (loading) of a measured variable (in object space) onto a latent variable gives quantitatively the part of variance it shares with the latent variable. Thus, the whole variance/covariance structure in the part of object space spanned by the latent variables, can be quantitatively reproduced in loading plots.

Rules for interpretation and use of both orthogonal and oblique score plots and loading plots have previously been published [5,6,15,23], and, will therefore not be discussed here. Illustrations of the aims and properties of such plots can be found in a tutorial paper by Wold et al. [26] and refs. 6, 14, 15, 23, 24 and 25.

For many latent-variable projection methods the scores and loadings can be simultaneously displayed in so-called biplot representation [13]. For examples of use and interpretation of such plots the reader should consult the tutorial paper by Birks [11].

3.4 Mathematical representation and interpretation of latent-variable projection methods

Many of the methods for directly projecting the rows and columns of a data matrix **X** onto latent variables can be represented mathematically by the following decomposition formula:

$$\mathbf{X}_{N \times M} = \mathbf{U}_{N \times A} \mathbf{G}_{A \times A}^{1/2} \mathbf{P}'_{A \times M} + \mathbf{E}_{N \times M} \qquad (4)$$

The data matrix **X** is assumed centered in eq. 4. The subscripts show the matrix dimensions (N, M and A are number of objects, variables and latent variables, respectively). The matrix **U** contains orthonormal column score vectors, while **G** is a diagonal matrix containing the squared length of each score vector (in variable space). The rows of the matrix **P'** are proportional to the loadings (covariances/correlations between the measured variables and a latent variable) associated with each latent-variable. The exact properties of **P'** are method-dependent. For singular-value decomposition (SVD) [2,27] (of which correspondence analysis is a special case [28]), the rows of **P**′ are orthonormal. The partial-least-squares (PLS) method [4] and canonical-variate analysis (CVA) (ref. 11 and references therein) result in rows in **P**′ that are neither normal nor orthogonal [6]. The differences between the modelled and the actual elements of **X** are accounted for by the residual matrix **E**. The decomposition formula is sometimes given without the residual matrix **E**, i.e. implying that **X** is reproduced exactly by the three matrices **U**, **G** and **P**′. This is equivalent to an orthogonal rotation of the columns of the data matrix **X**.

Eq. 4 departs from the formulation most commonly used in chemometrics, e.g., by Wold et al. [26] who combine the two matrices **U** and $\mathbf{G}^{1/2}$ into one matrix **T**. Although the two formulations are mathematically equivalent, the representation given by eq. 4, which is based on the basic-structure theorem of Eckart and Young [27], has advantages for interpretative and graphic purposes. Thus, post-multiplication of **U** with $\mathbf{G}^{1/2}$ gives the projections of the samples in the reduced variable space, while pre-multiplication of **P**′ with $\mathbf{G}^{1/2}$ quantitatively reproduces the covariance structure of the reduced object space. Thus, the reduced variable space and the reduced object space are in affect connected (coupled) through the matrix **G**. Since this is also true in the case of a decomposition representing the full rank of **X**, then variable and object space must be connected in general, independent of the method actually used for the decomposition of **X**.

Eq. 4 forms the basis for the so-called biplot formulation of SVD, a graphic representation displaying sample and variable relations (scores and loadings) in the same plot [11,13]. Although the symmetric relationship between score and loading vectors, both being sets of orthonormal vectors, holds only for SVD, biplot representations are possible for all methods that produce orthogonal score vectors. Thus, the result of, e.g., PLS may be graphically represented as biplots. This holds for both the *X*- and the *Y*-space PLS decomposition [21,29], but is probably most useful for displaying the reduced *X*-space.

An analysis of the slightly more complex mathematical situation of oblique score vectors is outside the scope of the present paper. Note, however, that in that case, eq. 4 is not longer valid. Mathematical details of relevance for the oblique projection methods discussed in this work can be found in ref. 6.

3.5 Aims and use of latent-variable projection methods

This section focuses on some particular direct projection methods which the author has found useful for the analysis of chemical and geochemical multivariate data. For the projection methods using marker variables or objects some discussion is devoted to methodological aspects and to the marker selection procedures. Otherwise, the description of methods focuses on their advantages and disadvantages for particular purposes and on aspects connected to properties and interpretation of the graphic representations of the reduced variable and object space.

Aims of multivariate analysis in general have recently been reviewed by Birks [11] and Mellinger [30]. These authors have also categorized methods according to their specific aims.

3.5.1 Principal-component projections

Principal component analysis (PCA), or its mathematically equivalent formulation, singular-value decomposition (SVD) decomposes a data matrix into latent-variables (components, factors, eigenvectors) which successively account for as much variance as possible within the constraint of orthogonal score and loading vectors. The two major advantages offered by this approach are that no a priori knowledge about samples or variables is required and that the data structure is represented on as few latent variables as possible. Score and loading plots may thus reveal underlying structure in data as patterns among samples or variables (data exploration, see ref. 14). Furthermore, in combination with cross validation, this technique can be used to separate noise from systematic intra-class variations in modelling of a priori groups (classes) of objects (ref. 26 and references therein). The major drawback of the method is the difficulty of giving chemical or geochemical meaning to the components extracted. This is due to the often mixed or average character of the principal components, being a result of the use of a maximum-variance criterion for their determination in the absence of any useful chemical or geochemical criteria (see ref. 24 for an example).

Principal component analysis in chemometrics has recently been thoroughly reviewed by Wold et al. [26] and in a more general context by Jolliffe [31].

3.5.2 Partial-least-squares projections

Decomposition by means of the partial-least-squares (PLS) method has so far been used mostly for calibration. For this purpose, PLS represents an alternative to established approaches to regression [21,29]. The method aims at a decomposition of X-space into components which better describe another block of variables (Y-space) than, for instance, principal components. Thus, the covariance between X- and Y-space is explicitly used in the decomposition. However, with some modifications, the method is also effective for decomposing one or more data matrices in the presence of a priori information. Thus, in combination with target-transformation [7], the PLS method has proved useful for revealing information in multi-block systems [8].

Since the PLS method produces orthogonal scores, the biplot representation is possible for representing PLS projections. This should provide a useful alternative if the PLS components are amenable to interpretation and the number of variables and objects are few.

The PLS method has also been proposed as an alternative to canonical variate analysis (CVA) (ref. 26 and references therein) for partitioning multi-class data. A priori information about the samples' group-membership represents the Y-block (binary coded) in such cases. However, as pointed out, e.g., by Birks [11], CVA gives the latent variables with optimal discriminatory ability, and should be used if maximum discrimination is actually wanted. Furthermore, there are now available projection methods which require no a priori information of group-membership, and which, in addition, are simpler and more optimal than the PLS method for the partition of a data set (see ref. 6 and Section 3.5.4). In addition, these methods are based on criteria that are less sensitive to atypical samples than CVA, PLS etc. [32].

3.5.3 Marker-variable projections

The part of a variable's variance shared with a latent variable represents a quantitative measure of its specificity for describing the latent variable. This variance is proportional to the projection of the measured variable onto the latent variable. The specificity is thus equal to $g_a^{1/2}|p_{ia}|/\|x_i\|$, i.e. the ratio between the absolute value of the variance-weighted loading p_{ia} of the measured variable i on the latent variable a and the norm of the variable vector. Thus, the specificity is a number between zero and one. Alternatively, inspection of loading plots may reveal variables that are specific for particular latent variables.

If the latent variables, required to appropriately represent the structure in the data, are orthogonal or only slightly oblique, the latent variables can be replaced by variables with a high degree of specificity. If the latent variables in addition are amenable to chemical or geochemical interpretation, the above statements suggest a procedure for systematic selection of markers, i.e. variables carrying chemical or geochemical information related to specific chemical or geochemical factors [5].

A procedure for obtaining and representing a data matrix by selected marker variables has recently been published [5,6]. The usefulness of the approach has been shown for GC–MS data acquired for several petroleum fractions [24]. An extension of the method to construct marker ratios has recently been published [33].

In short, the marker-variable projection technique is an R-mode technique [2], i.e. one focusing on the relationships between variables. The method is mathematically similar to PLS and has partly similar aims as the trial vector method used by Cattell [20] and other factor analysts, e.g., representation of a data matrix X in terms of interpretable ('real') factors (ref. 5 and references therein). Note, however, that markers are useful also in multi-block situations, since a block of marker variables represents an efficient tool for the decomposition of other matrices. Observe also that the markers are defined via object space since the latent-variable loadings are used for the marker selection.

As the score vectors for marker variables are columns in the data matrix **X**, the reduced variable space on marker variables can be displayed in ordinary bivariate variable plots.

Each axis in a marker-variable loading plot represents a row of the covariance matrix **X'X**. Thus, a marker-variable loading plot is a graphic representation of the covariances between the marker variables and all the other measured variables, each marker variable lying on the axis it spans. Furthermore, the correlation between pairs of columns in **X** representing marker variables defines exactly the obliqueness of the axes of the loading plots [6]. Thus, if the latent variables replaced by the markers are almost orthogonal, the procedure used for selecting marker variables assures that the marker-variable loading plot also has almost orthogonal axes.

Because marker-variable score plots have orthogonal axes, while the coresponding loadings are plotted on axes that are oblique, the biplot [13] representation is, unfortunately, not possible for the marker-variable projection technique.

3.5.4 Marker-object projections

The systematic selection of marker objects can proceed in two different ways, the choice between them depending partly on the purpose of the selection, partly on the available information. If the purpose is to find objects that can replace a latent variable found by using one of the other projection techniques described in this paper, the score plots can be used to detect marker-object candidates. The procedure is then analogous to the use of loading plots for revealing marker variables. Thus, if the scores of a sample are relatively small for all but one latent variable in a set of latent variables which appropriately represent the data matrix **X**, the sample may carry the same information as the particular latent variable, i.e. the sample is a marker candidate.

The angle ϕ in variable space between the object vector and the latent variable can be used to calculate quantitatively the specificity of a marker-object candidate. The cosine of the angle ϕ is equal to the scalar product of the object vector k and the row vector of coordinates of the latent-variable a, both vectors being normalized to unit length, i.e. a relation similar to eq. 1 is obtained:

$$\cos \phi = x'_k \cdot w_a / \| x'_k \| \| w_a \| \qquad (5)$$

The coordinate vector w_a contains the expansion coefficients linearly combining the measured variables into the latent variable a. Eq. 5 simplifies to $|u_{ka}| g_a^{1/2} / \| x'_k \|$ (see ref. 6). Thus, the specificity of the marker object k is equal to the ratio between the absolute value of the sample's score on the latent-variable a and the norm of its object vector. This relation can be used to check the result of a visual inspection of score plots. As for marker variables, the specificity is a number between zero and one.

Another approach to marker-object selection is to start directly from the relations between objects in variable space as defined by the scalar products calculated from eq. 1 in Section 3.2. Dissimilar pairs of samples are recognized by their scalar products being close to zero. By using this information together with the information about the norm of the samples in variable space, a strategy for selecting marker objects can be proposed. The procedure is based on three criteria: (i) select samples with cosine-association measure (eq. 1, Section 3.2) close to zero, (ii) if several samples have similar cosine-association measures with the other samples, select the one with maximum norm, and (iii) be careful not to include unique samples which might be outliers.

The described procedures for marker-object selection aim at providing subsets of samples that span the information in a data matrix. Such a subset of samples can be used to check prediction samples for outliers before calibration.

The interpretation of multi-block systems represents another field where marker objects are useful. Thus, projecting onto marker samples may produce interpretable latent variables without the rotation step involved in target-transformation of a PC or PLS decomposed matrix and without the restriction of a priori quantitative target factors [6]. In such cases, the marker selection is usually based on inspection of score plots. Marker objects revealed in one matrix are thus used to explore a whole set of matrices with respect to sample relations. Interpretation may then be possible by using independent information about samples, e.g., qualitative information such as the geographical location of markers. An example is given in ref. 6 of the use of marker objects detected in one matrix to produce a classification in terms of interpretable oblique axes in another matrix.

The marker-object projection technique is a Q-mode technique [2], i.e. it focuses on the relationships among objects. Independent of the selection procedure, i.e. whether scalar products, score plots or information of class-membership are used, object markers are defined via variable space. Selected object vectors in variable space define the latent variables on which the other samples are projected. Thus, the scores on a marker object display quantitatively the similarity between the marker and all the other samples. If the samples are normalized to unit length, the scores are the cosine-association measure of Imbrie and Purdy (see Section 3.2). Thus, for marker-object projections bivariate score plots are graphic representations of two columns (or rows) of a similarity table. The obliqueness between the axes in marker-object score plots reflects exactly the obliqueness between the marker-object vectors.

Similarly to what was found for marker-variable projections, the biplot [13] representation is not possible for marker-object projections.

3.5.5 Choice of methods

All the methods discussed above have been shown to have some potential for classification, calibration, and interpretation. The methods, however, have both strengths and weaknesses, a fact that has to be utilized in actual applications. For example, in exploratory situations a natural first step in the analysis is to project onto principal components. The maximum-variance property of principal components ensures that the most important sources of variation are represented in the graphic displays of sample and variable relations.

The maximum-variance property may be less useful if classification of samples or variables on chemically or geochemically interpretable factors is required. The mixed character of principal components may prevent an interpretation in terms of concepts that a chemist can relate to (see ref. 24 for an example) and is certainly in conflict with the common demand that each latent-variable should represent a single chemical or geochemical phenomenon. For such purposes, projections onto marker objects or marker variables may produce more useful results (see ref. 6 for an example). Thus, these latter techniques have their largest potential for investigations where the purpose is classification of samples and variables on chemical or geochemical factors.

In calibration situations, the marker-variable and marker-object approach may be useful in multiple-class situations where oblique axes may better represent the sample structure in X-space, thus, improving the ability for outlier detection compared to a partial-least-squares approach. However, such situations are rarely encountered in real problems, so that the computational simplicity due to orthogonal score vectors obtained by use of the partial-least-squares and principal-component approaches to calibration, makes these two latter methods more useful in the calibration context.

Consider again the equilateral triangle in Section 2.3 (Fig. 3). This figure defines the problem areas where methods for projecting onto latent variables have their major use. From the summary above, it is clear that principal-component projections are useful for all the purposes represented by Fig. 3. Thus, the method may be placed inside the triangle with equal distance to each of the three corners. Although the partial-least-squares (PLS) method is of some use for classification in the presence of a priori information of group-membership for objects and has been shown to have a potential for interpretative analysis when combined with target-transformation [8], the method is most useful for calibration. Thus, PLS falls inside the simplex, but closer to the calibration and interpretation corners than to the classification corner. Finally, the marker-variable and marker-object projection methods are located between the classification and interpretation corners in Fig. 3 close to the line connecting the two corners.

3.6 Quantitative comparison of latent variables by means of congruence coefficients

So far, the discussion has been limited to situations where the purpose is to reveal the information in one data matrix. For some interpretation problems, however, the use of a block of independently determined variables has been implied, thus, aiming at constraining the decomposition of a data matrix so as to obtain more interpretable components than those resulting from the use of the maximum-variance criterion implicit in a principal component decomposition. For example, a block of marker variables can be used to constrain the decomposition of another matrix by means of the PLS approach. To achieve chemical interpretation it will usually be necessary to rotate the PLS components with the markers as targets [8].

In exploratory multi-block analysis, the purpose is to establish the common and the unique information contained in a set of matrices, i.e. general factors vs. system- or sample-dependent factors (Fig. 5). Because the number of factors above noise level and their relative size may vary considerably from one matrix to another, the matrices are usually decomposed independently of each other. The common and unique information carried by the extracted latent-variables is subsequently revealed by means of congruence coefficients [23,24].

Multi-block data relating to exploratory data analysis can be divided into two basic categories: (i) data where several variable sets characterize a single sample set [24], and (ii) data in which several sample sets are characterized by a single variable set [23]. A sample set analyzed by different experimental techniques creates a multi-block structure of the first category. A data matrix partitioned into different classes of samples represents an example of the second category of multi-block structures.

A combination of the two problem categories defined above is also possible, e.g. when several sample sets are measured by several experimental techniques. Such multi-block problems, however, are probably best interpreted in a two-step procedure based on the division above.

Although the analysis of both categories of multi-block data defined above aims at inter-correlating blocks of latent variables, different congruence coefficients are required for the two categories. As shown below, a multi-block system of the first category uses the relations in object space for the inter-correlation, while the second category uses relations in variable space. Thus, it is tempting to use the R- and Q-mode terminology for distinguishing the two cases.

The congruence coefficients for the two categories of multi-block systems are easily defined after a straightforward generalization of the representations in variable and object space to accommodate multi-block situations. Fig. 6 displays the structure of R- and Q-mode multi-block systems. In both cases, each block defines its own variable and object space. Hence, the number of variable and object spaces is equal to the number of blocks.

For R-mode multi-block systems, all the object spaces are defined on the same object axes. Thus, the similarity between the latent variables obtained for two variable sets can be evaluated by comparison of pairs of score vectors, each score vector obtained for one variable set being compared with each score vector obtained for the other variable set. The correlation coefficients between the two sets of score vectors represent a quantitative measure of this similarity:

Fig. 5. General (common) factors $\{c_1, c_2, \ldots, c_A\}$ vs. sample/system-dependent (unique) factors $\{u_1, u_2, \ldots, u_B\}$.

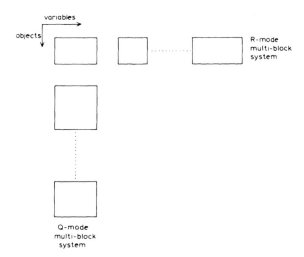

Fig. 6. Structure of R- and Q-mode multi-block systems.

$$r_{ab} = \cos \theta_{ab} = \boldsymbol{u}'_a \boldsymbol{v}_b,$$
$$a = 1, 2, \ldots, A; \quad b = 1, 2, \ldots, B \quad (6)$$

These correlation coefficients, also called direction cosines [19,20], differ only in normalization from the linear coefficients obtained by a least-squares fit of one set of latent-variables to the other [5].

The number of latent variables, denoted A and B in eq. 6, extracted for the two variable sets can be different because of different number of factors above noise level. This may happen if, for example, different experimental techniques are used to analyze the same sample set, or different compound classes are used to characterize, e.g., a set of crude oils [24].

If two variable spaces are decomposed on orthogonal score vectors, the similarity between each latent variable in one block with all the latent variables of another block may be quantitatively evaluated by means of the Squared Multiple Correlation Coefficient (SMCC) (ref. 5 and references therein). This similarity measure is obtained by squaring and summing correlation coefficients calculated from eq. 6:

$$\text{SMCC}_a = \sum_{b=1}^{B} r_{ab}^2, \quad \{a = 1, 2, \ldots, A\} \quad (7)$$

From the orthogonality between score vectors and the properties of correlation coefficients it follows that the congruence coefficient defined by eq. 7 varies from zero for a unique latent variable up to one for a latent variable that is completely accounted for by the other set of latent variables.

By summing the SMCCs in eq. 7 (over the index a), a quantitative measure is obtained for the congruence between the reduced object spaces spanned by the two sets of latent variables [5].

Note that any use of the congruence measures defined by eqs. 6 and 7 for interpretation and comparison is based on the assumption that similar object patterns implies the influence of the same chemical or geochemical factor.

Latent variables obtained for Q-mode multi-block systems are best compared by use of a congruence coefficient based on the coordinates of the latent variables in variable space (see Section 3.5.4 and ref. 6):

$$\cos \phi_{ab} = \boldsymbol{w}'_a \cdot \boldsymbol{w}_b / \| \boldsymbol{w}_a \| \| \boldsymbol{w}_b \| \quad (8)$$

In the special case of principal components this coefficient is the same coefficient of congruence introduced by Burt [34,35].

In complete analogy with the SMCC, eq. 8 can be used to obtain a congruence coefficient measuring the similarity between each latent variable in one block of samples with all the latent variables of another block of samples. As for the SMCC this coefficient varies between zero and one, and assumes orthogonal coordinate vectors.

The quantitative comparison of data structures offered by congruence coefficients is extremely useful for 'naming' factors in exploratory situations. An example is given in ref. 23, where principal components obtained for sample sets of oils and source rocks from various geographical areas are quantitatively compared by use of eq. 8, thus, revealing a source/maturity factor. This general factor is then used to correlate oils and sources, thus, eliminating sample/system-dependent parts such as the effect of oil alteration during migration from source to reservoir. Furthermore, as such factors describe variations in the distributions of classes of chemical molecules, inspection of loadings and scores on geochemically interpretable factors may produce useful hypotheses about the influence of the various geochemical factors governing the composition of crude oils and their sources. Thus, congruence coefficients may play an important role in an iterative approach to hypothesis generation and concept construction in fields where controlled experiments to give representative sample sets are only partly, if at all, possible [36].

Eq. 8 and its generalizations can also be used for the comparison of the intra-class structure of disjoint PC or PLS models. Again, this represent a way to separate the sample/system-dependent parts of the variation from the joint and general variation [36].

Another possible application of congruence coefficients is the validation of the significance of latent-variables used for correlative purposes. By means of a step-wise deletion procedure for variables and objects, the stability of the factors extracted from a data matrix X can be checked and outlying samples or variables revealed.

Graphic representation is possible for detailed quantitative comparison of the congruence between a pair of latent variables (see ref. 25 for an example).

4 CONCLUSION

The usefulness of latent-variable projection methods for the interpretation of spectroscopic and chromatographic multivariate data has been illustrated in numerous applications: (i) for qualitative interpretation, e.g., peak assignment in multicomponent spectra [15], (ii) for correlating samples in the presence of large interfering effects [23], and, (iii) for exploration of data and generation of models [14,33]. In combination with various congruence coefficients interpretation and correlation of multi-block systems are possible without assuming all blocks to be influenced by the same factors [5,24].

In addition to being a powerful tool for exploring and displaying multivariate data on meaningful axes, the methods of marker projections provide increased insight into the geometrical basis and the interpretation of latent-variable projection methods. Finally, these techniques represent a link to the work of the early factor analysts, e.g., Thurstone [19] and Cattell [20], who projected factor spaces onto marker variables to gain interpretation and understanding.

5 ACKNOWLEDGEMENTS

This tutorial is based on a talk given at the conference SCA – Scientific Computing and Automation, Amsterdam, 1987. I would like to thank Professor D.L. Massart, Chairman of the conference, for encouraging me to write this tutorial. John Birks, Odd Borgen, Rolf Manne and Richard Reyment are gratefully acknowledged for valuable discussions and comments on the manuscript, and the Norwegian Research Council for Science and Humanities (NAVF) for generous financial support and a grant for my travel to the SCA meeting.

REFERENCES

1 P. Horst, *Factor Analysis of Data Matrices*, Holt, Rinehart and Winston, Inc., New York, Chicago, San Francisco, Toronto, London, 1965.
2 K.G. Jöreskog, J.E. Klovan and R.A. Reyment, *Geological Factor Analysis*, Elsevier, Amsterdam, Oxford, New York, 1976.
3 R.J. Rummel, *Applied Factor Analysis*, Northwestern University Press, Evanston, IL, 1970, pp. 372–385.
4 H. Wold, Soft modeling: The basic design and some extensions, in K.G. Jöreskog and H. Wold (Editors), *Systems under Indirect Observation: Causality — Structure — Prediction*, Part II, Elsevier, Amsterdam, New York, London, 1982, pp. 1–55.
5 O.M. Kvalheim and N. Telnæs, Visualizing information in multivariate data: Applications to petroleum geochemistry. Part 1. Strategies for obtaining and representing information in complex petroleum fractions, *Analytica Chimica Acta*, 191 (1986) 87–96.
6 O.M. Kvalheim, Latent-structure decompositions (projections) of multivariate data, *Chemometrics and Intelligent Laboratory Systems*, 2 (1987) 283–290.
7 E.R. Malinowski and D.G. Howery, *Factor Analysis in Chemistry*, Wiley, New York, 1980.
8 O.M. Kvalheim, A partial-least-squares approach to interpretative analysis of multivariate data, *Chemometrics and Intelligent Laboratory Systems*, 3 (1988) 189–197.
9 T.V. Karstang and R. Eastgate, Multivariate calibration of an X-ray diffractometer by partial least square regression, *Chemometrics and Intelligent Laboratory Systems*, 2 (1987) 209–219.

10 A.A. Christy, R.A. Velapoldi, T.V. Karstang, O.M. Kvalheim, E. Sletten and N. Telnaes, Multivariate calibration of diffuse reflectance spectra as an alternative to rank determination by vitrinite reflectance, *Chemometrics and Intelligent Laboratory Systems*, 2 (1987) 199–207.
11 H.J.B. Birks, Multivariate analysis in geology and geochemistry: an introduction, *Chemometrics and Intelligent Laboratory Systems*, 2 (1987) 15–28.
12 B.P. Tissot and D.H. Welte, *Petroleum Formation and Occurrence. A New Approach to Oil and Gas Exploration*, Springer Verlag, Heidelberg, 1984.
13 K.R. Gabriel, The biplot graphic display of matrices with application to principal components analysis, *Biometrika*, 58 (1971) 953–967.
14 N. Telnæs, A. Bjørseth, A.A. Christy and O.M. Kvalheim, Interpretation of multivariate data: relationship between phenanthrenes in crude oils, *Chemometrics and Intelligent Laboratory Systems*, 2 (1987) 149–153.
15 O.M. Kvalheim, D.W. Aksnes, T. Brekke, M.-O. Eide, E. Sletten and N. Telnaes, Crude oil characterization and correlation by principal component analysis of ^{13}C nuclear magnetic resonance spectra, *Analytical Chemistry*, 57 (1985) 2858–2864.
16 C. Spearman, General intelligence, objectively determined and measured, *American Journal of Psychology*, 15 (1904) 201–293.
17 K. Pearson, On lines and planes of closest fit to systems of points in space, *Philosophical Magazine*, 6 (1901) 559–572.
18 L.L. Thurstone, Multiple factor analysis, *Psychological Review*, 38 (1931) 406–427.
19 L.L. Thurstone, *Multiple-Factor Analysis*, The University of Chicago Press, Chicago, London, 1947.
20 R.B. Cattell, *Factor Analysis: An Introduction and Manual for the Psychologist and Social Scientist*, Harpers and Brothers, New York, 1952.
21 H. Martens, *Multivariate calibration — Quantitative interpretation of non-selective chemical data*, Dr. techn. thesis, Technical University of Norway, Trondheim, 1985.
22 M.O. Eide, O.M. Kvalheim and N. Telnaes, Routine analysis of crude oil fractions by principal component modelling of gas chromatographic profiles, *Analytica Chimica Acta*, 191 (1986) 433–437.
23 O.M. Kvalheim, Oil–source correlation by the combined use of principal component modelling, analysis of variance and a coefficient of congruence, *Chemometrics and Intelligent Laboratory Systems*, 2 (1987) 127–136.
24 O.M. Kvalheim and N. Telnæs, Visualizing information in multivariate data: Applications to petroleum geochemistry. Part 2. Interpretation and correlation of North Sea oils using three different biomarker fractions, *Analytica Chimica Acta*, 191 (1986) 97–110.
25 E. Jantzen, O.M. Kvalheim, T.A. Hauge, N. Hagen and K. Bøvre, Grouping of bacteria by Simca pattern recognition on gas chromatographic lipid data: patterns among Moraxella and rod-shaped Neisseria, *Systematic and Applied Microbiology*, 9 (1987) 142–150.
26 S. Wold, K. Esbensen and P. Geladi, Principal component analysis, *Chemometrics and Intelligent Laboratory Systems*, 2 (1987) 37–52.
27 C. Eckart and G. Young, The approximation of one matrix by another of lower rank, *Psychometrika*, 1 (1936) 211–218.
28 M.J. Greenacre, *Theory and Applications of Correspondence Analysis*, Academic Press, London, 1985.
29 R. Manne, Analysis of two partial-least-squares algorithms for multivariate calibration, *Chemometrics and Intelligent Laboratory Systems*, 2 (1987) 187–197.
30 M. Mellinger, Multivariate data analysis: its methods, *Chemometrics and Intelligent Laboratory Systems*, 2 (1987) 29–36.
31 I.T. Jolliffe, *Principal Components Analysis*, Springer Verlag, Berlin, 1986.
32 R.A. Reyment, Multivariate analysis in geoscience: fads, fallacies and the future, *Chemometrics and Intelligent Laboratory Systems*, 2 (1987) 79–91.
33 O.M. Kvalheim, A.A. Christy, N. Telnæs and A. Bjørseth, Maturity determination of organic matter in coals using the methylphenanthrene distribution, *Geochimica et Cosmochimica Acta*, 51 (1987) 1883–1888.
34 C. Burt, The factorial study of temperamental traits, *British Journal of Psychology Statistical Section*, 1 (1948) 178–203.
35 H.H. Harman, *Modern Factor Analysis*, The University of Chicago Press, Chicago, 2nd ed., 1967, pp. 269–270.
36 O.M. Kvalheim, *Methods for the Interpretation of Multivariate Data — Examples from Petroleum Geochemistry*, Dr. Philos. thesis, University of Bergen, Bergen, 1987.

ND
Soft Modelling and Chemosystematics

NILS B. VOGT

Department of Analytical Chemistry, Senter for Industrial Research, P.B. 350, Blindern, 0314 Oslo (Norway)

(Received 3 June 1986; accepted 11 March 1987)

CONTENTS

1 Introduction 321
2 Multivariate data analyses in taxonomy and chemosystematics 322
   2.1 Clustering and ordinal methods 322
   2.2 Soft modelling 323
3 Partial least squares — multivariate regression/calibration 323
4 Chemical taxonomy and the multivariate approach 324
   4.1 The chemistry in taxonomy 324
   4.2 Multivariate chemotaxonomy 325
   4.3 Cladistic descriptions 327
5 Chemosystematics and the multivariate approach 327
   5.1 Chemosystematics 327
   5.2 Multivariate chemosystematics 328
   5.3 Samples and the choice of characters 328
6 Selected strategies for soft modelling 329
   6.1 Two blocks with different sets of variables 329
      6.1.1 Geno- and phenovariables together 329
      6.1.2 Composition activity regression (CARE) 330
   6.2 Two blocks with the same set of variables 332
      6.2.1 Phenovariable residual patterns 332
      6.2.2 Transposed partial least squares (TPLS) 333
   6.3 Hypothesis formulation and experimental design 335
7 Chemometrics and chemosystematics 336
8 Future developments of soft modelling in chemosystematics 336
9 Glossary of terms used 337
10 Acknowledgements 337
References 337

1 INTRODUCTION

Biological systems are in their nature complex. A simple way of understanding complex, multidimensional systems in nature is by recognising patterns.

Chemotaxonomy is concerned with classification using the chemical expression of the genetic code [1]. Traditionally, in chemotaxonomy, the combination of genotype and phenotype variations have been considered as a problem creating confusion (ref. 1, p. 96). Recently, through the use of discriminant analysis and SIMCA pattern recognition, the presence of chemical compound groups containing both genetic and phenotypic information have been used to chemically group taxa in multivariable systems [2–6].

Chemosystematics is concerned with analysing the chemical expression of type (classification) *and* condition using family- or organism-specific

0169-7439/87/$03.50   © 1987 Elsevier Science Publishers B.V.

chemical compounds. As such it is concerned with arrival at methods where chemical data may be used to describe both the genotype and the phenotype expression and to describe which part of the chemical and/or biological variation stems from the different sources. Chemosystematics thus encompasses the combined use of chemical and statistical/mathematical methods in biology and as such is part of the field of numerical taxonomy which was developed in the 1960s together with other fields of biological and chemical data analysis [1,7,8].

The ultimate question in chemosystematics is the connection between the chemical expression (biochemistry) of an organism and the connection with heredity (genetics), ecological, geographical and environmental regulators and the biological expression of these regulators.

Those interested in a more thorough discussion of topics in numerical taxonomy are adviced to refer to Sneath and Sokal, *Numerical Taxonomy* [1], an inspring book covering most aspects of systematics, including chemical methods.

The large number of variables (characters) available by chemical analysis and the complexity of this material suggests two important considerations. The one-variable-at-a-time approach is futile and time consuming and the a priori knowledge of the systematics, although a prerequisite for interpretation, must not be allowed to determine the outcome, or results, of the chemosystematic investigation. The use of computerised methods of data analyses applying multivariate methods which give interpretable output are necessary and greatly facilitate the task of analysing and interpreting complex material.

This article addresses some of the possibilities available in applying soft modelling using principal components and partial least squares (PLS) to reduce phenomenological dimensionality and allow interpretation of complex multivariate biological systems. The potential available in chemosystematics when using chemical compound groups, defined here as phenovariables and genovariables (see Fig. 1), in combination with multivariate data analytical methods using the SIMCA-PLS2 pattern recognition approach are indicated and paths for future research are outlined.

2 MULTIVARIATE DATA ANALYSES IN TAXONOMY AND CHEMOSYSTEMATICS

2.1 Clustering and ordinal methods

Methods of multivariate data analyses may be divided into two different types as described in Sneath and Sokal [1] and Johnson and Wichern [9]. These are clustering methods and ordinal methods [1,9–12].

Clustering methods may be viewed as being algorithmic approaches to divide or collect, in groups, samples (defined OTUs — operational taxonomic units [1]) depending on similarities. Although based on mathematic definitions the different algorithmic approaches are best described by differences in the criteria used to divide, or collect, groups of samples [12], i.e. the basic mathematical ideas defining the output information from the algorithm. The many different clustering methods available together with a wide variety of criteria and choices has made them popular. Clustering methods are also popular in taxonomic studies for the reason that based on the criteria used they give division or collection of groups of samples and presentation of results which allow interpretation in a straightforward way, e.g. dendrograms. By defining the algorithm and the measure that has been used the classification, e.g. taxonomic, may be described and discussed.

Ordinal methods, such as principal component analysis [10], factor analysis [11], principal coordinates and several other methods [9,13] are based on decomposing data matrices using mathematical properties [11]. The way ordinal methods work is to obtain new coordinates describing the variance,

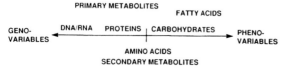

Fig. 1. Fingerprint interpretation of chemical compound groups. Patterns of different groups of compounds may be considered to contain varying amounts of information on the scale from genovariables to phenovariables.

or covariance, in the data set as linear combinations of the original data [9]. Most often a large data set may then be represented by only a few new ("latent") variables (linear combinations) may be needed to describe a large part of the variance in the data set. Thus, a complex material consisting of possibly hundreds of chemical variables analysed on many samples is reduced to a lower dimensionality allowing visual inspection [9]. Collections of samples ("clusters") may then be identified visually in the plots. An example of this is unsupervised principal component analysis. Verification of the groups as defining different entities, e.g. taxonomic groups, is then based on external data available describing the members of the visual classes, or by interpretation, again allowing for somewhat subjective considerations. Ordinal methods have therefore been popular for interpretation purposes, where the factors/components are assigned biological/ecological/geographical or related meening and less so for classification purposes.

2.2 Soft modelling

The concept of soft modelling is based on the ordinal method approach to data analyses. The "soft" part of the modelling approach stems from the fact that principal components/factors are calculated from the data and therefore describe structures in the data. This is opposed to traditional "hard" modelling, i.e. where a priori ideas of mathematical connections are used and models made from these. The SIMCA soft modelling is also different to other ordinal methods such as discriminant analysis and canonical variates, where the description is optimised with respect to either discriminating between groups of samples or some other preselected function. The class models in SIMCA-PCA are strict principal component descriptions of the data structure in each class.

The SIMCA "method" of data analyses [13–15] has been advocated in the chemical literature for two different reasons depending on level of interest. For those interested in computational aspects the advocation has been that the SIMCA method uses the NIPALS algorithm as described by Wold et al. [13]. This is an efficient method to compute principal components. This has been unique for the SIMCA package. The traditional method of obtaining principal components has been to use matrix inversion of the variance–covariance matrix. The NIPALS algorithm uses an iterative approach where the direction of a principal component is determined by (a) choosing some arbitrary vector direction and then (b) rotating this vector until the largest variance is explained.

The other major principle of advocation has been the conceptual description of "disjoint soft modelling". Disjoint soft modelling implies that principal component models are made on the separate (disjoint) groups identified in the unsupervised principal component analyses. In SIMCA the class boundaries (hyperspheres) are then used to determine if a sample is inside a class model or if the sample is an outlier, i.e. to determine class boundaries for discrete groups of samples. This is done by using approximate F-tests [14]. This concept greatly increases the use of ordinal methods beyond the pure data-analytical/interpretation use. By defining class boundaries a statistical verification of the visual clusters seen in the unsupervised principal component analysis is made possible. Using disjoint soft modelling therefore accomplishes the same as clustering methods, i.e. classification of groups of samples, but retains the basis of a mathematically defined, and thus objective, coordinate system which may be used to ease the interpretation of complex data sets.

3 PARTIAL LEAST SQUARES — MULTIVARIATE REGRESSION/CALIBRATION

Ordinal methods are similar to regression. The difference is that the components or factors (latent vectors) extracted in ordinal methods are themselves not a priori available such as in regular regression [10,11]. The properties of the most commonly used principal component and/or factor analytical solutions are such that they are ideally orthogonal vectors representing a subspace ($k < m$, k being the number of extracted components and m the number of original variables) which optimally describes a systematic variation in the data being analysed. If the components/factors may be assigned physical, chemical, biological or environmental and ecological meanings the factors of components extracted by using ordinal methods may be used for regression purposes. Principal component regression is an example of this [12].

Partial least squares (PLS) [16–18] is a term used to describe several related mathematical algorithms for use in multivariate calibration/regression. A recent tutorial by Geladi and Kowalski [19] describes the basic algorithm and gives examples of their use in chemistry. The PLS method has been implemented for use on microcomputers and is available in several different multivariate packages. The PLS method is an optimal method for determining predictive relations between two or more blocks of variables (chemical, biological, environmental etc.) analysed on the same samples (objects). Although the regression is linear in each dimension the use of many dimensions allows non-linear relationships to be analysed. The discussion in this paper is based on the concepts and ideas described in refs. 13–15 on the SIMCA-PLS2 package developed by Wold and co-workers at Umeå University in Sweden. The program and both geometric and mathematical descriptions of the algorithms used in principal component analysis, principal component regression and PLS may be found in the literature references. The interested reader is referred to these for in-depth description of the mathematics and computational aspects of the method.

In the PLS method a block of X (independent) variables are related to a block of Y (dependent) variables through a process where the variance structure in the Y block influence the calculation of the linear combination components in the X block and vice versa. This is accomplished by first analysing the **Y** block matrix obtaining what are termed "weights" according to the variance in this matrix. These weights are then used on the **X** block matrix to obtain the first set of scores for the X block. The scores from the X block are then used as weights in **Y** block matrix. An iteration of this "criss-cross" process gives what is described geometrically as a "tilting" of the components in both blocks so as to optimize inter-block regression of the latent variables. This is different from principal components analysis where the principal components are optimised with respect to variance/covariance description. Cross-validation [20], i.e. measuring description of systematic structure, is done by leaving out groups of data in a systematic manner, modelling the remainder, and predicting the omitted set. The factors/components or dimensions found in the X and Y blocks by PLS are no longer principal components in the strict sense and do not necessarily have the properties of principal components, e.g. orthogonality, but are components derived through influence from the structure in the other block (see ref. 19). These will therefore be referred to as PLS components. The PLS algorithm(s) may be viewed as "influencing" algorithms. The latent vectors that are calculated in PLS represent linear combinations of the original variables for the purpose of best describing the structure in the Y block from the structure in the X block for each dimension that is calculated.

It follows from the discussion above that the PLS vectors represent one type of linear solution to the problem of representing structure from one data matrix. Other solutions which might be more appropriate in some problems are canonical correlation [11] or some of the association and similarity functions described by Sneath and Sokal [1].

As shall be indicated later the PLS approach of obtaining optimal regressional relationship between two (or more) data matrices [19], deriving from a family of related methods using the same line of thought [21,22], greatly increases the possibilities of the biologically interested chemist.

4 CHEMICAL TAXONOMY AND THE MULTIVARIATE APPROACH

4.1 The chemistry in taxonomy

Chemotaxonomy is one of several methods of relating biological systems to each other in terms of similarity. Traditional chemotaxonomy has relied on identifying specific, most often genetically defined, biochemical macromolecules to specify common origin. The information in these molecules is at the "bit" level, i.e. univariate present or not present. Several methods have been formalised which use the DNA or protein sequence of these "information" macromolecules in defining taxonomic classification. Protein taxonomy, comparative serology, and the nucleic acid pairing technique are prominent examples of this approach [1]. The foundation of these methods is the close connection between genetic code and the biochemical expression of the code as present in these genovariables. These methods are therefore concerned with defining genotypic similarity or classification. Sneath and Sokal [1] distinguish between episemantic and semantic molecules. The episemantic molecules are defined as simple chemical substances which yield relatively little information about the organism. Semantic molecules are classes of biochemical compounds which are quite complex and yield a great deal of information. An example of the latter are proteins. The sequence of amino acids in proteins may be different in the same functional protein of different animals creating the possibility of organising (systemising) species according to similarity of the amino acid sequence similarities.

A totally different approach is to use chemical compounds much further removed from the genetic code. As noted by Bergan et al. [23] the distinction of fatty acid composition in an organism is subordinate to that of DNA pairing. The variability of fatty acid distributions in organisms, as analysed by simple univariate methods, has precluded the use of other than very special fatty acids, again either present or absent (episemantic molecules [1]), for taxonomic classification and identification [24].

Chemical compound groups which are secondary or tertiary, products of the genetic code and in which biosynthesis may be strongly influenced by conditions not related to heredity, e.g. fatty acids, amino acids, carbohydrates and hydrocarbon material, have therefore been considered applicable as taxonomic identifiers only when conditions of environmental differences have been controlled, such as in microbiological investigations, where preliminary culturing under controlled conditions may preclude variation from this source [23,24].

On the other hand lipids and fatty acids, together with amino acids and carbohydrates, are found in abundant amounts in all organisms. They are readily analysed by relatively simple polyphenic (multiple character) (ref. 1, p. 91) methods, and most importantly, they are present in a large number of different forms which opens the way for the use of chemometric methods, such as pattern recognition.

4.2 Multivariate chemotaxonomy

Chemical analysis in chemotaxonomy has been restricted to developing more specialised methods for isolating specific compounds.

The examples given in Fig. 2 are from some recent work on classification and interpretation of the fatty acid fingerprinting of five meiofauna taxonomic groups [2]. As may be seen the five meiofauna groups described by 20 non-specific chemical variables are separated into five visually distinct groups. The statistical evaluation of separate models of each class (Table 1) verifies the

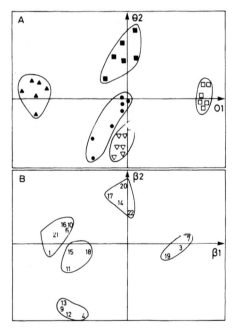

Fig. 2. (A) Object score plot from SIMCA principal components analysis of five meiofauna groups. ●, Nematodes; ■, Polychaetes; ▲, Oligochaetes; ▽, Ostracods; □, Copepods. (B) Variable loading plot for the same data set showing groups of correlated variables [2].

TABLE 1

Class distance between groups of samples (classes) in the meiofauna example (figs. 2A and B). A class distance above 3 is significant

Class	Nematodes	Polychaetes	Oligochaetes	Ostracods	Copepods
Nematodes	1	7.6	12.4	21.6	22.2
Polychaetes	7.6	1	7.4	14.6	43.2
Oligochaetes	12.4	7.4	1	42.2	50.8
Ostracods	21.6	14.6	42.2	1	13.7
Copepods	22.2	43.2	50.8	13.7	1

classification and shows that the five classes may be separated using the fatty acid fingerprinting [2]. These results have been used in an alternative approach as an attempt to identify the feeding patterns of juvenile (stage IV) lobsters (*Homarus gammarus*) [6].

A similar investigation using the fatty acid pattern available from chemical analysis of single fish eggs is given in Fig. 3. This figure shows how fish eggs from day 1, containing no morphological characteristics, and day 8, from two closely related species, cod (*Gadus morhua*) and haddock (*Melanogrammus aeglefinus*), are separated along principal component number 1 and thus may be classified using the pattern of 24 common fatty acids. By computing separate models for each of the classes, using cross validation to determine the number of statistically significant principal components, and using the approximate F-test to define class boundaries the class models have been found to be descriptive of the samples collected [3]. This shows that the fingerprint pattern of fatty acids may be used to taxonomically classify organisms. This has long been recognised in microbiology and is now increasingly being applied in taxonomy concerned with larger organisms.

The variable loading plots may be used to determine the relationship between the variable space and the object space. The variable loading plots in Figs. 2 and 3 show that several of the variables used in both analyses are correlated along the principal components. The circles around groups of variables are only used to indicate groups of correlated variables.

The second principal component in the fish egg example would seem to describe changes occurring during maturation from day 1 to day 8 [3]. The fact that the fatty acid pattern also contains additional information and that fatty acids also may be used to evaluate changes occurring during maturation shows that this compound group contains at least two types of information. A more thorough discussion of these plots has been presented in refs. 2 and 3.

Multivariate chemotaxonomy using phenovariables and pattern recognition methods resembles polythetic group organisation where organisms are placed together which have the greatest number of shared states, and no single state is either essential to group membership or is sufficient to make an organism a member of a group (ref. 1, p. 21). The term phenovariables may be used to describe chemical variables (compound types) which contain both genotypic and phenotypic information.

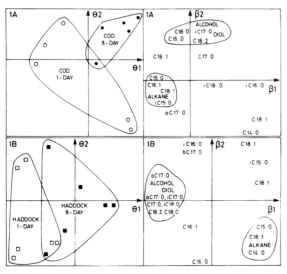

Fig. 3. (A) Object score and variable loading plots of SIMCA principal components analysis of cod eggs from day 1 and day 8. (B) Object score and variable loading plots of haddock eggs from day 1 and day 8 [3].

This definition follows the definition of Sneath and Sokal [1] of phenetic relationships as being similarities based on resemblance of phenetic characters.

Recently a DNA cutting technique combined with liquid chromatographic analysis and multivariate pattern recognition (SIMCA-PLS2) of DNA fragment peak patterns has been applied for taxonomic classification [15]. This shows that multivariate chemotaxonomy is not restricted to using phenovariables. The combined use of genovariables and phenovariables in a multivariate approach to resolving chemosystematic problems is suggested in the chapter on strategies for soft modelling.

4.3 Cladistic descriptions

Taxonomic investigations of biological systems are not only restricted to classification of different groups of organisms, but are as much concerned with determining the connection between families, or the evolutionary development of species [1]. Cladistic descriptions are concerned with genetic systematics and based on being able to resolve genetic similarities and describe evolutionary development. Changes in the DNA and amino acid sequences have been used as cladistic measures. Biochemical evidence is therefore often thought to be a "true indicator" of a cladistic relationship [1]. The development of cladistic models are very often based on phenetic data because these are the only data available. The use of chemical phenovariables, such as fatty acids, in cladistic investigations, i.e. to describe the pathway of evolution, will possibly run into problems because of the complex problem of resolving subtle differences in genetic fatty acid patterns from other influences causing variations. The changes resulting in development of evolutionary differences between species are however often reflected in changes in the patterns of biosynthesis and will therefore be present in the chemical fingerprint of a species. The concept of soft modelling is not restricted to describing discrete class models. Applying some measure of distance (as is available in some of the SIMCA "clones", i.e. class distance based on euclidean discrimination power measure) would allow trees describing similarity between classes to be calculated. Ref. 6 (part II) describes an attempt at ranking groups of potential prey species according to distance from the pattern of fatty acids found in juvenile lobster stomachs. Similar, more refined, approaches would possibly be usable in cladistic descriptions. The use of soft modelling in data analysis of "biochemical" data, e.g. the pattern of chemical compound groups closer to the genetic macromolecules such as secondary metabolites (e.g. alkaloids, sesquiterpenes etc.) to determine cladistic relationships should therefore be an interesting supplement to traditional methods of univariate biochemical analysis.

5 CHEMOSYSTEMATICS AND THE MULTIVARIATE APPROACH

5.1 Chemosystematics

The traditional approach in chemosystematic work where the interest is in investigating changes, differences or effects has been to seek out single chemical compounds containing information on either biochemical quality or fitness or, in pollution studies, some specific biochemical pathway developed by organisms to cope with pollutants [24]. Multiple statistical analysis with simple univariate methods is then used to identify differences between samples. The possible statistical difficulties of attacking chemosystematic problems by this approach are illustrated in Zar (ref. 26, p. 153) and in Wold et al. [13].

As pointed out in the previous section the possibility of using fatty acids in multivariable chemotaxonomic classification does not imply that the distribution, or pattern, of this compound group is not influenced by environmental factors or that all taxonomic divisions may be identified using only non-specific fatty acids. As has been suggested by Dore and Jaubert [27] and shown by Grahl-Nielsen et al. [28] and Grahl-Nielsen and Barnung [29] the fatty acid pattern of fish larvae (*Gadus morhua*), blue mussels (*Mytilus edulis*), shrimp (*Crangon crangon*) and the antenna of periwinkles (*Littorina littorea*) contain information on the environmental conditions to which an organism is subject.

The questions to be asked are how do different types of environments, ecological niches and geographical variations, or pollution regimes, effect the biochemical expression of organisms? In which way is this influence reflected in the fingerprint pattern of phenovariables, how can this pattern be determined and finally is it possible to resolve the genetic difference between species and interpret the different "external" factors (e.g. environment) in the patterns of the phenovariables?

5.2 Multivariate chemosystematics

The use of different compound groups to obtain information on different aspects of genotypic and phenotypic variations in populations of genetically identical or different organism groups reflects the levels of information available in the different groups of compounds.

The genetically controlled macromolecules, such as DNA and RNA, contain information on the *genetic* similarity or difference between taxonomic groups. The same is the case, to some degree, for some products of secondary metabolism, which are traditionally taxonomic identifiers, e.g. specific sesquiterpenes [30]. These compound groups may either be used on their own, as singular identifiers, to classify in genetic taxonomy or, if distribution are available, be analysed by pattern recognition methods.

The chemical products of enzymatic reactions, which are a function of both the genetic code, through the genetically controlled protein synthesis and ecological, environmental etc. induced influences, such as lipids and fatty acids and also the carbohydrates, or even some non-specific primary and secondary metabolites [27,30,31], of abundance, vary in distribution depending on environmental conditions. The patterns of these compound groups, in addition to containing information on the genetic origin, also contain information on the phenotypic variation, because the fluctuation in biosynthesis of these compound groups is influenced by the condition of the organisms and therefore by the environment in which the animal lives.

Resolving the complex influences of the different factors governing the synthesis of the phenovariables makes it necessary to use multivariate methods of data analyses. The essential part is that it is now possible to relate parts of the *pattern structure* of chemical compound groups to genome or phenome and it thus becomes possible to use the chemical compound groups described as phenovariables to identify chemical differences which were previously difficult to obtain.

Multivariate chemosystematics using phenovariables therefore leads to new possibilities in applying a chemical approach to problems which have previously been considered impossible to investigate.

5.3 Samples and the choice of characters

A recurring problem whenever sampling is involved is the question of sample size and/or number of operational taxonomic units (OTUs).

Depending on the level of resolution, i.e. the problem definition, either individuals or collected samples may be used. For multivariate purposes it is advantageous to use as large a number of samples for each class, i.e. type of sample, as possible and between 5–10 samples should be used if classes are to be statistically evaluated.

The data illustrated in Fig. 2 were obtained from samples containing from 1 to 100 organisms. In this meiofauna example the problem essentially was one of investigating feedig patterns of lobsters. The intention was therefore to be able to classify groups of potential prey organisms and not to investigate the variability between species within the groups. In the fish egg example (Fig. 3) the intention was to investigate the taxonomic potential of the approach used and the variability in the fatty acid pattern as representative of the phenovariables. Single fish eggs were used. A traditional chemotaxonomic approach will often require large numbers of individuals. The use of chemical phenovariables increases the sensitivity of taxonomic investigations and permits resolution of differences between samples consisting of single organisms.

Application of simple chemical derivatization and analyses on small amounts of material and whole organisms, e.g. fish eggs, gives organism specific data. If parts of organisms are used these must be representative [1] or have functions related to the problem, e.g. liver samples may be of interest in studying the effect of toxic substances. The important point in multivariate chemosystematics is to select chemical compounds which reflect both the genetic code and the phenome of the organisms being investigated.

Combining knowledge from problems solved by using ordinal methods in environmental, ecological and biochemical research [32] and especially SIMCA and SIMCA-PLS2 [13–18] there is now the possibility of obtaining chemosystematic interpretations by resolving the data from phenovariables into a genotypic and a phenotypic component. Strategies for such research will be discussed in the next section and areas of development in ecological and environmental biology and chemistry will be used to illustrate the possibilities.

6 SELECTED STRATEGIES FOR SOFT MODELLING

In biology multivariate data analysis has long been an integral part of systematic research. In chemistry the application of multivariate methods of data analysis is rapidly increasing.

The emphasis in this section is put on attempting to sort the area of soft modelling/PLS in chemosystematics into some of the type of problems which are typical in chemosystematics. The examples given and the suggestions for approaches represent possibilities which seem logical in the development of soft modelling in multivariate chemosystematics.

6.1 Two blocks with different sets of variables

In many biological investigations, such as in determining "quality" or "fitness" (e.g. biorearing) of organisms there is interest in decomposing and interpreting genotypic and phenotypic variation. Another area where an interest in biological response is important is toxicology. In these cases it is possible to define two blocks of variables. The first example describes an approach where data on both geno- and phenovariables are available. The interest in this case is to determine the part of the structure in the phenovariable matrix which may be considered to come from genetic variation. The second example is analogous to quantitative structure–activity relationships (QSAR) using SIMCA-PLS. Here the interest is in determining how a mixture of chemical compounds can be used to predict a biological effect and how the soft modelling approach allows interpretation of the way in which these variables influence the response measured.

6.1.1 Geno- and phenovariables together

One approach which may be used if data from both geno- and phenovariables are available is analogous to that which has been described by Esbensen et al. [33] in geochemical prospecting and Bisani et al. [34] in geochemical whole rock characterisation. These papers describe deconvolution of complex geochemical data into components with different geogenetic origin using PLS techniques.

In this approach the data from analysis of genovariables containing information on genotypic and phenovariables containing both the genotypic and the phenotypic variation may be decomposed. By using the different patterns of information available in the different groups and investigating the transfer of "information", i.e. in the sense of variance, between the different blocks of variables [10] the genetic variation, as expressed in patterns of genovariables, may be compared (calibrated/regressed) to a block of phenovariable compounds making it possible to investigate which part of the variance in the phenovariables comes from variance in the genovariables [33]. After the statistically significant number of PLS components have been determined the residuals of the phenovariables may be obtained. By subtracting the genetic structure present in the genovariable block from the total phenotypic variance present in the phenovariable block the variance which is not a function of genetic variance may be obtained and may be modelled for systematic structure and interpreted by a regular principal component/factor analysis.

A possible problem of such an approach, and thus a suggestion for caution whenever using the

PLS method, is where the two data matrices have variance dimensions which are not sequentially interrelated. An example of this may be where the environmental influence on the variation of the phenovariables is much larger than the genetical variance. As the PLS algorithms usually work cross-validation is made on the Y block. Choosing the genovariable block as the Y block, i.e. to be able to determine the statistically significant number of PLS vectors describing the genetic structure, and thus the dimensions of variance in the phenovariable (X block) which should be subtracted, seems logical. If the environmentally induced variance structure in the phenovariable matrix (X block) dominates, the first PLS vector(s) in the X block, even though they are "tilted" by the weights from the Y block, will not represent genetic variance, but be some form of "tilted" environmental vectors. A subsequent subtraction of this structure will therefore lead to residual matrices which do not represent the environmental variation, but some confounded intermediate structure. An alternative is then to switch around the referece set and the test set (see section on TPLS).

In all circumstances it is suggested, for the purpose of verification that the structure subtracted represents genetic variation, to apply a regular principal components analysis, or similar analysis, to both blocks of data. Interpretation of the phenovariable loading plots, from PLS and principal components analysis together, will suggest how well the genetic structure found in the phenovariable block by PLS is represented by the "objective" principal components analysis. A principal components regression, vector from Y block to vector from X block, will indicate which structure in the phenovariable block best conforms with the genovariable structure.

6.1.2 Composition activity regression (CARE)

It is generally accepted, although seldom admitted, that the relationship between environmental or ecological situations and biological condition is a function of many interrelated, often correlated, and seldom well described factors.

The traditional approach to such problems is to select one, or at most the sum of a few chemical constituents, as being representatives of a group [25]. The prime example is benzo[*a*]pyrene. Although it is highly active in several biological senses, the amount of research into this one compound reflects its relative importance to all other chemical and/or physical factors or even to all the other polycyclic aromatic hydrocarbons (PAHs) determining the biological activity of a mixture of pollutants.

Analogous to QSAR methods where the PLS approach has been used to determine quantitative relationships between chemical structure and biological activity [35–38], and as a natural extension of this, principal components analysis/regression and PLS may be used to find the connection between environmental/ecological situations and the biological condition expressed in the chemistry or biology of organisms.

Selecting as the Y block data on toxicity, carcinogenity or other related effects determined by using the actual mixtures from the effluent and as X block the concentrations of chemical variables in the mixture it is possible not only to determine which chemical compounds are important, but also to decide on how combinations of these compounds influence the measured effect. Including other types of variables in the X blocks makes it possible also to account for other factors than just chemical. Fig. 4 is taken from some work we are presently carrying out to investigate this approach on the connection between mutagenicity, measured by number of revertants of a TA98 bacterial strain, by a mixture of 26 PAHs analysed in air samples. In addition the amount of particulate material sampled has been included as a variable. The samples have been collected from above ovens where incineration of garbage and combustion of oil, coal, wood and pressed wood is done. The samples presented represent a selected group where only wood and pressed wood burning has been included. The analysis has been made on the concentrations with scaling to unit variance. Fig. 4 shows predicted mutagenity using two PLS components against measured mutagenity. Two statistically significant PLS components account for 99.1% of the variance in the Y block (mutagenicity) and 65.7% of the variance in the X block (PAH concentrations). The asterisks represent

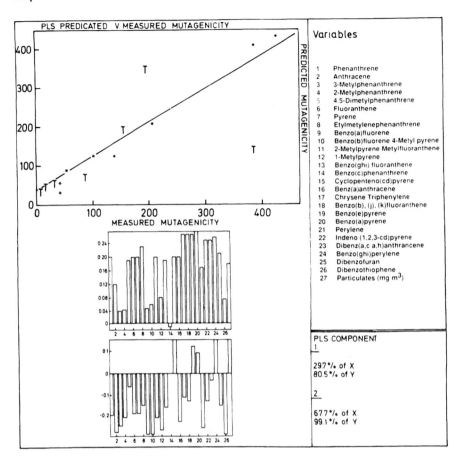

Fig. 4. The upper part of the figure shows PLS predicted versus measured mutagenicity together with the regression line. The equation for the line is: Mutagenicity(PLS pred.) = 33116 + 0.77 Mutagenicity(meas.). $R^2 = 0.74(++)$. The asterisks represent samples which were included in the modelling and the T's represent test samples. The bar diagrams show variable loadings for the 26 polycyclic aromatic hydrocarbons and the amount of particulates. The amount of variance explained by the two PLS components in each block used is given in the lower right corner.

samples that were part of the modelling (training set) and the T's represent samples that were left out of the modelling. The two T samples that are seen to deviate strongly from the regression line represent samples that are classified close to the 99% boundaries of the class model indicating that these samples differ in PAH composition from those on which the model has been based. The predicted values for these two samples are therefore expected to be anomalous. The lowest part of Fig. 4 shows two bar diagrams of the normalised variable loadings of the two PLS components from the X block (PAH block) together with the amount of variance accounted for in the Y block and the X block for each PLS component. The loadings show which variables are best correlated with the PLS components in the X block. A high positive or negative loading shows that the variance structure of this PAH correlates positively or negatively with mutagenicity.

The results show that the predictive relationship between mutagenicity as measured by number of revertants and the 26 PAHs is well defined.

The bar diagram in Fig. 4 show that although variable 20 (benzo[a]pyrene) has the highest positive loading along PLS component 1, and therefore is the most "important" variable, there are several other PAHs with almost equal importance. For the second PLS component (explaining 38% of the X block and 18.6% of the Y block) variables 15 (cyclopentene[cd]pyrene), 19 (benzo[e]pyrene), 20 (benzo[a]pyrene), 24 (benzo[ghi]perylene) and 27 (particulates) show positive correlation with mutagenicity whereas the others show negative correlation. Since there are two statistically significant PLS components both dimensions explain some systematic structure in the Y block, i.e. the mutagenicity block. The relatively low explanation power of the two PLS components in the PAH block compared to the mutagenicity block suggests that there are factors responsible for variation in the PAH block which are not represented in the mutagenicity block and thus the relation between the X block and the Y block is not fully described.

As pointed out in Section 6.1.1 the representation of variable structure (e.g. by loadings) in the **X** and the **Y** block matrices when using PLS might not allow straightforward interpretation of the importance of the variables. The separation of the amount of structure described by the PLS vectors in the X block which may be accounted to influence from the Y block and/or to structure from the X block is, with present programs, confounded. Geladi and Kowalski [19] describe an orthogonalisation of the PLS vectors subsequent to PLS analysis. This will give mutually independent scores for the objects in a PLS analysis, but will still not resolve the problem of confounding.

The strong positive correlation for all PAH variables along the first PLS component does suggest that this component in the X block describes a quantitative correlation, i.e. that the PAHs have a dominating variance structure which is a result of covariance of concentrations of PAHs in the samples. If the first PLS component is taken to represent the concentration dependence of the mutagenicity reaction the second component may be taken to describe that the five previously mentioned variables work by some additional mechanism to give a mutagenic reaction. Variable 14 (benzo[c]phenanthrene) is seen not be important for either of the two PLS dimensions. The interpretation of what the second PLS component might describe warrants a designed experiment where the factors which might be important may be controlled.

The traditional approach to determine which variables are important in cause and effect studies is to establish a "background", or reference, level so that quantitative differences may be determined. This is often difficult because of the many factors which must be controlled and the fact that a background, or reference situation might be different from the real situation. Using the CARE approach in analysis of cause and effect relationships allows the important variables to be determined from their influence on describing the relationship between the two blocks of variables in a real situation. From a preliminary analysis, such as the one described, subsequent hypothesis formulation is facilitated and design of experiments to accomplish a more detailed interpretation is necessary.

Returning to the implications put forward at the beginning of this section the CARE concept applied in environmental studies allows for a more detailed description, using all available information simultaneously, of the environmental conditions resonsible for the effect that is measured and thus will more accurately describe the real situation.

6.2 Two blocks with the same set of variables

The previous section discussed the situation where two different sets of variables were to be related. Often it is not possible to analyse two different sets of variables. In environmental analysis and pollution monitoring where the same set of chemical variables are measured at different locations or in biological laboratories where advanced instrumentation for analysis of DNA is not available the possibility to obtain specific information and to resolve differences between sample groups may still be of interest. In the following section two approaches to decomposing two blocks of the same variables describing two different conditions will be discussed. The first example suggests the possibility of being able to obtain residual structures of a test set matrix by comparing this to a model of a reference set matrix using a regular principal components analysis approach. The second example describes a similar situation and discussed the use of a transposed version of the PLS approach.

6.2.1 Phenovariable residual patterns

The situation in biological laboratories is quite

often that analytical data are only available for phenovariables, e.g. fatty acids. To be able to resolve the information available in phenovariable data is therefore important. An example may be investigation of the changes in patterns of phenovariables in groups of organisms which have been treated differently.

The simplest way to obtain a measure of what may be considered the reference set, e.g. genotype, is by calculating the average value for each variable. Assuming that the average represents the reference set, i.e. background, may have serious disadvantages. In most cases the reference set sample group will have structure because not all samples are from genetically identical populations. It is therefore necessary to model the genetic, or from other sources, derived structure in the phenovariables on a controlled set of samples (reference group) before subtracting this from the data set of affected samples. This may be done using soft models such as the SIMCA principal components analysis approach.

The data from the reference class are first modelled by principal components analysis using a statistically significant number of principal components to describe the structure in this data set. The objects from the treated set are then fitted to this model. The residual matrix (E) of the treated set will then contain any remaining structure which is not present in the reference set. Fig. 5 describes this procedure in blocks. When the genetic structure is subtracted from the treated data set any remaining systematic variance may be taken to describe the phenotypic variation specific for the condition which the test group has been subject to. Analysing this structure by principal components analysis will allow differences between phenotypic expressions to be modelled and interpreted. If some form of design matrix is available, e.g. by use of some form of experimental design, the effect of the variation of the variables may be described and verification of the subtraction of reference structure may be done by using principal components regression or PLS. Finally, assuming that interest is in determining the response of an organism to some external factor, by using multivariate calibration, the phenovariable structure which is specific for the treatment may be calibrated against intensity of environmental stress (CARE). Interpretation will allow investigation of the chemical mechanisms by which organisms adjust.

6.2.2 Transposed partial least squares (TPLS)

In the case when the same variables have been measured on different types (classes) of samples (objects) there will be two matrices with different objects, but the same variables. PLS is most often applied to two matrices with the same objects, but different variables [39], i.e. one set of independent variables (X block) and one set of dependent variables (Y block). To comply with a similar situation in the present case one possible way is to simply transpose the matrices involved (see Fig. 6), and operate the PLS straightforwardly on the objects as variables and the variables as objects. For those familar with cluster and/or factor analysis this may be viewed as an *R*-mode analysis

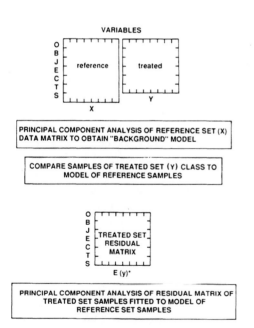

Fig. 5. Phenovariable resolution by principal components analysis. The structure in the reference set is modelled by principal components analysis. After the number of statistically significant principal components have been determined the test set data may be "fitted" to the class model. By subtracting the structure described by the principal components in the reference set remaining variance will represent structure specific for the test set.

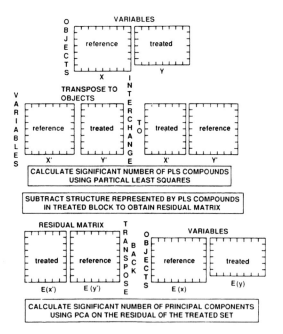

Fig. 6. Transposed PLS. Block description of the "inverse" TPLS2 analysis. The illustration uses reference set and treated set. The original data matrix is transposed, the significant number of PLS components is determined and the residual matrix is obtained. The transposed residual matrix is transposed back to normal. Analysis of the residual matrix by principal components analysis gives information on variance which has not been subtracted through TPLS.

where the characters (e.g. chemical variables) are classified according to the organisms that possess them.

Philosophically, if there is a pattern structure in the chemical data describing samples, the samples should reflect this. Describing the characters by using the samples is thus just a change in interest and should be as valid an approach in soft modelling as in other methods of multivariable data analysis. The operation and interpretation of this approach is not straightforward.

Depending on the objective of the investigation the TPLS approach may be used to obtain either a residual matrix or a predictive relationship.

Assuming that two sets of samples have been analysed for the same set of variables and that the interest is in determining the remaining variance structure in the test set which is not described by some structure in the data of the reference set the TPLS idea may be a way to obtain a regressional description giving a residual matrix which contains specific information. In this case the number of cross-validated significant PLS components (which with present implementations depends on the structure in the Y block) should be made to depend on the systematic structure in the matrix of the reference set (see Fig. 7). This may rather cryptically be described as an inverse transposed application of the PLS algorithm. If the remaining matrix, i.e. the residual (**E**), in the test set, after having constructed the cross-validated significant number of PLS components according to the structure in the matrix of the reference set, contains systematic structure then this may subsequently be analysed by regular principal components analysis/regression or CARE *after* having transposed the matrix back to normal. The same precautions exist here as were pointed to in the part on PLS applied to data sets containing both pheno- and genovariables. The TPLS approach should therefore be useful for obtaining residual matrices when the effect from the environmental conditions are smaller than the variance in the genotypic variance.

The suggested approach represents a relatively simple form of transposing the standard PLS method in order to cope with the problems posed by only having phenovariable data available. A similar concept is presently being investigated in environmental research, i.e. as a PLS approach to chemical mass balance modelling where the same set of pollutants are analysed at several sources in the environment and the aim is to determine the contribution from different sources. The interest here is to investigate the possible use of the predictive properties of PLS in determining the contributions from several sources to some recipient.

Wold has described the use of "discriminant PLS", i.e. adding a design matrix containing 1's and 0's as Y block, to model and characterize classes of samples. By then fitting samples of unknown source composition to the model the relative contribution from different sources may be obtained. Using this approach allows a straightforward determination of the relative contribution of different sources to a recipient (40) and then, knowing the mass of a sample, the mass balance contribution from different sources can be calculated. The TPLS approach, using the source profiles as Y block, would potentially allow the actual mass balance to be identified directly from the data.

Transposing the matrix results in a form which mathematically is equivalent to the regular matrix as long as the matrix has not been treated in some manner, e.g. average subtracted or data scaled. The interpretation of a transposed analysis is not straightforward, thus the suggested transposing back to an interpretable form in the case where the residual matrix is of interest.

6.3 Hypothesis formulation and experimental design

Experimental planning [41,42] should be an integrated and important part of chemosystematics. The importance of the design of an experiment becomes obvious whenever questions as to the validity of a data analysis are presented. This fact has long been recognised in agricultural research where cause and effect investigations are of interest. In environmental chemistry/biology relatively less emphasis seems to have been put on experimental planning.

The reluctance to acknowledge the fact that every experiment, intentionally or otherwise, is based on some hypothesis about the situation to be investigated leads to problems at two levels. When a hypothesis has not been clearly stated the factors which may be important for the investigation can not be determined. Mathematical and statistical data analysis, with validation of conclusions, is not possible without stating the assumptions and hypotheses which have been investigated.

Experimental designs have most often been associated with subsequent ANOVA [41] analysis. Principal components analysis and PLS may be viewed as extensions of the ANOVA approach including the description of covariance structure by linear combinations of the variations. Where ANOVA gives a number of the variance accounted for by a factor, or interaction between factors, principal components analysis provides for both visual and numerical interpretation of the variance struc tures.

Experimental planning is not only concerned with designs [43]. The process of determining which factors (e.g. chemical variables) are important, and how it is possible, by designing a relevant experiment, to obtain answers to the questions asked, may be the most important part of experimental planning.

With the use of soft models in data analysis, hypothesis formulation is made easier. The CARE example (Section 6.1.2) illustrates a typical situation. From the analysis of available data two statistically significant PLS components have been found. The previous discussion suggests the formulation of at least two hypotheses:

(1) Does the strong correlation of PAHs in PLS component 1 represent a quantitative covariance which is not related to mutagenicity (i.e. the PLS "confounding" effect)? The alternative hypothesis is that the quantitative correlation in the PAH matrix is represented by the first PLS component in the mutagenicity matrix, i.e. that there is, as is most often assumed, a connection between the amount of PAH and the mutagenic properties of a mixture, and:

(2) Does the covariance structure described by the second PLS component reflect an interaction, or secondary, mechanism by which the positively correlated PAHs give mutagenic reaction and the negatively correlated PAHs counteract this reaction? One alternative hypothesis here is that the second PLS component reflects two different types of mutagenic reaction pathways from two types of source profiles.

From the literature it is known that different bacterial strains respond differently to different types of PAHs. From a principal components analysis of the samples used we have indications that they do in fact represent different types of combustion profiles.

By designing an experiment where the samples represent different profiles and concentrations of PAHs the variance/covariance structures, represented by principal components analysis/PLS components, and the connection between these and the mutagenic effect could be interpreted. We are presently working to obtain more data on this particular problem and it is our intention to discuss this approach more thoroughly in a separate paper.

7 CHEMOMETRICS AND CHEMOSYSTEMATICS

The present strong increase in the use of multivariable data analysis in chemistry reflects the fact that chemists are interested in obtaining information from the large amount of data they are supplied with. The somewhat holistic approach suggested by using all available data, e.g. data analysis of mixtures instead of single compounds, is increasingly being appreciated by the chemist educated in the tradition of one-variable-at-a-time. The systematic biologist having been trained in multivariable data-analysis should appreciate this trend.

It is important to be aware of the fact that many of the applications and adjustments suggested in chemically related fields have already been proposed in other fields, e.g. biosystematics, where multivariable data analysis has long been accepted. Chemists interested in approaches to multivariable data analysis should use this source of information.

Multivariate chemosystematics and chemometrics are concerned with characterisation (classification and identification) and with the more general topic of relating studies of chemical expression to that of "extra"-chemical relationships. With the development of multivariable data-analytical methodology in chemistry the traditional approach in chemotaxonomy has been expanded to also include the use of genotypic less specific compound groups, such as lipids and fatty acids. This opens the way for interpreting chemical expressions of genotype and phenotype. Quantitative composition activity regression (QCARE) between amounts of variables in mixtures and measures of effect, e.g. toxicity or mutagenicity, greatly enhances the usefulness of chemistry in environmental studies. Depending on the objective of a study there are several ways to resolve the structure in data sets. Transposed PLS analyses of classes allows the variance in one class to be regressed and thus subtracted from another class on which the same variables have been measured. This gives the option of being able to analyse by principal components analysis only the variance which is of interest for specific chemosystematic questions.

8 FUTURE DEVELOPMENTS OF SOFT MODELLING IN CHEMOSYSTEMATICS

Since disjoint modelling by principal components analysis (SIMCA) is directed at optimal description of structures in single data matrices and PLS is directed towards optimal prediction/regression between two blocks of data the two methods have advantages in different situations in chemosystematics.

In regression analysis the use of residual analysis has been a popular means of interpreting if a model describes the systematic variation in data set [26,40]. This paper has suggested several areas where the application of residual matrices from soft models and PLS should be of interest for obtaining information in chemosystematics and environmental chemistry. To arrive at a form of approach for the PLS which may be more generally applicable, e.g. in chemosystematics and environmental chemistry/biology, various modifications of the standard manner in which the PLS algorithms are implemented are needed. Some of the problems (e.g. PLS "confounding") with the present implementations have been suggested. Cross-validation [20] in PLS is presently only made on the **Y** matrix. For interpretational purposes it would be of interest to be able to determine the amount of systematic structure explained by the PLS components in the **X** matrix also.

A future development of the principal components analysis/PLS family algorithms in chemosystematics and environmental chemistry will depend on the availability of a versatile "toolkit" which may be used to design procedures which may be used in turn to obtain the information of interest for a particular investigation.

A closer connection between experimental design and the use of disjoint modelling has been advocated by Wold for many years. Development of integrated procedures where experimental design is incorporated into the process of investigation is likely to increase in the future. The possibility of obtaining fuzzy classification in soft disjoint modelling would facilitate the use of these methods for chemosystematic purposes.

9 GLOSSARY OF TERMS USED

Cladistic	Evolutionary development of species
Genovariables	Characters describing genetic origin
Latent variable	Linear combination of original variables
Matrix	Ordering of data in tabular form
Orthogonal	Synonymous with independent, i.e. describing different variance structures
Phenovariables	Characters describing both genetic origin and phenotypic variation
Polyphenic	Analytical method yielding many characters
Polythetic	Classification based on many characters
Principal component	Linear combination of variables describing the largest variance in a data set
Principal components regression	Regression between principal components from analysis of separate data sets or between a principal component and a variable
Partial least squares	Algorithms for obtaining optimal regression between linear combinations of variables in two blocks of data
Residual matrix	Residual values of variables after model has been fitted
Transposed matrix	Switching treatment of variables and objects

10 ACKNOWLEDGEMENTS

I would like to thank Dr. K. Esbensen and Dr. P. Geladi from the Norwegian Computing Center for discussions on the PLS method and S. Nordenson and K. Kolset from the Senter for Industrial Research for discussions on the sorting of the different topics covered. O. Grahl-Nielsen was the one who introduced me to multivariate data analysis and has since then been a source of inspiration and ideas. The suggestions and results presented have been obtained by playing with the SIMCA-PLS2 package from Umeå University, The MAGIC package from Pattern Recognition Systems in Bergen and the UNSCRAMBLER package developed by Dr. H. Martens and available from CAMO A/S, Trondheim, Norway. Any mistakes or misconceptions are mine alone and information regarding these wil gratefully be accepted.

REFERENCES

1 P.H.A. Sneath and R.R. Sokal, *Numerical Taxonomy*, W.H. Freeman, San Francisco, CA, 1973.
2 N.B. Vogt and H. Knutsen, SIMCA pattern recognition classification of five infauna taxonomic groups using non-polar compounds analysed by high resolution gas chromatography, *Marine Ecology: Progress Series*, 26 (1985) 145–156.
3 N.B. Vogt, E. Moksness, S.P. Sporstøl and H. Knutsen, Multivariate chemotaxonomy on 1 and 8 day old cod

(*Gadus morhua*) and haddock (*Melanogrammus aeglefinus*) eggs using gas chromatography and SIMCA principal component analysis, *Marine Biology*, 92 (1985) 173-182.

4 F. Peladan, J.C. Turlot and H. Monteil, Discriminant analysis of volatile fatty acids produced in culture medium: A novel approach to the identification of *Pseudomonas* species, *Journal of General Microbiology*, 130 (1984) 3175-3182.

5 B. Søderstrøm, S. Wold and G. Blomquist, Pyrolysis-gas chromatography combined with SIMCA pattern recognition for classification of fruit bodies of some Ectomycorrhizal *Suillus* species, *Journal of General Microbiology*, 128 (1982) 1773-1784.

6 H. Knutsen and N.B. Vogt, A supplementary approach to identifying feeding patterns of juvenile lobsters (stage IV) using chemical analysis and pattern recognition by the method of SIMCA. Part I&II, *Journal of Experimental Marine Biology and Ecology*, 89 (1985) 109-131.

7 W.T. Williams and M.B. Dale, *Fundamental Problems in Numerical Taxonomy*, Academic Press, London, 1965, pp. 37-69.

8 D.L. Massart and G. Hoogewijs, The impact of chemometrics on microchemical analysis, *Pure and Applied Chemistry*, 55 (1983) 1861-1868.

9 R.A. Johnson and D.W. Wichern, *Applied Multivariate Statistical Analysis*, Prentice Hall, Englewood Cliffs, NJ, 1982, p. 594.

10 I.T. Jolliffe, *Principal Component Analysis*, Springer Verlag, New York, 1986, p. 271.

11 H.H. Harman, *Modern Factor Analysis*, University of Chicago Press, Chicago, IL, 1976, p. 485.

12 H.C. Romesburg, *Cluster Analysis for Researchers*, Lifetime Learning Publications, London, 1984, p. 334.

13 S. Wold, C. Albano, W.J. Dunn III, U. Edlund, K. Esbensen, P. Geladi, S. Hellberg, E. Johansson, W. Lindberg and M. Sjøstrøm, Multivariate data analysis in chemistry, in B.R. Kowalski (Editor), Proc. NATO Adv. Study Institute of Chemometrics, Cosenza, Italy, Reidel, Dordrecht, 1983, pp. 17-97.

14 S. Wold, C. Albano, W.J. Dunn III, K. Esbensen, S. Hellberg, E. Johansson, W. Lindberg and M. Sjøstrøm, Modelling data tables by principal components and PLS: class patterns and quantitative predictive relations. *Analusis*, 12 (1984) 477-485.

15 S. Wold, C. Albano, W.J. Dunn III, K. Esbensen, P. Geladi, S. Hellberg, E. Johansson, W. Lindberg, M. Sjøstrøm, B. Skageberg, C. Wikstrøm and J. Øhman, Multivariate data analysis: Converting chemical data tables to plots, *VIIth International Conference on Computers in Chemical Research and Education, Garmisch-Partenkirchen, June 10-14, 1985*, p. 166.

16 T. Naess and H. Martens, Comparison of prediction methods for multicollinear data. Communications in Statistics, part B. *Simulations and Computations*, 14 (1985) 545-576.

17 S. Wold, C. Albano, W.J. Dunn III, K. Esbensen, S. Hellberg, E. Johansson and M. Sjøstrøm, Pattern recognition: finding and using regularities in multivariate data, in H. Martens and H. Russwurm (Editors), *Food Research and Data Analysis*, Applied Science Publishers, London, 1983, pp. 147-189.

18 S. Wold, A. Ruhe, H. Wold and W.J. Dunn III, The collinearity problem in linear regression, the partial least squares (PLS) approach to generalised inverses, *SIAM Journal of Scientific and Statistical Computing*, 5 (1984) 735-743.

19 P. Geladi and B. Kowalski, Partial least squares: a tutorial, *Analytica Chimica Acta*, 185 (1986) 1-17.

20 S. Wold, Cross validatory estimation of the numbers of components in factor and principal component models, *Technometrics*, 20 (1978) 397-406.

21 H. Martens, Multivariate calibration—Quantitative interpretation of non-selective chemical data, Dr. Techn. Thesis, Technical University of Norway, Trondheim, 1985.

22 A. Lorber, L.E. Wangen and B.R. Kowalski, A theoretical foundation for the PLS algorithm, *Journal of Chemometrics*, 1 (1987) 19-31.

23 T. Bergan, P.A.D. Grimont and F. Grimont, Fatty acids of Serratia determined by gas chromatography. *Current Microbiology*, 8 (1983) 7-11.

24 N. Shaw, *Lipid Composition as the Guide to the Classification of Bacteria*, Academic Press, New York, 1974, pp. 63-105.

25 J.M. Neff, *Polycyclic Aromatic Hydrocarbons in the Aquatic Environment Sources, Fates and Biological Effects*, Applied Science Publishers, London, 1979, pp. 102-152.

26 J.H. Zar, *Biostatistical Analysis*, Prentice-Hall, Englewood Cliffs, NJ, 1974, p. 44.

27 J.C. Dore and J.N. Jaubert, Exemple d'étude systematique de chromatogrammes d'essences naturelles par analyse factorielle des correspondances, *Parfums, Cosmetiques, Aromes*, 61 (1985) 79-85.

28 O. Grahl-Nielsen, O.M. Kvalheim and K. Øygard, SIMCA multivariate data analysis of blue mussel components in environmental pollution studies, *Analytica Chimica Acta*, 150 (1983) 145-152.

29 O. Grahl-Nielsen and T. Barnung, Variations in the fatty acid profile of marine animals caused by environmental and developmental changes, *Marine Environmental Research*, 17 (1985) 218-221.

30 J. Mann, *Secondary Metabolisme*, Oxford University Press, Oxford, 1980, p. 99.

31 A.B. Smith III, A.M. Belcher, G. Epple, P.C. Jurs and B. Lavine, Computerised pattern recognition: A new technique for the analysis of chemical communication, *Science*, 228 (1985) 175-177.

32 A. Thielemans and D.L. Massart, The use of principal component analysis as a display method in the interpretation of analytical chemical, biochemical, environmental and epidemiological data, *Chimia*, 39 (1985) 236-242.

33 K. Esbensen, E. Johansson and S. Wold, Recognising polygenetic (geochemical) data patterns by multivariate soft bilinear modelling using selective partial least squares techniques, in D. Edwards and A. Høskuldson (Editors), *Symp.*

in Anvendt Statistikk, 26–27 January 1983, NEUCC, RECAU&RECKU, Copenhagen, 1983, pp. 111–133.

34 M.L. Bisani, D. Faraone, S. Clementi, K.H. Esbensen and S. Wold, Principal components and partial least squares analysis of the geochemistry of volcanic rocks from the aeolian archipelago, *Analytica Chimica Acta*, 150 (1983) 129–143.

35 B. Norden, U. Edlund and S. Wold, Carcinogeneity of polycyclic aromatic hydrocarbons studied by SIMCA pattern recognition, *Acta Chemica Scandinavica Series B*, 32 (1978) 602–608.

36 S. Hellberg, A multivariate approach to QSAR, *Dr. phil. Thesis*, Umeå University, Umeå, 1986, 222 pp.

37 S. Wold, W.J. Dunn III, U. Edlund, S. Hellberg and J. Gasteiger, Multivariate structure–activity relationships between data from a battery of biological tests and an ensemble of structure descriptors: The PLS method. *Quantitative Structure–Activity Relationships*, 3 (1984) 131–137.

38 W.J. Dunn III and S. Wold, Structure–activity analyzed by pattern recognition: The asymmetric case, *Journal of Medicinal Chemistry*, 23 (1980) 595–599.

39 R.W. Gerlach, B.R. Kowalski and H.O.A. Wold, Partial least squares path modelling with latent variables, *Analytica Chimica Acta*, 112 (1979) 417–421.

40 R. Vong, 1987, personal communication.

41 G.E.P. Box, W.G. Hunter and J.S. Hunter, *Statistics for Experimenters. An Introduction to Design, Data-analysis and Model Building*, Wiley, New York, 1978, p. 653.

42 G. Kateman and F.W. Pijpers, *Quality Control in Analytical Chemistry*, Wiley, New York, 1981, p. 276.

43 C.K. Bayne and I.B. Rubin, *Practical Experimental Designs and Optimization Methods for Chemists*, VCH Publishers, Deerfield Beach, FL, 1986, p. 205.

Chapter 24

Multivariate Analysis in Geology and Geochemistry: an Introduction

H.J.B. BIRKS

Botanical Institute, University of Bergen, P.O. Box 12, N-5027 Bergen (Norway)

CONTENTS

1 Introduction 340
2 What are multivariate data? 341
   2.1 Examples 341
3 What is multivariate data analysis? 342
   3.1 Types of statistical data analysis 342
   3.2 An example of exploratory data analysis 344
4 Why perform multivariate data analysis? 345
5 What is the geometrical basis of multivariate data analysis? .. 346
6 What types of methods are available for exploratory multivariate data analysis? ... 347
   6.1 Classification, partitioning or clustering methods 347
   6.2 Scaling, ordination or geometrical representation methods . 348
   6.3 Discrimination, group separation methods 348
   6.4 Multivariate correlation 348
   6.5 Coordinated uses 348
7 What types of questions can be answered by multivariate data analysis? ... 349
8 What is the role of multivariate data analysis in geological and geochemical research? ... 352
9 Acknowledgements 352
   References 352

1 INTRODUCTION

Geological and geochemical data are commonly quantitative, complex and multivariate, consisting of many variables or attributes recorded in a large number of samples or objects. Multivariate data analysis provides useful tools for the exploratory analysis of such data. This tutorial aims to provide a basic conceptual background and general introduction to exploratory multivariate data analysis in the earth sciences by considering seven questions: (1) What are multivariate data? (2) What is multivariate data analysis? (3) Why perform multivariate data analysis? (4) What is the geometrical basis of multivariate data analysis? (5) What types of methods are available for exploratory multivariate data analysis? (6) What types of questions can be answered by multivariate data analysis? (7) What is the role of multivariate data analysis in geological and geochemical research? I shall illustrate answers to some of these questions with simple examples using a small published data set.

There are now many excellent texts about multivariate data analysis. At an introductory level the books by Davis [1], Everitt [2,3], Marriott [4] and Manly [5] are all strongly recommended. The underlying mathematical theory for the various methods discussed here is covered in depth by, for example, Cooley and Lohnes [6], Everitt and Dunn [7], Gordon [8], Jöreskog, Klovan and Reyment [9], Lebart, Morineau and Warwick [10], Jambu and Lebeaux [11] and Pimental [12]. Useful sources of relevant computer programs include refs. 10 and 11. Recent important developments in the exploratory analysis of multivariate data were discussed by Barnett [13], whereas Reyment, Blackith and Campbell [14] provide an up-to-date appraisal of the subject, together with many fascinating examples of multivariate data analysis in, for example, geology, biology, morphometrics, linguistics, serology and palaeontology. Other useful sources of information and reviews of particular techniques of direct relevance to geologists and geochemists include the Sage University Paper Series on Quantitative Applications in the Social Sciences, the Concepts and Techniques in Modern Geography series, the journals *Mathematical Geology* and *Biometrics*, and the six volumes published to date of the *Encyclopedia of Statistical Sciences* [15].

This tutorial draws extensively on many of these sources. In the interests of readability, no attempt is made here to refer particular ideas or suggestions to specific sources. Several of the ideas and concepts discussed are presented in greater detail in refs. 8, 16 and 17. The tutorial is based on the introductory lecture at the Multivariate Statistical Workshop for Geochemists and Geologists held at Ulvik, Hardanger, Norway, in June 1986.

2 WHAT ARE MULTIVARIATE DATA?

2.1 Examples

The answer is: just about all complex data sets from the real world! Multivariate data sets consist of a set of many objects ($\equiv$ individuals, sampling units, samples), with each object described by several variables ($\equiv$ characters, attributes). Consider the following examples.

A micropalaeontologist estimates the frequencies of different types of microfossils in samples of sediment at different but known depths through a vertical borehole sequence. The data consist of counts, often expressed as percentages, for the various fossil types (variables) in the different stratigraphic samples (objects) (see ref. 16 for examples).

A geochemist analyses inorganic elements of organic compounds in a series of rock specimens. Here the data matrix consists of estimates of chemical composition (variables) in many rock specimens (objects) (see ref. 1 for examples).

A sedimentary petrologist analyses the grain-size classes of different sediments accumulating in contrasting environments, such as beaches, river channels, shallow-water bays and deep-water basins. The data here consist of grain-size classes (variables) for different environmental settings (objects) (see refs. 1 and 9 for examples).

Although seemingly different, these three examples have several features in common. Each data set consists of, say n, objects, described by several, say m, variables. We can represent these multivariate data as a data matrix $\mathbf{X}$ with m columns and n rows, with each matrix element, X_{ik}, representing the estimated value for variable k in object i.

The elements within matrix $\mathbf{X}$ can be values for different types of variables. There are four common types of variables (see refs. 3, 8, 11 and 18 for further details). Variables may be quantitative ($\equiv$ numerical, continuous), with values of real numbers, such as measurements on a continuous scale (e.g., weight, concentration, temperature) or on a constrained scale (e.g., percentages). A distinction is sometimes made between ratio and interval quantitative variables [8]. For ratio variables, zero is clearly defined and thus for a ratio variable with two values, X_1 and X_2, the expression X_1/X_2 is well defined. For interval variables (e.g., temperatures), zero is not well defined but the expression $(X_1 - X_2)$ is useful. The same data transformations may thus not be appropriate for ratio and interval variables.

Variables may be qualitative multi-state variables, namely variables with discrete states, so that each object belongs to one but only one variable state, e.g., eye colour or rank. Qualitative variables can be of two types. They can be nominal, where there is no ordering, so-called "disordered multistate variables", such as colour, or they can be ordinal, where there is an underlying order, so-called "ordered multistate variables", such as scaled or ranked variables (e.g., small, medium, large). Variables may be dichotomous or binary, so-called two-state variables. These include the presence ($+$) or absence ($-$) of a particular variable. Variables may be conditionally present variables. These are variables whose relevance in a particular study depends entirely on the presence of other variables. For example, the size of punctae in a microfossil carapace is clearly dependent on the presence of punctae. Considerable care must be taken in coding such variables (see refs. 8, 11 and 18 for details).

In many geological studies, variables of mixed type, so-called mixed data, are not uncommon. For example, in palaeoecology, a given set of objects may be most appropriately described by a combination of quantitative, qualitative and dichotomous variables. Such data are important and should be analysed in their entirety. Many investigators seem reluctant to analyse mixed data, and may either discard all the non-quantitative variables or reduce all the variables to dichotomous form, thereby losing potentially important geological information. Gower has developed powerful means of analysing mixed data that are of considerable potential in geology (see refs. 3, 8 and 18 for details).

An example of qualitative multivariate data that will be analysed later is given in Table 1. The data are the percentages of all households surveyed in 16 European countries (objects) that had 20 food types (variables) in the house at the time of the survey (from ref. 19). In addition, the area, population, annual percentage increase in population and the latitudes of the northernmost and southernmost parts of each country are given in Table 1.

3 WHAT IS MULTIVARIATE DATA ANALYSIS?

Until recently, statistical data analysis was viewed solely as something in which a specific hypothesis, often a null hypothesis, or some formal model was stipulated and defined, parameters were estimated for the model and the hypothesis was rejected or accepted. This is the basic approach of formal, classical or confirmatory data analysis.

Although this may be the ideal, with its emphasis on hypothesis testing and falsification, its close parallels to the ideas of Karl Popper about scientific progress being made by hypothesis rejection, and its roots in probability theory, there is an increasing acceptance that there is more to statistical data anaysis than hypothesis testing [8,13,14].

3.1 Types of statistical data analysis

Three broad activities can now be recognized within the general term of statistical data analysis [16]. First, there is hypothesis testing or confirmatory data analysis, with its emphasis on testing specific hypotheses, estimating parameters, studying specific properties of the data and studying data from well designed structured investigations such as controlled experiments. There are obvious limitations to this approach when considering complex multivariate data sets from the real world. Many of the assumptions of formal statistics, such as random and independent samples and multivariate normality of data, are difficult to sustain in geological studies.

Second, there is model building. This attempts to simulate patterns or processes in the real world in quantitative terms. Models can never prove anything: they can only indicate what models are inadequate descriptions of particular patterns or processes. Model building can be useful in polarizing thoughts, posing specific questions and generating testable hypotheses [16,17]. Models can be of an explanatory type that attempt to understand causal mechanisms and processes and of an empirical type where little or no attempt is made to understand the underlying processes and the emphasis is on modelling patterns. Both types can, in certain instances, be useful; in other instances,

TABLE 1

European food and geographical data

Food data (from ref. 19) showing the percentages of all households with various foods in house at time of questionnaire, and geographical data (compiled by C.J.A. Birks from various sources). International country abbreviations are used throughout.

	D	I	F	NL	B	L	GB	P	A	CH	S	DK	N	SF	E	IRL
Ground coffee	90	82	88	96	94	97	27	72	55	73	97	96	92	98	70	13
Instant coffee	49	10	42	62	38	61	86	26	31	72	13	17	17	12	40	52
Tea or tea bags	88	60	63	98	48	86	99	77	61	85	93	92	83	84	40	99
Sugarless sweeteners	19	2	4	32	11	28	22	2	15	25	31	35	13	20	–	11
Packaged biscuits	57	55	76	62	74	79	91	22	29	31	–	66	62	64	62	80
Packaged soup	51	41	53	67	37	73	55	34	33	69	43	32	51	27	43	75
Tinned soup	19	3	11	43	25	12	76	1	1	10	43	17	4	10	2	18
Instant potatoes	21	2	23	7	9	7	17	5	5	17	39	11	17	8	14	2
Frozen fish	27	4	11	14	13	26	20	20	15	19	54	51	30	18	23	5
Frozen vegetables	21	2	5	14	12	23	24	3	11	15	45	42	15	12	7	3
Fresh apples	81	67	87	83	76	85	76	22	49	79	56	81	61	50	59	57
Fresh oranges	75	71	84	89	76	94	68	51	42	70	78	72	72	57	77	52
Tinned fruit	44	9	40	61	42	83	89	8	14	46	53	50	34	22	30	46
Jam (shop)	71	46	45	81	57	20	91	16	41	61	75	64	51	37	38	89
Garlic cloves	22	80	88	15	29	91	11	89	51	64	9	11	11	15	86	5
Butter	91	66	94	31	84	94	95	65	51	82	68	92	63	96	44	97
Margarine	85	24	47	97	80	94	94	78	72	48	32	91	94	51	25	
Olive/corn oil	74	94	36	13	83	84	57	92	28	61	48	30	28	17	91	31
Yoghurt	30	5	57	53	20	31	11	6	13	48	2	11	2	–	16	3
Crispbread	26	18	3	15	5	24	28	9	11	30	93	34	62	64	13	9
Area (km² ×10³)	248	300	552	34	31	2.6	245	91	83	41	455	44	324	337	505	70
Population (×10⁵)	619	548	521	134	97	3.5	559	95	75	64	81	50	39	47	348	30
Annual % population increase	1.0	0.8	0.9	1.2	0.6	0.7	1.1	0.9	0.5	1.3	0.8	0.8	0.8	0.6	1.0	0.4
Latitude of northernmost part (°)	55	47	51	54	51	50	57	42	48	48	70	58	72	71	42	55
Latitude of southernmost part (°)	47	38	42	51	49	49	50	37	46	46	55	54	58	60	36	51

they can be misused. There is much debate currently about the potential value of model building. Pielou [20] summarized the present position in ecology as follows: "Models are ... displayed with little or no effort to link them with the real world. As a result, the whole body of ecological knowledge and theory has, I think, grown top-heavy with models ... Models are certainly not useless ... But too much should not be expected of them. Modelling is only a part, and a subordinate part, of ecological research".

The third major activity is hypothesis generation or exploratory data analysis. In this, an initial investigation of a data set is carried out with no explicit ideas about the patterns or "structure" within the data. Different data-summarizing techniques are used to detect what, if any, structure exists within the data. It is, in effect, "data detective work". No assumptions are made about random, independent sampling or multivariate normality. It is often useful in suggesting hypotheses about the underlying processes by which the patterns in the data arose by summarizing and aiding the interpretation of large data sets that are of interest in their own right.

This type of informal exploratory data analysis is often an invaluable first stage in multivariate data analysis, because multivariate data commonly have many different properties and it is clearly undesirable to concentrate too early on formal tests related to one or two specific properties of the data [8]. Techniques of exploratory multivariate data analysis can be safely used with almost all data sets as long as one realizes that what the methods are doing is only detecting and revealing structure or patterns in the data set at hand. If questions are asked about the underlying populations of which the data are a sample, formal statistical or confirmatory data analysis is required.

Confirmatory and exploratory analyses of multivariate data are therefore distinct, and have different underlying philosophies, working methods and texts (see ref. 17 for further details). Ideally, an exploratory analysis should be followed by a confirmatory analysis in which hypotheses are proposed about the underlying populations or causal processes. These hypotheses can then be tested using independent data collected in the same way but not used to formulate the hypotheses. In practice, this is rarely done because of the inherent complexity of many geological data and the inevitable difficulties of collecting further data in a truly identical way. In addition, geological data are often too complex for the formulation of realistic and useful null hypotheses that could lead to tractable test procedures. There is, however, growing interest and activity in establishing and strengthening links between exploratory and confirmatory analysis [8,17]. This should be beneficial to all, for as Tukey [21] commented, "we need both exploratory and confirmatory." Such links are not easy to establish, but they represent an area for potentially important future developments in multivariate data analysis [8,17].

The bulk of multivariate data analysis in geology and geochemistry thus falls within the broad category of exploratory data analysis, and is designed to assist in detecting patterns within complex data sets as an aid to generating hypotheses about underlying causal mechanisms.

3.2 An example of exploratory data analysis

As a simple example of exploratory data analysis, consider a data matrix of types of protein consumption from 9 types of food (variables) in 25 countries in Europe (objects). We can explore the patterns in this data set (from Gabriel in ref. 13) by means of a covariance biplot (Fig. 1) which permits a simultaneous analysis and geometrical display of the objects and the variables (see refs. 2, 8 and 22, Gabriel in ref. 13, and Gower and Digby in ref. 13 for details of biplots). Fig. 1 shows a two-dimensional biplot with a goodness-of-fit of 0.85. As it captures 85% of the total variability, the main features of the data should be accurately portrayed in this two-dimensional summary. It shows that in these two dimensions variables such as cereals and milk have high variances (long vectors), that animal protein sources are highly correlated (small angles between their vectors) and are thus grouped together, that cereals are negatively correlated with animal protein (angle between vectors about 180°) and that fruit and vegetables are poorly correlated with both cereals

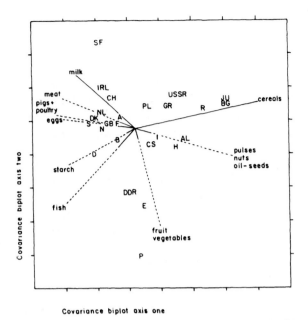

Fig. 1. Covariance biplot of protein consumption in 25 European countries. The positions of the 9 protein types (variables) are shown as vectors radiating from the centroid of the first and second axes. The 25 European countries are represented by their international car symbols. Redrawn from Gabriel in ref. 13.

and meat (angles of about 90°). The relationships between the variables and the objects are also clearly displayed, with countries in eastern Europe closely asociated with cereal protein sources, western and northern Europe countries associated with animal protein and Mediterranean and southern European countries correlated with nut, fruit and vegetable protein sources. This type of biplot is easily derivable from a standard principal components analysis [2,8,22] and permits a simple but powerful, combined graphical display of the objects and the variables in terms of their correlations, variances and interrelationships.

4 WHY PERFORM MULTIVARIATE DATA ANALYSIS?

Although multivariate data analytical techniques are applicable in many disciplines (e.g., geology, geochemistry, biology, ecology, archaeology, anthropology, geography, linguistics, medicine, astronomy, sociology, psychology), the aims of multivariate data analysis are often very similar in these different subjects. There are at least seven major reasons why exploratory data analysis can be useful in the geosciences.

(i) *Data simplification and reduction*. Multivariate geological data may be so complex that it is difficult for the brain to assimilate and comprehend them. This is particularly so in studies with automated methods for data collection and recording, for example in analytical geochemistry, and in digital processing of morphometric data. Multivariate data analysis aids the detection of relationships and patterns within complex data and allows you to "see the wood for the trees" and thus to isolate "signal" from "noise" in data sets.

(ii) *Display and description of complex data*. These can be aids in interpretation, can assist in detecting features in the data that might otherwise escape notice if the data were only tabulated and can point to new and potentially profitable lines of research.

(iii) *Hypothesis generation and prediction*. If the data are summarized and consistent patterns emerge, these results may allow predictions to be made and hypotheses generated to explain the detected patterns. The human mind is very flexible and can often think up explanations for any pattern, even if the patterns have no significance statistically. It is therefore important to test any explanatory hypothesis by subsequent confirmatory data analysis (see ref. 17 for details).

(iv) *Data exploration*. It can be useful to analyse an existing data set to see where new, further data could, most usefully, be collected.

(v) *Communication*. By simplifying and summarizing patterns in complex data, multivariate data analysis can aid the communication of results from large and often incomprehensible data sets in a form readily understandable to non-specialists.

(vi) *Repeatability*. Multivariate data analytical techniques are powerful and can analyse large data sets rapidly and precisely in a repeatable manner, assuming that different investigators use the same data and mathematical method.

(vii) *Explicit aims and assumptions*. Being

mathematical, such techniques force the investigator to be explicit about his or her aims, interests, methods and assumptions. Others can then follow the arguments precisely and this can assist in developing new ideas or in proving the initial arguments to be erroneous. It is better to be explicit but wrong than inexplicably right! As Walker [23] elegantly commented: "the more orthodox amongst us should at least reflect that many of the same imperfections are implicit in our own cerebrations and welcome the exposure which numbers bring to the muddle which words may obscure".

5 WHAT IS THE GEOMETRICAL BASIS OF MULTIVARIATE DATA ANALYSIS?

The major aim of exploratory multivariate data analysis is to detect the basic structure or patterns in complex data sets. Graphical representation is therefore essential and an important rule in all exploratory analysis is, never forget the graphs. Although all multivariate data analytical techniques are rooted in linear and matrix algebra, they provide a means of representing and summarizing features of complex data in a simple geometrical way (see refs. 2, 7 and 11 and Gower and Digby in ref. 13 for further details).

There are many graphical ways of presenting univariate or bivariate data, such as histograms, scatter plots and box and whisker plots. Given data for two variables (for example, ground coffee and instant coffee in Table 1), we can plot them geometrically with one axis representing ground coffee and the other axis representing instant coffee, and then position the 16 countries in this two-dimensional space depending on their values for ground coffee and for instant coffee (Fig. 2). The distance between any pair of object (country) points is a function of how similar the countries are in terms of their coffee types — the closer they are, the more similar they are; the further apart they are, the more dissimilar. We can readily calculate the squared Euclidean distances between, for example, West Germany and Italy in this two-dimensional "coffee" space, as (from Table 1)

$$d^2_{D,I} = (90 - 82)^2 + (49 - 10)^2 = 1585$$

Thus the Euclidean distance $\equiv \sqrt{1585} = 39.81$. This large distance contrasts, for example, with an Euclidean distance of 4.12 between Sweden and Denmark.

If we consider three variables (e.g., ground coffee, instant coffee and tea), the idea is the same and each country can be plotted in a three-dimensional space as a function of their values for these three variables. The relative positions are a measure of similarity between countries. For example, the squared distance between West Germany and Italy in these three variable dimensions is (from Table 1)

$$d^2_{D,I} = (90 - 82)^2 + (49 - 10)^2 + (88 - 60)^2 = 1729$$

whereas the squared Euclidean distance between Sweden and Denmark is

$$d^2_{S,DK} = (97 - 96)^2 + (13 - 17)^2 + (93 - 92)^2 = 18$$

The Euclidean distances of 41.58 and 4.24 between West Germany and Italy and between Sweden and Denmark, respectively, provide a geo-

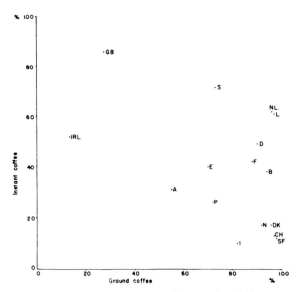

Fig. 2. Plot of the percentages of houses in 16 European countries with ground coffee and instant coffee at the time of survey. Data from Table 1. Country abbreviations follow the international car symbols.

metrical representation of the similarities between pairs of objects (countries).

Clearly, for m variables a geometrical representation with m axes is required. We cannot visualize an m-dimensional model, but the geometry for two or three dimensions is the same as for m dimensions. The Euclidean distance between object i and object j is thus

$$d_{ij} = \sqrt{\sum_{k=1}^{m} (X_{ik} - X_{jk})^2}$$

The concepts of similarity and dissimilarity and of distance or proximity are fundamental to almost all multivariate data analyses. Each pair of objects will have a measure, d_{ij}, of how close object i is to object j. Such a proximity measure can either be a similarity or a dissimilarity measure. In general, dissimilarities are considered because they can be regarded as distance measures in geometrical representations, as the more dissimilar a pair of objects are, the greater is the distance between them. Dissimilarities, d_{ij}, can be readily transformed into similarities, s_{ij}, and vice versa, by, for example, $s_{ij} = c - d_{ij}$, where c is a constant, $d_{ij} = \sqrt{(1 - s_{ij})}$, or $d_{ij} = 1 - s_{ij}$.

The concept of dissimilarity or distance is the basis, explicitly or implicitly, of nearly all exploratory multivariate analytical techniques such as principal components analysis, principal coordinates analysis, correspondence analysis, biplots, cluster analysis, canonical variates analysis, linear discriminant analysis and canonical correlation analysis. The implicit distance measure is readily and often unwittingly altered by data transformations, standardization, scalings or normalizations. It has been widely shown in ecology [24] and pollen analysis [25] that such data manipulations can greatly alter the results obtained. Much critical, a priori thought is therefore required before attempting any data transformations.

There is therefore a simple geometrical basis to multivariate data analysis, and the basic aim of exploratory data analysis is to allow us to explore this m-dimensional space. As we cannot do this directly, it is necessary to try to simplify this model into a small number of dimensions that can be studied. There are various such methods for attempting this, and they are discussed in the next section.

6 WHAT TYPES OF METHODS ARE AVAILABLE FOR EXPLORATORY MULTIVARIATE DATA ANALYSIS?

If we imagine the data set in Table 1 as a 20-dimensional (variable) geometrical space in which the 16 objects (countries) are positioned as a function of their values for the 20 variables, there are four broad types of methods available for exploring these multivariate data.

6.1 Classification, partitioning or clustering methods

Here the aim is to partition or classify the n objects into groups of broadly similar composition, namely to group objects that are close to each other and hence are similar to each other and dissimilar to other objects or groups of objects within the m-dimensional space. As used here, classification refers to the process where any group structure within the data set is sought and described. The basic question therefore is, do the objects fall into a number of groups ($\equiv$ clusters) so that objects within a group are, in some way, more similar to each other than they are to objects in different groups? This use of the term classification contrasts with its use in some mathematical literature to refer to the procedure of deciding to which of a number of groups a new object should be assigned. This latter procedure is here called identification or assignment. Classification, as used here, does not presuppose the number of groups within a data set and is therefore distinct from the process of dissection, where the aim is to dissect a data set into convenient sectors of, for example, approximately equal size (see refs. 3, 8, 11, 18 and 19 for further discussions of the concept of classification).

Many clustering methods are available, including agglomerative hierarchial techniques, iterative relocation methods, polythetic divisive techniques, density search techniques and mode analysis, overlapping clustering, methods of mixtures and fuzzy-set clustering (see refs. 3, 6, 8, 18 and 19 for details).

6.2 Scaling, ordination or geometrical representation methods

Here the aim is to represent the m-dimensional space in a low number of dimensions, usually two or three, so that the original distances between objects are reproduced as accurately as possible in the low-dimensional space and hence the inter-object dissimilarities are represented geometrically. These can then be examined visually. This is effectively a problem of dimension reduction and involves projecting the original object points on to the best fitting, in a mathematical sense, lines or planes. Many techniques are available for this, including principal components analysis, principal coordinates analysis, correspondence analysis, biplots, non-linear mapping and non-metric multidimensional scaling (see Gower and Digby in ref. 13 and refs. 1, 2, 6, 8–11 and 18 for details).

6.3 Discrimination, group separation methods

If we know, a priori, that some objects are from one group, others from another group and yet others from a third group, we may ask what is the most effective way, in a mathematical sense, of telling the three groups apart and how distinct are the groups within the m-dimensional space when allowance is made for the inherent within-group variation. Linear or quadratic discriminant analysis provides the mathematically optimal solution for two groups, whereas multiple discriminant analysis and its geometrical relative canonical variates analysis are appropriate for the analysis of three or more groups (see refs. 1, 4, 5, 12 and 14 for details). The identification, assignment and allocation of unknown objects to the known groups within the m-dimensional space are closely related to these discrimination procedures.

6.4 Multivariate correlation

If we have two sets of multivariate data, say food data with m_1 variables and geographical data with m_2 variables for the same n objects (see Table 1), these data can be visualized as two separate multidimensional spaces, one an m_1-dimensional food space and the other an m_2-dimensional geographical space. The questions then arise of what the correlation between these two spaces is and whether there are any common patterns between the two. Canonical correlation analysis and its related technique of redundancy analysis (see refs. 4, 5, 12 and 14 for details) provide powerful means of exploring patterns of multivariate correlation in two or more spaces.

6.5 Coordinated uses

The basic geometrical representation of the m-dimensional data discussed in the previous section provides a useful conceptual basis for much of exploratory multivariate data analysis. There are now a large number of useful and powerful methods available for exploratory analysis that are appropriate for particular problems in the geosciences. In the early days of the subject, great emphasis was placed on one or two methods such as Q-mode factor analysis or average-link cluster analysis, which, because of computational availability and mathematical limitations, were used on just about all data sets. Sometimes this use was appropriate, sometimes it was not. Now, thanks to the many mathematical developments in multivariate data analysis by, for example, J. Aitchison, J.-P. Benzécri, N.A. Campbell, K.R. Gabriel, A.D. Gordon, J.C. Gower, D.M. Hawkins, M.O. Hill, J.B. Kruskal, R.A. Reyment, R. Sibson and C.J.F. ter Braak [17], we have an arsenal of powerful multivariate data analytical tools appropriate to different problems and data sets in the geological sciences.

Data from the real world are commonly so different and there are often so many ways of looking at such complex multivariate data that any single line of attack may be inadequate. With the many recent developments in multivariate analysis, we are now at the stage of being able to take advantage of using different methods, each with their own strengths and weaknesses, to explore a given data set, to complement each other and hopefully to provide new insights into the data set of interest.

Reyment et al. [14] commented that, "it is by no means rare to see that one or other method is considered by a particular author to be in some respect superior to another when the capacities of the method considered as being inferior have been inadequately explored or understood. This is a transient phase of incomplete dissemination of the basic theories ... Unfortunately this transient phase is still with us ... A great deal of unnecessary polemics are devoted to the question of method superiority." The approach I recommend is a coordinated, critical use of several methods (see refs. 7, 11, 14 and 16 for examples), thereby exploiting different ways of exploring multivariate data in an attempt to provide insights into the data set and into the problem at hand.

7 WHAT TYPES OF QUESTIONS CAN BE ANSWERED BY MULTIVARIATE DATA ANALYSIS?

The questions that can be answered by exploratory data analysis primarily concern the patterns that exist within a complex data set, for example, the frequencies of microfossils in borehole sequences, geochemical data from oil samples, petrological data from rock collections and morphometric data for fossils. Answers to questions about what patterns exist in a given data set provide a basis for interpretation. However, it is necessary to check that the patterns detected by one method are not a mathematical artefact of that particular method. It is important, therefore, to use several methods and to look for consistency of results. Such consistency gives us confidence that the results are reflecting basic patterns or structure in the data.

A minimum-variance ($\equiv$ sum-of-squares) agglomerative hierarchical cluster analysis of the 16 European countries defined by the 20 food variables (Table 1) using Euclidean distance as the distance measure results in four major partitionings of countries (Fig. 3). These are (1) Finland, Norway, Denmark and Sweden; (2) Great Britain and Ireland; (3) West Germany, Belgium, Luxembourg, France, Switzerland and The Netherlands; and (4) Portugal, Austria, Spain and Italy.

Correspondence analysis (see refs. 8–11 and 26 for details) provides a low-dimensional geometrical representation of the objects and variables together and is therefore, for many purposes, one of the most powerful techniques for the exploratory analysis of multivariate data. Correspondence analysis of the European food data (Fig. 4) captures 58.6% of the total inertia in the first two dimensions. The Scandinavian countries (Sweden, Finland, Norway, Denmark) are positioned together along with crispbread, frozen vegetables, frozen fish, ground coffee and instant potatoes. Spain, Italy and Portugal contrast with the Nordic countries on the first axis and are characterized by garlic and olive oil. The second axis separates Great Britain, Ireland and The Netherlands from the Nordic and southern European countries, along

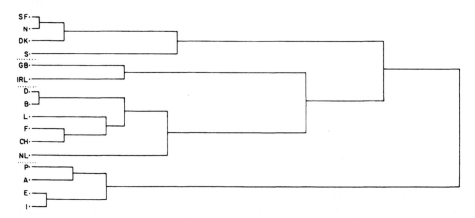

Fig. 3. Dendrogram showing the results of minimum-variance agglomerative cluster analysis of the 16 European countries defined by the 20 food variables listed in Table 1. Country abbreviations follow the international car symbols.

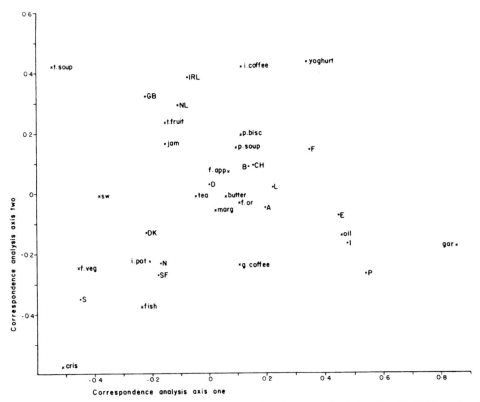

Fig. 4. Correspondence analysis results for the analysis of the European food data listed in Table 1, showing the positions of the 16 countries and 20 food variables on the first and second correspondence analysis axes. Country abbreviations follow the international car symbols. Food abbreviations: i. coffee = instant coffee; t. soup = tinned soup; t. fruit = tinned fruit; p. bisc = packaged biscuits; p. soup = packaged soup; f. app = fresh apples; f. or = fresh oranges; marg = margarine; sw = sweeteners; f. veg = frozen vegetables; i. pot = instant potatoes; g. coffee = ground coffee; fish = frozen fish; oil = olive/corn oil; gar = garlic; cris = crispbread.

with tinned soup, instant coffee, yoghurt and tinned fruit. In the centre of the plot, West Germany, Switzerland, Belgium, Luxembourg, Austria and France are closely positioned together, along with butter, tea, margarine, fresh oranges, fresh apples and packaged soups.

If a minimum-spanning tree (see refs. 2, 8 and 11 for details) is superimposed on the correspondence analysis results and the plot is rotated and reflected, there is some correspondence with the geography of Europe (Fig. 5; see also Gower and Digby in ref. 13). The Nordic countries are linked together and located in the north, Great Britain, Ireland and The Netherlands occupy a westward position, Portugal, Italy, Spain and, to a lesser extent, France are located in the south and Germany and adjacent countries are positioned together in the centre of the plot.

There are clearly some patterns that are consistently detected by both the partitioning and scaling methods, suggesting that these patterns represent the major structure within the data set. Hypotheses about the possible causes of these observed patterns, for example, between the foods found in kitchens in Nordic countries and Great Britain, can now be formulated.

Patterns within and between the four groups suggested by the minimum-variance cluster analysis can be explored by canonical variates analysis of the 4 groups defined by the 20 food variables. The results (not presented here) show no overlap between the groups and very little within-group variation.

Correlations and patterns between the 16 countries defined by 20 food variables and the same 16 countries defined by 5 geographical and popula-

Chapter 24

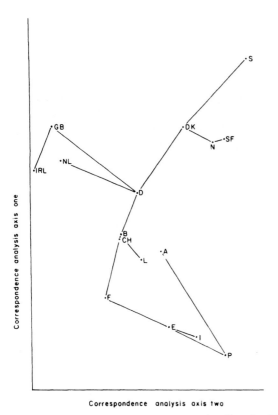

Fig. 5. Results of Fig. 4 rotated and reflected with a minimum-spanning tree fitted to the full 15-dimensional correspondence analysis results. Country abbreviations follow the international car symbols.

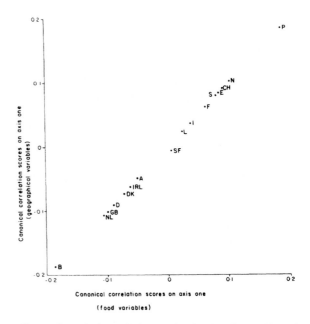

Fig. 6. Canonical correlation results showing the position of the canonical scores for the 16 European countries for the first canonical correlation of food variables and geographical variables. Country abbreviations follow the international car symbols.

tion variables (Table 1) can be explored by means of canonical correlation analysis. The first canonical correlation is 0.988 and is positively correlated with ground coffee, garlic, olive oil, and crispbread and negatively associated with instant coffee, sweeteners, tinned soup, biscuits, frozen vegetables, fresh apples, tinned fruit and jam amongst the food variables. Of the geographical variables, area of country and percentage annual population increase have positive correlations with the first canonical correlation, whereas population size and southernmost latitude have negative correlations. The pattern of scores for the 16 countries (Fig. 6) contrasts countries of large populations and small areas at southern or central latitudes (Belgium, The Netherlands, Great Britain, West Germany, Ireland) in which tinned soup, jam, sweeteners, biscuits, instant coffee, frozen vegetables, tinned fruit and fresh apples are common with countries of large areas, small populations and/or high annual population increase in which crispbread, garlic and/or ground coffee are frequent (Portugal, Norway, Sweden, France, Spain). The multivariate correlation between the two sets of variables is by no means simple, and appears to reflect, in part, patterns between population density, annual population increase and particular foods. Subsequent canonical correlations suggest further interactions between geography and food types, contrasts between population size and pre-prepared versus fresh food, and correlations between rate of population increase and certain food types.

These examples are presented here solely to illustrate the types of questions and analyses that can be attempted by exploratory data analysis. They do not attempt to explore fully the many interesting aspects of this data set, only to suggest possible approaches to detecting patterns within a multivariate data set.

8 WHAT IS THE ROLE OF MULTIVARIATE DATA ANALYSIS IN GEOLOGICAL AND GEOCHEMICAL RESEARCH?

There is a very great potential role for the careful and critical use of appropriate multivariate data analytical techniques in many areas of geological and geochemical research where data sets are commonly multivariate. However, like everything, these techniques should be used with care and thought. Such methods are a means to an end, not an end in themselves. The main intellectual challenge is in the interpretation and evaluation of the results. All that multivariate data analysis does is to help us by rotating or summarizing our data so that hopefully we can see any patterns within them more clearly. Multivariate data analysis is a tool, just like a petrological microscope, an X-ray diffractometer, a scanning electron microscope or a gas chromatograph. They are all tools that help us explore the immense, often bewildering, diversity of geological patterns in the real world. The role of multivariate data analysis is to help us explore complex data.

Despite this potentially important role in many branches of geology and geochemistry, many investigators seem reluctant either to adopt multivariate techniques or to exploit the many important recent mathematical developments that are of direct relevance to geologists and geochemists. Examples of such neglect include recent developments in the analysis of "closed" compositional data in petrology, geochemistry and palaeoecology (see ref. 27 for an important book on new and important methods for the statistical analysis of compositional data); robust time-series analysis; analysis of asymmetric relationships; boot-strapping, jack-knifing, and other validation techniques; randomization procedures; and size–shape analysis (see ref. 17 for a review of recent developments).

There is an immense potential for multivariate data analysis in geology and geochemistry. This role is only just beginning to be exploited to its fullest.

9 ACKNOWLEDGEMENTS

I am indebted to Christopher Birks, Hilary Birks, John Line and Annechen Ree for assistance, to Michel Mellinger and other participants at the Ulvik Multivariate Statistical Workshop for Geologists and Geochemists for comments and to Olav Kvalheim and Rolf Manne for encouraging me to prepare this tutorial.

REFERENCES

1 J.C. Davis, *Statistics and Data Analysis in Geology*, Wiley, New York, 2nd ed., 1986, 646 pp.
2 B.S. Everitt, *Graphical Techniques for Multivariate Data*, Heinemann Educational Books, London, 1978, 117 pp.
3 B.S. Everitt, *Cluster Analysis*, Heinemann Educational Books, London, 2nd ed., 1980. 136 pp.
4 F.H.C. Marriott, *The Interpretation of Multiple Observations*, Academic Press, London, 1974, 117 pp.
5 B.F.J. Manly, *Multivariate Statistical Methods. A Primer*, Chapman and Hall, London, 1986, 159 pp.
6 W.W. Cooley and P.R. Lohnes, *Multivariate Data Analysis*, Wiley, New York, 1971, 364 pp.
7 B.S. Everitt and G. Dunn, *Advanced Methods of Data Exploration and Modelling*, Heinemann Educational Books, London, 1983, 253 pp.
8 A.D. Gordon, *Classification: Methods for the Exploratory Analysis of Multivariate Data*, Chapman and Hall, London, 1981, 193 pp.
9 K.G. Jöreskog, J.E. Klovan and R.A. Reyment, *Geological Factor Analysis*, Elsevier, Amsterdam, 1976, 178 pp.
10 L. Lebart, A. Morineau and K.M. Warwick, *Multivariate Descriptive Statistical Analysis. Correspondence Analysis and Related Techniques for Large Matrices*, Wiley, New York, 1984, 231 pp.
11 M. Jambu and M.-O. Lebeaux, *Cluster Analysis and Data Analysis*, North-Holland, Amsterdam, 1983, 898 pp.
12 R.A. Pimental, *Morphometrics: The Multivariate Analysis of Biological Data*, Kendall Hunt, Dubuque, 1979, 276 pp.
13 V. Barnett (Editor), *Interpreting Multivariate Data*, Wiley, Chichester, 1981, 374 pp.
14 R.A. Reyment, R.E. Blackith and N.A. Campbell, *Multivariate Morphometrics*, Academic Press, London, 2nd ed., 1984, 233 pp.
15 S. Kotz and N.L. Johnson (Editors-in-Chief), *Encyclopedia of Statistical Sciences*, Wiley, New York, 1982–..., 6 volumes published so far.
16 H.J.B. Birks and A.D. Gordon, *Numerical Methods in Quaternary Pollen Analysis*, Academic Press, London, 1985, 317 pp.
17 H.J.B. Birks, Recent and possible future mathematical developments in quantitative palaeoecology, *Palaeogeography, Palaeoclimatology, Palaeoecology*, 50 (1985) 107–147.

18 G. Dunn and B.S. Everitt, *An Introduction to Mathematical Taxonomy*, Cambridge University Press, Cambridge, 1982, 152 pp.
19 J.A. Harigan, *Clustering Algorithms*, Wiley, New York, 1975, 351 pp.
20 E.C. Pielou, The usefulness of ecological models: a stocktaking, *Quarterly Review of Biology*, 56 (1981) 17–31.
21 J.W. Tukey, We need both exploratory and confirmatory, *American Statistician*, 34 (1980), 23–25.
22 C.J.F. ter Braak, Principal components biplots and alpha and beta diversity, *Ecology*, 64 (1983) 454–462.
23 D. Walker, Quantification in historical plant ecology, *Proceedings of the Ecological Society of Australia*, 6 (1972) 91–104.
24 I. Noy-Meir, D. Walker and W.T. Williams, Data transformations in ecological ordination. II. On the meaning of data standardization, *Journal of Ecology*, 63 (1975) 779–800.
25 I.C. Prentice, Multidimensional scaling as a research tool in Quaternary palynology: a review of theory and methods, *Review of Palaeobotany and Palynology*, 31 (1980) 71–104.
26 M.J. Greenacre, *Theory and Applications of Correspondence Analysis*, Academic Press, London, 1984, 364 pp.
27 J. Aitchison, *The Statistical Analysis of Compositional Data*, Chapman and Hall, London, 1986, 416 pp.

Multivariate Analysis in Geoscience: Fads, Fallacies and the Future

RICHARD A. REYMENT

Paleontologiska Institutionen, Box 558, S-751 22 Uppsala (Sweden)

CONTENTS

1 Introduction 354
2 Fads in multivariate analysis 355
3 Fallacies 355
4 Future of multivariate analysis 356
    4.1 The closure problem 357
    4.2 Atypical observations 357
    4.3 Robust estimation in multivariate analysis 360
5 From geochemistry to geology 361
    5.1 Multivariate analysis of strain 362
    5.2 Growth and shape as deformations 364
    5.3 Soft and hard modelling 365
6 Conclusions 365
7 Acknowledgements 365
    References 366

1 INTRODUCTION

Perhaps more than any other branch of mathematical statistics, multivariate analysis has been surrounded with an aura of cultism, such that various methods have attracted groups of devotees, and users of some particular procedure can often be heard to swear by its superior properties in relation to other methods. The claims and counter-claims for standard methods of R-mode and Q-mode analysis are oft-occurring features of this kind. One might well ask what the reason for this situation can be.

Although all statistical innovations must perforce have their roots in practice, the subject of multivariate statistics is, for several reasons, a special case. The earliest history of multivariate analysis is inextricably connected with the development of quantitative procedures in psychology; for example, the concept of "vectors of the mind". The method of factor anaysis grew out of this line of thought.

A consequence of the extensive lay-interest in the application of a complicated quantitative procedure to a diffusely defined problem is that interpretations are often made on grounds that are scientifically challengeable.

A source of uncertainty lies with the ready

availability of mammoth computer programs at professional computing centres. One the input requirements have been mastered, vast series of calculations, with or without direct relevance to a particular problem, are produced. It can indeed be unfortunate if the user of the program is poorly oriented with regard to what has been done and whether the program used is suitable for the data. It is frequently found that computer packages contain errors, which can be of various degrees of seriousness. In other cases, necessary theoretical requirements are not detected by the package and results of little scientific value may result.

2 FADS IN MULTIVARIATE ANALYSES

By fads are meant computing methods that catch on for a short time, are widely used and then are dropped from the repertoire. One might say that the linear discriminant function, when first introduced, bordered on fad status, but it has by no means been dropped. It is now one of the standard techniques of multivariate analysis, although the prominence once accorded the method has diminished significantly.

In the geological world, we have seen the rise and decline of tailored regressional techniques such as trend surface analysis and the wide and often inadmissible use of principal component factor analysis with all kinds of data. The 1960s saw a remarkable symbiosis between ambitious earth scientists and, in the U.S.A., the category known as "student programmers". There was, to a certain extent, an industry aimed at publishing computer programs of variable validity and insight.

A current fad involves correspondence analysis [28], a useful graphically combined numerical method for Q- and R-mode analyses of a contingency table.

A critical analysis of the fad situation reveals that many people, even professional statisticians, are unaware of how some new method is related to existing procedures. This problem becomes all the more acute if the new technique is embellished with a specialized terminology which masks familiar concepts and bestows upon the new method an aura of uniqueness. Correspondence analysis is just such a technique. It was developed with complete disregard for established nomenclature and for previous work along the same lines [28]. Thus, the fundamental contributions of the eminent statisticians Hirschfeld (Hartley) [1], Fisher [2] and, later in a similar field, Gabriel [3] were not noticed by the people promoting the method. Correspondence analysis has only recently begun to catch on in anglophone circles. In a fashion with which we are now familiar, it is being hailed in some quarters as the cure-all for all our problems. To statisticians, such as Greenacre [4], who has recently published the first English treatise on correspondence analysis, the method is considered to be a superior, graphically oriented technique.

Before leaving this topic, I should like to mention a malaise common to both biology and geology, notably, the tendency to surround some particular method with which one feels oneself to be particularly at home with sentiments amounting almost to religious fervour. Such is the case, in some quarters, with correspondence analysis, Q-mode factor analysis etc. The writing of the book *Geological Factor Analysis* [5] brought this home to me, and my co-authors, very clearly.

3 FALLACIES

This topic follows on directly from the last sentence, to wit, the revivalist intensity connected with some applications of multivariate analysis.

The main class of fallacies encountered in applied multivariate work are ascribable to misunderstandings about (1) what a method is designed for doing and (2) traditional misuse.

In many instances, the main cause of these misunderstandings lies with the way in which some new piece of theory or a new method was introduced. Often, this can be traced to the lack of familiarity of the statistician with biology or geology and his original wrong assessment of the situation that he was trying to quantify. In addition, exaggerated claims are not infrequently made for some method or other, claims for which sound bases in fact may be lacking.

A type of fallacy occurring in chemical work is

related to the problem of statistically analysing closed data sets, that is, tables of chemical analysis that have constant row-sums. It has long been appreciated by some workers that standard methods of multivariate (including bivariate) analysis cannot be applied to constant-sum data without some kind of preliminary step aimed at overcoming the distortion inherent in such data, if normal methods of calculation are used. Many attempts at solving the problem have been tried over the last 25 years. The most successful attempt so far is that of Aitchison [6–10], who has demonstrated that constant-sum data must be treated by methods applying to simplex space (S-space) as opposed to normal cartesian space (R-space), the space for which statistical methods in their usual form have been devised.

One of the standard procedures of applied multivariate analysis concerns the interpretation of the elements of the latent vectors of principal component analysis (or principal component factor analysis) in terms of equations in which the vector elements are the coefficients of the variables included in the analysis. It has struck me on more than one occasion over the years that success varies greatly in such enterprises. In principal component factor analysis, ad hoc ways have arisen of deciding which coefficients are meaningful and which of them are not.

It is a fallacy to assume that all analyses can yield usefully interpretable coefficients. The reason may lie with the unstabilizing effect of atypical values in a data set [29] — this is a commonly occurring problem and one that will be further examined in the next section of this text.

In the analysis of more than one statistical population, such as occurs in the method of canonical variate analysis, it is not logically possible to attempt interpretations of the canonical vectors along the same line as is done for principal component analysis. There are several reasons why this is not good policy (although it is widely done, even among professional statisticians). Suffice it to mention here that the distributional complications already noted as being a source of uncertainty in principal components are vastly compounded in the case of several populations.

Fallacious arguments may be produced if too much reliability is placed on canonical variate vectors, taken as in a "principal component" kind of interpretation, if the vectorial elements are unstable. It is therefore essential that the stability of the vectors be carefully examined.

In summary, it can be said that the sources of fallacies in multivariate analysis lie in the following situations:

(1) theoretical results introduced without implicit instructions for their use;

(2) a method employed in one field of research, and applied in another, without the realization that the model underlying the two is not compatible with both areas; the method may be in correct use in the original situation, but is not validly applicable in the second;

(3) the application of a statistical method to data that do not accord with the basic stipulations of that method; a common case is the use of multivariate methods in situations where the sample size, N, is less than the number of variables, p; the correlation or covariance matrix is then not defined in, for example, principal component analysis if $N - 1 < p$;

(4) a very common source of fallacious analyses in multivariate work is, however, the computer package program. Most users of these programs have no more than an anecdotal acquaintance with statistical theory in general and multivariate analysis in particular. The output from package programs is often voluminous and insufficiently annotated, in either the output or in the accompanying handbook, is there be one. Only a person with a good knowledge of statistics can sort out what is genuinely useful in a particular analysis and what must be discarded. In many instances in the biological and geological literature, the conclusions are based on computations of doubtful validity. It seems that provided the topic seems to be trendy enough, and the results sufficiently astounding, many will be tempted to make use of a fallacious statistical approach to their data. Thus a fallacy breeds a fad.

4 FUTURE ON MULTIVARIATE ANALYSIS

The above title implies a greater scope than I intend to cover. In the present connection, I place emphasis on future developments in multivariate analysis of particular applicability in the geosciences, such as methods aimed at improving standard estimation procedures of multivariate analysis with the aim of disclosing how well methods of statistical analysis portray what is really represented in the data.

4.1 The closure problem

As is well known to most analytical chemists, data consisting of proportions cannot be usefully analysed by the same graphical and statistical methods as apply in open systems. Closed data are such that add up to a constant. Examples are tables of chemical analyses and tables of rock analyses that add up to 100% and sedimentary compositional data. Less obvious closed systems are those which are produced in connection with frequency diagrams, such as in pollen analysis and foraminiferal distributional charts. Closed systems also arise in biological work; for example, tables of frequencies of blood polymorphisms, phytosociology and ecological distributions.

Several attempts at solving the closure problem have been presented by various workers, drawn both from geological and statistical areas. Up to the present, the most satisfactory approach is that of Aitchison [6,7]. Aitchison [9] has published an expositionary paper on the topic with geochemical studies particularly in mind, and, more recently, a book on the analysis of compositional data [10]. Aitchison's solution is clearly superior to previous suggestions based on the "open set concept", which is not a valid way to represent true compositions.

The mainspring of Aitchison's solution lies with the representation of a closed data system in simplex space. Thus, a real p-dimensional space R^p will be associated with a positive simplex space S^p. A vector of compositions (a set of analysis on one sample) has the property that $p + 1$ variables are wholly determined by a sub-set composed of p variables. With blood polymorphisms, as an example, we need to determine only seven of the eight Rhesus morphs; the values for the eighth morph can be obtained by simple subtraction.

Aitchison suggests that the best way of describing variations in compositional data is by means of the additive logistic normal distribution. Although logarithmic ratios in multivariate analysis offer a useful approach, they do not solve all riddles and much research remains to be done, especially with respect to the reduction of dimensionality and non-linear data. In many instances, compositional data display curvilinearity. Butler [11], for example, has demonstrated that such trends are usually no more than a patent consequence of simplex space and it is seldom possible to identify any real significance in such deviations from linearity. The solution proposed by Kork [12] is difficult to grasp, as it is presented in terms of crude covariances (cf., Table 9.3 in ref. 10).

Fig. 1 presents a comparative analysis based on artificial data for three variables in a closed relationship, showing correlations computed for S-space and R-space. The correlations in R-space are ordinary correlations and those in S-space are computed using Aitchison's logarithmic ratio transformation. We also present below a real example of the chemical analysis of Swedish glacial clays (data by courtesy of Dr. Per Wedel and Mr. Rodney Stevens, Gothenburg University).

Aitchison [8] has pointed out that it is not immediately obvious how to proceed with a multivariate analysis of a covariance matrix formed from logarithmic ratios. He has suggested a transformation for producing a matrix in suitable form for, say, a practical version of a compositional principal component analysis (cf., ref. 10, p. 88). Such an analysis normally yields latent roots that are positive [30].

It will be appreciated that the differences in the results yielded by analyses carried out in R-space and S-space are, in most instances, substantial. The question as to whether a principal component analysis of proportional data is really justifiably located in S-space is still a moot point. Many of the results published by Aitchison indicate that this should be so, whereas J.C. Gower (personal communication) does not believe this, if the sole aim of the investigation is to ordinate observations in a graphical display (see also ref. 8).

4.2 Atypical observations

How often are statistical analyses, let alone multivariate statistical analyses, preceded by an

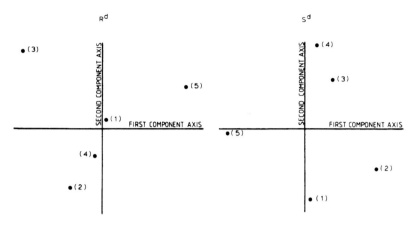

Correlations

Analysis in R-space (using ordinary correlations)				Analysis in S-space (using the logarithmic ratio transformation)			
	1	2	3		1	2	3
1	1.000	0.950	−0.086	1	1.000	0.581	−0.803
2		1.000	−0.237	2		1.000	−0.951
3			1.000	3			1.000

Principal component analyses
Latent roots

R-space			S-space		
(1) 2.0025	(2) 0.9600	(3) 0.0375	(1) 2.5668	(2) 0.4330	(3) 0.0002

Latent vectors

	1	2	3		1	2	3
1	−0.681	0.238	−0.693	1	0.532	−0.795	0.292
2	−0.698	0.077	0.712	2	0.575	0.592	0.565
3	0.223	0.968	0.114	3	−0.622	−0.133	0.772

Fig. 1. Simulated data for effects of closure. There are three chemical elements, determined in five samples and expressed in constant-sum for the two comparisons.

exploratory study of the properties of a data set? Most computerized applications do not examine the data for serious deviations from normality, nor do they take up the question of atypical observations, a subject which, of late, has to an increasing extent begun to capture the attention of mathematical statisticians [29]. Atypical values belong to a data set, generically, but for some good reason deviate from the distributional properties of most of the observations. The more sweeping terms outliers, rogue observations, etc., also encompass wrongly entered observations, incorrect measurements, misallocated specimens etc. [13]. These categories are not included under the present heading.

An atypical observation is, therefore, one that cannot be excluded from the analysis on formal grounds, although the investigator may suppress it for reasons of stability in his computations and the credibility of the statistical analysis. The problem of what to do with atypical values has led to augmented activity in the field of *robust estimation* [17], that is, estimates using the entire set of observations, but with some kind of reductional

TABLE 1

Correlations in R-space and S-space for 10 chemical determinations in a closed relationship for 16 samples

Correlation matrix (upper triangle) and logarithmic correlations (lower triangle). The variables determined in Pleistocene clays from southwestern Sweden are: (1) = Fe_2O_3, (2) = MnO_2, (3) = TiO_2, (4) = CaO, (5) = K_2O, (6) = P_2O_5, (7) = SiO_2, (8) = Al_2O_3, (9) = MgO, (10) = Na_2O. Data supplied by Dr. Per Wedel, Gothenburg.

	1	2	3	4	5	6	7	8	9	10
1	1.0000	0.6368	0.9443	−0.0536	0.3516	−0.1618	−0.9073	0.9451	0.1932	−0.0399
2	0.8280	1.0000	0.5206	−0.1423	0.1260	−0.3953	−0.5617	0.5947	0.0126	0.2363
3	0.9577	0.7871	1.0000	0.0858	0.2618	−0.0439	−0.8924	0.9318	0.1955	−0.1164
4	0.0918	0.0460	0.2962	1.0000	−0.8523	0.2014	−0.2430	−0.0331	0.6899	0.1397
5	0.6686	0.5994	0.7419	−0.0178	1.0000	0.1356	−0.0664	0.3024	−0.4612	−0.3857
6	0.4694	0.3486	0.6475	0.5756	0.7052	1.0000	0.1737	−0.1809	0.2243	−0.6483
7	0.5083	0.5287	0.6949	0.4779	0.8168	0.8333	1.0000	−0.8879	−0.4245	−0.1131
8	0.8469	0.7589	0.9390	0.3578	0.8274	0.7484	0.8752	1.0000	0.1050	−0.0986
9	−0.7246	−0.7774	−0.7997	−0.2607	−0.7316	−0.5044	−0.7882	−0.8620	1.0000	−0.0188
10	−0.7293	−0.5801	−0.8443	−0.4941	−0.7307	−0.9045	−0.7897	−0.8710	0.5590	1.0000

weighting arising from the data and applied to the outlying point or points. An application to a geological problem is given in ref. 14. A well rounded treatment of the subject is available in ref. 15, in which particular attention is given to robust estimation and atypical values in multivariate analysis.

Let us consider again the data analysed in Tables 1 and 2, the set of ten chemical determinations. The plot of the principal component scores of the data considered in S-space indicates that specimen 12 is an atypical observation. This deviating point was not picked up by the conventional principal component analysis. This observational vector deviates from all of the others in having almost zero entries for two of the elements.

The previous set of analyses was repeated, with sample 12 excluded from the data set (thus $N = 15$). The results obtained are as follows. The correlations displayed in Table 3 for both sets of computations display important differences for some of the pairs of variables, for example the correlations of variable four with all others (which are higher for the logarithmic ratio correlations). Least marked seem to be the correlations of the tenth variable with the others and there are no serious differences in the entries of the first and second matrices for these correlations. Of greater interest is the comparison between the matrices in Tables 1 and 3, the values for the entire set and the values after the atypical observational vector has been removed. It will be seen that the changes in the correlations computed in usual space are only exceptionally strongly different. The comparison of the logarithmic ratio correlations, on the other hand, shows much more strongly manifested shifts; see, for example, the column for $\hat{r}_{1i}$, the column for $\hat{r}_{4i}$, etc. Although a more detailed empirical analysis would be required for a firm conclusion, it does seem possible that the logarithmic ratio correlations are sensitive to atypical values, i.e., they are not robust estimators. This is a possible weakness in the method that should be kept in mind when conducting an investigation and one which necessitates particular care in monitoring data sets before entering into an analysis of interrelationships.

The principal component analyses corresponding to the new correlation matrices (Table 3) are summarized in Table 4. Here, the effects of the removal of the outlier are even more marked than in the case of the correlation coefficients.

The effects on the elements of the latent vectors computed in usual space are not great and the shifts are more of degree than of kind. In the case of the computations performed in simplex space,

TABLE 2

Principal component analyses of the correlation matrices of Table 3 in R-space and S-space

(1) Principal component analysis in R-space

Eigenvalues

| 4.33266 | 2.49003 | 1.89735 | 0.45979 | 0.39502 | 0.30374 | 0.05950 | 0.03097 | 0.02199 | 0.00895 |

Percentages of the total variance

| 43.32664 | 24.90029 | 18.97350 | 4.59786 | 3.95016 | 3.03742 | 0.59504 | 0.30967 | 0.21993 | 0.08948 |

	Eigenvectors									
	1	2	3	4	5	6	7	8	9	10
1	−0.4721	0.0402	−0.0514	0.0609	−0.0858	0.0239	−0.2829	0.3838	−0.7123	0.1629
2	−0.3384	0.0469	0.2735	−0.8391	0.2678	0.1136	−0.0114	−0.0961	0.0381	−0.0947
3	−0.4541	−0.0123	−0.1361	0.2450	0.1529	0.1180	−0.6338	−0.1535	0.4990	0.0591
4	0.0001	−0.6066	−0.1200	0.1227	0.3131	0.1242	−0.0432	−0.1862	−0.2724	−0.6152
5	−0.1296	0.5617	−0.1728	0.0001	−0.4332	0.1369	−0.0166	−0.2319	−0.0710	−0.6102
6	0.1073	−0.0247	−0.6462	−0.1907	0.0407	0.6704	0.1405	0.1574	0.0477	0.1898
7	0.4518	0.1640	0.0224	−0.1425	0.1732	−0.0640	−0.4494	0.6260	0.1139	−0.3276
8	−0.4621	0.0471	−0.0551	0.1937	0.1928	−0.1161	0.5361	0.5025	0.3037	−0.2467
9	−0.1021	−0.4954	−0.2319	−0.3123	−0.6525	−0.2949	−0.0825	0.1648	0.2072	−0.0561
10	−0.0093	−0.1936	0.6203	0.1594	−0.3424	0.6200	0.0089	0.1875	0.1113	−0.0541

(2) Principal component analysis in S-space

Eigenvalues

| 7.00904 | 1.47363 | 0.70798 | 0.51959 | 0.18737 | 0.06772 | 0.02379 | 0.00696 | 0.00391 | −0.00000 |

Trace = 9.999996

Percentages of the total variance

70.0904	14.7363	7.0798	5.1959
1.8737	0.6772	0.2379	0.0696
0.0391			

	Eigenvectors									
	1	2	3	4	5	6	7	8	9	10
1	−0.2844	0.3168	−0.2814	0.3415	−0.2958	0.0903	−0.0313	0.4441	−0.5287	0.1682
2	−0.3084	0.3788	−0.3034	−0.1169	0.7886	0.0622	0.0660	−0.0407	0.0742	0.1675
3	−0.3713	0.1321	−0.2196	0.2130	−0.3480	−0.0026	0.1665	−0.7530	0.1499	0.1544
4	−0.1311	−0.6574	−0.5302	−0.1748	−0.0425	0.0500	0.3617	0.2206	0.1198	0.1935
5	−0.3243	0.1343	0.5816	−0.0138	−0.0754	0.1319	0.5812	0.2594	0.2681	0.2007
6	−0.2951	−0.4084	0.2668	0.2302	0.1731	0.5904	−0.4430	−0.0943	−0.0477	0.1772
7	−0.3247	−0.2252	0.2835	−0.3832	0.0854	−0.4612	−0.0098	−0.1960	−0.5738	0.1542
8	−0.3795	0.0245	−0.0191	−0.0465	−0.1510	−0.4502	−0.5187	0.2629	0.5241	0.1414
9	0.3295	−0.1633	0.0932	0.6216	0.2410	−0.3882	0.0926	−0.0194	−0.0034	0.5061
10	0.3444	0.2090	−0.0245	−0.4531	−0.2095	0.2312	−0.1562	−0.0375	0.0012	0.7147

some of the changes are large as, for example, the entries for elements 4, 6, 7 and 9 in the first latent vector and the entries for elements 1, 4, 6, 7 and 10 in the second eigenvector.

4.3 Robust estimation in multivariate analysis

Another field that has attracted interest recently concerns methods of robust estimation in

TABLE 3

Correlations in R-space and S-space for 10 chemical determinations in 15 samples

Correlation matrix (upper triangles) and log-ratio (lower diagonal). Note: The logistic transformation has the property, assumed to be an advantage, of transforming positive values to spread throughout Euclidean space, seemingly a useful property if test statistics are required. The closure property for, say, three variables $x_1 + x_2 + x_3 = 1$ implies that points representing samples lie on a region of a plane. If a principal component analysis has as its aim to wish to provide a best fitting plane, then closure causes no problem; on the contrary, the fact that the points originally lie in a linear sub-space is irrelevant. A separate consideration is that logarithmic transforms may transform curved sub-spaces of samples to linear sub-spaces, although the empirical evidence for this is sketchy. It is possible for the opposite situation to arise, viz., that the original samples may lie in linear sub-spaces that are transformed into non-linear sub-spaces of doubtful interpretability.

	1	2	3	4	5	6	7	8	9	10
1	1.0000	0.6360	0.9520	−0.0629	0.3779	−0.1453	−0.9217	0.9534	0.1829	−0.1037
2	0.6950	1.0000	0.5410	−0.1689	0.1749	−0.3568	−0.5405	0.6178	−0.1109	0.1567
3	0.9570	0.6332	1.0000	0.0927	0.2570	−0.0673	−0.9384	0.9316	0.2674	−0.1065
4	−0.5882	−0.4842	−0.5077	1.0000	−0.8530	0.2637	−0.2217	−0.0267	0.7521	0.0877
5	0.2498	0.2448	0.2486	−0.8353	1.0000	0.0621	−0.1299	0.2976	−0.4183	−0.3260
6	−0.6763	−0.6504	−0.6923	0.1761	0.0746	1.0000	0.0894	−0.2166	0.5409	−0.5734
7	−0.4595	−0.0728	−0.3597	−0.0668	0.4812	0.1681	1.0000	−0.9354	−0.3510	0.0494
8	0.6598	0.6269	0.7017	−0.5998	0.5143	−0.6385	0.2844	1.0000	0.1628	−0.0806
9	−0.4176	−0.6042	−0.5128	0.2545	−0.4154	0.4741	−0.4758	−0.7805	1.0000	−0.5043
10	−0.0900	0.1439	0.0396	0.2820	0.0235	−0.2750	0.6025	0.4169	−0.7865	1.0000

some areas of applied multivariate statistical analysis. In general, the ideas stem from work in the univariate field, although the transitions are not always immediately obvious. Papers by Campbell [16,17,29] and the book by Hawkins [18] are useful references. Campbell and Reyment [14] applied robust estimation procedures to canonical variates, using an example drawn from micropalaeontology.

Table 5 presents a comparison of the latent vectors for the full data matrix and the data matrix formed by deleting the atypical value, both computed by robust estimation procedures [15]. The calculations were made using a program written in GENSTAT by Dr. J. Matthews (Department of Mathematics, University of Oxford), and were performed on the covariance matrix.

It will be seen that the effect of removing the atypical value has only a marginal effect on the first three latent vectors, but a more substantial structural influence on the subsequent latent vectors.

Thus, in the present example, the outlying observation is not of special significance if interest lies in attempting an interpretation of the first three principal components of the covariance matrix.

5 FROM GEOCHEMISTRY TO GEOLOGY

Up to this point, I have couched this presentation largely in terms of geochemistry. Let me now take a new direction. As far as I can see, future developments in the application of multivariate statistical analysis in analytical chemistry and geochemistry will be oriented towards improved methods of estimation and stability, as outlined in the foregoing examples. There is, however, more to the geological field than that.

Among the many possible fields likely to undergo expansion in multivariate geostatistics, mention may be made of the statistics of orientational data, for example, the analysis of dips and strikes, palaeomagnetic data and sedimentational orientations. Advances here are likely to be made in augmenting the number of applications to geological data, but even theoretical developments can be expected to occur.

Stratigraphical data generate, naturally, material that should be suitable for time series analyses. There is a considerable body of theory from the world of econometrics that could, with a little work, be brought into a state suitable for geological application. I have used aspects of this theory in studies of polarity reversals and volcanic

TABLE 4

Principal component analyses of the correlation matrices for R-space and S-space with an atypical value removed ($N = 15$)

(1) Principal component analysis in R-space
Eigenvalues

4.40084	2.65208	1.87408	0.48919	0.38696	0.08299	0.06043	0.03279	0.01732	0.00331

Percentages of the total variance

44.00839	26.52075	18.74077	4.89194	3.86964	0.82994	0.32788	0.17324	0.03312	

Eigenvectors	1	2	3	4	5	6	7	8	9	10
1	0.4698	−0.0095	0.0341	−0.0900	0.0195	−0.2733	0.2430	−0.3348	0.7152	−0.1110
2	0.3290	−0.1715	−0.1715	0.8586	0.2762	−0.0044	0.0083	0.0935	−0.0805	−0.0648
3	0.4575	0.0768	0.0034	−0.2210	0.0600	0.2716	0.6898	0.1390	−0.3689	0.1656
4	−0.0134	0.5387	−0.3436	−0.0592	0.0869	0.0733	0.0396	0.2018	0.0184	−0.4719
5	0.1568	−0.3647	0.5159	−0.1648	0.1468	−0.4213	−0.0426	0.2306	−0.2690	−0.4719
6	−0.0810	0.3695	0.4566	−0.0076	0.7421	0.2095	−0.0974	−0.1513	0.0767	0.1242
7	−0.4511	−0.1322	0.0786	0.1960	−0.0298	0.0117	0.4755	−0.6230	−0.1783	−0.2956
8	0.4653	−0.0061	−0.0206	−0.1185	−0.1311	0.2946	−0.4740	−0.5528	−0.3271	−0.1666
9	0.0911	0.5788	0.0823	0.1585	−0.1843	−0.6430	0.0016	−0.1536	−0.3125	0.2371
10	−0.0463	−0.2287	−0.6028	−0.3124	0.5365	−0.3528	−0.0390	−0.1604	−0.1760	0.1188

(2) Principal component analysis in S-space
Eigenvalues

4.85524	2.41920	1.74950	0.43023	0.33582	0.11681	0.06392	0.01698	0.01230	0.00000

Trace = 9.999995

Percentages of the total variance

48.5524	24.1920	17.4950	4.3023
3.3582	1.1681	0.6392	0.1698
0.1230			

Eigenvectors	1	2	3	4	5	6	7	8	9	10
1	−0.3820	0.3184	−0.0236	0.2156	−0.1414	0.0088	−0.1402	0.0015	−0.7821	0.2287
2	−0.3693	0.0805	−0.0861	−0.7896	−0.3615	−0.0777	0.0716	−0.0078	0.1272	0.2626
3	−0.3935	0.2451	−0.0805	0.4144	−0.1153	0.1127	0.5475	0.3448	0.3770	0.1499
4	0.3006	−0.0814	−0.5474	0.0904	−0.1173	0.0742	0.3907	−0.5589	−0.0945	0.3202
5	−0.2279	−0.2159	0.5831	0.1094	−0.0556	0.4826	−0.0440	−0.4720	0.1369	0.2635
6	0.3307	−0.1237	0.3739	0.1978	−0.6678	−0.4645	0.0725	0.0807	−0.0215	0.1531
7	−0.0201	−0.6107	0.1495	−0.1657	0.2755	−0.0049	0.4791	0.3258	−0.3797	0.1441
8	−0.4154	−0.1445	0.0129	0.1491	0.3425	−0.7145	−0.0350	−0.3495	0.1513	0.1003
9	0.3488	0.3504	0.1696	−0.0794	0.3942	−0.0636	−0.1043	0.1844	0.0982	0.7109
10	−0.1289	−0.4969	−0.3934	0.2074	−0.1568	0.1113	−0.5234	0.2777	0.1589	0.3530

eruptions [19,20]. Birks and Gordon [21] have studied the question in connection with the quantification of pollen analysis.

These, and other areas, are doubtless of developmental potential in geostatistics, but they do not represent any really new thinking and in no way break with the traditional or conventional path taken in multivariate work. In order to present some really innovative approaches, it will be necessary to consider palaeontological aspects.

5.1 Multivariate analysis of strain

Multivariate statistical analysis has long been applied to fossil material and, in fact, some of the

TABLE 5

Latent roots and vectors for the geochemical data, computed by a robust estimation procedure applied to the covariance matrix [14–16]

Computed by means of a GENSTAT program written by Dr. John Matthews, Oxford.

	U (1)	U (2)	U (3)	U (4)	U (5)	U (6)	U (7)	U (8)	U (9)	U (10)
Full sample (N = 16)										
	−0.3856	−0.2994	−0.7918	−0.0326	0.1392	0.2177	0.2533	0.0500	−0.0128	−0.0007
	−0.0053	−0.0025	−0.0223	0.0129	−0.0480	−0.0717	−0.0339	0.4112	0.7310	0.5360
	−0.0294	−0.0187	−0.0365	−0.0018	−0.0184	0.1667	−0.1216	−0.8551	0.4720	0.0233
	−0.0494	0.5382	−0.2265	−0.1184	−0.1390	0.5220	−0.5650	0.1284	−0.1054	0.0624
	−0.0203	−0.4946	0.1977	−0.4910	0.5066	0.1063	−0.4430	0.0524	−0.0483	0.0747
	0.0015	−0.0037	−0.0010	−0.0405	0.0177	0.0404	−0.0838	0.2569	0.4714	−0.8373
	0.8606	−0.1854	−0.3624	0.2485	0.1107	0.0603	−0.1252	0.0102	−0.0119	0.0050
	−0.3196	−0.2748	0.0753	0.7708	0.0348	−0.0501	−0.4630	0.0381	−0.0527	−0.0101
	−0.0636	0.2440	−0.3746	−0.1668	0.0913	−0.7898	−0.3530	−0.1015	−0.0147	−0.0330
	−0.0268	0.4542	0.0589	0.2413	0.8246	0.0661	0.2066	−0.0114	0.0625	0.0178
Latent roots	15.2810	2.1157	0.1901	0.1615	0.0877	0.0024	0.0016	0.0008	0.0004	0.0002
Censored sample (N = 15); atypical value removed										
	−0.3994	0.3501	−0.5409	−0.4523	0.1157	0.4509	−0.0090	0.0632	−0.0013	0.0041
	−0.0051	0.0053	−0.0191	−0.006	−0.0621	−0.0142	−0.0368	−0.0873	−0.3618	0.9250
	−0.0311	0.0068	−0.0202	−0.0211	0.0028	0.0613	0.1969	−0.9661	−0.0908	−0.1184
	−0.0440	−0.6342	−0.2884	−0.1858	0.2852	−0.1548	0.5907	0.1006	0.0857	0.0806
	−0.0330	0.4923	0.4453	−0.3306	0.4923	−0.3336	0.2893	0.0245	0.0721	0.0760
	0.0008	−0.0010	0.0075	−0.0325	0.0516	−0.0537	0.1079	0.1472	−0.9194	−0.3386
	0.8470	0.2566	−0.4204	0.0310	0.1848	0.0321	0.0606	−0.0045	0.0064	0.0120
	−0.3418	0.2687	−0.3681	0.7143	0.2535	−0.3002	0.1029	0.0053	0.0115	0.0110
	−0.0458	−0.1424	−0.2889	−0.3355	0.1100	−0.6333	−0.5886	−0.1378	−0.0191	−0.0518
	0.0134	−0.2728	0.1718	0.1623	0.7388	0.4030	−0.3935	−0.0598	−0.0494	0.0204
Latent roots	15.1070	0.8696	0.2606	0.1580	0.0402	0.0319	0.0122	0.0010	0.0002	0.0001

earliest applications of methods, now standard, were to organic remains. Original interest lay with ordinating samples of fossils from different locations in time and space, that is, allocating collections to established groups. Further considerations led to the elements of the latent vectors being used to interpret growth relationships, such that all-positive elements for measures on dimensions of an organism have been taken as indicating that such a vector is an indicator of size. A latent vector, the elements of which bear positive and negative signs, has been interpreted as exhibiting variability in shape. Now, there is no doubt that size and shape are encapsulated in such vectors and there may be several vectors, connected with latent roots of almost equal status, each of which can be given some kind of interpretation in terms of shape. The method of principal component analysis lends itself most easily to this procedure. Less easily interpreted are the latent vectors of canonical variate analysis. Reyment et al. [22] have considered several reasonably founded examples, but with some unease.

A more direct and logically sounder approach is that of Mosimann [23], who established a method based on a theorem which states that at most one size variable is independent of the entire space of shape variables. For many purposes, the method advocated by Mosimann is a useful way of indicating difference in shape. If, however, one desires to study in detail how shape is changing from arbitrarily chosen point to arbitrarily chosen point through time, then a graphically oriented method is clearly necessary.

The reasons for wanting to carry out such an investigation are obvious to the evolutionary biologist, but probably not so to the statistician. One is required to abandon the ordered world of conventional multivariate analysis, with its arsenal of tests for correct methods and its convenient distributional theory, for a semi-statistical environment in which approximate estimations abound and heresy may not be far away.

5.2 Growth and shape as deformations

As early as 1917, the remarkably clairvoyant English biologist D'Arcy Wentworth Thompson published his celebrated transformation grids for the transformational between the fish *Diodon* and *Mola*. Easy enough in perception, it was to baffle mathematicians for 60 years before somebody arrived at an approximate way of tackling the quantification of Thompson's stroke of genius.

Bookstein [24] proposed a solution based on the recognition of homologous landmarks and their mappings from one form to the other, whereby the "strain" produced by the transformation was to be interpreted in terms of the strain tensor. Notwithstanding the fact that much remains to be done with respect to the formal statistical development of the concept, it is clear to me that the geometrical approach of Bookstein represents a real step forward in multivariate morphometrics and the analysis of how changes in shape are manifested over an organism. For the purposes of evolutionary biology, such a representation has decided advantages over a more statistically enunciated interpretation of shape. Bookstein [24] called his graphical procedure the method of "biorthogonal grids" (see also ref. 25). It would lead too far to go into all aspects of geometric morphometry, so I shall simply outline an application.

The method presents change in shape as a smooth deformation of points in one coordinate system and aims at finding and quantifying the principal directions of change in shape in a manner that will ensure evenness of areal coverage. It is a solution of the problem of representing the Cartesian transformational grids of D'Arcy Wentworth Thompson in a quantitative and graphically constructable form, and analyses the shift between the

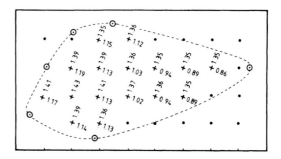

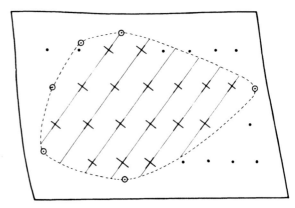

Fig. 2. *Brizalina mandoroveensis*. Top: biorthogonal analysis of the Thompsonian transformation for a sample from one level to a sample from a higher level in a borehole (Ikang, Cameroun, Miocene). The figure shows selected curves and dilatations (strains along the principal axes of deformation), superimposed on reference mesh points. Bottom: this figure is deformed in relation to the first, as reference, in that the rectangular basis of the first has been strained to form a parallelogram, whereby the shape of the younger material is, on the average, squatter.

forms in terms of the dilatations at comparable points in the grid.

For the example indicated in Fig. 2, six landmarks were measured around the test of *Brizalina mandoroveensis* (from the Miocene of Cameroun, West Africa), viz., at the proloculus, the location of the aperture and at sutural contacts of the last two chambers. The diagrams were constructed by placing a mesh of points over the average form arbitrarily selected as the starting point for the graphical display. The interpolation yielded by homologous data was then used to compute the mapping of all of these points into the figure of the second form, an average for a sample of the species taken at a stratigraphically higher level. The orientations of the biorthogonal axes were drawn in at the mesh points lying inside the outlines.

These axes are oriented along the directions of maximum and minimum directions of rate of change of length. The curves and dilatations that can be superimposed on the mesh of crosses describe the change in shape of the average outline by a tensor field, which expresses a deformation of the interior by particular rates in particular directions.

The results summarized in Fig. 2 indicate the biologically significant fact that the deformational intensity is not evenly spread over the entire surface of the test and that the difference in the two rates tends to be greater along the medial axes, and proximal to the proloculus.

The present application is of biological interest. It is, however, easy to see how this kind of analysis can be applicable to the study of strain in structural geology if the measurable objects are mineral grains with observable crystal properties, or some regularly formed object occurring in the deformed rock.

5.3 Soft and hard modelling

An area that should receive more attention in the geosciences and biosciences is one that involves causal interpretations of events. The original ideas lie with the path analysis concept of Wright [26] and later with Wold's approximate modelling technique which he called "soft modelling" (see, for example, ref. 27). Wright's ideas were, for natural scientists, unfortunately eclipsed by the simultaneously formulated multivariate method of principal component analysis and by the subsequent multivariate school with which geostatisticians are closely familiar (see ref. 25).

Ease of computing, the availability of multipurpose computer packages for other methods, etc., have had the effect that the modelling efficiency of path analysis has not been realized. For example, growth and palaeocological (ecological) studies, etc., can often be more realistically developed in terms of path analysis. The reason for this lies with the fact that the conventional arsenal of multivariate techniques has, as its prime motivation, the desire to explain or optimize variance, whereas path analysis explains covariances and interprets certain combinations of variables that often do not correspond to what is being optimized by the usual methods. Doubtless, part of the problem is due to the computational complications that sometimes arise in a path analysis, which makes it difficult to produce general methods of solution amenable to computer packaging.

Presumably, people will eventually tire of analysing their material in a stereotyped manner and start to want to have more control over determining causal effects and interaction. It is to be expected, then, that path analysis and soft modelling will prove attractive to those who wish to use statistics as an interactive tool in their research, just as much as are scanning electron microscopes, atomic absorption spectrometers, etc.

6 CONCLUSIONS

The wide availability of computer programs often leads to the uncritical use of techniques that may achieve the status of fads, despite their unsuitability for many of the applications made of them. Fallacious interpretations of multivariate results can also derive from misunderstandings on the parts of both the statistician who developed some new procedure as well as the users of the method. Moreover, a method which may be suitable for solving one kind of multivariate problem may not be validly applicable in another situation. Closed data sets (tables of proportions and percentages) require specially adapted multivariate procedures and pose a particular problem in chemometrics. Another source of error in multivariate analysis is caused by the presence of atypical values in a sample. Robust estimation procedures are often useful in such instances. The significance of these aspects is illustrated here using a set of chemical analyses on Swedish glacial clays. Recent advances in the field of shape analysis, as applicable to geological problems, have been briefly reviewed.

7 ACKNOWLEDGEMENTS

I am grateful to Rolf Manne and Olav Kvalheim for their encouragement in preparing this paper.

John Birks has given me the benefit of his wide experience in the applied multivariate field. I am particularly grateful to John Aitchison for a penetrating analysis of the manuscript. Per Wedel and Rodney Stevens generously supplied the chemical data used in part of the paper.

REFERENCES

1. H.O. Hirschfield (H.O. Hartley), A connection between correlation and contingency, *Cambridge Philosophical Society Mathematical Proceedings*, 31 (1935) 520–524.
2. R.A. Fisher, The precision of discriminant functions, *Annals of Eugenics*, 10 (1940) 422–429.
3. K.R. Gabriel, The biplot graphical diplay of matrices with application to principal components analysis, *Biometrika*, 58 (1968) 453–467.
4. M.J. Greenacre, *Theory and Applications of Correspondence Analysis*, Academic Press, London, 1984, 364 pp.
5. K.G. Jöreskog, J.E. Klovan and R.A. Reyment, *Geological Factor Analysis*, Elsevier, Amsterdam, 1976, 178 pp.
6. J. Aitchison, A new approach to null correlations of proportions, *Journal of Mathematical Geology*, 13 (1981) 175–189.
7. J. Aitchison, The statistical analysis of compositional data, *Journal of the Royal Statistical Society B*, 44 (1982) 139–177.
8. J. Aitchison, Principal component analysis of compositional data, *Biometrika*, 70 (1983) 57–65.
9. J. Aitchison, Reducing the dimensionality of compositional data sets, *Journal of Mathematical Geology*, 16 (1984) 617–635.
10. J. Aitchison, *The Statistical Analysis of Compositional Data*, Chapman & Hall, London, 1986.
11. J.C. Butler, Principal component analysis using the hypothetical closed array, *Journal of Mathematical Geology*, 8 (1976) 25–36.
12. J.O. Kork, Examination of the Chayes–Kruskal procedure for testing correlations between proportions, *Journal of Mathematical Geology*, 9 (1977) 543–562.
13. V. Barnett and T. Lewis, *Outliers in Statistical Data*, Wiley, New York, 1978.
14. N.A. Campbell and R.A. Reyment, Robust multivariate procedures applied to the interpretation of atypical individuals of a Cretaceous foraminifer, *Cretaceous Research*, 1 (1980) 207–221.
15. N.A. Campbell, Canonical variate analysis: practical aspects, *Ph.D. Thesis*, Imperial College, University of London, 1979.
16. N.A. Campbell, Shrunken estimators in discriminant and canonical variate analysis, *Applied Statistics*, 29 (1980) 5–14.
17. N.A. Campbell, Robust procedures in multivariate analysis. II. Robust canonical variate analysis. *Applied Statistics*, 31 (1982) 1–8.
18. D.M. Hawkins, *Identification of Outliers*, Chapman & Hall, London, 1980, 188 pp.
19. R.A. Reyment, Statistical analysis of some volcanologic data regarded as series of point events, *Pure and Applied Geophysics*, 74 (1969) 57–77.
20. R.A. Reyment, Trends in Cretaceous and Tertiary geomagnetic reversals, *Cretaceous Research*, 1 (1980) 27–48.
21. H.J.B. Birks and A.D. Gordon, *Numerical Methods in Quaternary Pollen Analysis*, Academic Press, London, 1985, 317 pp.
22. R.A. Reyment, R.E. Blackith and N.A. Campbell, *Multivariate Morphometrics*, Academic Press, London, 2nd ed., 1984, vii + 233 pp.
23. J.E. Mosimann, Size allometry: size and shape variables with characterizations of the lognormal and generalized gamma distributions, *Journal of the American Statistical Association*, 65 (1970) 930–945.
24. F.L. Bookstein, The measurement of biological shape and shape change, *Lecture Notes in Biomathematics 24*, Springer Verlag, New York, 1978, 191 pp.
25. F.L. Bookstein, B. Chernoff, R. Elder, J. Humphries, G. Smith and R. Strauss, *Morphometrics in Evolutionary Biology*, Special Publications of the Academy National Sciences, Philadelphia, No. 15, 1985, 277 pp.
26. S. Wright, General, group and special size factors, *Genetics*, 17 (1932) 603–619.
27. H. Wold, Systems under indirect observation using PLS, in C. Fornell (Editor), *A Second Generation of Multivariate Analysis*, Praeger Publishers, New York, 1982, pp. 325–347.
28. J.P. Benzécri, *L'Analyse des Données. II, L'Analyse des Correspondances*, Dunod, Paris, 1973, 619 pp.
29. N.A. Campbell, The influence function as an aid to outlier detection in discriminant analysis, *Applied Statistics*, 27 (1978) 251–258.
30. W.W. Cooley and P.R. Lohnes, *Multivariate Data Analysis*, Wiley, New York, 1971, 364 pp.

Interpretation of Lithogeochemistry using Correspondence Analysis *

MICHEL MELLINGER

Saskatchewan Research Council, 15 Innovation Blvd., Saskatoon, Saskatchewan S7N 2X8 (Canada)

CONTENTS

1 Introduction 367
2 Major and trace elements: Data analysis strategy 368
3 Analysis of the raw data 369
   3.1 Major element patterns 369
   3.2 Trace element patterns 371
4 Analysis of recoded data 375
5 Conclusion 380
6 Acknowledgements 381
References 381

1 INTRODUCTION

Lithogeochemical data describe the chemical composition of rock samples in terms of many elements, some of which are abundant and are called major elements, the others being less abundant and called trace elements. Such data are usually organized in a table whose rows are the rock samples and columns the chemical elements. The number of samples analyzed may vary from about 50 or 100 to several thousands; the number of elements analyzed for typically involves from about 8 to 12 major elements and up to 20 or more trace elements. Thus, we usually are dealing with a large data table and many variables, which suggests the following remarks: (1) one better define a useful data analysis strategy, that is, a strategy based on geological and geochemical knowledge, before analyzing the data table; (2) multivariate statistics will provide useful tools for the analysis of such data.

As a first step towards setting an adapted data analysis strategy, let us note that a rock is comprised of several minerals, each of a usually well-defined chemical composition. Thus, the chemical composition of a rock is a linear combination of the chemical compositions of its constituting minerals. Unfortunately, due to the nature of the analytical procedures used in the geochemical laboratories (crushing → grinding → digestion → elemental analysis), only element concentrations are returned to the user (this is the appropriately-named 'ultimate analysis' of petroleum geochemists). The mineral information is lost in that process, but it can be regained by multivariate data analysis.

Multivariate data analysis techniques which are well adapted to the study of data patterns are those of the factor analysis type (see ref. 1 for a review of multivariate techniques). Correspon-

---

* SRC Publication No. R-851-3-A-87.

dence analysis is one such technique, and is preferred here to other factor analysis techniques because: it treats the rows and columns of a data table in a symmetrical manner, it produces output parameters which give good control to the user during interpretation, and it is flexible with respect to the type of data investigated and in particular with respect to data recoding (see refs. 2 and 3 for a review of this technique and its usage; refs. 4–6 provide further details about correspondence analysis).

In this paper, a data analysis strategy suggested for the study of lithogeochemical data is explained, and is then applied to the interpretation of raw concentration data in several examples which illustrate a range of geological applications. In the last section, examples of the usage of data recoding for the enhancement of specific data features are given.

2 MAJOR AND TRACE ELEMENTS: DATA ANALYSIS STRATEGY

A general strategy for the application of correspondence analysis to lithogeochemical data was explained in ref. 7. Its basis is that rocks contain two categories of elements, not in terms of elemental abundances but, as suggested by Korzhinskii [8], based on thermodynamic and mineral phase stability considerations. Indeed, major and trace elements do not play the same role in a rock with respect to the constituting minerals (Table 1). Major elements are mineral-forming elements, or "framework" elements, whereas trace elements substitute for major elements in existing minerals (for chemical data on common rock-forming minerals, see for example ref. 9).

As a consequence, the behaviour of elements from each category will be controlled by different parameters. The behaviour of major elements is determined by their chemical potential and the mineral phase stability relationships into play at given pressure and temperature conditions (see for example refs. 10 and 11). The behaviour of trace elements is determined by their availability or abundance, and by the nature of their interaction with existing crystal structures as governed by ionic radii, electronegativity, as well as the crystal-field stabilization energy for transition elements (see for example refs. 12 and 13).

From the point of view of data analysis, one is interested in what causes variations in the data, because our statistical tools are built to detect such variations. For major elements, causes for data variations can be found very simply in variations in mineral porportions. This assumes that variations in the chemical composition of a given mineral are negligible, which is usually the case for a specific data set; if this assumption is wrong, then variations will likely become large enough to be explained in terms of two or more end-members which should be identifiable in the data patterns. For trace elements, however, causes for data variations are more complex and involve essentially three components: variations in mineralogy (and thus in major element data), variations in the availability of the trace elements in the geological situation(s) under study (the 'rock specialization' of Russian geochemists, see ref. 14), and variations in the crystallization history of the rocks investigated (which relates to the geochemical signature of specific geological processes).

The data analysis strategy that results from these considerations is summarized in Fig. 1, and consists in carrying out separately, but in parallel, the analysis and interpretation of major element data and trace element data. The purpose of this

TABLE 1

Distinguishing between major and trace elements in rocks (after ref. 7)

Major elements	Trace elements
Definition	
Mineral-forming elements	Substituting elements
Behaviour parameters	
Chemical potential	Availability
Mineral phase stability	Interaction with crystal structures
Causes for data variations	

Variations in mineralogy	Variations in:
	 –mineralogy (general)
	 –availability
	  (rock specialization)
	 –crystallization history
	  (geological process signature)

Chapter 26

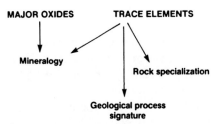

Fig. 1. Data analysis strategy for the study of lithogeochemical data; see also Table 1 (after ref. 7).

strategy is to recover from the observed data patterns, the causes for data variations just explained and which constitute the core of the geological information of interest to the geologist and geochemist. This is done by first calculating a factor space for the major elements (active variables), with trace elements being simply projected into that space (supplementary variables); secondly, a factor space is calculated for the trace elements (active variables), with major elements being projected as supplementary variables; one must be careful to treat outlier samples as supplementary rows in order to ensure that the matching observed between both sets of elements has general applicability for the data investigated. The data analysis will mostly involve analyzing the raw concentration data; however, as illustrated in the latter section of this paper, several data recoding schemes can also be used in addition, in order to enhance particular features of interest in the data patterns and provide further information of interest to the user.

3 ANALYSIS OF THE RAW DATA

3.1 Major element patterns

Let us first examine several examples which illustrate that major element patterns are mineralogical patterns.

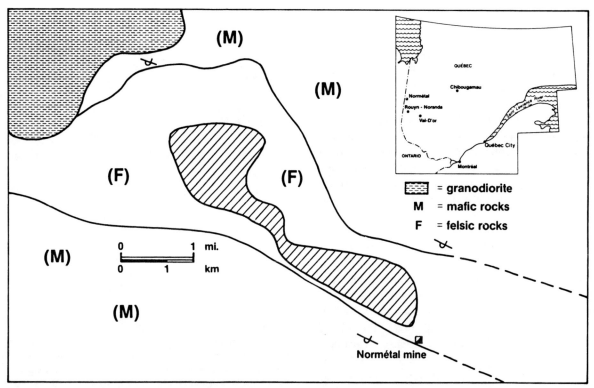

Fig. 2. The Normétal project: location (inset) and generalized geological map of the study area. The hatched area indicates the distribution of anomalous zones as defined at the end of the project (after ref. 15).

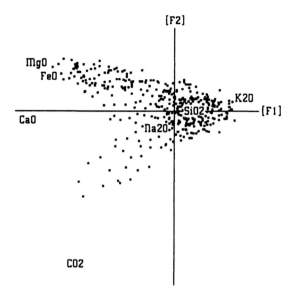

Fig. 3. The Normétal project: correspondence analysis of the major oxide data, showing the general petrographic characteristics of rock samples (see text for discussion). Only those variables which contribute to either factor or are close to this factorial plane as displayed; each closed square symbol represents one rock sample (after ref. 15).

The first example deals with an Archean metavolcanic pile which has good potential for volcagenic massive sulphide deposits, located in the Normétal area of northwestern Québec, Canada (Fig. 2). The Normétal Cu-Zn deposit produced a large amount of ore until the early 1970's, and was enclosed in a sequence of mainly felsic metavolcanic rocks. These rocks were sampled systematically on a regional scale in order to define zones that may have been affected by the hydrothermal alterations characteristically associated with volcanogenic massive sulphide deposits [15]. Correspondence analysis of the raw major oxide data shows patterns that are easily interpretable. On factor plot F1-F2 (Fig. 3), two sample trends are apparent: one is an expression of the presence of mafic (MgO, FeO, CaO) to felsic (SiO_2, K_2O, Na_2O) rocks in the sampled area; the other, smaller in size, results from the extensive carbonatization (CaO, CO_2) of some of these rocks during greenschist metamorphism. Of interest is that the direction of this carbonatization trend is towards a mixture of CaO and CO_2, which indicates that calcite (CaO · CO_2) is the mineral involved; also, the degree of carbonatization is progressively higher for increasingly mafic rocks, indicating that most likely, CaO has been remobilized from these rocks during metamorphism to form calcite, while only CO_2 was introduced by the metamorphic fluids (note that the two trends observed in Fig. 3 are correlated because they are non-orthogonal). When factor plot F2-F3 is examined (Fig. 4), it is apparent that factor F3 discriminates between rocks that have been affected by potassic hydrothermal alteration stabilizing minerals such as sericite (K_2O), and unaffected rocks still rich in albitic plagioclase (Na_2O). This factor is most useful from the point of view of mineral exploration, because potassic alteration is indicative of the presence of hydrothermal centres favorable for the deposition of massive sulphides. Sample coordinates along factor F3 can be used as a measure of the degree of alteration reached (in this case higher alteration towards more positive coordinates) and plotted on the sample location map.

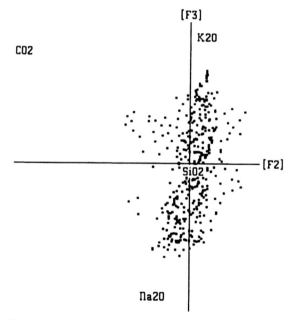

Fig. 4. The Normétal project: correspondence analysis of the major oxide data, showing hydrothermal alteration characteristics of the rock samples, essentially along factor F3 (see text for discussion). Same comments and notations as in Fig. 3 (after ref. 15).

This was done, indicated that potassic alteration is indeed consistently distributed in the field (Fig. 2), and several drilling targets could be further identified in that area after taking mapping information into account.

The next example deals with Proterozoic sandstone from the Dawn Lake area in the Athabasca Basin of northern Saskatchewan, Canada (Fig. 5). The sandstone is essentially comprised of detrital quartz grains, precipitated (red-bed) hematite, and a clay mineral matrix consisting of major kaolinite and illite, and of some magnesian chlorite. In this project, a large number of samples (about 2700 drill core samples) were treated, which means that about all ranges of possible mineralogy for sandstone of that area were probably sampled in the process. Indeed, factor plot F2–F3 obtained by correspondence analysis of the major oxides (Fig. 6) is equivalent to a ternary plot kaolinite (upper left)–illite (K_2O)–chlorite (MgO), with the majority of samples occurring along the kaolinite–illite edge (see ref. 16 for a full discussion). Of interest is the fact that the 'kaolinite' pole is observed on factor plot F2–F3, even though the associated oxide (H_2O^+) was not analyzed for (see the next example for the role of H_2O^+ in such paterrns): actually, Al_2O_3 projects close to the origin in the direction of this pole, and does have a relatively large weight, thus contributing to the extraction of the full pattern.

The last two examples also deal with sandstone from the Athabasca area, and illustrate further the well-defined mineralogical information that is obtained from major oxide data. Samples from the Midwest Lake property (see location on Fig. 5) display several separate alteration trends which emerge from a background population characterized independently [17]: a siderite trend (FeO and CO_2) and magnesian–chlorite trend (MgO), both displayed on factor plot F3–F4 (Fig. 7); a kaolinite-rich subpopulation (H_2O) discriminated along factor F3 (Figs. 7 and 8); and an illite trend extending along factor F5 (Fig. 8). It was an easy task in this case to use factor coordinates along F3 to F5 to isolate samples from each alteration trend for further data analysis [17]. In the case of the Maurice Bay property (see location on Fig. 5), although many samples are altered (i.e. distinct from background), only a continuous population is observed on factor plot F3–F4 (Fig. 9). Petrographical and clay mineralogy studies of the samples have shown that this pattern has resulted from a strong magnesian–chlorite overprinting of much sandstone in that area, some of which had been affected earlier by illitization [18]. Hence the continuous spectrum of altered rocks, from essentially pure illitized samples (K_2O, to the bottom left of Fig. 9) to completely overprinted (chloritized) samples (MgO, to the bottom right). Selection of sample subpopulations in that case was also carried out in the factor space, but by including other mineralogical information (see ref. 17 and Fig. 9).

3.2 Trace element patterns

The following examples illustrate the fact that trace element patterns are more than mineralogical patterns.

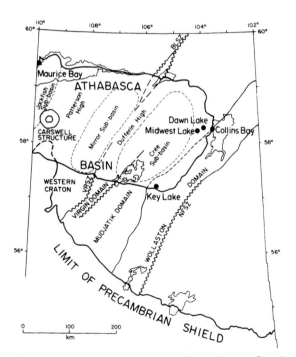

Fig. 5. General geology of northern Saskatchewan (Canada) with location of some of the Athabasca unconformity-type uranium deposits. The Proterozoic Athabasca Basin is essentially flat-lying on the deformed Archean and Aphebian basement (from ref. 17).

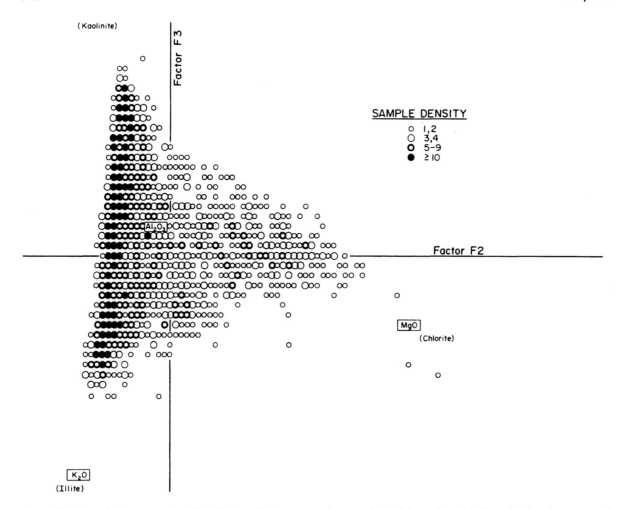

Fig. 6. The Dawn Lake property (see Fig. 5 for location): correspondence analysis of the major oxide data, showing the nature and distribution of the clay mineralogy of the sandstone samples (see text for discussion); note the equivalence of this factor plot to a ternary mineralogical diagram. Only Al_2O_3, K_2O and MgO contribute significantly to the displayed factors; the interpreted clay minerals are shown in brackets; samples are displayed using density symbols as per the scale shown (from ref. 16).

In the first case, a regional lithogeochemical survey was carried out to try to characterize areas which could warrant further prospecting. The questions at hand are: what are the major lithologies encountered? which mineral commodity has good potential in the survey area? and what lithogeochemical prospecting strategy should be followed at a more detailed scale? Preliminary answers to these questions could be given [19] using correspondence analysis in the following way:

(a) analyze major element patterns to identify lithologies,

(b) project the trace elements into the major element factor space as supplementary variables (i.e. after calculation of the factors using only the major element data) to evaluate their correlation with lithological trends (using the proximity output parameter, see ref. 2), and

(c) analyze trace element patterns to identify trace elements which have a univariate behaviour and to cross-validate the findings of step (b).

Fig. 10 is an example of lithologies identification in the major oxide factor space: carbonate rocks containing essentially calcite (CaO), and silicate rocks (MgO, FeO, Al_2O_3, SiO_2), are the

Chapter 26

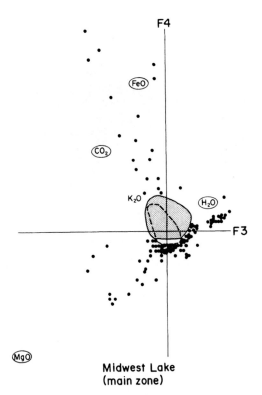

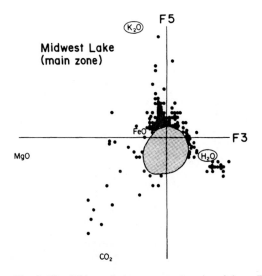

Fig. 7. The Midwest Lake property (see Fig. 5 for location): correspondence analysis of the major oxide data, showing anomalous mineralogical trends of samples (see text for discussion). The circled oxides contribute significantly to the displayed factors; others are projected for reference (see Fig. 8); the dotted area represents the expected compositional range of background sandstone; the stippled outline represents the projected illite trend of Fig. 8 and located along factor F5; each closed circle symbol represents one sandstone sample (from ref. 17).

Fig. 8. The Midwest Lake property (continued from Fig. 7): correspondence analysis of the major oxide data, showing the anomalous illite trend of samples (see text for discussion). Same notations as for Fig. 7 (from ref. 17).

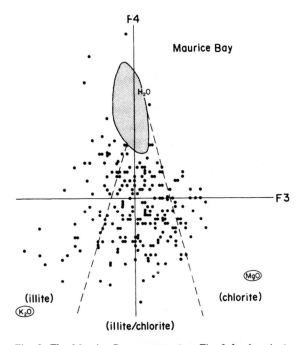

Fig. 9. The Maurice Bay property (see Fig. 5 for location): correspondence analysis of the major oxide data, showing anomalous mineralogical trends of samples (see text for discussion). Same notation as for Fig. 7; the anomalous sample population has been divided according to a mineralogical and petrographical study of the same samples (stippled lines towards bottom; from ref. 17).

main categories; factor F3 (not shown here), when combined with F2, helps distinguish further end-members of the silicate rocks. Fig. 11 is an example of patterns in the trace element factor space, which illustrates the essentially univariate behaviour of some trace elements, here of Hg and Zr. In fact, Zr could be considered a major element (actually an 'indifferent' element from a thermodynamic viewpoint, see ref. 8) because it occurs only in the accessory mineral zircon, and its univariate behaviour is a result of zircon occurring only with sediments of a distinct sedimentary basin located in the survey area. After the correlation between the major and trace element factor

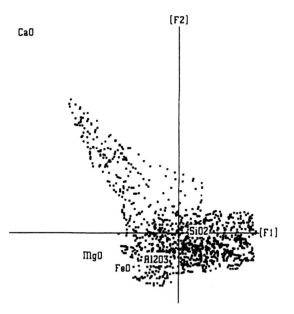

Fig. 10. A regional lithogeochemical survey: correspondence analysis of the major oxide data, showing major lithologies (see text for discussion). Only those oxides which contribute significantly to either factor or are close to the displayed factorial plane are shown; each closed square symbol represents one rock sample (from ref. 19).

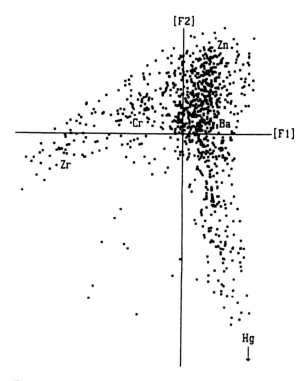

Fig. 11. A regional lithogeochemical survey: correspondence analysis of the trace element data, showing an example the univariate behaviour of some elements, here of Hg and Zr (see text for discussion). Same notations as in Fig. 10 (from ref. 19).

spaces was examined as suggested for step (b) above, recommendations for further geochemical prospecting in that area for the detailed interpretation of the data can be made, as summarized in Table 2. This example clearly shows that trace element variations are more complex than major element variations.

TABLE 2

Recommendations for further geochemical prospecting and guidelines for interpretation, based on a preliminary study of a regional lithogeochemical survey (after ref. 19)

Group 1: elements which can be plotted directly on a concentration map due to their univariate behaviour:
Pb, Hg, Sn, Cu, Mo, Zr

Group 2: elements which will require some correction for lithological variations before plotting:
Zn, Ni

Group 3: elements which will require careful study of lithological controls before their geographical distribution is interpreted:
Mn, Ba, Ce, Co, Cr, La, Nb, Rb, Sc, Sr, Th, V, Y

The second example deals with the study of metamorphosed cratonic rocks, and illustrates the complementary nature of the major and trace element factor spaces. In Fig. 12, obtained for the major elements, sodic and potassic felsic gneisses can be discriminated from each other and form very homogeneous populations; meta-arkoses and meta-pelites overlap largely with each other and have a wide-ranging chemical (mineralogical) composition. One may notice that this factor plot F2–F4 is actually equivalent to a classical KCN (K_2O–CaO–Na_2O) petrological diagram, but that it displays more variability information than a normal ternary diagram would. In Fig. 13, trace element patterns for the same rocks are shown and are quite different from the patterns of Fig. 12: sodic felsic gneisses are easily discriminated from other rocks due to their characteristic Sr content; the three other sample populations share an essentially linear pattern resulting from a nega-

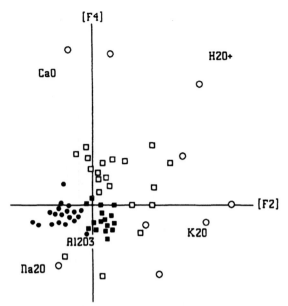

Fig. 12. Geochemistry of Precambrian metamorphic rocks: correspondence analysis of the major oxide data, showing petrological characteristics of the main rock types identified in the field (see text for discussion). Note the equivalence of this factor plot to the classical $K_2O-CaO-Na_2O$ (KCN) petrological diagram. Only those oxides which contribute significantly to either factor or are close to the displayed factorial plane are shown; each symbol represents one rock sample (from ref. 3). ● = Sodic felsic gneisses; ■ = potassic felsic gneisses; □ = meta-arkoses; ○ = meta-pelites.

tive correlation between Ba and the other trace elements of significance here (i.e. Rb and the incompatible elements); the potassic felsic gneisses show more heterogeneity in trace elements than their mineral composition suggested in Fig. 12. Clearly, some more effort must be spent understanding the trace element geochemistry of these rocks in relationship with their geological history.

4 ANALYSIS OF RECODED DATA

Data recoding adds much flexibility to the toolbox of the user of multivariate data analysis. Data recoding involves modifying the data in one of the following ways (see ref. 2 for more details):

(1) changing the scale of a variable using a continuous transform function (see Table 3a for examples);

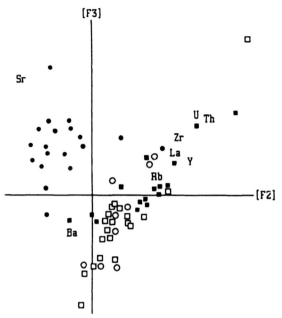

Fig. 13. Geochemistry of Precambrian metamorphic rocks: correspondence analysis of the trace element data. Compare the sample patterns to those of Fig. 12 (see text for discussion). Same notation as in Fig. 12 (from ref. 3).

TABLE 3

Examples of data recoding schemes

(a) Rescaling a continuous variable:
from: $X = \{x_{min} \ldots x_i \ldots x_{max}\}$
to: $X' = \log X$
or to: $X' = \dfrac{X - \text{mean}(X)}{\sigma(X)}$
or to: etc.

(b) Recoding a continuous variable to an ordinal variable:
from: $X = \{x_{min} \ldots x_i \ldots x_{max}\}$
to: $X' = 1$, if $x_{min} \leqslant X < x_1$
$= 2$, if $x_1 \leqslant X < x_2$
$=$ etc.
$= n$, if $x_{n-1} \leqslant X \leqslant x_{max}$
giving: $X' = \{1 \ldots n\}$

(c) Recoding a nominal variable to a logical variable:
from: $X = \{1 \ldots n_1 \ldots n_2 \ldots n\}$
to a binary variable: $X' = 0$, if $1 \leqslant X < n_1$
$= 1$, if $n_1 \leqslant X \leqslant n$
or to a complete disjunctive variable:
$X' = \{X'_1, X'_2, X'_3\}$
$X' = \{1, 0, 0\}$, if $1 \leqslant X \leqslant n_1$
$= \{0, 1, 0\}$, if $n_1 \leqslant X < n_2$
$= \{0, 0, 1\}$, if $n_2 \leqslant X \leqslant n$
or to: etc.

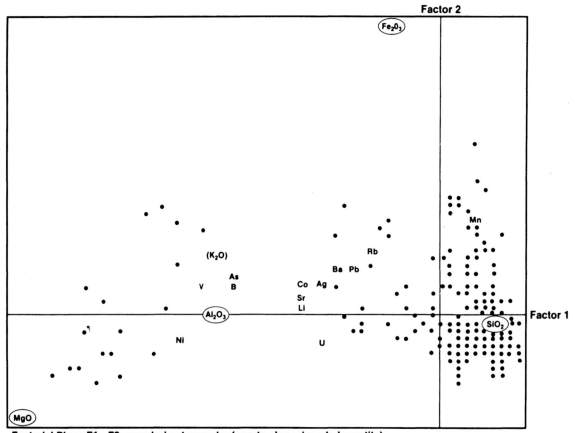

Factorial Plane F1 - F2: samples' petrography (quartz-clay minerals-hematite)

Fig. 14. The Maurice Bay property (see Fig. 5 for location): correspondence analysis of the major oxide data, showing the main petrographical types. The factors displaying clay matrix mineralogy were shown in Fig. 9 (see text for discussion). Circled oxides contribute significantly to either factor, while K_2O is close to the factorial plane displayed; all trace elements were treated as supplementary variables and only those close enough to this factorial plane are shown here. Each closed circle symbol represents one sandstone sample (from ref. 7).

(2) changing the nature of the variable from continuous to ordinal and/or from ordinal to logical (see Table 3b and 3c for examples).

Data recoding must have a purpose, related either to the type of data or to the project under study. For example, one may have to take into account the nature of the data: calculating the mean of a nominal variable does not make much sense, and such a variable will be recoded as a complete disjunctive variable. In cases where the measured variables are expressed on very different scales, it will be wise either to rescale the data through some linear or other type of transform functions, or to recode the continuous variables to ordinal variables. This latter type of coding can also be used when data distributions are very skewed or heterogeneous, and it then helps increase the robustness of the analysis (i.e. decrease its susceptibility to small changes in the data). Finally, one may wish to radically change the nature of the initial variables, to enhance some features of interest in the data such as when continuous variables are recoded into ordinal variables and then into complete disjunction variables: in that case, one will be able to investigate non-linear patterns of variation present in the continuous variables [20]. The first example used here illustrates how non-linear lithogeochemical

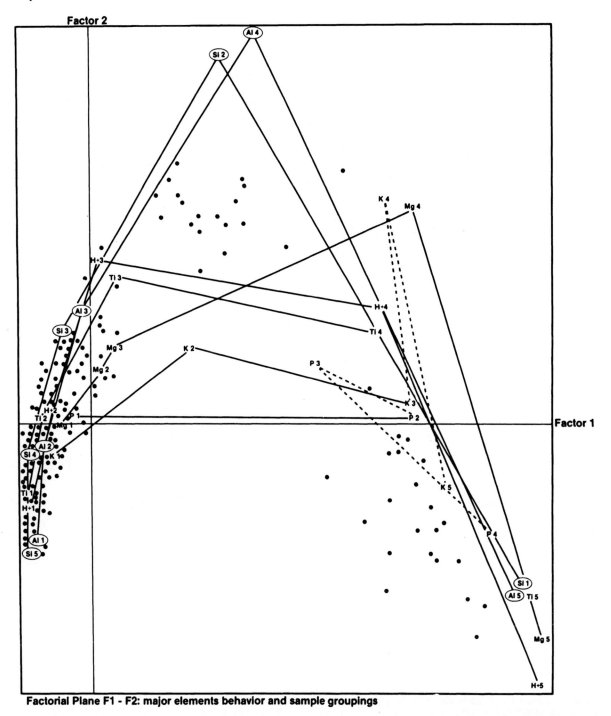

Fig. 15. The Maurice Bay property (see Fig. 5 for location): correspondence analysis of the major oxides recoded as complete disjunctive variables, showing non-linear variation patterns (see text for discussion). Note the various shapes of the non-linear trends, the presence in them of some discontinuities, and the better discriminant power of this analysis for sample subpopulations (compare with factor 1 of Fig. 14). Circled variables contribute significantly to either factor, others are close to the factorial plane displayed; solid lines outline continuous non-linear trends, whereas stippled lines indicate discontinuities in such trends; each closed circle symbol represents one sandstone sample (from ref. 7).

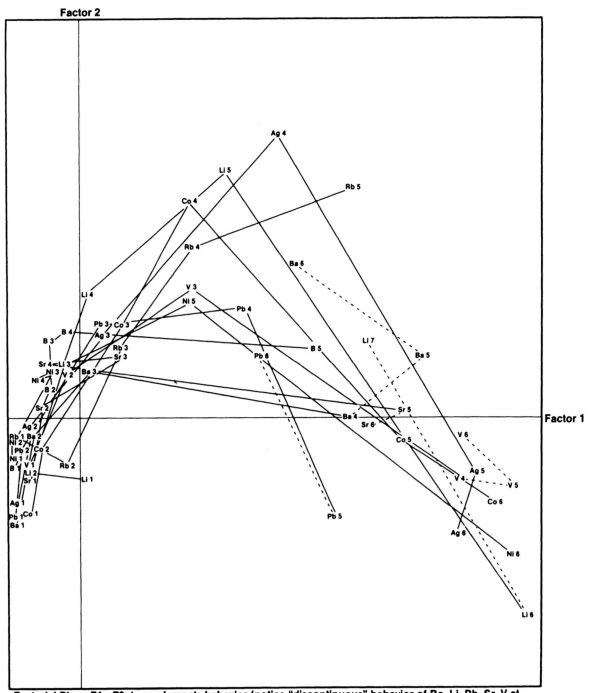

Factorial Plane F1 - F2: trace elements behavior (notice "discontinuous" behavior of Ba, Li, Pb, Sr, V at higher concentration ranges)

Fig. 16. The Maurice Bay property (see Fig. 5 for location): projection of the trace elements recoded as complete disjunctive variables onto the factorial plane of Fig. 15, all treated as supplementary variables (see text for discussion). Same scale and variable notations as for Fig. 15 (from ref. 7).

patterns can be examined, while the second example illustrates the enhancement of ore element topologies in a mineralized suite of samples.

The first example, discussed in more detail in ref. 7, uses the data from Maurice Bay (see location on Fig. 5). First, correspondence analysis of the raw major elements resulted in mineralogical patterns; those related to the clay minerals were shown in Fig. 9, while those related to the general petrography of the sandstones and of interest to us in this section, are displayed in factor plane F1–F2 on Fig. 14. The first factor results from variations in the quartz/clay minerals ratio of the sandstone (quartz-rich rocks at right, clay-rich rocks at left); factor 2 is essentially a pure hematite (Fe_2O_3) factor. In a second stage, the major elements were recoded first into ordinal variables ($SiO_2 = 1$ to 5, ...) and then into complete disjunctive variables (Si1 to Si5, ...) as explained in Table 3c; trace elements were recoded in the same manner; up to 5 and 7 intervals were used for the recoding of major and of trace elements, respectively. Correspondence analysis of the recoded major elements produces non-linear patterns for these complete disjunctive variables (Fig. 15): silica (Si1 to Si5) and alumina (Al1 to Al5) show a nice negative correlation on the plot factor 1–factor 2, while some other major elements follow a similar but not quite identical non-linear pattern. A detailed interpretation will not be carried out here. Of special interest, however, are those elements which show discontinuities in their patterns, such as potassium (for range K4) and phosphorus (for range P3); the overprinting chloritization mentioned earlier may be invoked for explaining the 'unaligned' location of K4 (which is 'pulled' towards Mg4). Of interest also is the fact that complete disjunctive variables appear to be more discriminating than the initial continuous variables: two essentially distinct sample subpopulations can been seen on the factor plot of Fig. 14, while three sample subpopulations are clearly distinguished on the factor plot of Fig. 15. When the complete disjunctive trace elements are projected as supplementary variables (i.e. without being taken into account when calculating the factors) on this same factor plot (Fig. 16), their non-linear patterns can tentatively be correlated with those of Fig. 15.

Several of the trace elements also have a discontinuous behaviour at higher ranges of concentration: Ba, Li, Pb, Sr, and V. In this example, the information that is extracted by factor 1 alone for the continuous variables (Fig. 14), that is, the

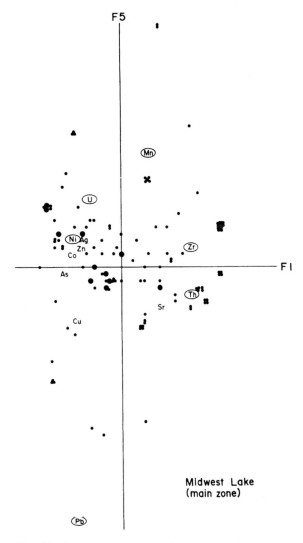

Fig. 17. Ore/trace element topology at the Midwest Lake property (see text for explanation): Zr and Th are related to detrital minerals, whereas Mn is related to siderite and Sr to phosphates; elements to the left of factor F5 form the mineralization suite. Circled elements contribute significantly to either factor, others are close to the factorial plane displayed; each closed circle symbol represents one altered sandstone sample of the illite trend (smaller circles) or the chlorite trend (larger circles). See Figs. 7 and 8 for identification and separation of these alteration trends (from ref. 17).

mineralogical relationships between detrital quartz and the clay mineral matrix, is now expressed along the first three factors extracted from the complete disjunctive data (factor 3 is not shown here) and in non-linear form: the pattern obtained from the continuous variables is thus 'dissected' or 'exploded' into several non-linear patterns when one recodes the continuous variables to complete disjunctive variables.

In the final example, explained more fully in ref. 17, trace and ore element data for selected alteration trends in sandstone are first separated (see Section 3.1, examples shown in Figs. 7–9), and then recoded into a binary form, as explained in Table 3c, so that concentrations higher than a given threshold for an element are recoded as '1' while concentrations lower than the threshold are recoded as '0'. The purpose of this coding scheme is to enhance the data in such a way that only the 'high' (here: mineralized) and 'low' (here: non-mineralized) character of the element concentrations is kept. Correspondence analysis of these recoded data produces a topology of ore and trace element associations for the alteration trends observed at various uranium deposits in the Athabasca Basin. In Fig. 17, the result for the Midwest Lake property are shown, where a mineralization episode involving Pb (negative side of F5) is separate from the main U–Ni–Co–(As) mineralization phase (negative side of F1); at Maurice Bay (Fig. 18), it is a Cu–Zn mineralization episode (lower right of the projection) that can be discriminated from the main U–Ni–Co–As episode (positive side of F4). The purpose of that recoding exercise was achieved in the study: several mineralization phases could be detected for the Athabasca unconformity-type uranium deposits as a group, while local variations along that theme were identified at each of the deposits investigated.

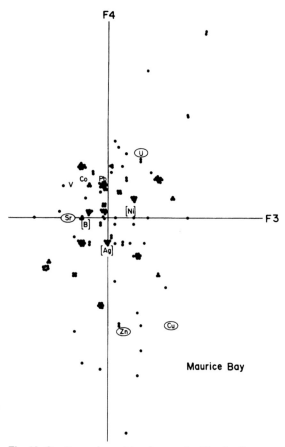

Fig. 18. Ore/trace element topology at the Maurice Bay property (see text for explanations): Sr is related to phosphates and B to tourmaline and clay minerals; other elements form the mineralization suite. Same element notations as for Fig. 17, with elements in square brackets being ubiquitous at Maurice Bay. See Fig. 9 for the identification of altered sandstone samples, all represented with the same symbol here (from ref. 17).

5 CONCLUSION

The success obtained to date in the application of correspondence analysis to the investigation of lithogeochemical data is seen as having two components. Firstly, a data analysis strategy that is based on geological and geochemical knowledge has been established, which guarantees that information useful to the geologist and geochemist will be produced. The key aspects of this strategy are:

(1) the selection of sets of chemical elements which are homogeneous with respect to the information sought,

(2) the complementary nature of the parallel treatment of these sets of variables,

(3) the geological character of the interpretation of the results, and

(4) when required, the implementation of data

recoding (data enhancement) schemes which are adapted to specific aspects of each project under study.

Secondly, the method used for the investigation of the data, correspondence analysis, is both powerful and flexible, and, with a minimum of practice, the user can learn to avoid some of the interpretational pitfalls of multivariate statistical techniques. The key features of correspondence analysis in this respect are:

(1) its geometrical approach (a feature shared with other factor analysis methods such as principal components analysis),

(2) its complete set of output parameters which allow the user to assess precisely the status of each row, column and factor, and

(3) its flexibility ('universality') with respect to data types. Of course, the more flexible a method is, the more responsibility the user will have to assume in making decisions; however, because these decisions are related to the user's field of endeavour and not to the field of statistics, this will not cause any difficulty.

Also, it is evident that correspondence analysis can be applied successfully to geological data other than lithogeochemical data, inasmuch as the user has established a data analysis strategy that is well adapted to the type of data being investigated.

6 ACKNOWLEDGEMENTS

Each one of us is first a learner: the ideas underlying French data analysis philosophy were learned from the several sources cited and integrated during a summer school lead by French colleagues in Orsay some time ago. In addition, one does not become a user of multivariate statistics overnight: the application of these ideas to lithogeochemical data has been carried out over the past few years within the scope of the project "Geochemistry and Data Analysis" at the Saskatchewan Research Council, and has benefited from the cooperation of various mineral exploration companies active in the Province of Saskatchewan.

I wish to thank the organizers and sponsors of the workshop "Multivariate Statistics for Geochemists and Geologists" for inviting me to present my work. R.G. Brereton, G. Nickless, and an anonymous reviewer are gratefully acknowledged for their constructive criticisms and their suggestions for improving an earlier manuscript.

REFERENCES

1 M. Mellinger, Multivariate data analysis: its methods, *Chemometrics and Intelligent Laboratory Systems*, 2 (1987) 29–36.

2 M. Mellinger, Correspondence analysis: the method and its application, *Chemometrics and Intelligent Laboratory Systems*, 2 (1987) 61–77.

3 M. Mellinger, Statistical analysis and modelling of geochemical data: what can multivariate data analysis do for you?, *Geochemistry and Data Analysis, Proceedings of a Symposium held in Luleå, Sweden, October 15–16, 1985*, Swedish Geological Co., Luleå, 1986, pp. 61–73.

4 J.P. Benzécri, *L'Analyse des Données: 2. L'Analyse des Correspondances*, Dunod, Paris, 1980, 632 pp.

5 L. Lebart, A. Morineau and K.M. Warwick, *Multivariate Descriptive Statistical Analysis: Correspondence Analysis and Related Techniques for Large Matrices*, Wiley, New York, 1984, 231 pp.

6 M.J. Greenacre, *Theory and Applications of Correspondence Analysis*, Academic Press, London, 1984, 364 pp.

7 M. Mellinger, Correspondence analysis in the study of lithogeochemical data: general strategy and the usefulness of various data-coding schemes, *Journal of Geochemical Exploration*, 21 (1984) 455–469.

8 D.S. Korzhinskii, *Physico-chemical Basis of the Analysis of the Paragenesis of Minerals*, Consultants Bureau Inc., New York, 1959, 142 pp.

9 W.A. Deer, R.A. Howie and J. Zussman, *An Introduction to the Rock-Forming Minerals*, Longman, Edinburgh, New York, 1966, 528 pp.

10 W.G. Ernst, *Petrologic Phase Equilibria*, Freeman, San Francisco, 1976, 333 pp.

11 H.G.F. Winkler, *Petrogenesis of Metamorphic Rocks*, Springer, Berlin, New York, 1974, 320 pp.

12 R.G. Burns, *Mineralogical Applications of Crystal Field Theory*, Cambridge University Press, 1970, 224 pp.

13 D. Crerar, S. Wood and S. Brantley, Chemical controls on solubility of ore-forming minerals in hydrothermal solutions, *Canadian Mineralogist*, 23 (1985) 333–352.

14 A.A. Beus and S.V. Grigorian, *Geochemical Exploration Methods for Mineral Deposits*, Applied Publishing Ltd., Wilmette, IL, 1977, 287 pp.

15 M. Mellinger, Recherche des métallotectes dans la région de Normétal: pétrographie et géochimie, *Rapport Intérimaire*, Ministère des Richesses Naturelles du Québec, 1978, DPV-582, pp. 33–160.

16 M. Mellinger, Evaluation of lithogeochemical data by use of multivariate analysis: an application to the exploration

for uranium deposits in the Athabasca Basin of Saskatchewan (Canada), *Proceedings of the APCOM '84 conference, held in London, U.K., March 1984*, Institute of Mining and Metallurgy, London, 1984, pp. 21–27.

17 M. Mellinger, D. Quirt and J. Hoeve, Geochemical signatures of uranium deposition in the Athabasca Basin of Saskatchewan (Canada), in B. Poty and M. Pagel (Editors), Concentration Mechanisms of Uranium in Geological Environments — Conference Report, *Uranium*, 3 (1987) 187–209.

18 M. Mellinger, The Maurice Bay uranium deposit (Saskatchewan): geology, host-rock alteration and genesis, in T.I.I. Sibbald and W. Petruk (Editors), *Geology of Uranium Deposits*, CIMM Special Volume 32, Canadian Institute of Mining and Metallurgy, Montreal, pp. 140–150.

19 M. Mellinger, oral communication, Workshop: Geochemical Anomaly Recognition, *11th International Geochemical Exploration Symposium, held in Toronto, Canada, April 28–May 2, 1985*.

20 J.M. Bouroche and G. Saporta, *L'Analyse des Données*, Presses Universitaires de France, Paris, 1980, 127 pp.

Multivariate Analysis of Stratigraphic Data in Geology: A Review

H.J.B. BIRKS

Botanical Institute, University of Bergen, P.O. Box 12, N-5027 Bergen (Norway)

ABSTRACT

Birks, H.J.B., 1987. Multivariate analysis of stratigraphic data in geology: a review. *Chemometrics and Intelligent Laboratory Systems*, 2: 109–126.

Stratigraphic multivariate data are common in geology, for example as biostratigraphic, lithostratigraphic, geochemical, geophysical, morphometric or isotopic variations in stratigraphic sections and boreholes. All such stratigraphic data are multivariate in character and have a known sequence or temporal ordering of samples, so-called stratigraphic constraints. Multivariate data-analytical techniques that take specific account of this stratigraphic ordering permit (1) partitioning or zonation of the stratigraphic data, (2) comparison and correlation of the stratigraphic data, (3) detection of stratigraphic patterns and (4) the establishment of relationships between different stratigraphic variables in the same sequence. They can also aid in the interpretation of stratigraphic patterns. Recent developments in these topics are reviewed. In addition, problems of analysing proportional data and of data transformations, and possible future developments aimed towards an analytical phase in the analysis of stratigraphic data, are outlined.

INTRODUCTION

Stratigraphic multivariate data abound in historical geology in the form of biostratigraphic, lithostratigraphic, geochemical, geophysical, morphometric and isotopic variations through geological time sequences, as preserved in boreholes and stratigraphic sections. Examples of such data include: (1) abundance estimates of different types of microfossils (e.g., pollen, ostracods, foraminifers) at different but known depths in boreholes or sections, so-called biostratigraphic data; (2) estimates of sediment composition (e.g., particle size classes) at different but known depths in boreholes or sections, so-called lithostratigraphic data; (3) abundance estimates of different chemical variables (e.g., elements, organic compounds) at different but known depths in stratigraphic sequences, so-called geochemical data; (4) physical properties of sediments in boreholes such as magnetic orientation and susceptibility, neutron logging (induced radiation), gamma radiation, electrical self-potential, electrical resistivity, sonic logs and redox logs, so-called geophysical data; (5) measurements of morphometry (e.g., length, breadth, shell thickness, coiling angles) in the same fossil taxon at different but known depths in stratigraphic sequences, so-called morphometric data; and (6) stable isotope ratios in one or more fossil taxa at different but known depths in stratigraphic sequences, so-called isotopic data.

Although seemingly rather different, these data types have certain features in common. They are multivariate, consisting of estimates or measurements of many variables in a large number of samples, and there is a known stratigraphic sequence or temporal ordering of the samples. The samples may form a continuous observational series, as in many geophysical data, or a discontinuous series with discrete sampling levels, as in many biostratigraphic, geochemical, morphometric and isotopic data. This ordering of samples with depth and hence age, so-called stratigraphic constraints, is of paramount importance in historical geology. It provides the basis for (1) presenting the data as stratigraphic diagrams of variables plotted against depth and thus through time, (2) partitioning the data into sections of broadly similar composition or physical properties, (3) recognizing stratigraphic changes within the entire sequence, within and between sequences of individual variables, and between entire sequences, and (4) interpreting the data as a temporal record of changes in past biota, sedimentary conditions or regional environment.

Questions commonly arising in the handling, analysis, summarization and interpretation of stratigraphic data include: (1) can the sequence be partitioned into sections of more-or-less similar composition or physical properties, namely problems in segmentation or zonation?; (2) how similar are two or more sequences and how do they relate to each other, namely problems in sequence comparison and correlation?; (3) what stratigraphic trends and changes occur within the sequence as a whole, namely problems in pattern detection?; (4) what relationships exist between different types of variables (e.g., morphometric and geochemical) in the same sequence, namely problems in correlation between time series?; and (5) how can observed stratigraphic patterns be interpreted in terms of underlying processes, namely problems in geological interpretation?

Standard multivariate data-analytical techniques such as principal components analysis (PCA), correspondence analysis (CA) and hierarchical cluster analysis do not take account of stratigraphic constraints and hence may discard important geological information when applied to the analysis of stratigraphic data. Standard time series analysis techniques such as auto-correlation, cross-correlation and power-spectral analysis commonly assume that sample observations are made at equal time intervals. Clearly there are great difficulties in meeting this requirement with almost all stratigraphic data. In recent years, several robust multivariate data-analytical methods have been developed, particularly in the field of stratigraphic Quaternary pollen analysis, that take account of stratigraphic constraints and do not require or assume equal time intervals between samples. These were reviewed by Birks and Gordon [1]. Many of these methods are directly applicable to other types of stratigraphic data from earlier time periods, as shown, for example, by Reyment's [2] Cretaceous studies.

The aims of this paper are (1) to present briefly methods of potentially wide utility in stratigraphic geology that can assist in answering the types of questions presented above; (2) to outline new developments that have occurred since Birks and Gordon's [1] text was completed; and (3) to suggest possible future developments in analysing stratigraphic data. Before considering data-analytical methods, it is important to consider problems of analysing proportional compositional data, one of the commonest types of geological data, and of appropriate data transformations for different types of quantitative data. Solutions, complete or partial, to these problems are fundamental to subsequent multivariate data analysis.

The contents of the paper are therefore (1) proportional data, closure problems and data transformations; (2) zonation of stratigraphic data; (3) comparison and correlation of stratigraphic data; (4) detection of stratigraphic patterns; (5) establishment of relationships between different stratigraphic variables; (6) interpretation of stratigraphic patterns; and (7) possible future developments.

PROPORTIONAL DATA, CLOSURE PROBLEMS AND DATA TRANSFORMATIONS

Proportional data are common in geology and geochemistry, for example as proportions of different fossil taxa in stratigraphic samples or proportions of different elements in geochemical analyses. The mathematical properties of proportional data with their constant sum create the widely discussed "closure" problems and the resulting dangers of interpreting correlations between variables when their values are expressed as proportions [3]. Proportions are constrained within a closed simplex space of $m - 1$ dimensions, where there are m variables, whereas almost all statistical and multivariate data-analytical methods assume open Euclidean space. Log-ratio [4] and log-linear contrasts [5] transform proportional data confined within a closed simplex into open Euclidean space. Recent applications in geochemistry [6,7], serology [8] and sedimentology [9] illustrate their value in avoiding the inevitable "spurious correlations" [3] and minimizing the curvatures that commonly occur in PCA of proportional data [5]. Log-ratio transformations also provide means of modelling and analysing different sources of measurement and observer errors and of assessing effects of such errors on statistical inferences from geochemical data [10]. Problems arise in these transformations with very small proportions that inevitably are not estimated accurately and with zero proportions [1,4,6,7]. Ad hoc techniques such as log-normal or conditional modelling can help, but further work is needed to resolve these difficulties [6]. Despite these limitations, Aitchison has opened up a new and important area of statistical methodology for analysing closed, proportional compositional data, and has rightly tried [7] "to persuade geologists... to abandon use of "standard methods" quite inappropriate to "nonstandard" data sets such as compositions."

Concentration, flux density or other non-relative data often require variance-stabilizing transformation prior to multivariate analysis. Many transformations exist in the literature [11]. Gordon's [12] arc-tan transformation is particularly useful for biostratigraphic data. Standardization of variables to zero mean and unit variance, called autoscaling in chemometrics [13], increases the contribution of minor variables and increases the "noise" relative to "signal" in multivariate data [14]. Standardization or autoscaling should therefore only be used when absolutely necessary, for example when variables are estimated in different units. In other situations it should be avoided and other, more appropriate, transformations [14] used. Abundance coefficients [15], based on geometric density categories, approximate to the Poisson parameter, λt, and represent a natural coding of abundances with useful mathematical and predictive properties. They could usefully be used with many types of biostratigraphic and geochemical data.

ZONATION OF STRATIGRAPHIC DATA

It is often convenient to partition sequences of complex multivariate stratigraphic data into smaller units for purposes of (1) description, (2) discussion and interpretation and (3) correlation in time and space of different sequences [1]. The most useful stratigraphic unit for subdividing the vertical, temporal dimension of a sequence is a zone, informally defined here as a sediment body with a broadly similar composition (for example, fossils, sedimentary components or chemistry) or similar physical properties (for example magnetic, isotope or electrical-logging parameters), and that differs from underlying and overlying sediment bodies in the kind and/or amount of its contained stratigraphic variables [1].

Various partitioning methods exist that either take account of the stratigraphic ordering of the samples or the concept of a stratigraphic zone as a body of sediment with a broadly similar composition or homogeneous physical properties [16]. Unless both are considered, important geological information is discarded in zonation. All methods outlined here take account of stratigraphic ordering and the above concept of a zone.

Sequential digitized data such as univariate electrical logs, seismic traces or magnetic profiles or discontinuous univariate stratigraphic data such as a stable isotope ratio curve can be partitioned by means of split moving-window techniques [17] if the sequence shows low overall variability but sharp changes. Piecewise regression [18] is appropriate for sequences containing trend, whereas maximum level variance techniques [19–22] are robust and effective for almost all types of trend-free sequences (see refs. 23 and 24 for comparisons of these techniques).

Birks and Gordon [1] reviewed different methods for zoning multivariate stratigraphical data. The basic idea is to partition the sequence of samples on the basis of overall dissimilarity, however defined, into zones but only to classify together samples that are stratigraphically adjacent. The simplest procedure is constrained single-linkage cluster analysis. Dissimilarities between all pairs of samples are calculated on the basis of total composition using an appropriate dissimilarity measure [14,25]. Stratigraphically adjacent samples of low dissimilarity and hence similar composition are grouped together using the single-linkage criterion. Hence only contiguous samples or groups of samples are permitted to fuse. The process continues until all samples are classified. Results are conveniently represented as a dendrogram, and groups of broadly similar, contiguous samples delimited as zones. Its principal disadvantage is that with gradual stratigraphic changes, zones may be difficult to delimit because of chaining [1]. Recently, Grimm [26] developed a similar approach but instead of using single-linkage criteria, he used minimum sum-of-squares criteria. This is an intense clustering criterion that minimizes chaining. Its disadvantage is that it tends to produce compact, "spherical shaped" groups of samples, often of roughly similar size.

In constrained binary divisive analysis [1], the aim is to partition a sequence into t zones by placing $t-1$ boundaries so as to minimize within-zone variation, as measured by information content or sum-of-squares deviations about t centroids. This partitioning is computationally possible because of the stratigraphic ordering of samples and can be implemented in various ways. For stratigraphic purposes a hierarchical solution is preferable (see refs. 1 and 16 for examples). Initially, total information (or total sum-of-squares) is calculated, and a boundary placed in all possible places ($n-1$, where n is the number of samples). The position giving the largest reduction in within-zone variability and hence the most homogeneous pair of zones is selected as the first boundary. This binary division is repeated by partitioning existing subdivisions so as to produce maximal decreases in variability until little further reduction in variability occurs. The pathway of divisions can be represented as a "block" dendrogram, and the resulting partitions regarded as zones [1]. The principal advantage of these and other divisive procedures is that data for a whole zone are considered in partitioning and locating zone boundaries. Their main disadvantage is that there is no guarantee that the nested partitions are the optimal divisions that give the overall total within-group variability. This is because by positioning the first boundary at position n_1, the second boundary, n_2, is located by binary divisions in a mathematically optimal position subject to the constraint that the first boundary is fixed at n_1.

Partitions from optimal divisive analysis [1] need not be hierarchically nested. For example, optimal partitioning into three zones is not necessarily obtained by dividing one of two pre-existing groups into two. It is computationally very time consuming to examine all possible partitionings into 3, 4, ..., zones especially for larger data sets. Optimal least-squares partitionings of stratigraphic data can be obtained efficiently using the optimality principle of dynamic programming (see refs. 1 and 27–29 for details of algorithms). As the analysis is non-hierarchical, the results cannot be presented as a dendrogram but are most conveniently given in a triangular format (see refs. 1, 12 and 27 for examples). Although optimal divisive analysis has, to date, only been used with sum-of-squares criteria, it can utilize any criterion for which total variation of the partitions is additive. Of the various partitioning methods currently available, it merits widest use. A FORTRAN program for optimal and binary divisive analysis and constrained single-linkage cluster analysis is given in ref. 1.

Birks and Gordon [1] discussed an additional method, the variable-barriers approach. This is more restrictive and attempts to distinguish groups of transitional samples with intermediate composition or properties from groups of samples with broadly similar composition of properties, and hence to delimit two zone types — those of relatively constant character and those of abrupt or systematic changes in character (see ref. 30 for a recent application). It can only be implemented by an iterative procedure that is computationally expensive. Its use is restricted to comparatively small sequences (< 50 samples).

Because all these partitioning procedures make different assumptions with regard to the data, use different concepts and measures of sample dissimilarity and sequence variability and utilize different partitioning criteria, no single method is the most appropriate for a given data set. As in other exploratory data analysis, several methods should be used, their results compared and any consistent patterns viewed as the most effective partitions. As in other multivariate data analysis, great care is needed in selecting appropriate dissimilarity measures [14], data transformations and differential weighting of variables [1,31–33]. When used carefully and critically, these procedures provide a repeatable, useful and unambiguous zonation of stratigraphic multivariate data. They have recently been applied to problems in lithostratigraphy [34], foraminifer stratigraphy [35] as well as pollen stratigraphy [36,37].

COMPARISON AND CORRELATION OF STRATIGRAPHIC DATA

Comparison and correlation of sequences are important for constructing regional stratigraphies, matching of sequences and boreholes and interpreting, for example, within-basin events. For many stratigraphic problems it is important to use methods that do not attempt a direct matching of a sample in one sequence with a sample in another sequence, as in cross-association techniques [38], because a sample in one sequence may simply not be directly matchable with a sample in another sequence, but may only resemble it to a lesser or greater extent and because gaps commonly occur in sedimentary sequences. Howell [39] described a robust form of cross-association that allows for gaps and the matching of one sample in a sequence to several samples in another sequence. It has obvious potential in lithostratigraphy. Cross-correlation techniques assume that consecutive groups of samples in one sequence can be correlated with consecutive groups of samples in another sequence with no gaps or repeats in the sequences and that the samples are at regular time intervals [40]. Modifications designed to maximize cross-correlations between sequences include shifting, stretching or shrinking of sequences [41,42], Holgate "fine-tuning" [43] and cross-correlations between power spectra [44]. Cross-correlation is useful in matching continuous varve, isotope, magnetic and geophysical logging sequences, but is less appropriate and rarely useful in matching biostratigraphic, lithostratigraphic, geochemical or other discontinuous sequences. More robust methods of sequence comparison are required for such data (see refs. 1, 2, 16, 45 and 46).

The simplest and, in many ways, the most satisfactory approach is combined geometrical scaling of the sequences of interest, using PCA, principal coordinates analysis (PCoA), CA or biplots [1,47]. All these techniques seek to represent, in low-dimensional space, similarities defined mathematically between samples. Samples of similar composition or physical properties, irrespective of the sequence from which they are derived, will, if the low-dimensional scaling is an accurate and efficient representation of the original multidimensional data, be positioned near to each other in the scaling plot. Although PCA and PCoA are most commonly used for comparing sequences (see ref. 1 for references), CA and biplots are probably the most useful for general comparative purposes because samples from all sequences of interest and their associated variables can be displayed together on the same scaling plot, thereby facilitating ease of interpretation [12,47,48]. The use of PCA, PCoA, CA and biplots allows scope for various data transformations [1,12,14,16], whereas PCoA allows a choice of different measures of sample similarity [1,14].

An alternative approach for comparing sequences is sequence slotting [1,45,49]. In combined scalings, no stratigraphic constraints are imposed during numerical analysis, whereas in sequence slotting sample ordering is an essential constraint. Sequences are slotted together into a single sequence so that parts of sequences with similar composition or properties are placed close together in the joint sequence, subject to the stratigraphic ordering in the separate sequences. A discordance measure (Ψ) is defined for the slotting,

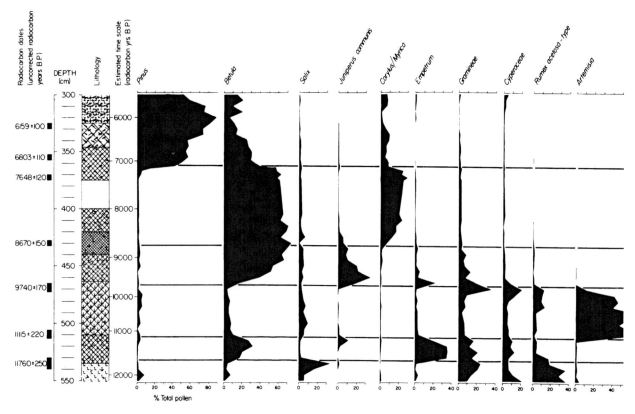

Fig. 1. Stratigraphic pollen diagram from Abernethy Forest, Inverness-shire, Scotland [62]. The curves represent the amounts of the nine pollen types shown expressed as percentages of the sum of these nine types. The radiocarbon dates [as years before present (B.P.)], sediment lithology, estimated time scale, and pollen sum (ΣP) are also shown, together with the sample scores on the first three principal component axes derived from a PCA of the correlation matrix between the nine taxa, after log-linear contrast transformation [5] of the pollen proportions. For further details of the site, sediment lithology and pollen stratigraphy, see ref. 62.

and the optimal slotting that minimizes Ψ is found by a dynamic programming algorithm [49,50]. As the method always produces a slotting, the value of Ψ must be examined to see how reliable the slotting is, although, as yet, no formal statistical evaluation can be made for a given value of Ψ [51].

Gordon and Reyment [45] introduced constraints into sequence slotting. When comparing long geophysical logs from boreholes, it can be useful to specify that samples are situated above or below an independent marker horizon such as a tephra layer, bone bed or coal lens, or a well marked event such as a biotic extinction, all of which are assumed to be regionally synchronous.

A FORTRAN program for unconstrained and constrained sequence slotting is available [49].

The principal advantages of sequence slotting are [1,45,46] that it is multivariate and considers all stratigraphic variables, it condenses similar parts of a sequence and expands dissimilar parts and it does not assume continuous sedimentation or regular time intervals between samples. Its main disadvantage is that it is moderately sensitive to different stratigraphic patterns within sequences. "Blocking" [1] can occur if there are marked differences between sequences or if there are very minor differences between consecutive samples within a sequence (see ref. 52 for striking examples). In the former instance Ψ is usually large,

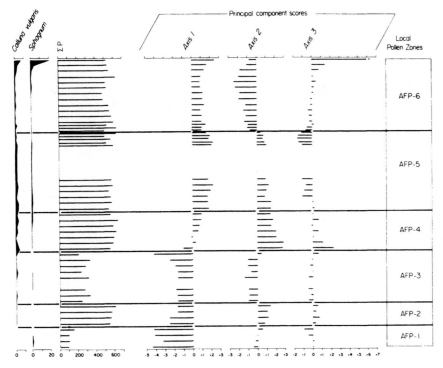

Anal. R.W. Mathewes, 1974

whereas in the latter Ψ is often low. Slotting is most robust when comparing sequences that are moderately similar but not too similar or dissimilar. Recent applications include slotting of pollen sequences [36,53], diatom profiles [54], Cretaceous boreholes [2] and frequencies of foraminifers and their proloculi diameters in boreholes [55]. Recent developments include extending slotting to three or more sequences [56] and adopting a minimal combined path length as the criterion of optimal slotting [57]. Both developments extend the potential utility and improve the robustness of sequence slotting as a technique for comparing and correlating multivariate stratigraphical data as a basis for establishing a common stratigraphy.

DETECTION OF STRATIGRAPHIC PATTERNS

For many purposes, particularly to aid interpretation, it is useful to detect major stratigraphic changes and trends within complex multivariate data and to summarize these principal patterns of variation through time. The simplest approach is to apply geometric scaling procedures such as PCA, PCoA or CA, paying due attention to questions of data transformations and dissimilarity measures. Results are most usefully presented in a stratigraphic context as stratigraphic plots of sample scores or coordinates on each scaling axis (see Figs. 1 and 2 and refs. 1 and 58–61 for examples). This representation is appropriate if the axes effectively summarize trends in the data, as the stratigraphic plots are thus "composite curves" for several variables. They commonly portray patterns of change within the sequence and the nature (gradual, abrupt) and rates of change, and maximize the "signal-to-noise" ratio in stratigraphic data.

The first three principal components of the stratigraphic pollen data shown in Fig. 1 (data

ABERNETHY FOREST, INVERNESS - SHIRE.

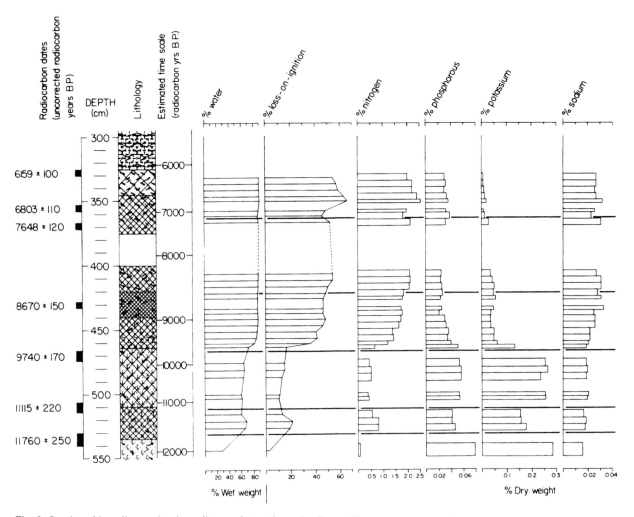

Fig. 2. Stratigraphic sediment chemistry diagram from Abernethy Forest. The curves represent the amount of water expressed as percentages of wet sediment weight and of N, P, K, Na, Ca and Mg and loss-on-ignition at 550°C expressed as percentages of dry sediment weight. Radiocarbon dates, sediment lithology and estimated time scale are also shown, together with the sample scores on the first four principal component axes derived from a PCA of the correlation matrix between the eight variables.

from ref. 62) represent 34, 16 and 12% of the original variability, respectively. Together the stratigraphic plot of the sample scores on these axes summarizes 62% of the total variation in these data. The first three components of the sediment-chemical data from the same profile (Fig. 2) represent 79, 12 and 4%, respectively. Axis one thus provides an extremely effective summary of the major stratigraphic patterns in the chemical data.

Biplots and CA [12,47] are particularly useful for detecting patterns within stratigraphic data because of the simultaneous analysis and geometric representation of samples and variables, thereby providing a firm basis for interpretation. Sample coordinates on the first two or three axes can be plotted and the samples joined up in stratigraphic order. This time course can then be traced through the different variables within the same low-dimensional, geometrical space [12]. Examples of applying CA to stratigraphic data include refs. 12, 30

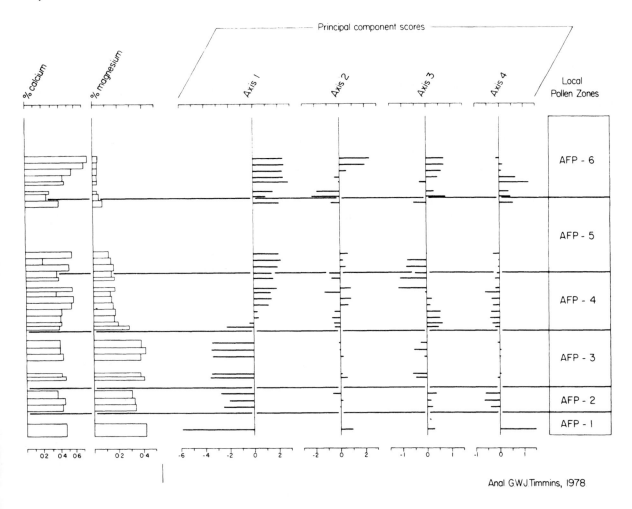

Anal. G.W.J. Timmins, 1978

and 63–65 (see refs. 47, 48, 66 and 67 for a fuller discussion of its theory, advantages and potential).

Geometric scaling methods, particularly PCA, canonical variates analysis (CVA) [68] and growth-invariant CVA [2], can be used to summarize major patterns of morphometric change in a given fossil taxon through time in borehole sequences and to attempt a partition of these patterns into size and shape components [2,46,69,70]. Plots of sample scores against stratigraphic position provide chronoclines of morphometric variation through time, so-called "biologs" [2,46] that can be compared with geochemical data or geophysical logs for the same stratigraphic interval. Different data sets, such as biostratigraphic and geochemical data from the same sequence, can be analysed together by, for example, PCA or PCoA to detect major stratigraphic patterns and temporal changes and to derive "ecologs" [2,71]. Problems arise when establishing relationships between stratigraphic variables by comparing, for example, biologs of morphometric change through time with stratigraphic logs of other variables, such as geophysical properties. This problem is considered in the next section.

ESTABLISHING RELATIONSHIPS BETWEEN STRATIGRAPHIC VARIABLES

In considering stratigraphic patterns within pollen-stratigraphic and geochemical data from the same sequence (Figs. 1 and 2), an important question prior to interpretation is whether patterns in the two data sets are similar or not. Visual inspection of the two data sets is extremely difficult and no firm conclusions can be reached. It is simplest to concentrate attention on the first few principal components or "composite curves" for the two sets. Sample scores on component one of the chemical (79%) and pollen (34%) data are superimposed in Fig. 3. Strong similarities are immediately apparent, with the two sets of scores running in close parallel. Major stratigraphic patterns in both sets contrast the late-glacial (465–550 cm) with the Holocene (300–465 cm) and highlight the strong co-variance between K, Mg and P and herbaceous taxa (e.g., *Artemisia, Gramineae*) in the late-glacial and between N and loss-on-ignition and trees (e.g., *Betula, Pinus*) in the early Holocene. Both sets reflect the major environmental change at the end of the late-glacial (about 9700 B.P.) from an open, unstable landscape with mineral soils, erosion and treeless vegetation to a closed stable landscape with organic soils, humus accumulation and forest vegetation.

After visual inspection, tentative statistical relationships between two or more time series from the same sequence can be explored by constructing an "oscillation log" [27], a record of the number of swings to the right (+) and left (−) in the two time series. The hypothesis that the oscillations in the series are significantly different can be simply tested by a χ^2 test or a likelihood ratio G-test [2]. For Fig. 3 this hypothesis is rejected.

More sophisticated and rigorous correlations between time series from the same sequence are beset with problems of "nonsense correlations" [72]. These arise because samples in a stratigraphic sequence are not independent observations but have Markovian auto-correlative properties. Hence samples at time t or depth k are dependent, to some extent, on the composition of samples $t-1$ or $k-1$. In addition, sequences may not be stationary and hence have trends in means and variances of particular variables. Spuriously high cross-correlations arise between non-stationary, auto-correlated sequences [73,74].

Sequences of different stratigraphic variables can be dichotomized into above (+) or below (−) the series median, and relationships between sequences explored by tabulating +/− runs [2,69] or tested by rank correlations [75]. Modified runs tests to test for randomness in two ordered series of +s and −s for paired observations and logistic and χ^2 tests applied to binary data can also be useful in comparing patterns between sequences and to test the hypothesis of randomness in the sequences [2,76]. More powerful methods for

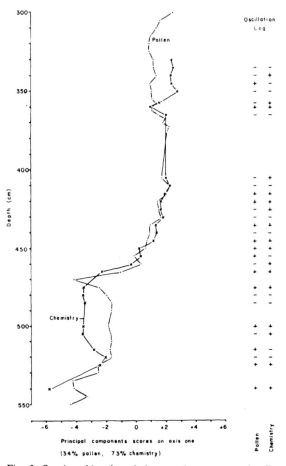

Fig. 3. Stratigraphic plot of the sample scores on the first principal component of the pollen (Fig. 1) and chemical (Fig. 2) data. An oscillation log [2] for shifts to the right (+) or left (−) in the curves is shown for directly analogous stratigraphic samples only.

quantifying relationships between stratigraphic variables include partial correlation, concordance and discordance probability tests for direction of change in variables and piecewise regression [77]. Techniques using regressions and lagged regressions with parameters based on power- and cross-spectra of time series [78] assume equal time intervals and are of limited applicability to many, if not all, geological sequences.

If one sequence is a time series of a continuous variable (e.g., self-potential log) and another is based on discontinuous, discrete sampling (e.g., fossil composition), it is important only to compare stratigraphically analogous levels [2].

Canonical correlation analysis (CCA) [68] provides a useful tool for establishing and quantifying the strength of relationships between two sets of stratigraphic variables, such as geochemical and biostratigraphic data [2,71,79]. Stratigraphic plots of CCA sample scores for the two sets provide a useful graphical summary or "ecolog" of those relationships — in parts they may closely covary, indicating a strong relationship, and in other parts they may deviate, suggesting a weak relationship [2,46,71]. CCA and related techniques of redundancy analysis and multiple correlation analysis [80] are powerful methods for establishing and modelling relationships between stratigraphic variables. They warrant wider use in geology and chemometrics.

If many stratigraphic variables are studied in the same sequence (e.g., foraminifers, inorganic chemistry, sediment particle sizes), it is valuable to establish if the major patterns in the different variables correspond stratigraphically. With many variables, the easiest approach is to partition independently each data set into zones, construct correlation diagrams for the sequence showing the zonation for each data set, compare the stratigraphic overlap between zone boundaries and, if required, test statistically the hypothesis of stratigraphically independent boundaries in the different data sets using tests based on Maxwell–Boltzman statistics [81].

Emphasis, so far, has been on detecting patterns within or between stratigraphic variables and sequences, using numerical techniques. Quantitative methods can also aid in interpreting observed patterns in terms of underlying processes, as discussed in the next section.

INTERPRETATION OF STRATIGRAPHIC PATTERNS

Stratigraphic data are time series, namely sequences of variables (for example, fossil abundances, sediment geochemistry, isotope ratios) varying through time. Time series can be uni-, bi- or multivariate and it is often difficult to interpret the patterns within long time series. Time series analysis provides many numerical and statistical techniques for investigating and modelling patterns within temporal data (see ref. 82 for an excellent introduction). It is natural that geologists have explored these techniques to help interpret observed patterns of stratigraphic variables and to investigate and model relationships between variables with reference to time and frequency domains (see ref. 83 for details).

In time-domain analysis, auto-correlations provide a means of exploring temporal patterns within a variable, whereas cross-correlations are useful for detecting temporal patterns between variables. In frequency-domain analysis, power-spectral analysis highlights different frequencies of temporal variation that account for the major patterns of variability within the time series and helps to detect periodicities. In almost all time-series analytical techniques, observations are assumed to be equally spaced in time. Clearly this is not the case with most stratigraphic data [1]. Interpolation between adjacent samples is commonly used to provide data points with constant inter-sample time intervals. However, interpolation may be of little value and unreliable when there is no reliable independent absolute chronology, when the temporal resolution of the samples is imprecise and when there is a paucity of data points within the time series (see ref. 84 for a review of handling irregularly spaced or missing data in time series analysis). Exceptions are stratigraphic data sampled at numerous, equal time intervals from varved sediments or from dated, rhythmically deposited sediments. In these cases, periodicity patterns can be confidently and reliably estimated using time-series analysis [85,86].

Additional problems can arise in time series analysis of stratigraphic data because of the statistical requirement that the time series be stationary over time, i.e., show no trends in means or variances. Although techniques exist for determining if time series are stationary and for transforming or filtering data to ensure that this is met, there is frequently a paucity of data points (30–50 samples) within stationary sections of geological sequences. Reliable estimates of power spectra, on the other hand, require stationary sequences with 100–200 data points. Only very well marked features are therefore likely to be detected by time-series techniques from the short stationary sequences common in stratigraphic studies.

Recent discussions about temporal patterns and possible periodicities in "mass extinctions" over the last 250 million years highlight the considerable problems of applying statistical time-series analysis to sparse, incomplete geological data and of using time-series techniques as a tool for interpreting patterns in stratigraphic data (see refs. 83 and 87–91 for contrasting analyses and conclusions). As Connor [83] emphasizes, "the unique challenge of the fossil record to the time-series analyst is to develop techniques to deal with unequally and irregularly spaced observations from a sparse temporal record."

The robustness and interpretative value of time-series analysis of most stratigraphic data, with small numbers of samples at irregular time intervals, can usefully be strengthened by using periodic regression techniques [92], randomization and bootstrap procedures [83] and the logically structured analytical approach developed for analysing time series of events [93].

Periodic regression analysis provides a means of statistical frequency domain analysis when the time intervals between samples are unequal [92]. It decomposes a stationary time series into a series of non-harmonic waveforms that correspond to any actual periodicities within the time series. The statistical significance of these can be evaluated in relation to stochastic effects by Monte Carlo simulation. Recent applications include refs. 93 and 94. It deserves wider use.

When data points are sparse, it is important to supplement and evaluate statistically the results from frequency-domain analysis by randomization [89, 95] and bootstrap [96,97] procedures [83]. In essence, these are robust data resampling techniques for estimating standard errors or probability distributions of test statistics. In time series analysis they involve repeated random resampling from the observed distribution of the stratigraphic variable of interest to generate time series of the same length as the observed series [83]. Each series is then analysed, the power distribution at each frequency recorded and the process repeated very many times (> 1000) to derive by "brute force" [87] empirical probability distributions for the relevant test statistics (see refs. 83 and 87–89 for examples). In randomization procedures, data are resampled without replacement and thus probability distributions are generated solely for random permutations of the samples in the observed sequence, whereas in bootstrapping resampling occurs with replacement and produces probability distributions based on broader rearrangements of the data because the same samples may be used more than once to construct a simulated time series [83]. In practice, both are useful in validating results of time series analysis of sparse geological data sets.

There is a common tendency in time series analysis of stratigraphic data to concentrate on seeking periodicities, without considering or testing for other equally interesting temporal patterns in the sequence of interest. These include linear trends with changing rates of occurrences in time, independent successive occurrences and renewal processes and Poisson distributions. The logically structured approach developed by Cox and Lewis [93] for analysing series of events viewed as point processes has considerable potential and could usefully be extended and adopted in the critical interpretation of stratigraphic data [98] (see refs. 99 and 100 for examples).

A second, potentially valuable approach to the interpretation of stratigraphic patterns involves modelling quantitatively observed trends either within the entire sequence or in sections of roughly homogeneous means and variances within the sequence. The entire multivariate sequence can be partitioned into zones or sections using optimal divisive methods [1,27] or sequences of individual variables can each be split into homogeneous sections characterized by their means, variances and mean/variance ratios [1,20,23,101–103]. In the latter instance, the observed number and distribution of splits for all variables within a sequence can be tested statistically [16,81] against the hypothesis that the splits are randomly distributed in time, thereby distinguishing between linked, grouped and independent, individualistic changes through time. This distinction can be of geological interest and important in interpretation [1].

Within each section, individual data points can also be modelled as linear, quadratic or cubic functions of time using regression techniques in an attempt to characterize temporal trends within sections [102,103]. Often sample numbers within a section are too small to justify polynomial equation fitting [1]. Piecewise regression [18,77,104,105] can be useful for modelling trends within an entire sequence as, in effect, it implements sequence splitting and regression modelling in a single analysis. Spline functions [53,106] can also be valuable. Observed biostratigraphic patterns can be modelled in terms of exponential and logistic models [1,107] using regression analysis (see ref. 108 for recent examples). Overall biotic change through time can also be quantified by, for example, CA of stratigraphic data [30,109]. Rates and patterns of biotic change in one or more sequences can be compared in time and space, thereby aiding interpretation of the observed biostratigraphic patterns [30]. More elaborate modelling is possible using procedures such as discriminant analysis [110] and "excess after regression" analysis [111]. For example, Hilton et al. [112] modelled concentrations of different trace metals (Zn, Cd, Fe, Ni, Cu, Pb, Mn) in lake sediments in terms of background, erosion and pollution components using this regression technique.

A third general, quantitative approach to interpreting stratigraphic patterns involves transfer functions, first developed by Imbrie and Kipp [113] to calibrate biostratigraphic patterns in terms of environmental change [114]. In the parlance of chemometric modelling [13], transfer functions are predictive models developed from a training set of modern samples using unsupervised methods to predict or estimate properties of interest from fossil samples whose properties of interest are unknown (e.g., temperature, salinity). Techniques of transfer function derivation and calibration are well developed and are largely variants of the basic multiple linear regression model (see refs. 115 and 116 for current techniques). Multivariate calibration (= chemometric modelling sensu Jurs [13]) is a widespread problem in applied statistics, and Brown [117] reviewed current methods. Important new developments in transfer function and calibration methodology include using CA prior to regression analysis [118] and a combination of CCA, PCA, multiple regression and Kalman filters [119]. These developments overcome some, but not all, of the limitations of previous transfer function techniques [1] and extend the potential use of transfer functions to past situations lacking modern analogues and to areas where modern data sets are of limited applicability to the fossil record [120].

In conclusion, it should be emphasized that numerical analysis and statistical testing are only tools to assist the geologist in the task of interpreting observed patterns in terms of underlying processes. Numerical methods must not be viewed as a substitute for geological, chemical or biological knowledge [1]. Their critical use requires, however, an understanding of their assumptions, strengths and weaknesses and thus a familiarity with the underlying theory. Without this knowledge, to paraphrase Faegri [121], numerical analysis of any data set is bound to become at its best a lifeless computing exercise, at its worst useless altogether!

POSSIBLE FUTURE DEVELOPMENTS

I recently proposed [48] that quantitative palaeoecology could develop an analytical phase (sensu Ball [122]) if more attention was directed to generating and testing falsifiable hypotheses. Existing techniques for the quantitative analysis of stratigraphic data, as discussed above, have contributed to the development of descriptive and narrative phases [122] in quantitative historical geology, just as they have in quantitative palaeoecology [48]. Possible future developments could usefully be directed towards attaining an analytical phase in the quantitative analysis of stratigraphic data. In this section, I suggest possible ways of developing existing exploratory methods of data analysis into confirmatory data analytical techniques that can aid in generating and testing falsifiable hypotheses, the essential components of an analytical phase in quantitative stratigraphic geology.

One problem in analysing stratigraphic data is that they are often so complex that it is difficult to formulate realistic and useful hypotheses that could lead to analytically tractable test procedures [1]. On the other hand, many exploratory techniques, such as optimal divisive analysis, PCA, CA and PCoA, produce "results that give the appearance of things going on even when applied to data simulated to be completely random" [123]. It is thus necessary "to test ... 'nothing is going on' (as a null hypothesis) first" [123]. The simplest testable null hypothesis is that there are no patterns within a particular stratigraphic data set and hence that any observed patterns are indistinguishable from random expectations. It proposes that nothing in a given data set need be explained other than by chance [124,125]. This type of null hypothesis is probably the easiest to formulate in a non-experimental science such as historical geology [125].

For a given data set, a reference point or "control" for comparing observed patterns with random expectations is provided by repeated randomizations of the data within defined constraints of, for example, constant marginal totals [124]. In such a control any geologically determined structure is, in theory, removed by the randomization procedures and thus provides a useful "null" yardstick with which to compare observed patterns.

There are several methods developed in related fields that are directly applicable to the analysis of stratigraphic data and permit the null hypothesis of no patterns in the data to be tested. These include (1) probabilistic dissimilarity coefficients for comparing samples [126], (2) randomness tests for patterns in stratigraphically ordered dissimilarity matrices [127,128] that allow hypotheses of no change, continuous change or discontinuous change in stratigraphic sequences to be tested and (3) probabilistic partitioning of ordered, stratigraphic data [129]. In addition, there are randomization procedures that can usefully be employed to evaluate the statistical significance of (1) patterns of non-probabilistic dissimilarity coefficients [130], (2) patterns detected by partitioning methods [131,132], (3) patterns detected by scaling methods such as PCA and CA [133] and (4) patterns detected by time series analysis [83,134]. Patterns of stratigraphic boundaries in different data sets within the same sequence can be tested statistically using methods developed for ecological transects [81,135-137], geological successions [138] or geographical gradients [139,140]. Approximate tests for the statistical significance of sequence partitionings are available for some techniques [101,102], although excessive claims for the "statistical tested accuracy" of the partitions [102] may not be justified [1], particularly for sequences containing small numbers of samples. Careful, critical and imaginative use of randomization [95,124] and bootstrapping [96,97] procedures with stratigraphic data could, hopefully, provide some basis for answering some of the criticisms of existing methods for partitioning stratigraphic data, namely that they have "no direct reference to statistics and, consequently, make it impossible to verify the significance of the given division" [24]. These and related techniques can provide falsifiable null hypotheses for complex and otherwise intractable stratigraphic patterns and hence could contribute to developing an analytical phase in the quantitative multivariate analysis of stratigraphic data. Another potentially important future development would be to link multivariate analysis of stratigraphic data with methods devised by stratigraphers for determining most likely sequences of events based on observations from many profiles [141,142]. This could result in a more unified approach to stratigraphic analysis, and would represent a major "refinement (1980-1989)" (sensu Mann [143]) in quantitative stratigraphic analysis.

ACKNOWLEDGEMENTS

I am grateful to Christopher Birks, Hilary Birks, Kari Eeg, Sylvia Peglar and Geoffrey Timmins for assistance, to John Boyle, Edward Connor, Francis Gilbert, Allan Gordon, Eric Grimm and John Line for discussions, correspondence or preprints and to Richard Brereton, Michel Mellinger and Richard Reyment for comments during the Ulvik Multivariate Statistical Workshop for Geologists and Geochemists organised by Olav Kvalheim and Rolf Manne.

REFERENCES

1 H.J.B. Birks and A.D. Gordon, *Numerical Methods in Quaternary Pollen Analysis*, Academic Press, London, 1985, 317 pp.
2 R.A. Reyment, *Morphometric Methods in Biostratigraphy*, Academic Press, London, 1980, 175 pp.
3 K. Pearson, Mathematical contributions to the theory of evolution — on a form of spurious correlation which may arise when indices are used in the measurement of organs, *Proceedings of the Royal Society of London*, 60 (1897) 489–498.
4 J. Aitchison, The statistical analysis of compositional data, *Journal of the Royal Statistical Society B*, 44 (1982) 139–177.
5 J. Aitchison, Principal component analysis of compositional data, *Biometrika*, 70 (1983) 57–65.
6 J. Aitchison, Reducing the dimensionality of compositional data sets, *Mathematical Geology*, 16 (1984) 617–635.
7 J. Aitchison, The statistical analysis of geochemical compositions, *Mathematical Geology*, 16 (1984) 531–564.
8 R.A. Reyment, Moors and Christians: an example of multivariate analysis applied to human blood-groups, *Annals of Human Biology*, 10 (1983) 505–522.
9 R.A. Reyment, personal communication.
10 J. Aitchison and S.M. Shen, Measurement error in compositional data, *Mathematical Geology*, 16 (1984) 637–650.
11 R. Gnandasikan, *Methods for Statistical Data Analysis of Multivariate Observations*, Wiley, New York, 311 pp.
12 A.D. Gordon, Numerical methods in Quaternary palaeoecology. V. Simultaneous graphical representation of the levels and taxa in a pollen diagram, *Review of Palaeobotany and Palynology*, 37 (1982) 155–183.
13 P.C. Jurs, Pattern recognition used to investigate multivariate data in analytical chemistry, *Science*, 232 (1986) 1219–1224.
14 I.C. Prentice, Multidimensional scaling as a research tool in Quaternary palynology: a review of theory and methods, *Review of Palaeobotany and Palynology*, 31 (1980) 71–104.
15 R.M. Forester, Abundance coefficients, a new method for measuring microorganism relative abundance, *Mathematical Geology*, 9 (1977) 619–633.
16 H.J.B. Birks, Numerical zonation, comparison and correlation of Quaternary pollen-stratigraphical data, in B.E. Berglund (Editor), *Handbook of Holocene Palaeoecology and Palaeohydrology*, Wiley, Chichester, 1986, pp. 743–774.
17 R. Webster, Automatic soil-boundary location from transect data, *Mathematical Geology*, 5 (1973) 27–37.
18 D.M. Hawkins, On the choice of segments in piecewise approximation, *Journal of Institute of Mathematical Applications*, 9 (1972) 250–256.
19 D. Gill, Application of a statistical zonation method to reservoir evaluation and digitized log analysis, *American Association of Petroleum Geologists Bulletin*, 54 (1970) 719–729.
20 D.M. Hawkins and D.F. Merriam, Optimal zonation of digitized sequential data, *Mathematical Geology*, 5 (1973) 389–395.
21 T.R. Bement and M.S. Waterman, Locating maximum variance segments in sequential data, *Mathematical Geology*, 9 (1977) 55–61.
22 R. Webster, Optimally partitioning soil transects, *Mathematical Geology*, 29 (1978) 380–402.
23 D.M. Hawkins and D.F. Merriam, Segmentation of discrete sequences of geologic data, *Geological Society of America Memoir*, 142 (1975) 311–315.
24 A.J. Krawczyk, A critical evaluation of some methods of profile division into homogeneous sections, *Bulletin de l'Academie Polonaise des Sciences*, 30 (1983) 111–120.
25 P.H.A. Sneath and R.R. Sokal, *Numerical Taxonomy — the Principles and Practice of Numerical Classification*, Freeman, San Francisco, 1973, 573 pp.
26 E.C. Grimm, CONISS: A FORTRAN 77 program for stratigraphically constrained cluster analysis by the method of incremental sum of squares, *Computers and Geosciences*, in press.
27 D.H. Hawkins and D.F. Merriam, Zonation of multivariate sequences of digitized geologic data, *Mathematical Geology*, 6 (1974) 263–269.
28 D.H. Hawkins, FORTRAN IV program to segment multivariate sequences of data. *Computers and Geosciences*, 1 (1976) 339–351.
29 D.H. Hawkins and J.A. ten Krooden, Zonation of sequences of heteroscedastic multivariate data, *Computers and Geosciences*, 5 (1979) 189–194.
30 G.L. Jacobson, Jr. and E.C. Grimm, A numerical analysis of Holocene forest and prairie vegetation in central Minnesota, *Ecology*, 67 (1986) 958–966.
31 J.C. Brower, S.A. Millendorf and T.S. Dyman, Methods for the quantification of assemblage zones based on multivariate analysis of weighted and unweighted data, *Computers and Geosciences*, 4 (1978) 221–227.
32 S.A. Millendorf, J.C. Brower and T.S. Dyman, A comparison of methods for the quantification of assemblage zones, *Computers and Geosciences*, 4 (1978) 229–242.
33 J.C. Brower, The relative biostratigraphic values of fossils, *Computers and Geosciences*, 10 (1984) 111–131.
34 A.G. Brown, Traditional and multivariate techniques applied to the interpretation of floodplain sediment grain size variations, *Earth Surface Processes and Landforms*, 10 (1985) 281–291.

35 B. Denngård, Late Weichselian and early Holocene stratigraphy in southwestern Sweden with emphasis on the Lake Vänern area, *Geologiska Institutionen, Göteborgs Universitet Publications*, A 48 (1984) 1–187.
36 J.C. Ritchie, Late-Quaternary climatic and vegetational change in the Lower Mackenzie Basin, Northwest Canada, *Ecology*, 66 (1985) 612–621.
37 B.W. Leyden, Late Quaternary aridity and Holocene moisture fluctuations in the Lake Valencia basin, Venezuela, *Ecology*, 66 (1985) 1279–1285.
38 M.J. Sackin, P.H.A. Sneath and D.F. Merriam, ALGOL program for cross-association of nonmetric sequences using a medium-size computer, *Kansas Geological Survey Special Distribution Publication*, 23 (1965) 1–36.
39 J.A. Howell, A FORTRAN 77 program for automatic stratigraphic correlation, *Computers and Geosciences*, 9 (1983) 311–327.
40 A.J. Rudman and R.W. Lankston, Stratigraphic correlation of well logs by computer techniques, *American Association of Petroleum Geologists Bulletin*, 57 (1973) 577–588.
41 J.R. Southam and W.W. Hay, Correlation of stratigraphic sections by continuous variables, *Computers and Geosciences*, 4 (1978) 257–260.
42 F. Kemp, An algorithm for the stratigraphic correlation of well logs, *Mathematical Geology*, 14 (1982) 271–285.
43 D.R. Matuszak, Stratigraphic correlation of subsurface geologic data by computer, *Mathematical Geology*, 4 (1972) 331–343.
44 B.D. Kwon and A.J. Rudman, Correlation of geologic logs with spectral methods, *Mathematical Geology*, 11 (1979) 373–390.
45 A.D. Gordon and R.A. Reyment, Slotting of borehole sequences, *Mathematical Geology*, 11 (1979) 309–327.
46 R.A. Reyment, Biostratigraphical logging methods, *Computers and Geosciences*, 4 (1978) 261–268.
47 A.D. Gordon, *Classification: Methods for the Exploratory Analysis of Multivariate data*, Chapman and Hall, London, 1980, 193 pp.
48 H.J.B. Birks, Recent and possible future mathematical developments in quantitative palaeoecology, *Palaeogeography, Palaeoclimatology, Palaeoecology*, 50 (1985) 107–147.
49 A.D. Gordon, SLOTSEQ: a FORTRAN IV program for comparing two sequences of observations, *Computers and Geosciences*, 6 (1980) 7–20.
50 A. Delcoigne and P. Hansen, Sequence comparison by dynamic programming, *Biometrika*, 62 (1975) 661–664.
51 A.D. Gordon, An investigation of two sequence-comparison statistics, *Australian Journal of Statistics*, 24 (1982) 332–342.
52 K-D. Vorren and T. Alm, An attempt at synthesizing the Holocene biostratigraphy of a "type area" in northern Norway by means of recommended methods for zonation and comparison of biostratigraphical data, *Ecologia Mediterranea*, 11 (1985) 53–64.
53 K.J. Edwards and R. Thompson, Magnetic, palynological and radiocarbon correlation and dating comparisons in long cores from a Northern Irish lake, *Catena*, 11 (1984) 83–89.
54 N.J. Anderson, Diatom stratigraphy and comparative core correlation within a small lake basin, *Hydrobiologia*, 143 (1986) 105–112.
55 R.A. Reyment, Phenotype evolution in a Cretaceous foraminifer, *Evolution*, 36 (1982) 1182–1199.
56 D.H. Hawkins, A method for stratigraphic correlation of several boreholes, *Mathematical Geology*, 16 (1984) 393–406.
57 R.M. Clark, A FORTRAN program for constrained sequence-slotting based on minimum combined path length, *Computers and Geosciences*, 11 (1985) 605–617.
58 G.P. Lohmann and J.J. Carlson, Oceanographic significance of Pacific Late Miocene calcareous nannoplankton, *Marine Micropaleontology*, 6 (1981) 553–579.
59 L.V. Sergeeva, Trace element associations as indicators of sediment accumulation in lakes, *Hydrobiologia*, 103 (1983) 81–84.
60 M.W. Binford, Ecological history of Lake Valencia, Venezuela: interpretation of animal microfossils and some chemical, physical, and geological features, *Ecological Monographs*, 52 (1982) 307–333.
61 B.A. Malmgren and J.P. Kennett, Principal component analysis of Quaternary planktic foraminifera in the Gulf of Mexico: paleoclimatic applications, *Marine Micropaleontology*, 1 (1976) 299–306.
62 H.H. Birks and R.W. Mathewes, Studies in the vegetational history of Scotland. V. Late Devensian and early Flandrian pollen and macrofossil stratigraphy at Abernethy Forest, Inverness-shire, *New Phytologist*, 80 (1978) 455–484.
63 M. Melguen, Facies analysis by "correspondence analysis": numerous advantages of this new statistical technique, *Marine Geology*, 17 (1974) 165–182.
64 J.K. Elner and C.M. Happey-Wood, The history of two linked but contrasting lakes in North Wales from a study of pollen, diatoms and chemistry in sediment cores, *Journal of Ecology*, 68 (1980) 95–121.
65 S.M. Peglar, S.C. Fritz, T. Alapieti, M. Saarnisto and H.J.B. Birks, Composition and formation of laminated sediments in Diss Mere, Norfolk, England, *Boreas*, 13 (1984) 13–28.
66 M.O. Hill, Correspondence analysis: a neglected multivariate method, *Applied Statistics*, 23 (1974) 340–354.
67 M.J. Greenacre, *Theory and Applications of Correspondence Analysis*, Academic Press, London, 1984, 364 pp.
68 R.A. Reyment, R.E. Blackith and N.A. Campbell, *Multivariate Morphometrics*, Academic Press, London, 2nd ed., 1984, 233 pp.
69 R.A. Reyment, Spectral breakdown of morphometric chronoclines — a paleogenetic problem, *Mathematical Geology*, 2 (1970) 365–376.
70 J.O.R. Hermelin and B.A. Malmgren, Multivariate analysis of environmentally controlled variation in *Lagena*: Late Maastrichtian, Sweden, *Cretaceous Research*, 1 (1980) 193–206.

71 R.A. Reyment, Chemical components of the environment and Late Campanian microfossil frequencies, *Geologiska Föreningen i Stockholm Förhandlingar*, 98 (1976) 322–328.
72 G.U. Yule, Why do we sometimes get nonsense-correlations between time-series? — a study in sampling and the nature of time-series, *Journal of the Royal Statistical Society*, 89 (1926) 1–69.
73 B.A. Malmgren, Comparison of visual and statistical correlation in time series curves, *Mathematical Geology*, 10 (1978) 103–106.
74 J.H. Schuenemeyer, Reply to Comparison of visual and statistical correlation in time series curves, *Mathematical Geology*, 10 (1978) 106–108.
75 B.A. Malmgren, Morphometric studies of planktonic foraminifers from the type Danian of southern Scandinavia, *Stockholm Contributions in Geology*, 29 (1974) 1–126.
76 R.A. Reyment, Correlating between electrical borehole logs in paleoecology, in J.M. Cubitt and R.A. Reyment (Editors), *Quantitative Stratigraphic Correlation*, Wiley, Chichester, 1982, pp. 233–240.
77 A.D. Gordon, On measuring and modelling the relationship between two stratigraphically-recorded variables, in J.M. Cubitt and R.A. Reyment (Editors), *Quantitative Stratigraphic Correlation*, Wiley, Chichester, 1982, pp. 241–248.
78 B.W. Hamon and E.J. Hannan, Estimating relations between time series, *Journal of Geophysical Research*, 68 (1963) 6033–6041.
79 H. Ivert, Relationship between stratigraphical variation in the morphology of *Gabonella elongata* and geochemical composition of the host sediments, *Cretaceous Research*, 1 (1980) 223–263.
80 W.W. Cooley and P.R. Lohnes, *Multivariate Data Analysis*, Wiley, New York, 1971, 364 pp.
81 F.P. Gardiner and R.L. Haedrich, Zonation in the deep benthic megafauna — application of a general test, *Oecologia*, 31 (1978) 311–317.
82 J.M. Gottman, *Time-series Analysis, a Comprehensive Introduction for Social Scientists*, Cambridge University Press, Cambridge, 1981, 400 pp.
83 E.F. Connor, Time series analysis of the fossil record, in D.M. Raup and D. Jablonski (Editors), *Pattern and Process in the History of Life*, Springer-Verlag, Berlin, 1986, pp. 119–147.
84 W, Dunsmuir, Estimation for stationary time series when data are irregularly spaced or missing, in D.E.F. Findley (Editor), *Applied Time Series Analysis II*, Academic Press, New York, 1981, pp. 609–649.
85 I. Renberg, U. Segerström and J-E. Wallin, Climatic reflection in varved lake sediments, in N-A. Mörner and W. Karlén (Editors), *Climatic Changes on a Yearly to Millenial Basis*, Reidel, Boston, 1984, pp. 249–256.
86 T.D. Herbert and A.G. Fischer, Milankovitch climatic origin of mid-Cretaceous black shale rhythms in central Italy, *Nature (London)*, 321 (1986) 739–743.
87 D.M. Raup and J.J. Sepkoski, Jr., Periodicity of extinctions in the geologic past, *Proceedings of National Academy of Sciences U.S.A.*, 81 (1984) 801–805.
88 D.M. Raup and J.J. Sepkoski, Jr., Periodicity of extinctions of families and genera, *Science*, 231 (1986) 833–836.
89 J.J. Sepkoski, Jr. and D.M. Raup, Periodicity in marine extinction events, in D.K. Elliott (Editor), *Dynamics of Extinction*, Wiley, Chichester, 1986, pp. 3–36.
90 A. Hoffman and J. Ghiold, Randomness in the pattern of 'mass extinctions' and 'waves of origination', *Geological Magazine*, 122 (1985) 1–4.
91 J.J. Sepkoski, Jr., D.M. Raup, N.L. Gilinsky, J.A. Kitchell, G. Estabrook and A. Hoffman, Was there a 26-Myr periodicity of extinctions?, *Nature (London)*, 321 (1986) 533–536.
92 M. Briskin and J. Harrell, Time series analysis of the Pleistocene deep-sea paleoclimatic record, *Marine Geology*, 36 (1980) 1–22.
93 D.R. Cox and P.A.W. Lewis, *The statistical analysis of series of events*, Methuen, London, 1966, 285 pp.
94 T.A. Wijmstra, S. Hockstra, B.J. de Vries and T. van der Hammen, A preliminary study of periodicities in percentage curves dated by pollen density, *Acta Botanica Neerlandica*, 33 (1984) 547–557.
95 E.S. Edgington, *Randomization tests*, Marcel Dekker, New York, 1980, 287 pp.
96 B. Efron and G. Gong, A leisurely look at the bootstrap, the jackknife and cross-validation, *American Statistician*, 37 (1983) 36–48.
97 P. Diaconis and B. Efron, Computer-intensive methods in statistics, *Scientific American*, 248 (1983) 96–108.
98 R.A. Reyment, Some applications of point processes in geology, *Mathematical Geology*, 8 (1976) 95–97.
99 R.A. Reyment and J.D. Collinson, Periodicity in Namurian sediments of Derbyshire, England (a quantitative sedimentary analysis), *Sedimentary Geology*, 5 (1971) 23–36.
100 H.J.B. Birks, Long-distance pollen in Late Wisconsin sediments of Minnesota, U.S.A.: a quantitative analysis, *New Phytologist*, 87 (1981) 630–661.
101 A.J. Krawczyk and T. Stomka, A method of analysing linear trends in geological profiles, *Bulletin de l'Academie Polonaise des Sciences*, 30 (1983) 105–110.
102 D. Walker and S.R. Wilson, A statistical alternative to the zoning of pollen diagrams, *Journal of Biogeography*, 5 (1978) 1–21.
103 D. Walker and Y. Pittelkow, Some applications of the independent treatment of taxa in pollen analysis, *Journal of Biogeography*, 8 (1981) 37–51.
104 D.M. Hawkins, Point estimation of the parameters of piecewise regression models, *Applied Statistics*, 25 (1976) 51–57.
105 V.W. McGee and W.T. Carleton, Piecewise regression, *Journal of the American Statistical Association*, 65 (1970) 1109–1124.
106 J.G. Hayes, Numerical methods for curve and surface fitting, *Bulletin of the Institute of Mathematics and its Applications*, 10 (1974) 144–152.
107 A. Hallam, Models involving population dynamics, in T.J.M. Schopf (Editor), *Models of Paleobiology*, Freeman, Cooper & Company, San Francisco, 1972, pp. 62–80.
108 K.D. Bennett, Postglacial population expansion of forest

trees in Norfolk, UK, *Nature (London)*, 303 (1983) 164–167.
109 H.J.B. Birks, Quantitative estimation of changing human impact through time on cultural-landscape development: a multivariate approach, *University of Bergen Botanical Institute Report*, 41 (1986) 16.
110 T.P. Burnaby, The palaeoecology of the foraminifera of the Chalk Marl, *Palaeontology*, 4 (1961) 599–608.
111 S. Chatterjee and B. Price, *Regression Analysis by Example*, Wiley, New York, 1977, 228 pp.
112 J. Hilton, W. Davison and U. Ochsenbein, A mathematical model for analysis of sediment core data: implications for enrichment factor calculations and trace-metal transport mechanisms, *Chemical Geology*, 48 (1985) 281–291.
113 J. Imbrie and N.G. Kipp, A new micropaleontological method for quantitative paleoclimatology: application to a late Pleistocene Caribbean core, in K.K. Turekian (Editor), *The Late Cenozoic Glacial Ages*, Yale University Press, New Haven, 1971, pp. 71–181.
114 J. Imbrie and T. Webb, III, Transfer functions: calibrating micropaleontological data in climatic terms, in A. Berger (Editor), *Climatic Variations and Variability: Facts and Theories*, Reidel, Boston, 1981, pp. 125–134.
115 H.M. Sachs, T. Webb, III and D.R. Clark, Paleoecological transfer functions, *Annual Review of Earth and Planetary Sciences*, 5 (1977) 159–178.
116 S. Howe and T. Webb, III, Calibrating pollen data in climatic terms: improving the methods, *Quaternary Science Reviews*, 2 (1983) 17–51.
117 P.J. Brown, Multivariate calibration, *Journal of the Royal Statistical Society B*, 44 (1982) 287–321.
118 F. Gasse and F. Tekaia, Transfer functions for estimating palaeoecological conditions (pH) from East African diatoms, *Hydrobiologia*, 103 (1983) 85–90.
119 J. Guiot, A method for palaeoclimatic reconstruction in palynology based on multivariate time series analysis, *Geographie Physique et Quaternaire*, 39 (1985) 115–125.
120 J. Guiot, Late Quaternary climatic change in France estimated from multivariate pollen time series, unpublished manuscript.
121 K. Faegri, Some problems of representativity in pollen analysis, *Palaeobotanist*, 15 (1966) 135–140.
122 I.R. Ball, Nature and formulation of biogeographical hypotheses, *Systematic Zoology*, 24 (1975) 407–430.
123 R.H. Green, *Sampling Design and Statistical Methods for Environmental Biologists*. Wiley, New York, 1979, 257 pp.
124 E.F. Connor and D. Simberloff, Interspecific competition and species co-occurrence patterns on islands: null models and the evaluation of evidence, *Oikos*, 41 (1983) 455–465.
125 E.F. Connor and D. Simberloff, Competition, scientific method, and null models in ecology, *American Scientist*, 74 (1986) 155–162.
126 J.F. Grasle and W. Smith, A similarity measure sensitive to the contribution of rare species and its use in investigation of variation in marine benthic communities, *Oecologia*, 25 (1976) 13–22.
127 E.C. Pielou, Spatial and temporal change in biogeography: gradual or abrupt?, in R.W. Sims, J.H. Price and P.E.S. Whalley (Editors), *Evolution, Time and Space; The Emergence of the Biosphere*, Academic Press, London, 1983, pp. 29–56.
128 E.C. Pielou, Interpretation of paleoecological similarity matrices, *Paleobiology*, 5 (1979) 435–443.
129 E.D. McCoy, S.S. Bell and K. Walters, Identifying biotic boundaries along environmental gradients, *Ecology*, 67 (1986) 749–759.
130 J. Rice and R.J. Belland, A simulation study of moss floras using Jaccard's coefficient of similarity, *Journal of Biogeography*, 9 (1982) 411–419.
131 C.W. Harper, Jr., Groupings by locality in community ecology and paleoecology: tests of significance, *Lethaia*, 11 (1978) 251–257.
132 R.E. Strauss, Statistical significance of species clusters in association analysis, *Ecology*, 63 (1982) 634–639.
133 D.F. Stauffer, E.O. Garton and R.K. Steinhorst, A comparison of principal components from real and random data, *Ecology*, 66 (1985) 1693–1698.
134 L.C. Cole, Biological clock in the unicorn, *Science*, 125 (1957) 874–876.
135 E.C. Pielou and R.D. Routledge, Saltmarsh vegetation: latitudinal gradients in the zonation patterns, *Oecologia*, 24 (1976) 311–321.
136 A.J. Underwood, The detection of non-random patterns of distribution of species along a gradient, *Oecologia*, 36 (1978) 317–326.
137 M.R.T. Dale, The contiguity of upslope and downslope boundaries of species in a zoned community, *Oikos*, 49 (1984) 92–96.
138 C.W. Harper, Jr., Inferring succession of fossils in time: the need for a quantitative and statistical approach, *Journal of Paleontology*, 55 (1981) 442–452.
139 E.C. Pielou, The statistics of biogeographic range maps: sheaves of one-dimensional ranges, *Bulletin of the International Statistical Institute*, 47 (1977) 111–122.
140 E.C. Pielou, The latitudinal spans of seaweed species and their patterns of overlap, *Journal of Biogeography*, 4 (1977) 299–311.
141 W.W. Hay and J.R. Southam, Quantifying biostratigraphic correlation, *Annual Review of Earth and Planetary Science*, 6 (1978) 353–375.
142 J.C. Brower, Quantitative biostratigraphy, 1830–1980, in D.F. Merriam (Editor), *Computer Applications in the Earth Sciences: An Update of the 70's*. Plenum, New York, 1981, pp. 63–103.
143 C.J. Mann, Stratigraphic analysis: decades of revolution (1970–1979) and refinement (1980–1989), in D.F. Merriam (Editor), *Computer Applications in the Earth Sciences: An Update of the 70's*. Plenum, New York, 1981, pp. 211–242.

Fuzzy Theory Explained

MATTHIAS OTTO

Department of Chemistry, Bergakademie Freiberg, Leipziger Strasse, 9200 Freiberg (G.D.R.)

(Received 18 August 1987; accepted 27 January 1988)

CONTENTS

1 The management of vagueness and uncertainty 401
2 Mathematical tools 402
   2.1 Specification of fuzzy sets 402
   2.2 Elementary operations with fuzzy sets 404
   2.3 Computing with fuzzy sets 405
3 Optimization with several objectives 406
4 Handling fuzzy and incomplete data patterns 407
   4.1 Unsupervised learning 407
   4.2 Supervised learning 408
5 Calculations with fuzzy numbers 410
   5.1 How much is twenty minus about fourteen? 410
   5.2 Spectra are fuzzy functions 412
   5.3 Mixture component analysis 413
6 Fuzzy modelling 414
7 To come: fuzzy logic 416
   7.1 Linguistic variables 416
   7.2 Multi-valued logic 417
   7.3 The compositional rule of inference 418
8 Conclusions 419
References 419

1 THE MANAGEMENT OF VAGUENESS AND UNCERTAINTY

It happened when I sat at my PC and tried to implement a decision algorithm for optimizing an analytical procedure with respect to several objective functions. I knew that one could handle multi-criteria decisions by combining the different objective functions in a proper way, e.g. by summing the individual criteria with a proper weighting factor. What I did not know, however, was the "proper weighting factor". I only could state that the sensitivity of the method should be "high enough", the costs "as low as possible", and the speed of the analysis should be "less than 5 minutes".

How to represent that kind of knowledge in a computer? There was something needed that enables representation and handling of vague statements and uncertain information in order to describe terms such as "high enough" or "as low as possible".

In principle, this problem was already solved in 1965 by Lofti A. Zadeh of the University of California (U.S.A.) with his introduction of the concept of fuzzy sets [1]. With the theory of fuzzy sets the term "high" can be represented by a characteristic function over the variable of interest X (e.g. sensitivity) that increases monotonically with the variable as depicted in Fig. 1. This function — renormed normally to the interval [0, 1] — is the membership function to the variable (also called the Universe X), $m(x)$, and it represents, according to Zadeh's definition, a fuzzy set over X. To express, for example, the term "the sensitivity should be high enough" would mean to specify all those values of sensitivity as important which have a membership value $m(x)$ greater than, say, 0.5.

The initial problem of combining the different criteria can also easily be solved if these criteria are aggregated in the domain of their membership functions rather than in the original variable space.

Since 1965 the theory of fuzzy sets has been extensively developed and, of course, besides solving multi-criteria decisions the theory is used in areas as diverse as pattern recognition, data analysis, approximate reasoning and robot control. In general, with this theory problems can be solved that overcome idealization of real world problems in order to adapt them to conventional mathematical techniques. No assumptions are needed for defining the meaning of notions. Imprecise data or data with uncertain error bounds can be handled likewise, and even incomplete or vague knowledge can be managed by an exact mathematical theory.

Monographs by Dubois and Prade [2], by Zimmermann [3], by Kaufman [4] and by Kandel [5] are available. Nowadays there is almost no branch of science that has not benefited from applications of fuzzy set theory and, due to the proliferation of powerful computers, the methods are increasingly used.

This tutorial is aimed at providing fundamental tools of fuzzy set theory, at introducing preliminary applications within the field of chemometrics, at stimulating the reader to solve his own problems by fuzzy theory and at featuring future applications in chemometrics and with intelligent laboratory systems.

2 MATHEMATICAL TOOLS

2.1 Specification of fuzzy sets

To understand terminology and mathematical notation in fuzzy set theory we start with conventional set theory: A conventional or crisp (the opposite to fuzzy) subset A of a given universe X of elements is commonly defined by specifying for every element of the universe whether it is a member of A or it is not a member of A. In Fig. 2 this is demonstrated for the subset of all real numbers in the interval 1 to 4. Mathematically the containment of elements to a set can be expressed by a characteristic function, the membership function (m.f.), $m(x)$, that assigns a value of 1 to every

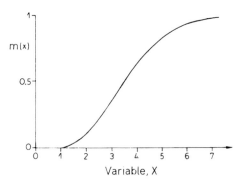

Fig. 1. A fuzzy set over the variable X (sensitivity) expressed by a characteristic function $m(x)$, the so-called membership function.

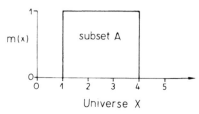

Fig. 2. All points in the interval [1, 4] of the real line X form the elements of the (crisp) subset A of the universe X.

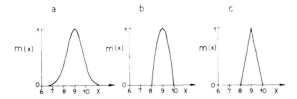

Fig. 3. Membership functions $m(x)$ for characterizing the imprecision of the position of a spectroscopic line. (a) Exponential m.f. (eq. 3); (b) quadratic m.f. (eq. 4); (c) linear m.f. (eq. 5).

element x of the Universe X ($x \in X$) that is a member of the subset A ($x \in A$) and a membership value of 0 to elements that are not members of A ($x \notin A$):

$$m(x) = \begin{cases} 1 & \text{if } x \in A \subseteq X \\ 0 & \text{if } x \notin A \subseteq X \end{cases} \quad (1)$$

Fuzzy sets can be derived by generalizing the concept of a membership function allowing membership values between 0 and 1. The m.f. for "high" in Fig. 1 could be specified, then, by the following characteristic function:

$$m(x) = 1 - \exp(-|x-a|^2/b^2) \quad (2)$$

where the constants a and b are fixed at 1 and 3, respectively. Note that the constant $1/b^2$ normalizes the m.f. to the interval [0, 1].

Other types of membership functions need to be defined. For characterizing the imprecision of the position of a spectroscopic line on the wavelength axis X a symmetrical function, e.g. a bell-shaped function might be more adaquate (Fig. 3a):

$$m(x) = \exp(-|x-a|^2/b^2) \quad (3)$$

with $b = 1$ and $a = 9$.

It has been found in practice that the functional type of the m.f. is less important than the fact that it is monotonic. Therefore, approximation of the bell-shaped m.f in eq. 3 by the functions in eq. 4 (Fig. 3b) and in eq. 5 (Fig. 3c) will have little influence on the final conclusions to be drawn.

$$m(x) = \left[1 - |x-a|^2/b^2\right]^+ \quad (4)$$

$$m(x) = \left[1 - |x-a|/b\right]^+ \quad (5)$$

The + sign denotes truncation of membership values to 0 at negative values.

In general, there are no restrictions to specify m.f.s. The functions could be also constructed from individual points on the universe and interconnected by a cubic spline (Fig. 4). Even the specification of higher dimensional m.f.s is important if fuzzy relations or multivariate problems are to be handled. Fig. 5 gives an example of m.f. over 2 variables (x–y-relation). Care has to be taken, however, to make the intended fuzzy operations feasible. Hence, the computational restrictions will dictate the possible complexity of the m.f.s. An example of how to speed up fuzzy calculations will be demonstrated in Section 2.3.

Sometimes there is no sense in specifying a continuous m.f.; e.g. if the suitability of a given physical state of samples has to be represented by

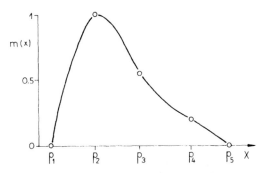

Fig. 4. Membership function constructed from five points on the variable axis X connected by a cubic spline. $p_1(x)$, $p_5(x)$ are the vanishing points with $m(p_i(x)) = 0$, $p_2(x)$ is the most possible x value with $m(p_2(x)) = 1$, and $p_3(x)$ and $p_4(x)$ have membership values of $m(p_3(x)) = 0.55$ and $m(p_4(x)) = 0.2$, respectively.

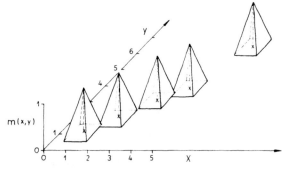

Fig. 5. A two-dimensional membership function representing the fuzziness of points in an x–y functional relationship.

a fuzzy set. Fig. 6 gives an example for characterizing the suitability of powdered (1), gaseous (2), liquid (3) and compact (4) samples for spectrophotometric determinations.

Where do membership functions come from

The m.f. that should be chosen for solving a given problem by using fuzzy set theory depends on the problem at hand. The following guidelines can be given:
— If enough information about the shape of the m.f. is available, e.g. by orientation on statistical material in case of fuzzy data analysis or in case of optimization with known objective functions, the most appropriate function can be defined;
— If no information about the m.f. exists but learning steps are inherent in the method, such as with supervised pattern recognition or with calibration techniques, different m.f.s could be tried and subsequently refined to derive that m.f. which gives the best solution to the question posed;
— Knowledge from different sources, e.g. statistical data, subjective experience and literature data, can be incorporated into the specification of the m.f.;
— Some instruments and data evaluating techniques produce fuzzy data by themselves. An early example was shown for particle shape analysis [6]. To derive the most parsimonious shape of a particle (circle, ellipse, square) the particle is imaged (photographically) giving a picture with blurred contours at the edges of the particle. These blurred (imaged) edges are then taken as fuzzy sets and compared to prototype of shapes, such as circles, ellipses etc. and the most common shape is evaluated.

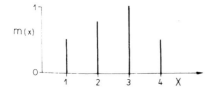

Fig. 6. Membership function characterizing the suitability of the "physical state" of a sample for spectrophotometric determinations.

2.2 Elementary operations with fuzzy sets

In this section we explain the principal operations with fuzzy sets. It should be mentioned, however, that fuzzy theory is not limited to set-theoretic operations; methods for handling fuzzy numbers, functions, relations, measures, integrals and fuzzy logic have been well established. Examples of such operations will be given in the applications sections.

The *intersection* of two (crisp) sets is known to contain all elements belonging to both of the sets simultaneously. The intersection of fuzzy sets A and B ($A \cap B$) over the universe X is given by the minimum of both of the m.f.s $m_A(x)$ and $m_B(x)$, respectively:

$$m_{A \cap B}(x) = \min(m_A(x), m_B(x)) \qquad (6)$$

Fig. 7a further explains this operation.

The *union* of two (crisp) sets involves all elements belonging to at least one of the sets. In case of fuzzy sets the maximum of both of the m.f. can be taken (Fig. 7b):

$$m_{A \cup B}(x) = \max(m_A(x), m_B(x)) \qquad (7)$$

The *complement* of a crisp set contains all elements of the universe that do not belong to a

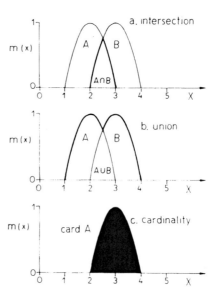

Fig. 7. Set-theoretic operations on fuzzy sets.

subset A. For fuzzy sets the complement has then the m.f.:

$$m_{\bar{A}}(x) = 1 - m_A(x) \tag{8}$$

Very often a fuzzy set is to be valued, giving a measure in the form of a single number. Here the *cardinality* of a set can be used. With common finite sets the cardinality of a set is the number of all elements in this set. By analogy, the cardinality with respect to a fuzzy set is defined by (cf. Fig. 7c):

$$\text{card } A = \sum_{x \in X} m(x) \quad \text{or} \quad \text{card } A = \int_X m(x)\,dx \tag{9}$$

Normalization of the cardinality of a fuzzy set to the interval [0, 1], e.g. for comparing the (absolute) cardinality of different fuzzy sets, can be performed by defining the *relative cardinality*. The relative cardinality is the cardinality of set A (card A) divided by the cardinality of a standard set S, normally the joint universe X, i.e.

$$\text{rel}_S \text{ card } A = (\text{card } A)/(\text{card } S) \tag{10}$$

Example: Calculate the relative cardinality for the set in Fig. 7c. This is done by integrating over the m.f. used, here $m(x) = [1 - |x - a|^2]^+$ was chosen (cf. eq. 4), in the interval 2 to 4:

$$\text{rel}_S \text{ card } A = \int_2^4 [1 - |x - 3|^2]\,dx/(4 - 2)$$
$$= 0.666$$

2.3 Computing with fuzzy sets

If one considers fuzzy set operations as demonstrated in Fig. 7 it may not be straightforward for the reader to see how to compute the intersection or union of a fuzzy set. Of course, there are no problems in performing these operations if at least one of the sets is taken to be crisp. For example, to calculate the intersection of the fuzzy set A in Fig. 8 with the (crisp) set B the actual x-value of B would have to be inserted into the m.f. for A,

$$m_A(x) = [1 - |x - a|^2/b^2]^+ \tag{11}$$

with $a = 2$, $b = 1$ and $x = 2.7$. Then the minimum is formed, i.e.

$$m_{A \cap B}(x) = \min[m_A(x), m_B(x)]$$
$$= \min[0.51, 1] = 0.51$$

Likewise, all operations on discrete representations of fuzzy sets, e.g. those in Fig. 6, can be handled by the same formalism: The fuzzy set A in Fig. 6 can be expressed by

$$A = |(1, 0.5), (2, 0.78), (3, 1), (4, 0.5)|$$

where the first element of the ordered pairs denotes the element of the universe and the second its degree of membership. To perform an intersection of the fuzzy set A with a fuzzy set B,

$$B = |(1, 1)|$$

e.g. in order to decide on the usefulness of analysing a powdered sample (1) by spectrophotometry would result in a fuzzy set C

$$C = A \cap B = |(1, 0.5)|$$

Very often, too, continuous m.f.s are transformed to discrete membership values. This is automatically done if graphics or image processing equipment is used for the computations. Sometimes the operations can be performed analytically if simple types of m.f.s are used.

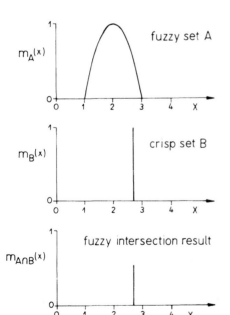

Fig. 8. Intersection of a fuzzy set A with a crisp set B.

Programming languages that can handle fuzzy-set manipulations have been proposed by several authors, e.g. FSTDS (fuzzy set-theoretic data structure) or FUZZY (pp. 265–267 in ref. 2). A software package for fuzzy computations including many-valued logic is offered in ref. 7.

Implementation of fuzzy logic hardware systems may greatly speed up fuzzy computations, giving us the possibility of producing fuzzy expert systems on a single chip [8].

3 OPTIMIZATION WITH SEVERAL OBJECTIVES

Fuzzy set theory has been extensively used in economics for multi-criteria decision making (ref. 3, p. 213). The same principles can be successfully applied in chemometrics as will be demonstrated with the following example.

For optimization of the enzymatic determination of the serum protein ceruloplasmin two objective functions were of importance: sensitivity and cost of analysis. The aim of optimization, then, was to determine the enzyme at maximum sensitivity and at minimum cost. Both criteria depend on the experimental variables pH, substrate and buffer concentration. Fig. 9 gives a plot of the objective area for both criteria "sensitivity" and "costs". This area has been computed on the basis of an empirical model relating the measured sensitivity and the estimated costs to the experimental variables, $x \in X$, by reduced second order polynomials. The compromise set of optimum sensitivity and cost is labeled by a bold curve in Fig. 9. Fuzzy set theory can now be used to derive the global optimum as a compromise between high sensitivity and low costs. With respect to cost (criterion $c_1(x)$) it is required that the costs do not exceed 80 currency units ($m(c_1(x)) = 0$ for $x \in X$) and, of course, the costs should be as low as possible. A minimum of cost is specified at 10 currency units ($m(c_1(x)) = 1$). Fig. 9 shows the resulting m.f.

The m.f. for the criterion sensitivity ($c_2(x)$) can be designed from three finite numbers on the variable axis: a minimum sensitivity required in order to fulfill clinical needs ($m(_2(x)) = 0$, here 30 sensitivity units); a maximum sensitivity of 98

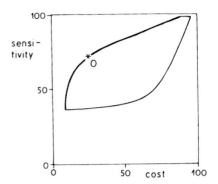

Fig. 9. Optimization of ceruloplasmin determination with respect to cost and sensitivity as objective functions.

units ($m(c_2(x)) = 1$) caused by chemical considerations and a membership value of $m(c_2(x)) = 0.5$ in the middle of the useful sensitivity range, i.e. at 64 sensitivity units. A continuous m.f. is constructed by a cubic spline function as mentioned in Section 2.2.

What remains to be solved is the problem of combining the membership values by suitable aggregation operators, such as the arithmetic or the geometric mean, the minimum or maximum operator, the algebraic product or algebraic sum, etc. The arithmetic mean is most frequently used and the optimum conditions are then found by evaluating the maximum membership value $m^*(x)$ over all conditions of the experimental variables in the compromise set:

$$m^*(x) = \max\left\{1/p \sum_{i=1}^{p} m_j(c_j(x))\right\} \quad (12)$$

with x being the vector $x = (x_1, x_2, x_3)'$ and $p = 2$, i.e. two objective functions. The maximum is found by inserting the model function for the criteria

$c_j(x)$ in dependence on the vector of the three variables pH (x_1), buffer (x_2) and substrate concentration (x_3) into the equations for the m.f. of cost and sensitivity (cf. Fig. 9) and searching for the maximum, e.g. by the method of steepest ascent. As optima for cost and sensitivity we obtain values of 25 and 71 units, respectively, which is demonstrated in Fig. 9 by the point labeled 0.

In conclusion, fuzzy set theory helps us to reach a decision for optimizing problems with several objective functions that can be characterized in broad categories rather than by distinct weighting factors, as is common with conventional multi-criteria decision making.

4 HANDLING FUZZY AND INCOMPLETE DATA PATTERNS

One of the earliest applications of fuzzy set theory has been in the area of pattern recognition. This is true for both unsupervised and supervised learning techniques of pattern recognition.

4.1 Unsupervised learning

With unsupervised pattern recognition, e.g. cluster analysis, the objects are assigned to different clusters on the basis of a set of characteristic features, such as elemental concentrations or physical and chemical parameters. In classical (crisp) cluster analysis [9] every object is assigned to exactly one cluster.

This is not an adequate solution to many real problems because very often there will also be objects lying between two clusters. Overlapping clusters are better described by fuzzy clustering algorithms. Here, to every object a membership value is extracted from the algorithm that characterizes the assignment of the object to a definite cluster. Fuzzy clustering has been applied in chemometrics to the partitioning of malt samples characterized by 9 chemical/physical analyses [10].

To explain the method reference is made to the well-known "Butterfly" cluster (cf. ref. 11). Fig. 10A shows this data set, consisting of 15 objects in

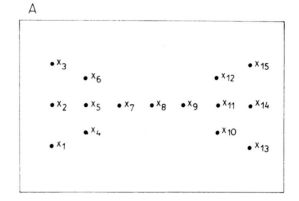

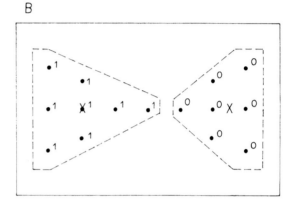

Fig. 10. Unsupervised learning: The Butterfly-cluster (A) and crisp clusters of the Butterfly (B).

a 2-feature space (plane). When using a crisp objective function based on the Euclidean distance [9] two clusters are obtained (Fig. 10B) where a membership value of 1 is assigned to the left cluster and a value of 0 to the right one. The centers of the clusters are labeled by X. Note that, apart from the symmetrical nature of the data pattern, the resulting clusters are not symmetrical. This is due to point 8 (Fig. 10A) which lies in between the two data patterns. This point in the middle of the two clusters should obviously be described as belonging equally to the left and right clusters. Fuzzy clustering enables this kind of partitioning of the data to be performed as shown in Fig. 11A and B were a membership value is assigned to every object, characterizing its membership to cluster 1 and 2, respectively. As expected, the point 8 "between" the clusters gets a membership value of 0.5.

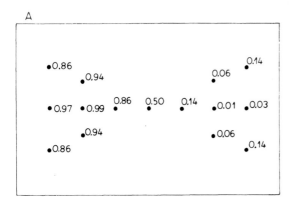

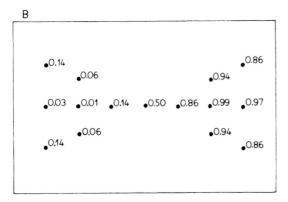

Fig. 11. Fuzzy clustering of the Butterfly. Left cluster (A) and right cluster (B).

As a consequence the clusters obtained are not only found as crisp sets, as with classical supervised pattern recognition techniques, but all the objects in the clusters are labeled by a possibility measure describing the membership to each of the clusters. In this example a 2-means algorithm has been used to evaluate the partitioning of the clusters. Details of the clustering procedure can be found in ref. 11.

4.2 Supervised learning

Classification of objects to known classes (clusters) is carried out by supervised learning techniques of pattern recognition, e.g. the k-nearest neighbor method, discriminant analysis, individual class modeling techniques (SIMCA) or potential methods of pattern recognition [12]. Typical examples are classification of medieval glass finds by their metal patterns as the features, and identi-

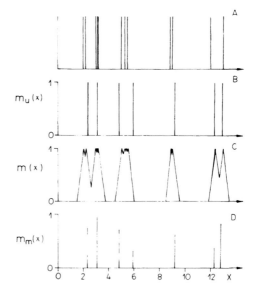

Fig. 12. Matching a class pattern (A) versus a crisp unknown sample pattern (B) based on fuzzifying the class pattern (C) and intersecting the membership functions of both patterns giving (D).

fication of chemical species based on spectral comparison or on chromatographic data patterns.

Problems with classical supervised learning techniques arise if non-randomly missing data have to be handled, for example due to the high variability of natural systems. In pyrolysis gas chromatography of tissues the number of peaks in the chromatogram can vary by up to ± 15 peaks. This high degree of variability in the data patterns is not caused by badly performed chromatography but results from the nature of the system and has to be taken as being an inherent data uncertainty. This is the point, then, where fuzzy set theory comes in. With fuzzy sets even feature data that have no definite order within the object vectors can be handled, since data sets rather than vectors are compared.

The principle is exemplified in the classification of data patterns that may originate from chromatograms or from spectroscopic lines.

Fig. 12A gives a data pattern for one sample class. The pattern is formed from individual data patterns that are obtained from the analysis of, e.g., the retention times on all samples of the class. For classifying the data pattern given in Fig. 12B with respect to its degree of membership to the class pattern in Fig. 12A, at first, the class pattern is fuzzified by using the membership type of eq. 5, i.e.

$$m_i(x) = [1 - |x - 0.5|]^+ \qquad (13)$$

where $i = 1, \ldots, n$, and n represents the number of peaks in the training class pattern, here 13 (Fig. 12A). The m.f. of the whole class, $m(x)$, is now defined by the maximum over all overlapping m.f.s from the individual peaks:

$$m(x) = \max_i m_i(x) \qquad (14)$$

Secondly, the unknown sample pattern, being crisp with the m.f. $m_u(x) = 1$, is classified by intersecting the data patterns in Fig. 12B and C giving the following m.f.

$$m_m(x) = \min[m(x), m_u(x)] \qquad (15)$$

As shown in Fig. 12D, a vector of sympathy values is obtained, $m_m(x_j)$ with $j = 1, \ldots, p$ and $p = 7$.

In order to derive one final value from the resulting membership values, the relative cardinality is usually calculated according to eq. 10, i.e.

$$\mathrm{rel}_S \text{ card } A = 1/7 \sum_{i=1}^{7} m_m(x)$$

$$= 1/7(0.8 + 0.96 + 0.78 + 0.26$$
$$+ 0.51 + 0.15 + 0.87) = 0.597 \qquad (16)$$

This measure represents the grade of containment of the unknown pattern in Fig. 12B to the class pattern in Fig. 12A. Note that, in the special case of unifying membership values as in Fig. 12D by calculating the cardinality (eq. 8) and refering to the joint universe $X = p$ (eq. 9), the relative cardinality is identical with the arithmetic mean as used in eq. 12.

A practical application of the method can be found in ref. 13 where chromatographically analysed urine samples had to be classified according to three "classes" of nephritis. The retention times were taken as the universe X and the classes were constructed from 5 to 8 training samples in each class. If samples are matched against the class from which they originate, values between 0.9 and 1 were derived for the degree of containment according to eq. 16. On the other hand, samples that did not belong to a class are returned with degrees of containment between 0.45 and 0.81, enabling classification of unknown samples even when the number of peaks is different in the samples and when the retention times vary to a reasonable degree (Table 1A).

Other applications of the method are found in library search in IR spectroscopy [14], classification of gasolines based on capillary gas chromatography [15], and quality control of tablets by measuring their UV spectra [13].

In principle the method is not limited to the consideration of a single variable but can also be used in cases of several variables. In chromatography, for instance, the retention position and the signal of a peak could be used for classifying unknown samples with the fuzzy method. In that case a two-dimensional m.f. would be needed, for example one based on a circle or ellipse as the domain of influence for each observation. Over these domains of influence (also called the "supports" in fuzzy theoretical terms [16]) the m.f.s are specified as surfaces of suitable structure, e.g. paraboloids or ellipsoids. If one variable is the absorbance y and the second variable is the retention time x the two-dimensional m.f. reads:

$$m_i(x, y)$$
$$= \left\{ 1 - \left[(x - x_i)^2 / u_i^2 + (y - y_i)^2 / v_i^2 \right] \right\}^+ \qquad (17)$$

The principles of building the m.f. of the training class from all individual y, x-patterns are the same as in the case of using only one variable.

TABLE 1

Classification of samples based on their chromatographic patterns assuming fuzzified class patterns and crisp sample patterns

Class origin	Grade of containment (eq. 16) obtained by matching against class number		
	1	2	3
A. Using retention data only			
1	1	0.55	0.32
2	0.41	1	0.84
3	0.56	0.73	1
B. Using retention time and signal response data			
1	1	0.33	0.085
2	0.083	1	0.39
3	0.26	0.35	1

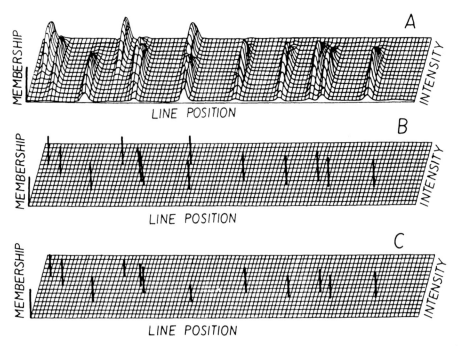

Fig. 13. A two-dimensional membership function specified for identification of components from their IR spectrum. Every peak is characterized by its line position (abscissa) and intensity (ordinate). The unknown component spectrum is fuzzified (A), the candidate spectrum is taken as crisp (B) and comparison is made by fuzzy intersection with the result as given in (C).

Overlapping m.f.s from neighboring peaks are aggregated by the union of the m.f., and classification of an unknown crisp pattern is performed by intersecting the actual values for absorbance y and retention time x with the class pattern according to eq. 17.

In the example of classifying different patients with respect to nephritis disease mentioned above, the classification was much improved, as is outline in Table 1B: the samples with different class origin are found to have degrees of containment less than 0.4 so that, for example, a sample from class 2 gives a degree of containment to class 3 of 0.39 instead of 0.84 when only the retention time is used for discriminating between the samples.

If still more information on the m.f. in its dependence on the variables is available, a suitable m.f. can be defined over the whole space of variables. This has been demonstrated by Blaffert [14] in the case of the identification of IR spectra with the fuzzy method. The line position and intensity of every spectroscopic peak is evaluated and that of the sample spectrum is fuzzified as given in Fig. 13. The same principles of comparing the candidate reference spectrum (A) with an unknown spectrum (B) by fuzzy intersection can be applied, leading to the membership values as given in Fig. 13C.

5 CALCULATIONS WITH FUZZY NUMBERS

5.1 How much is twenty minus fourteen?

One of the possibilities in fuzzy theory is the computation with fuzzy numbers rather than by means of (crisp) conventional numbers. Fuzzy arithmetic can be carried out by theoretically well-defined rules [2] and is important if, for instance, functions such as spectra or depth profiles are to be compared. Imagine a situation where the similarity of two blurred spectra is to be judged. One could in the first instance subtract one spectrum from the other one and use the

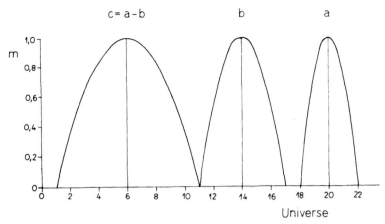

Fig. 14. Fuzzy subtraction of "about 20" (a) minus "about 14" (b) giving "about 6" (c).

resulting difference (or squared differences) as a measure of their similarity. In terms of fuzzy theory this means substracting two fuzzy functions or — if discrete values are measured — subtracting fuzzy numbers.

A fuzzy number a (for mathematical definitions see ref. 16, p. 196), such as "about 20" can be specified by the following m.f. (cf. Fig. 14 and eq. 4):

$$m_a(x) = \left[1 - |x - a|^2/\text{const}\right]^+ \quad (18)$$

$$m_{\tilde{20}}(x) = \left[1 - |x - 20|^2/2^2\right]^+ \quad (19)$$

The tilde stands for a "fuzzy 20".

Arithmetic operations with fuzzy numbers are based on the so-called extension principle (ref. 2, p. 47) and the difference of the fuzzy numbers $a - b$ would reveal the following m.f.:

$$m_{a-b}(z) = \sup_{z=x-y} \min\left[m_a(x), m_b(y)\right] \quad (20)$$

where sup represents the supremum (maximum) and $m_a(x)$, $m_b(y)$ are the m.f.s of the fuzzy numbers a and b.

The uncertainty of the second fuzzy number b, here 14, is characterized by the m.f.

$$m_{\tilde{14}}(y) = \left[1 - |y - 14|^2/3^2\right]^+ \quad (21)$$

then the m.f. for the difference c of both numbers is computed in the following way: take all combinations of $x - y$ that give a certain z_0, evaluate the minimum of $m_a(x)$ and $m_b(y)$ and store the maximum (supremum) of all the minima computed. For example (cf. Fig. 14),

$z_0 = 7$ if $x = 21$ and $y = 14$,
i.e. min(0.75, 1) = 0.75

$z_0 = 7$ if $x = 19$ and $y = 12$,
i.e. min(0.75, 0.56) = 0.56

⋮

$z_0 = 7$ then finally

$$\sup_{z=x-y} \min(m_{\tilde{20}}(x), m_{\tilde{14}}(y)) = 0.96$$

To compute the remaining membership values for the result of the subtraction, i.e. fuzzy 6, the procedure would have to be continued for all z_0-values which have non-zero membership values.

You will agree that this procedure is not extremely user-friendly and might put some people off computing with fuzzy numbers. Therefore a much more practicable version should be used that is based on so-called fast operations (cf. ref. 2, p. 53). With these operations a fuzzy number is characterized by its LR (left/right) representation, where the m.f. over the mean value of the fuzzy set splits into a left and a right branch. In case of the fuzzy number a the LR representation is as follows:

$$m_a(x) = \begin{cases} L((a-x)/\alpha_0) & \text{for } x \leq a \\ R((x-a)/\beta_0) & \text{for } x \geq a \end{cases} \quad (22)$$

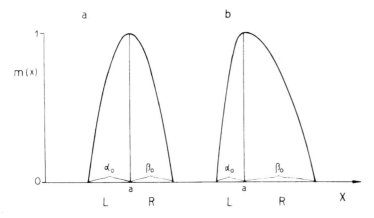

Fig. 15. LR (left/right) representation of a fuzzy number in case of a symmetric (a) and an asymmetric (b) membership function.

Here, $\alpha_0 > 0$ and $\beta_0 > 0$ are the left and right spreads of the domain of influence, respectively. Fig. 15 gives examples of fuzzy numbers with symmetric and asymmetric m.f.s.

As a preliminary to subtracting the two fuzzy numbers a and b the fuzzy number b is likewise specified by

$$m_b(x) = \begin{cases} L((b-x)/\alpha) & \text{for } x \leqslant b \\ R((x-b)/\beta) & \text{for } x \geqslant b \end{cases} \quad (23)$$

Now fuzzy subtraction can be performed quite easily by

$$m_{a-b}(x) = \begin{cases} L((a-b-x)/(\alpha_0 + \beta)) \\ R((x-(a-b))/(\beta_0 + \alpha)) \end{cases} \quad (24)$$

Computation is greatly simplified with this representation of the m.f. of the difference of two fuzzy numbers since the arguments of L and R can be directly inserted into the equation of the m.f. If we reconsider the type of the m.f. used in Fig. 14, i.e.

$$m(z) = [1 - z^2]^+ \quad (25)$$

then the m.f. $m_{a-b}(x)$ is calculated according to

$$m_{a-b}(x) = \begin{cases} 1 - ((a-b-x)/(\alpha_0 + \beta))^2 \\ \quad \text{for } x \leqslant (a-b) \\ 1 - (x-(a-b)/(\beta_0 + \alpha))^2 \\ \quad \text{for } x \geqslant (a-b) \end{cases} \quad (26)$$

The result has already been considered in Fig. 14.

Addition of two fuzzy numbers $a + b$ reveals:

$$m_{a+b}(x) = \begin{cases} L((a+b-x)/(\alpha_0 + \beta_0)) \\ \quad \text{for } x \leqslant (a+b) \\ R((x-(a+b))/(\alpha + \beta)) \\ \quad \text{for } x \geqslant (a+b) \end{cases} \quad (27)$$

Finally, we also mention multiplication of two fuzzy numbers $a * b$ (ref. 2, p. 55):

$$m_{a*b}(x) = \begin{cases} L((ab-x)/(a\alpha + b\alpha_0)) \\ R((x-ab)/(a\beta + b\beta_0)) \end{cases} \quad (28)$$

5.2 Spectra are fuzzy functions

As we mentioned above, typical applications of the use of fuzzy arithmetic is the comparison of spectra. Particularly, comparison of spectra with only few spectral features, such as ultraviolet spectra, may be difficult if solvent influence and other experimental parameters blur them.

Fig. 16 shows two spectra that represent a sample spectrum (a) and a candidate spectrum (b) that may originate form a library of reference spectra. Uncertainties due to experimental conditions are modeled by means of an m.f. of the type given in eq. 25. Thus the sample spectrum can be described as a family of fuzzy numbers $a(x)$, with x being the channel or wavelength axis. If the adequate notation is applied to the candidate reference spectrum, $b(x)$, then the difference be-

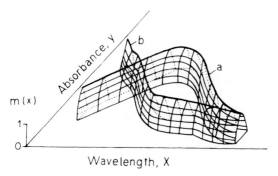

Fig. 16. Fuzzy functions used to characterize a blurred sample spectrum (a) and a blurred candidate reference spectrum (b).

TABLE 2

Component identification in the ultraviolet spectral range with use of the degree of similarity, m_S (eq. 31)

Candidate reference spectrum			
Phenacetin against phenacetin in % methanol (v/v)	Fuzzy criterion	Phenobarbital against	Fuzzy criterion
10	0.950	phenobarbital	0.823
20	0.914	barbital	0.236
30	0.889	codeinium	
40	0.879	phosphate	0.356

tween sample and reference spectrum gives the following m.f.:

$$m_{a-b}(y; x) = \begin{cases} L((a(x) - b(x) - y)) \\ /(\alpha_0(x) + \beta(x)) \\ R((y - (a(x) - b(x))) \\ /(\beta_0(x) + \alpha(x)) \end{cases} \quad (29)$$

where y is the absorbance and the spreads $\alpha_0(x)$, $\beta_0(x)$, $\alpha(x)$, $\beta(x)$ can be specified anew for every wavelength x.

In order to characterize the difference of the spectra quantitatively, a criterion N has been proposed, having the following form [17]:

$$N = \int_X \int_Y |y| m_{a-b}(y; x) \, dy \, dx \quad (30)$$

The criterion characterizes up to which degree the difference fits the "zero-function" in a fuzzy sense: it is formed by integrating the product of the residual value y and its corresponding membership value m_{a-b} over the whole wavelength range X and over the absorbance range Y likewise.

To normalize the result to the interval [0, 1] a degree of similarity, m_S, can be defined, i.e.

$$m_S = [1 - N/N_{max}]^+ \quad (31)$$

where N_{max} is the criterion N computed for the case where only the sample spectrum is taken into account without overlaying the candidate reference spectrum.

Table 2 shows records for identification of phenacetin and phenobarbital. Phenacetin is matched against spectra that were monitored with a photodiode array detector in mobile phases of varying methanol content. In spite of the fact that the UV spectra are heavily altered when the methanol content increases from 10 to 40% (v/v) (cf. ref. 17) the degree of similarity remains at values higher than 0.8. This enables the component phenacetin to be recognized despite the uncertain shape of the spectrum. The identification of phenobarbital (columns 3 and 4 in Table 2) among components with quite similar spectra (cf. Fig. 2 in ref. 17) is also possible because the different entities barbital and codeinium phosphate are returned with degree of similarity values of 0.236 and 0.35, respectively.

Further applications of comparing fuzzy functions are known for peak tracking in high-performance liquid chromatographic separations [18] and for depth-profiling in secondary ion mass spectrometry [19].

5.3 Mixture component analysis

A logical extension of the method for comparing fuzzy functions can be applied to spectra that consists of a combination of several component spectra. Usually, the mixture spectrum can be considered as a linear combination of the component spectra weighted by their concentrations c_j and Beer's law is valid, i.e.

$$a(x) = \sum_{j=1}^{r} c_j b_j(x) \quad (32)$$

with r being the number of components in the

mixture. This is the basic idea of multicomponent analysis and the procedure has been successfully applied to mixture analysis of pain relieving tablets [17].

The same criterion N as outlined in eq. 30 can be used. The m.f. for the candidate reference spectrum is now written for the concentration-weighted sum spectrum as follows:

$$m_{\Sigma c_j b_j(x)}(y; x)$$

$$= \begin{cases} L\left(\left(\Sigma c_j b_j(x) - y\right)\big/\left(\Sigma c_j \alpha_j(x)\right)\right) \\ R\left(\left(y - \Sigma c_j b_j(x)\right)\big/\left(\Sigma c_j \beta_j(x)\right)\right) \end{cases} \quad (33)$$

If one introduces the difference between the sample and reference spectrum as $d = a(x) - \Sigma c_j b_j(x)$ then the m.f. of the residuals becomes:

$$m_d(y; x) = \begin{cases} L\left(\left(a(x) - \Sigma c_j b_j(x) - y\right)\right) \\ \big/\left(\alpha_0(x) + \Sigma c_j \beta_j(x)\right) \\ R\left(\left(y - \left(a(x) - \Sigma c_j b_j(x)\right)\right)\right) \\ \big/\left(\beta_0(x) + \Sigma c_j \beta_j(x)\right) \end{cases} \quad (34)$$

By inserting this m.f. into eq. 30 for computing the criterion N the degree of similarity can be evaluated according to eq. 31. The optimum values for the sought concentrations are found by a simplex search routine minimizing N or maximizing m_S.

Typical relative prediction errors for analyzing mixtures of caffeine, propyphenazone and phenacetin based on recording their pure spectra and data reduction by the fuzzy method were found to be 2.4%, 13.88% and 1.05%, respectively. In comparison, application of ordinary least squares (OLS) analysis gives much worse prediction errors, namely 52.38%, 33.32% and 29.79%, respectively [17].

This is due to the fact that the non-additivity of component spectra, which is valid in the example studied, cannot be accounted for in OLS regression analysis and that, in addition, strongly overlapping spectra (ill-conditioned systems) are more easily handled by the fuzzy method.

It also should be mentioned that the fuzzy method is not limited to the assumptions of the validity of Beer's law, but any model — even a non-linear one — could be used.

6 FUZZY MODELLING

For controlling industrial, chemical or biotechnological processes adequate models of the systems are needed. Very often, however, the variables/parameters to be controlled are difficult to measure; they sometimes cannot be easily described, or their mutual interdependence may be very complex. In addition, knowledge about the process model may be incomplete or the elaboration of a precise model might be inadequate from an economic point of view. Furthermore, even the subjective experience of a skilled operator should be taken into consideration for controlling the process, which might be difficult to express linguistically in exact terms.

Such kinds of modelling can be carried out by means of fuzzy methods that include fuzzy implications and reasoning, as has been shown by Tanaka and Sugeno [21].

Modelling, of course, is not restricted to process control, but is necessary for calibrating or optimizing analytical procedures, for predicting environmental or pharmaceutical data and is used for deconvolution of spectra based on non-linear models.

To explain the feasibility of fuzzy modelling let us consider the well-known straight line model:

$$y = a_0 + a_1 x \quad (35)$$

This model is to be used for fitting the data shown in Fig. 5. These data can be considered as fuzzy observations and their uncertainty can be described by specifying a suitable m.f. The m.f. could be of the type given in eq. 5, as demonstrated in Fig. 5. In ref. 20 we used m.f.s based on elliptical supports revealing, for every observation i, an ellipsoid of the type given in eq. 17. The parameters were set at $u_i = 0.25$ and $v_i = 0.40$. For combining the i fuzzy observations the union M of the m.f.s of the observations is used according to eq. 7

$$m_M(x, y) = \max_i m_i(x, y) \quad (36)$$

For evaluation of the optimum parameters (a_0, a_1) to fit our observations M to the straight line the following measure can be applied [20]:

$$\text{rel}_{X_0} \text{ card } M = \int_{X_0} m_M(x, a_0 + a_1 x) \, dx \bigg/ \int_{X_0} dx \quad (37)$$

Here, X_0 denotes the set where (fuzzy) information concerning the modelling problem exists, i.e. where at least one point (x, y) with $m_M(x, y) > 0$ exists, and the cardinality card M characterizes the intersection of the actual straight line with the ellipsoids as the m.f.s of the fuzzy observations. As this relative cardinality is directly related to the parameter set (a_0, a_1) it can be taken as a fuzzy set over the parameter space and we write for its m.f.

$$m_{E_0}(a_0, a_1) = \int_{X_0} m_M(x, a_0 + a_1 x) \, dx \bigg/ \int_{X_0} dx \quad (38)$$

In order to enable different families of functions to be applied to the model, a dependence of the computed relative cardinality should be compared to local approximations of the functional relationship [20]. These local approximations have the following m.f.

$$m_{E_0}(f) = \int_{X_0} \sup_{y \in Y} m_M(x, y) \, dx \bigg/ \int_{X_0} dx \quad (39)$$

This means that at every x-position those y-values are considered where the function $m_M(x, y)$ assumes the supremum.

From eqs. 38 and 39 the degree of approximation of the straight line compared to the local approximation is derived as

$$m_C(a_0, a_1; f) = m_{E_0}(a_0, a_1)/m_{E_0}(f) \quad (40)$$

For explanation, consider the straight lines with the parameter sets (1) $a_0 = 0.020$, $a_1 = 0.92$; (2) $a_0 = 0$, $a_1 = 1.12$; and (3) $a_0 = -0.03$, $a_1 = 1.32$. For these parameter sets the relative cardinality is computed for every straight line according to eq. 38. Then they are compared to the previously calculated relative cardinality of the local approximations (eq. 39). As degree of approximation one gets from these calculations m_C-values (eq. 40) of 0.77, 0.58 and 0.32, respectively.

Since, by our definition, the optimum parameter set gives the best approximation of the straight line to the fuzzy observations the degree of approximation must be maximized by varying the parameter set based on a grid search or by means of a Monte-Carlo technique. In the present example a maximum membership value is found to be $m_C = 0.77$, with the parameters $a_0 = 0.20$ and $a_1 = 0.92$ [20].

The fuzzy set m_C over the parameter space can be visualized as shown in Fig. 17. These fuzzy estimates of the parameters can also be used to compute fuzzy predictions at a certain y_0-value. The basic idea is derived from the extension principle (cf. ref. 2) and the m.f. for the predicted x-value at the measured y_0-value, $m_0(x; y_0)$, is obtained as follows:

$$m_0(x; y_0) = \sup_{\{(a_0, a_1) : y_0 = a_0 + a_1 x\}} m_C(a_0, a_1; f) \quad (41)$$

That is, the supremum (maximum) is to be computed for every x taking into account in every case all those $(a_0, a_1) \in A$ that belong to the same value y_0 with $y_0 = a_0 + a_1 x$.

In practice the maximum is found by evaluating the maximum m_E-value for x at a given y_0 along the graph of $a_0 = y_0 - x a_1$, with a_0 being

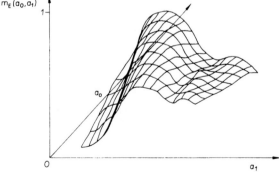

Fig. 17. Plot of the degree of approximations $m_C(a_0, a_1; f)$ according to eq. 40 as a function of the parameters (a_0, a_1).

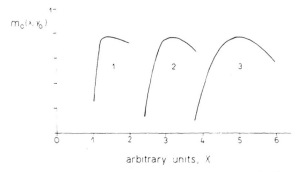

Fig. 18. Fuzzy predictions of x-values with the straight line model in eq. 35 at y_0-values of 1.48 (1), 3.17 (2) and 4.68 (3).

the dependent and a_1 the independent variable (cf. ref. 20).

Typical predictions are given in Fig. 18. As seen from the figure the predictions provide an evaluation of the predicted x-values along the x-axis. The broader membership curve of the 5th observation is due to the fact that this point is an obvious outlier compared to the other y-values (cf. Fig. 5). The main advantages with this kind of fuzzy modelling can be summarized as

(i) there are no restrictions concerning the number of experiments available for computing fuzzy estimates of the parameters and of fuzzy predictions,
(ii) all information on the uncertainty of the observations can be used for characterizing the fuzziness of the observations either based on data-oriented knowledge or on subjective aspects,
(iii) linear and non-linear models can be used likewise, even whole families of functions could be used for modelling a dependence, and
(iv) the method is not limited to one variable as in the straight line example but responses in dependence on several variables could be modelled.

7 TO COME: FUZZY LOGIC

Utilization of fuzzy set theory in the field of reasoning and logic dates back to the late seventies when Zadeh himself introduced a concept for fuzzy reasoning [22] in order to enable conclusions or inferences to be drawn from imprecise, vague and uncertain premises. Meanwhile fuzzy logic has been already squeezed onto a single chip [8] giving us a basis for the use of expert systems in faster processes, such as robotics or instrumental and industrial control.

7.1 Linguistic variables

A basic tool in fuzzy logic is the feasibility of defining linguistic variables that may be verbal units, words or sentences in a natural or artificial language. The most cited example refers to the linguistic variable "young" which can be labeled by a fuzzy set on the universe "age". Fig. 19 demonstrates a proposal for "young" that assigns membership values of 1 to persons aged less than 20 and shows a decreasing m.f. with increasing age. The m.f. for "young", m_Y, chosen here has the following analytical form [22]:

$$m_Y(x) = \begin{cases} 1 & \text{for } x \leq a \\ 1 - 2((x-a)/(c-a))^2 & \text{for } a \leq x \leq b \\ 2((x-c)/(c-a))^2 & \text{for } b \leq x \leq c \\ 0 & \text{for } x \geq c \end{cases} \quad (42)$$

where the parameters are $a = 20$, $b = 30$, and $c = 40$.

What can also be seen from Fig. 19 is the modification of all the labels by suitable transformation of the m.f. so that it is not always necessary to consider each label anew. The following labels are commonly deduced from the m.f. $m_Y(x)$,

very young $\qquad m_{vY} := m_Y^2$
not young $\qquad m_{nY} := 1 - m_Y \qquad (43)$
more or less young $\qquad m_{mY} := m_Y^{1/2}$

The membership function for "old" had been constructed by changing the parameters of the m.f. in eq. 40 to $a = 40$, $b = 55$, and $c = 70$.

Likewise, linguistic variables can be introduced, such as truth, probability and possibility whose base variable takes values in the unit interval.

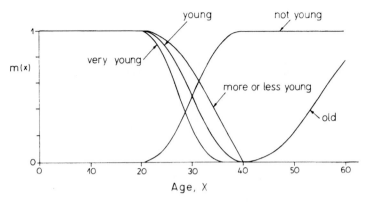

Fig. 19. Fuzzy sets representing the linguistic variable "young" over the universe "age".

Approximate reasoning can be performed on the basis of different propositions of fuzzy logic. In the following section the ideas of multi-valued logic and of Zadeh's compositional rule of inference is explained.

7.2 Multi-valued logic

Classical logic is based on Boolean, i.e. two-valued logic. Multi-valued logic means that truth-values are not restricted to "true" and "false" but gradually modified truth-values, e.g. "very true" or "rather false" are possible. This is achieved by assigning a truth-value T as degree of membership in the interval [0, 1].

To be able to reason about the logical formulas the connectives AND, OR and the negation NOT are applied. Let $T(p)$ be a logical formula "x is p" or simply p then the connectives mentioned have the corresponding fuzzy cases:

NOT p: $T(\text{NOT } p) = 1 - T(p)$
p AND q: $T(p \text{ AND } q) = \min[T(p), T(q)]$
p OR q: $T(p \text{ OR } q) = \max[T(p), T(q)]$ (44)

For explanation, consider the test of verifying the presence of the element bromine (atomic number 35) in an X-ray fluorescence spectrum. Here the atomic formula $p = $ "bromine is present" would be given a truth value of $T(p) = 1$ as a starting point for checking its presence against the actual data base. To combine this information, e.g. with the rule that refers to the detector used (proportional detector — prop — or scintillation detector — scin) the truth value would be sought that resolves, for instance,

if bromine is present AND the available detector

is prop (p AND q);   (45)

if bromine is present AND the available detector

is scin (p AND q);   (46)

if bromine is present AND (the available detector

is prop OR scin)   (47)

For $T(p) = 1$; $T(q) = 0.3$ and $T(r) = 0.65$ the latter expression (eq. 47) translates into

p AND (q OR r)

$= \min\{T(p), \max[T(q), T(r)]\}$

$= \min\{1, \max[0.3, 0.65]\} = 0.65$

In words, bromine can be assumed to be present with a truth-value of 0.65 and this only in case the scintillation detector has been used. Otherwise, with the proportional detector its presence can be validated only with $T(q) = 0.3$. These truth-values are taken in our case from a fuzzy logical database a part of which is illustrated in Table 3 and originates from a PROLOG based expert system for the interpretation of X-ray fluorescence spectra.

Incorporation of fuzzy logic into expert systems is one of the most active fields of using fuzzy logic. Examples of incorporating fuzzy theoretical

TABLE 3

Part of a fuzzy PROLOG-based database for characterizing the detectability of elements by use of the proportional detector (prop) or the scintillation detector (scin)

The arguments of the predicate "detector" represent (atomic number, detector, truth-value).

---

detector(X, prop, Y) :- Y is 1, X < 20
detector(X, scin, Y) :- Y is 1, X > 50
detector(20, prop, 1)
detector(30, prop, 0.5)
detector(35, prop, 0.3)
detector(40, prop, 0.2)
detector(50, prop, 0.03)
detector(25, scin, 0.1)
detector(30, scin, 0.5)
detector(35, scin, 0.65)
detector(40, scin, 0.8)
detector(50, scin, 0.97)

---

tools in an expert system are known from medical diagnosis (CADIAC-2 [23]) or in engineering (Speril [24]). It is expected that the use of fuzzy reasoning and the feasibility of incorporating imprecise and uncertain knowledge into today's expert systems will provide a new generation of those systems that might approach human thinking more closely than the current commonly used methods of artificial intelligence.

7.3 The compositional rule of inference

In order to characterize a relationship (or dependence) between linguistic variables a fuzzy conditional statement can be used. For instance, if t (temperature) and c (concentration) are linguistic variables taking on fuzzy values $A \overset{c}{\sim} X$ and $B \overset{c}{\sim} Y$ respectively, ("$A \overset{c}{\sim} X$" reads: A is contained in the fuzzy set X), then a dependence between t and c may be given by a fuzzy conditional statement as:

IF ($t = A$) THEN ($c = B$), or for short

IF A THEN B (48)

This simple fuzzy implication is assumed to be the Cartesian product $A \times B$ being a fuzzy relation of the form

IF A THEN $B = A \times B$ (49)

TABLE 4

Discrete membership values for "medium temperature" and "high concentration"

$t(°C)$	250	260	270	280	290	300	310
$m_A(x)$	0	0.6	0.8	1	0.7	0.4	0
$c(M)$	0.5	0.6	0.7	0.8			
$m_B(y)$	0.2	0.5	0.8	1			

The fuzzy set on the Cartesian product has the following m.f.:

$$m_{A \times B}(x, y) = \min[m_A(x), m_B(y)]; \quad x \in X, y \in Y \quad (50)$$

In terms of the conditional statement between temperature and concentration we are able now to describe a relationship, such as

IF temperature is medium

THEN concentration is high (51)

Let us assume for the fuzzy sets "medium temperature" and "high concentration" the discrete membership values as given in Table 4. Then we get for the Cartesian product $A \times B$ the membership values given in Table 5. These are obtained by, e.g.

$$\left.\begin{array}{l} m_A(260) = 0.6 \\ m_B(0.7) = 0.8 \end{array}\right\} \min(0.6, 0.8) = 0.6$$

We are now able to ask the question "if the temperature takes on a certain value what is the value of concentration implied by the relationship between t and c?".

The answer is provided by the compositional

TABLE 5

The Cartesian product $A \times B$ representing the fuzzy implication "IF temperature is medium THEN concentration is high"

$c(M)$	$t(°C)$						
	250	260	270	280	290	300	310
0.5	0	0.5	0.5	0.5	0.5	0.4	0
0.6	0	0.6	0.6	0.6	0.6	0.4	0
0.7	0	0.6	0.7	0.7	0.7	0.4	0
0.8	0	0.6	0.8	0.8	0.7	0.4	0

TABLE 6

Fuzzy concentrations corresponding to given fuzzy sets of temperature evaluated by the compositional rule of inference (eq. 52)

$t(°C)$	250	260	270	280	290	300	310
$m_{A'}(x)$	0	0	1	0	0	0	0
$c(M)$	0.5	0.6	0.7	0.8			
$m_{B'}(y)$	0.5	0.5	0.7	0.8			

rule of inference which states: if $R \subseteq X \times Y$ is a fuzzy relation that represents a dependence between t and c (a fuzzy conditional statement) and t assumes a value A', then the induced value of c is:

$$m_{B'}(y) = \max_{x \in X} \min[m_{A'}(x), m_{A \times B}(x, y)]$$

for each $y \in Y$ \hfill (52)

Returning to our example, if we want to know the concentrations that correspond to a temperature of, say, 270°C the compositional rule of inference reveals the fuzzy set over concentrations as given in Table 6. From the resulting fuzzy set it can be deduced that an 0.8 M concentration has the highest degree of membership ($m_{B'}(y) = 0.8$) and lower degrees of membership are assigned to decreasing concentrations.

Compared to the original fuzzy set "high concentration" (cf. Table 4) the fuzzy set over concentration is more uncertain, taking into account the dependence between temperature and concentration.

The case presented represents, of course, only a simple example. Real systems such as process controllers, are not described by one implication only. More complicated relations are therefore constructed on the basis of eq. 50, as may be found in the work by Zadeh [22].

Especially in the field of fuzzy logic, many important developments of the theory have been introduced since Zadeh's innovations in 1973. For example, recent investigations deal with relationships between fuzzy and probabilistic reasoning and can be learned from ref. 25.

8 CONCLUSIONS

The use of fuzzy theoretical methods within the area of chemometrics and in the laboratory is as yet in its infancy. However, the methods are nowadays being very actively explored by scientists from different fields. Initial applications in multi-criteria optimization, in pattern recognition, in fuzzy modelling and in the field of artificial intelligence have demonstrated with certainty that fuzzy theory will be a valuable tool to extend the possibilities of interpreting and modelling systems by mathematical techniques and therefore should be used as a supplementary technique in chemometrics and related disciplines.

It should be mentioned that many of the definitions used in this tutorial are not the only rational choices for a theory of fuzzy set operations, matching and reasoning, yet they are the most frequently used. Other choices that have been the focus of important developments over the last few decades can be found in, e.g., refs. 2, 3, 21, 22, 24 or 25.

For the future, we may expect that the proposed methods of fuzzy data handling will be validated and their performance will be investigated in many practical situations. On the other hand, the possibility of incorporating artificial intelligence tools in daily laboratory practice will create many new fields for the application of fuzzy logic and fuzzy reasoning in order to match human thinking as closely as possible.

The tutorial should also have made it clear that application of fuzzy theory very often represents a real advance by comparison with common (crisp) mathematics: if arithmetic is performed with fuzzy numbers, then "crisp arithmetic" is included in the operations, or if fuzzy logic is applied, classical logic is always inherent in fuzzy reasoning.

REFERENCES

1 L.A. Zadeh, Fuzzy sets, *Information and Control*, 8 (1965) 338.
2 D. Dubois and H. Prade, *Fuzzy Sets and Systems: Theory and Application*, Academic Press, New York, 1980.
3 H.J. Zimmermann, *Fuzzy Set Theory and Its Applications* (International Series in Management Sciences/Operations

Research), Kluwer Nijhoff, Boston, Dordrecht, Lancaster, 1986.
4 A. Kaufman, *Introduction to the Theory of Fuzzy Subsets, Vol. 1: Fundamental Theoretical Elements*, Academic Press, New York, 1975.
5 A. Kandel, *Fuzzy Mathematical Techniques with Applications*, Addison-Wesley, Reading, 1986.
6 H. Bandemer and A. Kraut, On the fuzzy theory based computer-aided particle shape description, *Fuzzy Sets and Systems*, in press.
7 FLOPS — A Fuzzy-logic Production System, Dr. William Siler, Kemp-Carraway Heart Institute, Birmingham, AL 35234, U.S.A.
8 Expert on a Chip, *Scientific American*, 254 (1986) 50.
9 J. Zupan, *Clustering of Large Data Sets*, Wiley, Chicester, 1982.
10 T. Jacobsen, K. Kolset and N.B. Vogt, Partial least squares regression and fuzzy clustering — a joint approach, *Mikrochimica Acta (Wien)*, II (1986) 125–138.
11 J.C. Bezdek, *Pattern Recognition with Fuzzy Objective Function Algorithms*, Plenum Press, New York, 1982.
12 D.L. Massart, B.G.M. Vandeginste, S.N. Deming, Y. Michotte and L. Kaufman, *Chemometrics: A Textbook*, Elsevier, Amsterdam, 1988.
13 M. Otto and H. Bandemer, Pattern recognition based on fuzzy observations for spectroscopic quality control and chromatographic fingerprinting, *Analytica Chimica Acta*, 184 (1986) 21–31.
14 T. Blaffert, Computer-assisted multicomponent spectral analysis with fuzzy data sets, *Analytica Chimica Acta*, 161 (1984) 135–148.
15 E. Stottmeister, H. Hermann, P. Hendel, D. Feiler, M. Nagel and H.-J. Dobberkau, Spurenbestimmung von Vergaser- und Dieselkraftstoffen in Wasser mittels Kapillargaschromatographie/automatischer Mustererkennung, *Fresenius Zeitschrift fuer Analytische Chemie*, 327 (1987) 709–714.
16 H. Bandemer and M. Otto, Fuzzy theory in analytical chemistry, *Mikrochimica Acta (Wien)*, II (1986) 93–124.
17 M. Otto and H. Bandemer, A fuzzy method for component identification and mixture evaluation in the ultraviolet spectral range, *Analytica Chimica Acta*, 191 (1986) 193–204.
18 M. Otto, W. Wegscheider and E.P. Lankmayr, A fuzzy approach to peak tracking in chromatographic optimizations, SCA (Europe), 13–15 May 1987, Abstracts, p. 63.
19 M. Otto, Chemometrics for material analysis, *Mikrochimica Acta (Wien)*, I (1987) 445–453.
20 M. Otto and H. Bandemer, Calibration with imprecise signals and concentrations based on fuzzy theory, *Chemometrics and Intelligent Laboratory Systems*, 1 (1986) 71–78.
21 T. Tanaka and M. Sugeno, Fuzzy identification of systems and its applications to modelling and control, *IEEE Transactions on Systems, Man and Cybernetics*, Vol. SMC-15, 1 (1985) 116.
22 L.A. Zadeh, A theory of approximate reasoning, in J.H. Hayes, D. Michie and L.I. Mikulich (Editors), *Machine Intelligence 9*, Wiley, New York, 1979, p. 49.
23 K.P. Adlassnig, in M.M. Gupta and E. Sanchez (Editors), *Approximate Reasoning in Decision Analysis*, North-Holland, Amsterdam, 1982, p. 203.
24 M. Ishizuka, K.S. Fu and J.T.P. Yao, in M.M. Gupta and E. Sanchez (Editors), *Approximate Reasoning in Decision Analysis*, North-Holland, Amsterdam, 1982, p. 261.
25 J.F. Baldwin, Evidential support logic programming, *Fuzzy Sets and Systems*, 24 (1987) 1–7.

Author Index

Berridge, J.C.
　Chemometrics and method development in high-performance liquid chromatography. Part 1: Introduction 139
Berridge, J.C.
　Chemometrics and method development in high-performance liquid chromatography. Part 2: Sequential experimental designs 153
Birks, H.J.B.
　Multivariate analysis in geology and geochemistry: an introduction 340
Birks, H.J.B.
　Multivariate analysis of stratigraphic data in geology: a review 383
Brereton, R.G.
　Fourier transforms: use, theory and applications to spectroscopic and related data 166
Buchanan, B.G.
　-, Feigenbaum, E.A. and Lederberg, J.
　　On Gray's interpretation of the Dendral project and programs: myth or mythunderstanding? (Invited Comments) 48
Burton, K.W., see Morgan, E. 104
Burton, K.W.C.
　- and Nickless, G.
　　Optimisation via Simplex. Part I. Background, definitions and a simple application 124
Christie, O.H.J.
　Some fundamental criteria for multivariate correlation methodologies 286
Church, P.A., see Morgan, E. 104
Dessy, R.E.
　Scientific word processing 1
Esbensen, K., see Wold, S. 209
Feigenbaum, E.A., see Buchanan, B.G. 48
Flerackers, E.
　Scientific programming with GKS: advantages and disadvantages 21
Geladi, P., see Wold, S. 209
Gray, N.A.B.
　Dendral and Meta-Dendral – the myth and the reality 26
Gray, N.A.B.
　Response to comments by Buchanan, Feigenbaum and Lederberg 51
Kateman, G.
　Sampling theory 196
Kleywegt, G.J.
　-, Luinge, H.-J. and Schuman, B.-J.P.
　　PROLOG for chemists. Part 1 67

Kleywegt, G.J.
　-, Luinge, H.-J. and Schuman, B.-J.P.
　　PROLOG for chemists. Part 2 92
Kvalheim, O.M.
　Interpretation of direct latent-variable projection methods and their aims and use in the analysis of multicomponent spectroscopic and chromatographic data 306
Lederberg, J., see Buchanan, B.G. 48
Lee, T.V.
　Expert systems in synthesis planning: a user's view of the LHASA program 53
Lewi, P.J., see Thielemans, A. 262
Lewi, P.J.
　Spectral map analysis: factorial analysis of contrasts, especially from log ratios 250
Luinge, H.-J., see Kleywegt, G.J. 67, 92
McDowall, R.D.
　-, Pearce, J.C. and Murkitt, G.S.
　　The LIMS infrastructure 12
Marshall, A.G.
　Dispersion vs. absorption (DISPA): a magic circle for spectroscopic line shape analysis 181
Massart, D.L., see Thielemans, A. 262
Mellinger, M.
　Multivariate data analysis: its methods 225
Mellinger, M.
　Correspondence analysis: the method and its application 233
Mellinger, M.
　Interpretation of lithogeochemistry using correspondence analysis 367
Morgan, E.
　-, Burton, K.W. and Church, P.A.
　　Practical exploratory experimental designs 104
Murkitt, G.S., see McDowall, R.D. 12
Nickless, G., see Burton, K.W.C. 124
Otto, M.
　Fuzzy theory explained 401
Pearce, J.C., see McDowall, R.D. 12
Reyment, R.A.
　Multivariate analysis in geoscience: fads, fallacies and the future 354
Schuman, B.-J.P., see Kleywegt, G.J. 67, 92
Thielemans, A.
　-, Lewi, P.J. and Massart, D.L.
　　Similarities and differences among multivariate display techniques illustrated by Belgian cancer mortality distribution data 262

Vogt, N.B.
   Soft modelling and chemosystematics 321
Windig, W.
   Mixture analysis of spectral data by multivariate methods 293

Wold, S.
–, Esbensen, K. and Geladi, P.
   Principal component analysis 209

Subject Index

Absorption
    Dispersion vs. absorption (DISPA): a magic circle for spectroscopic line shape analysis 181
Artificial intelligence
    Dendral and Meta-Dendral – the myth and the reality 26
Artificial intelligence
    PROLOG for chemists. Part 2 92
Bit-mapping
    Scientific word processing 1
Calibration
    Interpretation of direct latent-variable projection methods and their aims and use in the analysis of multicomponent spectroscopic and chromatographic data 306
Calibration
    Soft modelling and chemosystematics 321
Cancer mortality distribution
    Similarities and differences among multivariate display techniques illustrated by Belgian cancer mortality distribution data 262
Chemical composition of rock samples
    Interpretation of lithogeochemistry using correspondence analysis 367
Chemosystematics
    Soft modelling and chemosystematics 321
Chemotaxonomy
    Soft modelling and chemosystematics 321
Classification
    Multivariate data analysis: its methods 225
Classification
    Some fundamental criteria for multivariate correlation methodologies 286
Classification
    Interpretation of direct latent-variable projection methods and their aims and use in the analysis of multicomponent spectroscopic and chromatographic data 306
Classification
    Multivariate analysis in geology and geochemistry: an introduction 340
Closure problem
    Multivariate analysis in geoscience: fads, fallacies and the future 354
Clustering
    Multivariate analysis in geology and geochemistry: an introduction 340
Clustering methods
    Soft modelling and chemosystematics 321
Cole–Cole plot
    Dispersion vs. absorption (DISPA): a magic circle for spectroscopic line shape analysis 181

Component identification
    Fuzzy theory explained 401
Composition activity regression
    Soft modelling and chemosystematics 321
Computer graphics standards
    Scientific programming with GKS: advantages and disadvantages 21
Computer programming
    PROLOG for chemists. Part 1 67
Computer programming
    PROLOG for chemists. Part 2 92
Congruence coefficients
    Interpretation of direct latent-variable projection methods and their aims and use in the analysis of multicomponent spectroscopic and chromatographic data 306
Constraints
    Optimisation via Simplex. Part I. Background, definitions and a simple application 124
Contrasts
    Spectral map analysis: factorial analysis of contrasts, especially from log ratios 250
Control
    Sampling theory 196
Correlation
    Multivariate analysis of stratigraphic data in geology: a review 383
Correlation analysis
    Some fundamental criteria for multivariate correlation methodologies 286
Correspondence analysis
    Correspondence analysis: the method and its application 233
Correspondence analysis
    Interpretation of lithogeochemistry using correspondence analysis 367
Correspondence factor analysis
    Similarities and differences among multivariate display techniques illustrated by Belgian cancer mortality distribution data 262
Data analysis
    Multivariate data analysis: its methods 225
Data analysis
    Correspondence analysis: the method and its application 233
Data analysis
    Interpretation of lithogeochemistry using correspondence analysis 367

Data transformations
  Multivariate analysis of stratigraphic data in geology: a review 383
Dendral
  Dendral and Meta-Dendral – the myth and the reality 26
Description
  Sampling theory 196
Design
  Practical exploratory experimental designs 104
Desk top publishing
  Scientific word processing 1
Dielectric relaxation
  Dispersion vs. absorption (DISPA): a magic circle for spectroscopic line shape analysis 181
Dimensionality reduction
  Principal component analysis 209
Dimensionality reduction
  Similarities and differences among multivariate display techniques illustrated by Belgian cancer mortality distribution data 262
Discriminant analysis
  Soft modelling and chemosystematics 321
DISPA
  Dispersion vs. absorption (DISPA): a magic circle for spectroscopic line shape analysis 181
Dispersion
  Dispersion vs. absorption (DISPA): a magic circle for spectroscopic line shape analysis 181
Documentation
  The LIMS infrastructure 12
Experiment design
  Soft modelling and chemosystematics 321
Expert systems
  Expert systems in synthesis planning: a user's view of the LHASA program 53
Expert systems
  PROLOG for chemists. Part 1 67
Expert systems
  PROLOG for chemists. Part 2 92
Exploratory
  Practical exploratory experimental designs 104
Factor analysis
  Multivariate data analysis: its methods 225
Factor analysis
  Correspondence analysis: the method and its application 233
Factor analysis
  Mixture analysis of spectral data by multivariate methods 293
Factor analysis
  Interpretation of direct latent-variable projection methods and their aims and use in the analysis of multicomponent spectroscopic and chromatographic data 306
Factor analysis
  Interpretation of lithogeochemistry using correspondence analysis 367
Factorial
  Practical exploratory experimental designs 104
Factorial data analysis
  Spectral map analysis: factorial analysis of contrasts, especially from log ratios 250
Factor plots
  Correspondence analysis: the method and its application 233
Factors
  Optimisation via Simplex. Part I. Background, definitions and a simple application 124
Fourier spectrometer
  Fourier transforms: use, theory and applications to spectroscopic and related data 166
Fourier transform, discrete
  Fourier transforms: use, theory and applications to spectroscopic and related data 166
Fourier transforms
  Fourier transforms: use, theory and applications to spectroscopic and related data 166
Frequency domain
  Fourier transforms: use, theory and applications to spectroscopic and related data 166
Fuzzy data analysis
  Fuzzy theory explained 401
Fuzzy functions
  Fuzzy theory explained 401
Fuzzy logic
  Fuzzy theory explained 401
Fuzzy modelling
  Fuzzy theory explained 401
Fuzzy set operations
  Fuzzy theory explained 401
Geoscience
  Multivariate analysis in geoscience: fads, fallacies and the future 354
GKS
  Scientific programming with GKS: advantages and disadvantages 21
Graphical Kernel System
  Scientific programming with GKS: advantages and disadvantages 21
Graphic standards
  Scientific word processing 1
Hard modelling
  Multivariate analysis in geoscience: fads, fallacies and the future 354
High-performance liquid chromatography
  Chemometrics and method development in high-performance liquid chromatography. Part 1: Introduction 139
High-performance liquid chromatography
  Chemometrics and method development in high-performance liquid chromatography. Part 2: Sequential experimental designs 153
Intelligent databases
  PROLOG for chemists. Part 1 67
Intelligent databases
  PROLOG for chemists. Part 2 92
Laboratory information management system
  The LIMS infrastructure 12

Latent variables
    Interpretation of direct latent-variable projection methods and their aims and use in the analysis of multicomponent spectroscopic and chromatographic data 306

LHASA program
    Expert systems in synthesis planning: a user's view of the LHASA program 53

LIMS
    The LIMS infrastructure 12

LIMS user group
    The LIMS infrastructure 12

Logic programming
    PROLOG for chemists. Part 1 67

Logic programming
    PROLOG for chemists. Part 2 92

Log ratios
    Spectral map analysis: factorial analysis of contrasts, especially from log ratios 250

Many-way tables
    Principal component analysis 209

Maximization
    Optimisation via Simplex. Part I. Background, definitions and a simple application 124

Meta-Dendral
    Dendral and Meta-Dendral – the myth and the reality 26

Method development
    Chemometrics and method development in high-performance liquid chromatography. Part 1: Introduction 139

Minimization
    Optimisation via Simplex. Part I. Background, definitions and a simple application 124

Mixture analysis
    Mixture analysis of spectral data by multivariate methods 293

Modified Simplex
    Optimisation via Simplex. Part I. Background, definitions and a simple application 124

Monitoring
    Sampling theory 196

Multiblock analysis
    Interpretation of direct latent-variable projection methods and their aims and use in the analysis of multicomponent spectroscopic and chromatographic data 306

Multicomponent analysis
    Interpretation of direct latent-variable projection methods and their aims and use in the analysis of multicomponent spectroscopic and chromatographic data 306

Multicomponent analysis
    Fuzzy theory explained 401

Multi-criteria optimization
    Fuzzy theory explained 401

Multivariate analysis
    Multivariate analysis in geology and geochemistry: an introduction 340

Multivariate analysis
    Multivariate analysis in geoscience: fads, fallacies and the future 354

Multivariate analysis of stratigraphic data
    Multivariate analysis of stratigraphic data in geology: a review 383

Multivariate correlation
    Some fundamental criteria for multivariate correlation methodologies 286

Multivariate data
    Principal component analysis 209

Multivariate data analysis
    Multivariate data analysis: its methods 225

Multivariate data analysis
    Correspondence analysis: the method and its application 233

Multivariate data analysis
    Interpretation of direct latent-variable projection methods and their aims and use in the analysis of multicomponent spectroscopic and chromatographic data 306

Multivariate data analysis
    Soft modelling and chemosystematics 321

Multivariate data analysis
    Interpretation of lithogeochemistry using correspondence analysis 367

Multivariate data analysis, exploratory
    Multivariate analysis in geology and geochemistry: an introduction 340

Multivariate display techniques
    Similarities and differences among multivariate display techniques illustrated by Belgian cancer mortality distribution data 262

Multivariate regression
    Soft modelling and chemosystematics 321

Optimisation
    Optimisation via Simplex. Part I. Background, definitions and a simple application 124

Optimisation
    Chemometrics and method development in high-performance liquid chromatography. Part 2: Sequential experimental designs 153

Ordinal methods
    Soft modelling and chemosystematics 321

Partial least squares
    Soft modelling and chemosystematics 321

Partial least squares, transposed
    Soft modelling and chemosystematics 321

Partitioning
    Multivariate analysis in geology and geochemistry: an introduction 340

Pattern recognition
    Fuzzy theory explained 40

Pattern recognition
    Soft modelling and chemosystematics 321

PC graphic adapters
    Scientific word processing 1

Phasing
    Fourier transforms: use, theory and applications to spectroscopic and related data 166

Pixels
    Scientific word processing 1
Principal component analysis
    Principal component analysis 209
Principal component analysis
    Similarities and differences among multivariate display techniques illustrated by Belgian cancer mortality distribution data 262
Principal component analysis
    Mixture analysis of spectral data by multivariate methods 293
Principal component analysis
    Similarities and differences among multivariate display techniques illustrated by Belgian cancer mortality distribution data 262
Principal components analysis
    Spectral map analysis: factorial analysis of contrasts, especially from log ratios 250
Principal elements
    Correspondence analysis: the method and its application 233
Printing
    Scientific word processing 1
Programming languages
    PROLOG for chemists. Part 1 67
Programming languages
    PROLOG for chemists. Part 2 92
Projection methods
    Interpretation of direct latent-variable projection methods and their aims and use in the analysis of multicomponent spectroscopic and chromatographic data 306
PROLOG
    PROLOG for chemists. Part 1 67
PROLOG
    PROLOG for chemists. Part 2 92
Proportional data
    Multivariate analysis of stratigraphic data in geology: a review 383
Pulsed fourier methods
    Fourier transforms: use, theory and applications to spectroscopic and related data 166
Q-mode analysis
    Multivariate analysis in geoscience: fads, fallacies and the future 354
Rank
    Principal component analysis 209
Reflection
    Optimisation via Simplex. Part I. Background, definitions and a simple application 124
Regression
    Chemometrics and method development in high-performance liquid chromatography. Part 2: Sequential experimental designs 153
Response
    Optimisation via Simplex. Part I. Background, definitions and a simple application 124
Response-surface
    Practical exploratory experimental designs 104

R-mode analysis
    Multivariate analysis in geoscience: fads, fallacies and the future 354
Sampling
    Sampling theory 196
Sampling strategy
    Fourier transforms: use, theory and applications to spectroscopic and related data 166
Scientific word processing
    Scientific word processing 1
Self-modelling
    Mixture analysis of spectral data by multivariate methods 293
Separation optimisation
    Chemometrics and method development in high-performance liquid chromatography. Part 1: Introduction 139
Signal-to-noise ratio
    Fourier transforms: use, theory and applications to spectroscopic and related data 166
SIMCA
    Some fundamental criteria for multivariate correlation methodologies 286
SIMCA
    Soft modelling and chemosystematics 321
Similarity models
    Principal component analysis 209
Simplex
    Optimisation via Simplex. Part I. Background, definitions and a simple application 124
Simplex
    Chemometrics and method development in high-performance liquid chromatography. Part 2: Sequential experimental designs 153
Singular-value decomposition
    Interpretation of direct latent-variable projection methods and their aims and use in the analysis of multicomponent spectroscopic and chromatographic data 306
Soft modelling
    Some fundamental criteria for multivariate correlation methodologies 286
Soft modelling
    Multivariate analysis in geoscience: fads, fallacies and the future 354
Soft modelling
    Soft modelling and chemosystematics 321
Spectral map analysis
    Spectral map analysis: factorial analysis of contrasts, especially from log ratios 250
Spectral map analysis
    Similarities and differences among multivariate display techniques illustrated by Belgian cancer mortality distribution data 262
Spectroscopic line shape analysis
    Dispersion vs. absorption (DISPA): a magic circle for spectroscopic line shape analysis 181
Step size
    Optimisation via Simplex. Part I. Background, definitions and a simple application 124

Subject Index

Stratigraphic data
  Multivariate analysis of stratigraphic data in geology: a review 383
Structure elucidation
  Dendral and Meta-Dendral – the myth and the reality 26
Supplementary elements
  Correspondence analysis: the method and its application 233
Synthesis planning systems
  Expert systems in synthesis planning: a user's view of the LHASA program 53
System manager
  The LIMS infrastructure 12
Taxonomy
  Soft modelling and chemosystematics 321
Time averaging
  Fourier transforms: use, theory and applications to spectroscopic and related data 166
Time domain
  Fourier transforms: use, theory and applications to spectroscopic and related data 166
Time series
  Fourier transforms: use, theory and applications to spectroscopic and related data 166
Two-block regression
  Principal component analysis 209
User training
  The LIMS infrastructure 12
Validation
  The LIMS infrastructure 12
Variable-size Simplex
  Optimisation via Simplex. Part I. Background, definitions and a simple application 124
Windows
  Scientific word processing 1
Word processors
  Scientific word processing 1
WYSIWYG
  Scientific word processing 1

CPSIA information can be obtained at www.ICGtesting.com
Printed in the USA
BVOW05s2308301215

431229BV00019B/20/P